Lippincott
Illustrated Reviews:
Neuroscience

Second Edition

Lippincott Illustrated Reviews: Neuroscience

Second Edition

Claudia Krebs, MD, PhD

Professor of Teaching
Department of Cellular and Physiological Sciences
The University of British Columbia
Vancouver, British Columbia, Canada

Joanne Weinberg, PhD

Professor and Distinguished University Scholar, Emerita
Department of Cellular and Physiological Sciences
The University of British Columbia
Vancouver, British Columbia, Canada

Elizabeth Akesson, MSc

Professor Emerita
Department of Cellular and Physiological Sciences
The University of British Columbia
Vancouver, British Columbia, Canada

Esma Dilli, MD, FRCPC

Clinical Assistant Professor
Department of Medicine, Division of Neurology
The University of British Columbia
Vancouver, British Columbia, Canada

. Wolters Kluwer

Philadelphia • Baltimore • New York • London
Buenos Aires • Hong Kong • Sydney • Tokyo

Not authorised for sale in United States, Canada, Australia, New Zealand, Puerto Rico, and U.S. Virgin Islands.

Acquisitions Editor: Crystal Taylor
Development Editor: Andrea Vosburgh
Editorial Coordinator: Emily Buccieri
Marketing Manager: Michael McMahon
Production Project Manager: Kim Cox
Design Coordinator: Stephen Druding
Artist/Illustrator: Jennifer Clements
Manufacturing Coordinator: Margie Orzech
Prepress Vendor: SPi Global

Second edition

Library of Congress Cataloging-in-Publication Data
Names: Krebs, Claudia, 1974- author. | Weinberg, Joanne, author.
Title: Neuroscience / Claudia Krebs, Joanne Weinberg.
Other titles: Lippincott's illustrated reviews.
Description: Second edition. | Philadelphia : Wolters Kluwer Health, [2018] |
   Series: Lippincott's illustrated reviews | Includes index.
Identifiers: LCCN 2017029418 | ISBN 9781496367891
Subjects: | MESH: Nervous System Physiological Phenomena | Outlines | Examination Questions
Classification: LCC RC343.6 | NLM WL 18.2 | DDC 616.8—dc23 LC record available at https://lccn.loc.gov/2017029418

CCS0917

In Memoriam

Richard A. Harvey, PhD

1936–2017

Co-creator and series editor of the Lippincott Illustrated Reviews series, in collaboration with Pamela C. Champe, PhD (1945–2008).

Illustrator and co-author of the first books in the series: *Biochemistry, Pharmacology, and Microbiology and Immunology.*

Dedication

To our families who love and support us in all that we do.

To our teachers and colleagues who have mentored us.

To our students who have inspired this book.

Acknowledgments

This book would not have been possible without the help and support of our families, our friends, our colleagues, and the fantastic team at Wolters Kluwer.

In particular, we would like to thank:

Mark Fenger, Monika Fejtek, Angela Krebs, and Ole Radach whose creative input for photography and visualization was essential throughout the project.

Crystal Taylor and her wonderful Wolters Kluwer editorial team who supported this project and made it come to life.

Emily Buccieri and Andrea Vosburgh of the Wolters Kluwer team whose patience and attention to detail moved this second edition forward on target.

Prof. Kyeong-Han Park from the Kangwon National University in South Korea for his thorough feedback on the first edition.

Prof. Anne M. R. Agur from the University of Toronto for her feedback, advice, and support in the writing of this book.

Preface

"Men ought to know that from nothing else but the brain come joys, delights, laughter and sports, and sorrows, griefs, despondency, and lamentations. And by this, in an especial manner, we acquire wisdom and knowledge and see and hear and know what are foul and what are fair, what are bad and what are good, what are sweet and what are unsavory... And by the same organ we become mad and delirious, and fears and terrors assail us... All these things we endure from the brain when it is not healthy... In these ways I am of the opinion that the brain exercises the greatest power in the man."

—Hippocrates, On the Sacred Disease (Fourth century BC)

The brain has fascinated mankind for centuries and only in recent decades have we begun to unravel some of the mysteries of its function. Neuroscience has been a rapidly evolving field that continues to bring us new insights into the human brain. In the first edition of this book, our mission was to streamline this complex information and make it accessible to newcomers while still including new and exciting developments. We have built on and extended this mission in our second edition. We continue to endeavor to make the complex information in both foundational and clinical neuroscience accessible to readers at all levels. We have updated the neuroscience information according to the latest research and have expanded the clinical examples to highlight the functional importance and applicability of the basic neuroanatomy and neurophysiology that are at the core of the book. An exciting addition is the inclusion of clinical case studies that guide the readers to apply their knowledge to a clinical scenario. Finally, we have revised many of our figures and added new ones to illustrate all of the key concepts presented.

Overall, this book provides the most up-to-date information within an integrated framework, bringing together neuroanatomy, neurophysiology, and the clinical context in which they are applied. Undergraduate and graduate science students, medical and dental students, students in rehabilitation sciences and nursing, residents, and practitioners will find that foundational science concepts are brought from bench to bedside.

Contents

Introduction to the Nervous System and Basic Neurophysiology

1

I. OVERVIEW

The nervous system is what enables us to perceive and interact with our environment. The brain regulates voluntary and involuntary function, enables us to be alert and responsive, and enables us to respond physically and emotionally to the world. Brain function is what makes us the people we are.

The nervous system can be divided into a **central nervous system (CNS)**, composed of the brain and the spinal cord, and a **peripheral nervous system (PNS)**, composed of all nerves and their components outside of the CNS (Figure 1.1). Information can flow in two general directions: from the periphery to the CNS **(afferent)** or from the CNS to the periphery **(efferent)**. Afferent, or sensory, information includes input from sensory organs (eye, nose, ear, and taste buds) as well as input from the skin, muscles, joints, and viscera. Efferent, or motor, information originates in the CNS and travels to glands, smooth muscle, and skeletal muscle (see Figure 1.1).

II. CELLULAR COMPONENTS OF THE NERVOUS SYSTEM

The cells of the nervous system are the basic building blocks for the complex functions it performs. An overview of these cellular components is shown in Figure 1.2. Over 100 billion **neurons** populate the human nervous system. Each neuron has contacts with more than 1,000 other neurons. Neuronal contacts are organized in **circuits** or **networks** that encode for the processing of all conscious and nonconscious information

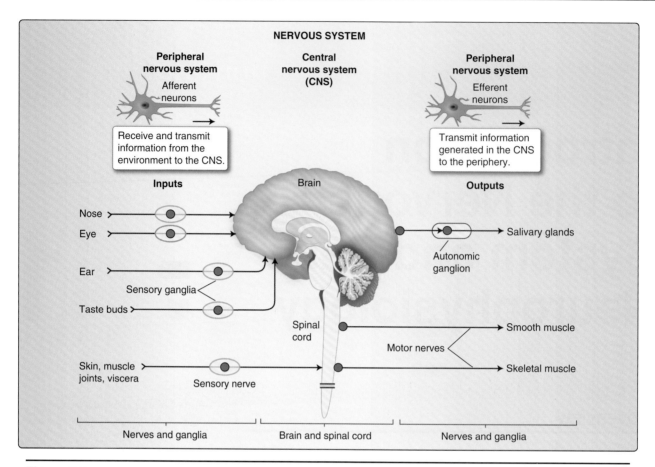

Figure 1.1
Overview of inputs and outputs of the central nervous system.

in the brain and spinal cord. A second population of cells called **glial cells** functions to support and protect the neurons. Glial cells, or **glia**, have shorter processes than do neurons and outnumber neurons by a ratio of 10:1. The function of glia goes beyond a simple supporting role. They also participate in neuronal activity, provide the stem cell pool within the nervous system, and provide the immune response to inflammation and injury.

A. Neurons

Neurons are the excitable cells of the nervous system. Signals are propagated via **action potentials**, or electrical impulses, along the neuronal surface. Neurons are connected to each other via **synapses** to form functional networks for the processing and storage of information. A synapse has three components: the axon terminal of one cell, the dendrite of the receiving cell, and a glial cell process. The **synaptic cleft** is the space between these component parts.

1. **Functional organization of neurons:** There are many types of neurons within the nervous system, but all neurons have structural components that enable them to process information. An overview of these components is illustrated in Figure 1.3. All neurons have a cell body, or **soma** (also called **perikaryon**), that contains the cell

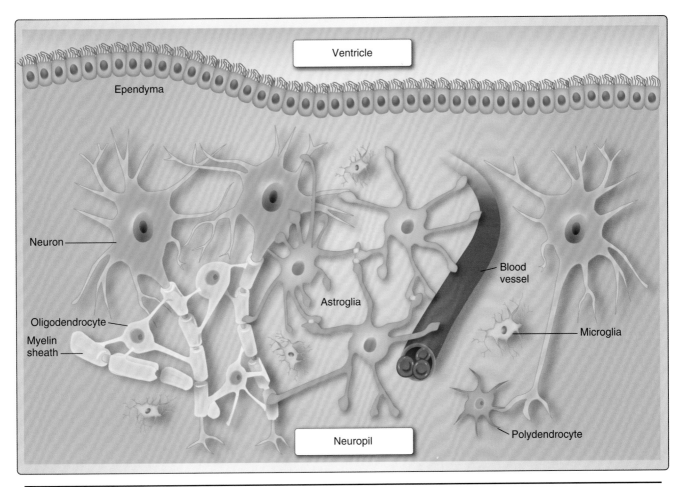

Figure 1.2
Overview of cellular components of the central nervous system.

nucleus and is where all proteins, hormones, and neurotransmitters are produced. A halo of endoplasmic reticulum (ER) can be found around the nucleus and is a testament to the high metabolic rate of neurons. This ER stains intensely blue in a Nissl stain and is commonly referred to as **Nissl substance**. Molecules produced in the soma are transported to the peripheral synapses via a network of **microtubules**. Transport from the perikaryon along the axon to the synapse is termed **anterograde transport**, which is how neurotransmitters needed at the synapse are transported. Transport along the microtubules can also be from the synaptic terminal to the perikaryon—this is termed **retrograde transport**. Retrograde transport is critical for shuttling trophic factors, in particular neurotrophin, from a neuron's target in the periphery to the soma. Neurons depend on the trophic substances supplied by their peripheral targets for survival. It is a sort of feedback mechanism to let the neuron know that it is innervating a "live target." Some viruses that infect neurons, such as the herpes virus, also take advantage of this retrograde transport mechanism. After they are taken up by the nerve ending, they are transported by retrograde transport to the perikaryon where they can lie dormant until activated. Synaptic input to a neuron occurs primarily at the **dendrites**.

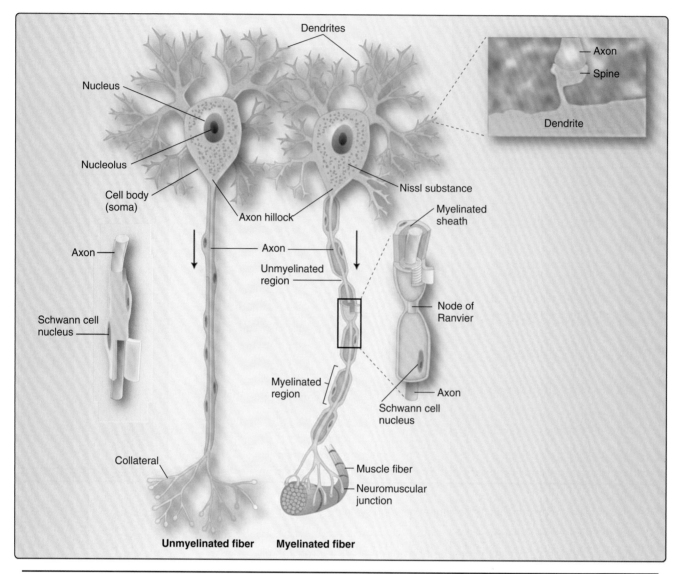

Figure 1.3
Neuron histology.

Here, small **spines** are the protrusions where synaptic contacts with axons are made. **Postsynaptic densities** in the spines serve as the scaffolding that holds and organizes neurotransmitter receptors and ion channels as shown in Figure 1.3. Every neuron also has an **axon**, whose terminals make synaptic contacts with other neurons. These cylindrical processes arise from a specialized area called an **axon hillock** or **initial segment,** and can be enwrapped by a protective layer called **myelin.** The initial segment of an axon is where all input to a neuron, both excitatory and inhibitory, is summed up and the decision to propagate an action potential to the next synapse is made.

2. **Types of neurons:** There are many types of neurons in the CNS. They can be classified according to their size, their morphology, or the neurotransmitters that they use. The most basic classification relates to morphology, as shown in Figure 1.4.

a. **Multipolar neurons:** Multipolar neurons are the most abundant type of neuron in the CNS and are found in both the brain and spinal cord. Dendrites branch directly off the cell body, and a single axon arises from the axon hillock.

b. **Pseudounipolar neurons:** Pseudounipolar neurons are found mainly in spinal ganglia. They have a dendritic axon that receives sensory information from the periphery and sends it to the spinal cord via an axon, *bypassing* the cell body along the way. Pseudounipolar neurons relay sensory information from a peripheral receptor to the CNS without modifying the signal.

c. **Bipolar neurons:** Bipolar neurons are found primarily in the retina and the olfactory epithelium. Bipolar neurons have a single main dendrite that receives synaptic input, which is then conveyed to the cell body and from there via an axon to the next layer of cells. Bipolar neurons integrate multiple inputs and then pass that modified information to the next neuron in the chain. The difference between a pseudounipolar and a bipolar neuron is the amount of processing that occurs in the neuron.

3. **Types of synapses:** A synapse is the contact between two neuronal cells. Action potentials encode the information that is processed in the CNS, and it is through synapses that this information is passed on from one neuron to the next (Figure 1.5).

a. **Axodendritic synapses:** The most common synaptic contacts in the CNS are between an axon and a dendrite called **axodendritic synapses**. The dendritic tree of any given multipolar neuron will receive thousands of axodendritic synaptic inputs, which will cause this neuron to reach threshold (see below) and to generate an electrical signal, or **action potential**. The architecture of the dendritic tree is a key factor in calculating the convergence of electrical signals in time and in space (called **temporospatial summation**, see below).

b. **Axosomatic synapses:** An axon can also contact another neuron directly on the cell soma, which is called an **axosomatic synapse**. This type of synapse is much less common in the CNS and is a powerful signal much nearer to the axon hillock where a new action potential may originate.

c. **Axoaxonic synapses:** When an axon contacts another axon, it is called an **axoaxonic synapse**. These synapses are often on or near the axon hillock where they can cause very powerful effects, potentially producing an action potential or inhibiting an action potential that would have otherwise been fired.

B. Glia

For many years, **glia** (Greek for "glue") were considered to be merely the scaffolding that holds neurons together, with no particular function of their own. However, our understanding of the function of glial cells has grown dramatically over the past decades. Glial cells outnumber neurons, and the ratio of glial cells to neurons is higher in vertebrates than in invertebrates. Humans and dolphins have a particularly high

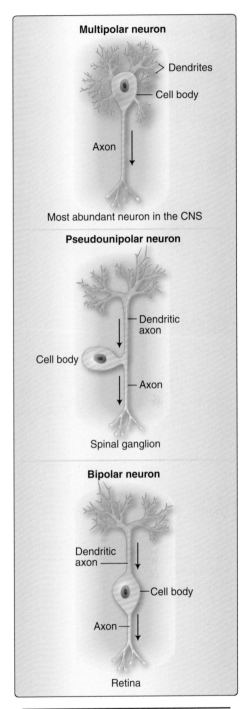

Figure 1.4
Types of neurons. CNS = central nervous system.

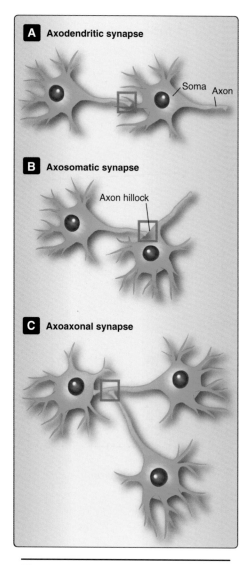

A Axodendritic synapse

Soma Axon

B Axosomatic synapse

Axon hillock

C Axoaxonal synapse

Figure 1.5
Types of synapses.

ratio of glia to neurons (10:1 or higher). When scientists were studying Albert Einstein's brain, one of the only morphological differences they found was a higher-than-usual ratio of glial cells to neurons. Far from being just "nerve glue," we now know that glial cells are an essential component of CNS function. **Oligodendroglia** and **Schwann cells** help provide the myelin sheath around axons in the CNS and PNS, respectively. **Astroglia** are involved in ion homeostasis and nutritive functions. Glia also have unique signaling and signal modification functions. **Polydendrocytes** are another type of glial cell that constitute the stem cell pool within the CNS, with the capability of generating both new glial cells and neurons. Finally, **microglia** are the immune cells within the brain, because the blood–brain barrier separates the brain from the blood-borne immune cells (Figure 1.6).

1. **Astroglia:** Astroglia, or **astrocytes**, can be subdivided into fibrous and protoplasmic astrocytes (found in white and gray matter, respectively) and Müller cells (found in the retina). Their main function is to support and nurture neurons. They take up and recycle excessive neurotransmitter at the synapse and maintain ion homeostasis around neurons. For example, at excitatory synapses, astrocytes take up glutamate and convert it to glutamine. Glutamine is then shuttled back to the neurons as a precursor of glutamate. All of this allows for efficient signal transduction at the synapse. Astrocyte end-feet line the blood vessels in the brain and are an important part of the **blood–brain barrier**, which separates the blood from the nervous tissue. They play an important role in maintaining homeostasis by shuttling excess ions into the blood stream. Besides this supportive role, astrocytes also have a signaling and signal modification role. Astrocytes are now known to be the third partner at the synapse. To value the importance of the astrocyte at the synapse, the term **tripartite synapse** (presynaptic neuron, postsynaptic neuron, and astrocyte) has been introduced. Astrocytes can release neurotransmitter into the synaptic cleft and strengthen the signal at that synapse. They also have neurotransmitter receptors and can communicate with each other through waves of intracellular calcium propagated from one astrocyte to another through gap junctions. During development, **radial glia**, a subpopulation of astrocytes, provide the direction and scaffolding for axon migration and targeting.

2. **Oligodendroglia:** Oligodendrocytes are the **myelinating cells** within the CNS. They can wrap their cellular processes around axons to provide an insulating and protective layer. One oligodendrocyte can myelinate multiple axons. The myelin sheath has important interactions with the axons it surrounds: it provides trophic support (promotes cell survival), protects the axon from the surroundings, and organizes the distribution of ion channels along the axon. The thickness of the myelin sheath is closely related to the diameter of the axon. Gaps in the myelin sheath occur at regular intervals to allow the passage of ions and are called **nodes of Ranvier** (see Figure 1.6).

3. **Schwann cells:** These are the myelinating cells in the PNS. Their function is similar to that of both oligodendrocytes and astrocytes in the CNS. In contrast to oligodendrocytes, however,

one Schwann cell can myelinate only a single axon. It also can enwrap several unmyelinated axons as a protective layer. At the **neuromuscular junction**, or the contact between a motor nerve and a muscle fiber, the Schwann cell will take up excessive neurotransmitter and maintain ion homeostasis to facilitate efficient signal transduction.

4. **Microglia:** These glial cells are derived from the monocyte–macrophage lineage and migrate into the CNS during development. Microglia are the **immune cells** in the CNS. They are small with numerous processes and are distributed throughout the CNS. When activated, microglia release inflammatory molecules such as cytokines, similar to the activation pathways of blood-borne macrophages. They are recruited to areas of neuronal damage, where they phagocytose cell debris and are involved in antigen presentation, again similar to blood-borne macrophages.

5. **Polydendrocytes:** Also referred to as glial precursor cells, polydendrocytes are the **stem cells** within the brain, and they can generate both glia and neurons. They are of particular interest in demyelinating disorders because their activation and recruitment as oligodendrocyte precursor cells are the first steps in remyelination. Glial precursor cells can also receive direct synaptic input from neurons, making this glial cell a direct link between the neuronal signaling network and the glial network. The finding that glial cells receive direct synaptic input revolutionized our understanding of how networks are organized in the CNS. It appears that there is significant cross talk between the neuronal networks and the parallel glial networks. The functional implications of this remain speculative.

6. **Ependymal cells:** These epithelial cells line the ventricles and separate the cerebrospinal fluid (CSF) from the nervous tissue, or **neuropil**. On their apical surface, they have numerous cilia. Some ependymal cells have a specialized function within the ventricles as part of the **choroid plexus**. The choroid plexus produces CSF. See Chapter 2, "Overview of the Central Nervous System," for details.

C. Blood–brain barrier

The CNS needs a perfectly regulated environment to function properly. This homeostasis must be preserved and cannot be influenced by fluctuations in nutrients, metabolites, or other blood-borne substances. The blood–brain barrier, illustrated in Figure 1.7, effectively isolates and protects the brain from the rest of the body. Endothelial cells in the CNS are linked to each other via tight junctions. In addition, astrocyte processes (end-feet) contact the vessel from the neuropil side. This effectively separates the blood compartment from the neuropil compartment. Transport across the blood–brain barrier can be by diffusion of small lipophilic molecules, water, and gas. All other substances must use active transport. Clinically, this is relevant because it limits the drugs that can be given to treat disorders in the brain to those that can cross the blood–brain barrier.

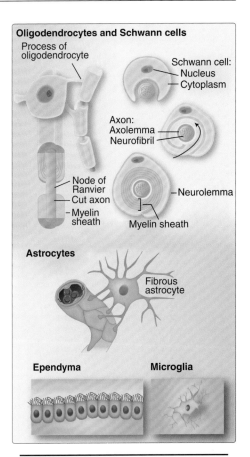

Figure 1.6
Types of glia.

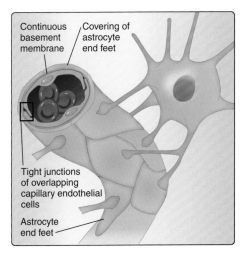

Figure 1.7
The blood–brain barrier.

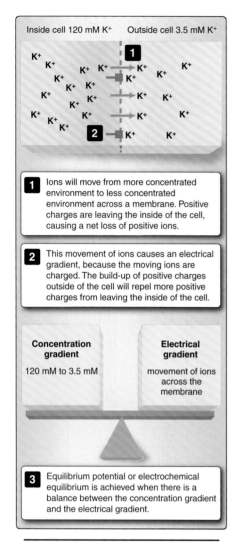

Figure 1.8
Ion movements.

III. BASIC NEUROPHYSIOLOGY

Neurophysiology is the study of ion movements across a membrane. These movements can initiate signal transduction and the generation of action potentials. The study of neurophysiology also includes the action of neurotransmitters.

A. Ion movements

A neuron is surrounded by a phospholipid bilayer membrane, which maintains differential ion concentrations on the inside versus the outside of the cell as shown in Figure 1.8. The movement of ions across this membrane generates an electrical gradient for each ion. The sum of all ion gradients is the **membrane potential**, also called the **electrical potential**. It is important to note that all cells are surrounded by a phospholipid bilayer, and all cells maintain a different ion concentration on the inside versus the extracellular space. Neurons and muscles are, however, the only cells that are able to send signals along their surface or to harness these ion differentials to generate electrical signals (Table 1.1).

1. **Equilibrium potential—the Nernst equation:** The differential ion concentrations on the inside versus the outside of the cell are maintained by the function of the cell membrane. Figure 1.8 illustrates the movement of ions until they reach a steady state. Ions move along a concentration gradient. Uncharged particles will move across the membrane until the same concentration is achieved on both sides (**diffusion**). Because ions are charged, their movement causes an **electrical gradient**. Charged ions repel ions with the same charge at the membrane. As ions move across the membrane along the concentration gradient, there is a buildup of charge preventing more ions from moving across the membrane. For example, the K⁺ concentration inside (i) the cell (120 mM) is higher than outside (o) the cell (3.5 mM). K⁺ will move along the concentration gradient across the membrane to the outside of the cell and take the positive charge with it. The potential of the inside of the cell is negative because it is constantly losing K⁺ to the outside of the cell. The result of this net flow is that the inside of the cell loses K⁺ ions. As these K⁺ ions move, they generate a **potential gradient**, or electrical gradient, across the membrane. At some point, this electrical gradient will prevent the further movement of K⁺ as positive charge buildup on the other side of the membrane will repel positive charges from crossing over. **Equilibrium potential** (also called **electrochemical equilibrium**)

Table 1.1: Intracellular and Extracellular Ion Concentrations

Ion	Extracellular (mM)	Intracellular (mM)	Equilibrium Potential (mV)
Na⁺	140	15	+60
K⁺	3.5	120	−95
Ca²⁺	2.5	0.0001 (unbound Ca²⁺)	+136
Cl⁻	120	5	−86

is thus achieved. This equilibrium potential can be expressed by the **Nernst equation**. The Nernst equation takes several physical constants and the ion gradient, or ion concentration inside the cell and outside the cell, to determine the potential at which there will be no more net movement of ions. As shown in Figure 1.9, the equilibrium potential for K⁺ is at −95 mV.

2. **Resting membrane potential—the Goldmann equation:** Of course, a neuron does not contain only K⁺ but other ions as well. Each ion has a different intracellular and extracellular concentration, and the permeability of the membrane is different for each ion. The **permeability** of the membrane determines how easily an ion can cross the membrane. In order to determine the resting membrane potential, we need to take into account the concentration of different ions inside and outside the cell as well as the permeability of the membrane for each ion. This resting membrane potential can be described by the **Goldmann equation** as shown in Figure 1.10. The different intracellular and extracellular ion concentrations are maintained by membrane proteins that act as **ion pumps**. The most prominent of these ion pumps is the **Na⁺/K⁺ ATPase**, which pumps Na⁺ (sodium) out of the cell in exchange for K⁺. This activity of the Na⁺/K⁺ exchanger is shown in Figure 1.11. As the name implies, these ion pumps depend on energy in the form of adenosine triphosphate (ATP) to function. The pump can only function in the presence of ATP, which is hydrolyzed to adenosine diphosphate (ADP) in order to release energy.[1] **Ion channels** are membrane proteins that allow ions to pass through them, which causes current flow. Ion channels are selective: The size of the **channel pore** and the amino acids in the pore will regulate which ion can pass through. The opening or closing of ion channels is regulated by different mechanisms as detailed in Figure 1.12.

Biological membranes can actively change their permeability for different ions. This changes the membrane potential, and it is the underlying mechanism of the action potential.

a. **Voltage-gated ion channels:** These channels are regulated by the membrane potential. A change in membrane potential opens the **channel pore**. The most prominent of these channels is the voltage-gated Na⁺ channel. Its opening underlies the initiation of an action potential (see Figure 1.12A).

b. **Ligand-gated ion channels:** These channels are regulated by a specific molecule that binds to the ion channel. This opens the pore, and ions can pass through. Postsynaptic neurotransmitter receptors are ligand-gated ion channels (see Figure 1.12B).

c. **Mechanically gated ion channels:** The pore in these channels is mechanically opened. Touch receptors in the skin and receptor cells in the inner ear are examples of mechanically gated ion channels. These channels open through the mechanical deflection that pries or pulls the channel open (see Figure 1.12C).

 [1]See pp. 72 and 73 in *Lippincott Illustrated Reviews: Biochemistry*.

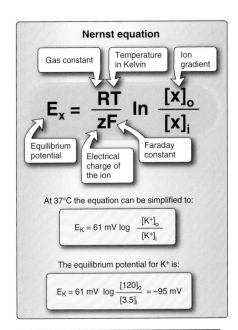

Figure 1.9
The Nernst equation.

$$E_x = \frac{RT}{zF} \ln \frac{[x]_o}{[x]_i}$$

At 37°C the equation can be simplified to:

$$E_K = 61\ mV \log \frac{[K^+]_o}{[K^+]_i}$$

The equilibrium potential for K⁺ is:

$$E_K = 61\ mV \log \frac{[120]_o}{[3.5]_i} = -95\ mV$$

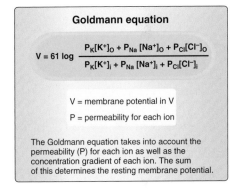

Goldmann equation

$$V = 61 \log \frac{P_K[K^+]_o + P_{Na}[Na^+]_o + P_{Cl}[Cl^-]_o}{P_K[K^+]_i + P_{Na}[Na^+]_i + P_{Cl}[Cl^-]_i}$$

V = membrane potential in V

P = permeability for each ion

The Goldmann equation takes into account the permeability (P) for each ion as well as the concentration gradient of each ion. The sum of this determines the resting membrane potential.

Figure 1.10
The Goldmann equation.

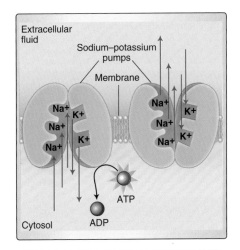

Figure 1.11
Ion pumps. ATP = adenosine triphosphate; ADP = adenosine diphosphate.

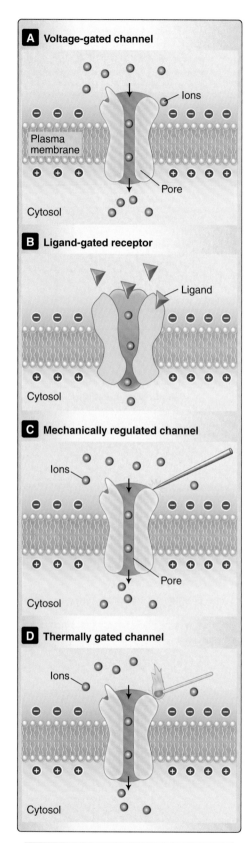

Figure 1.12
Ion channels.

d. Thermally gated ion channels: These channels are regulated by temperature. The channel protein acts as a thermometer, and a change in temperature opens the channel pore (see Figure 1.12D).

B. Action potential

Action potentials (APs) are electrical impulses, or changes in membrane potential, that travel along the surface of a neuron. The underlying mechanism for APs is the change in membrane permeability for different ions, first Na⁺ when initiating an AP and then K⁺ in the recovery phase. APs are the means of communication between neurons.

1. **Generation of an action potential:** The sequential changes in membrane permeability for Na⁺ and K⁺ that cause changes in membrane potential and underlie the AP are illustrated in Figure 1.13. A change in membrane potential during an AP is due to an increased permeability of the membrane to Na⁺. This temporary increase in Na⁺ permeability is due to the opening of Na⁺ channels and causes a **depolarization** of the cell membrane. When Na⁺ channels open, Na⁺ flows into the neuron and the resting membrane potential shifts from being close to K⁺ equilibrium to being close to Na⁺ equilibrium, that is, in the positive range. This Na⁺ permeability is short lived, as the Na⁺ channels close again, and the membrane becomes more permeable to K⁺, even more permeable than at rest (known as **undershoot** or **after-hyperpolarization**).

The Na⁺ channels that open when an AP is generated are voltage-gated channels. They will open only when the membrane depolarizes to a **threshold potential**. Once this threshold has been reached, the AP is generated as an all-or-none response. Because there is no gradation for the AP, it can only be either "on" or "off." Signal transduction via action potentials is like a binary system that computers use to encode information (1 or 0). All APs in a given population of neurons are of the same magnitude and duration.

After each AP, the Na⁺ channels involved are in a **refractory period**. This corresponds with the undershoot phase (increased K⁺ permeability) and has two main effects: the number of APs that can travel along an axon is limited, and the direction of the action potential is determined. The AP will not go toward refractory channels ("backward") but forward to the next set of channels. The next AP is generated when the channels are ready, or "reset."

The generation of action potentials has an energy price tag: ATP is needed to restore ion homeostasis. At the end of an action potential, ion pumps (e.g., the **Na⁺/K⁺ ATPase**) restore ion homeostasis through active transport. The activity of these pumps depends on the hydrolyzation of ATP to ADP to release energy.

2. **Propagation of action potentials:** The effective transmission of an electrical signal along an axon is limited by the fact that ions tend to leak across the membrane. The impulse will dissipate as charge is lost over the membrane. The AP has a way of circumventing the leakiness of the neuronal membrane. Electrical signals

along an axon are propagated through both passive current flow and active current flow, as illustrated in Figure 1.14.

a. Current: **Current** is measured in amperes (A) and describes the movement of charge or movement of ions. The amount of work necessary to move that charge is described as the **voltage** and measured in volts (V). The difficulty of moving ions is referred to as **resistance** and measured in ohms (Ω). **Conductance** is the ease of moving ions and measured in siemens (S). The current of a specific ion depends on the membrane permeability (**conductance**) and the electrochemical driving force for that ion. This can be expressed by the **Ohm law**, which is summarized in Figure 1.15.

Passive current is a shuttling of charge, much like the flow of electricity along a wire. Passive current is not a movement of Na⁺ ions. **Active current**, by contrast, does involve the flow of ions (Na⁺) through ion channels (see Figure 1.14). The propagation of an AP depends on both passive current flow and the opening of Na⁺ channels. Passive current will cause a change in membrane potential, which will open the voltage-gated Na⁺ channels. This causes the generation of another AP. Passive current generated by this AP will travel along the membrane to the next set of Na⁺ channels. By constantly regenerating the AP, the leakiness of the neuronal membrane is effectively avoided.

b. Continuous conduction: In **unmyelinated** axons, passive current flows along the axon and continuously opens Na⁺ channels (active current) that are inserted along the entire length of the axon. Continuous regeneration of APs along the entire length of axons is called **continuous conduction** and is illustrated in Figure 1.16.

c. Saltatory conduction: In myelinated axons, Na⁺ channels are accumulated at the gaps in the myelin sheath (nodes of Ranvier). Passive current is shuttled along a long segment of the myelinated axon. At the node of Ranvier, the change in membrane potential causes the opening of Na⁺ channels and with that, the regeneration of the AP. The action potential seems to "jump" from node to node, which is called **saltatory conduction** and is illustrated in Figure 1.17.

3. Velocity of the action potential: The velocity of an action potential is determined by both the passive and the active current flow. In order to increase the velocity of APs, you need to facilitate passive and active current flow. The two major obstacles to overcome are the resistance of the axon and the capacitance of the membrane, summarized in Figure 1.18.

a. Resistance: Resistance describes the difficulty of moving ions and is measured in ohms (Ω). Large diameter axons have a low resistance and a fast passive current flow. The larger the diameter of an axon, the easier it is to move ions. Unfortunately, the body cannot increase axon diameter indefinitely to increase the speed of conductance. The axon diameter needed to

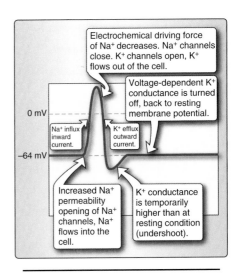

Figure 1.13
The action potential.

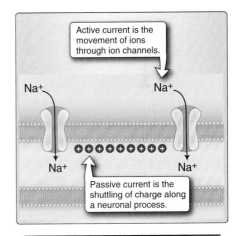

Figure 1.14
Passive current and active current.

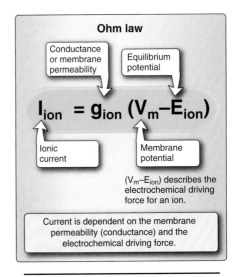

Figure 1.15
Ohm law.

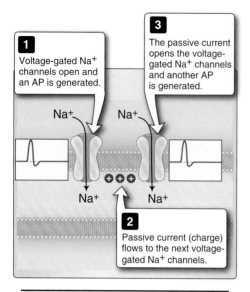

Figure 1.16
Continuous conduction.

accommodate the high speed of conductance needed over long distances would be too large to fit into our peripheral nerves.

b. **Capacitance:** A capacitor is composed of two conducting regions separated by an insulator. In neurons, the extracellular and intracellular fluids are separated by the cell membrane. The cell membrane is the insulator, and the extracellular and intracellular fluids are the two conducting regions. Charge is accumulated on one side of the membrane, which repels the same charge on the other side and attracts the opposite charge. In this way, the capacitor separates and accumulates charges. Every time an AP travels down an axon, passive current opens the Na⁺ channels and Na⁺ flows into the cell (active current). However, before this can happen, the capacitance of the cell membrane (or repelling charge accumulated at the cell membrane) must be overcome.

c. **Velocity of passive current:** The velocity of passive current flow depends on the resistance it encounters in the axon. Increasing axon diameter will decrease the resistance and speed up passive current. Another way to speed up the velocity of the passive current is to insulate the membrane with myelin, which would lessen the dissipation of current (**leak current**) through the membrane.

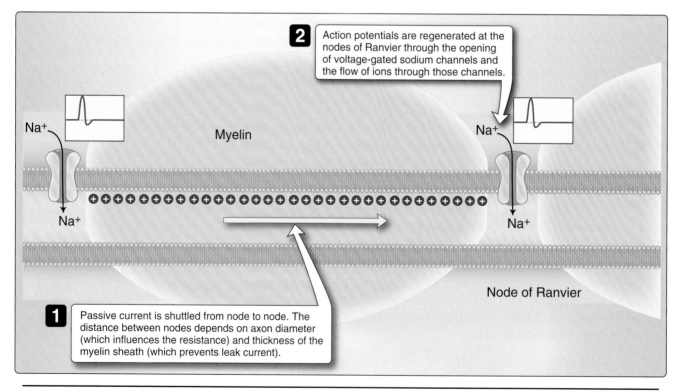

Figure 1.17
Saltatory conduction.

d. Velocity of active current: The velocity of active current flow depends on the capacitance of the membrane. The easier it is to overcome the accumulated repelling charge at the cell membrane, the quicker the ions can move across the membrane. A reduction in axon diameter would reduce capacitance by reducing the surface area of the membrane or the total net area where repelling charges can be accumulated. Decreasing the axon size, however, would increase the resistance for passive current flow. Another approach to reduce capacitance is through myelination of the axon. Myelin is such an effective insulator that once myelinated, the membrane can no longer act as a capacitor, and it no longer accumulates charge. The downside is that the membrane is no longer permeable to ions either, and Na^+ ions cannot pass through it. In order to have active flow, there need to be gaps (called **nodes of Ranvier**) in the myelin where Na^+ can pass through the cell membrane. Voltage-gated Na^+ channels are clustered at these nodes. The capacitance of the membrane must be overcome at every node of Ranvier, but this is just a small area compared to the large surface area of the entire axon (see Figures 1.16 and 1.17). Passive current flows the distance between the nodes and opens the voltage-gated Na^+ channels at the next gap in the myelin. The distance between nodes (**internode distance**) depends on the axon diameter, which determines the resistance the passive current encounters.

In summary, APs are the currency of communication in the CNS. As an AP travels along an axon, it must be

1. *Unidirectional*: This is achieved through the refractory period.
2. *Fast*: A decrease in both membrane capacitance (from myelin) and resistance (from increased axon diameter) helps speed the AP along.
3. *Efficient*: APs are generated only at nodes of Ranvier, not along the entire length of the axon, saving energy.
4. *Simple*: The AP is an all-or-none response, essentially a binary system.

C. Synaptic transmission

Synaptic transmission can occur via either electrical or chemical synapses. All synapses contain both presynaptic and postsynaptic elements.

1. **Electrical synapses:** Two neurons can be coupled electrically to each other via **gap junctions**. A gap junction is a protein pore complex (connexon) that lets ions and other small molecules move between cells (Figure 1.19A). Gap junction–coupled neurons are found in areas where populations of neurons need to be synchronized with each other, for instance, in the breathing center or in hormone-secreting regions of the hypothalamus.

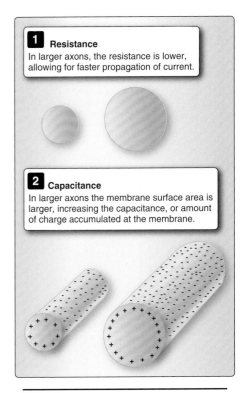

1 Resistance
In larger axons, the resistance is lower, allowing for faster propagation of current.

2 Capacitance
In larger axons the membrane surface area is larger, increasing the capacitance, or amount of charge accumulated at the membrane.

Figure 1.18
Resistance and capacitance.

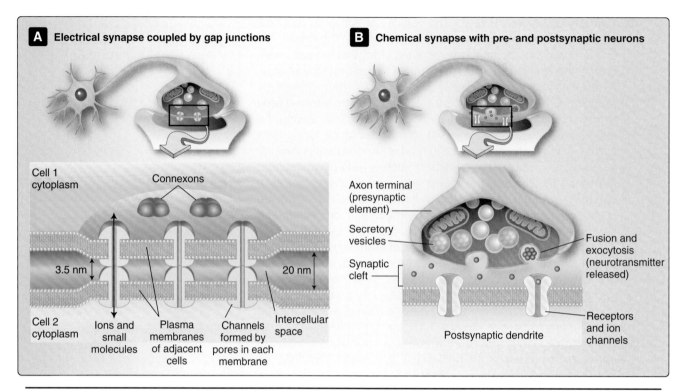

A Electrical synapse coupled by gap junctions

Cell 1
cytoplasm

Connexons

3.5 nm

20 nm

Cell 2
cytoplasm

Ions and
small
molecules

Plasma
membranes
of adjacent
cells

Channels
formed by
pores in each
membrane

Intercellular
space

B Chemical synapse with pre- and postsynaptic neurons

Axon terminal
(presynaptic
element)

Secretory
vesicles

Synaptic
cleft

Fusion and
exocytosis
(neurotransmitter
released)

Postsynaptic dendrite

Receptors
and ion
channels

Figure 1.19
Electrical synapses and chemical synapses.

2. **Chemical synapses:** A chemical synapse is composed of a presynaptic terminal, a synaptic cleft, and a postsynaptic terminal (see Figure 1.19B). Charge and ions do not move directly between cells. Communication is achieved via neurotransmitters (see below).

3. **Synaptic signal transduction:** The cascade of events leading to signal transduction at the synapse is shown in Figure 1.20. An AP arrives at the presynaptic terminal, which causes voltage-gated Ca^{2+} (calcium) channels to open. This influx of Ca^{2+} causes neurotransmitter-filled vesicles to fuse with the membrane and diffuse the neurotransmitter across the synaptic cleft. The neurotransmitter binds to postsynaptic receptors, and ion channels open. The type of ion channel that opens will determine whether an **inhibitory postsynaptic potential** (**IPSP**) or an **excitatory postsynaptic potential** (**EPSP**) is elicited. An influx of Na^+ causes an **EPSP** and brings the membrane closer to reaching threshold, as seen in Figure 1.21A. An influx of Cl^- (chloride) causes an **IPSP** and moves the membrane potential away from threshold, as seen in Figure 1.21B. Neurotransmitters are specific to producing either IPSPs or EPSPs (see Table 1.2).

   a. **Temporospatial summation:** The postsynaptic neuron will fire an AP when the threshold potential has been reached. One individual synapse does not have the power through a single **synaptically evoked potential** to bring a postsynaptic neuron closer to threshold. Only the cumulative effect of thousands of synapses on any given postsynaptic neuron will elicit an AP.

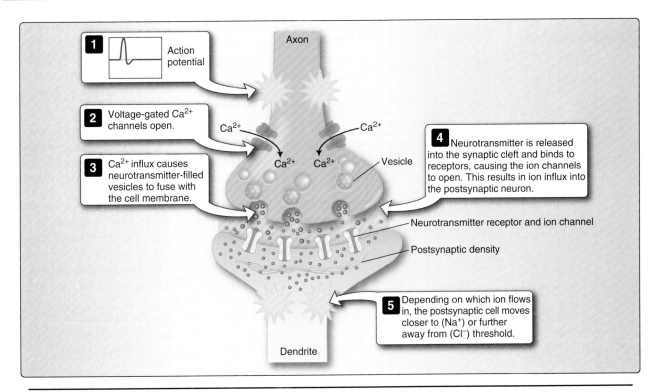

Figure 1.20
Synaptic signal transduction.

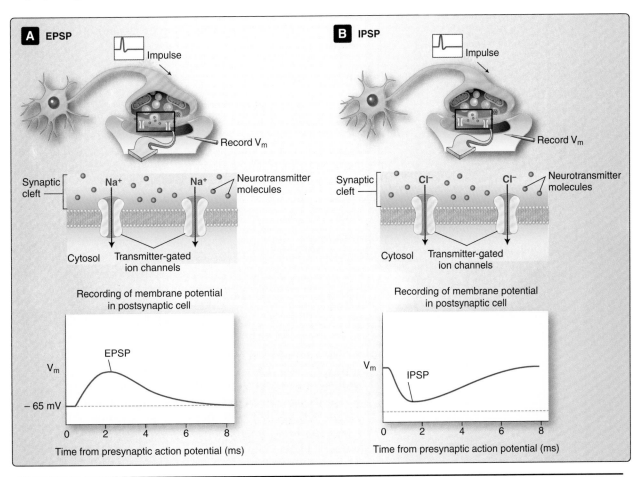

Figure 1.21
Excitatory (EPSP) and inhibitory postsynaptic potentials (IPSP). V_m = membrane potential.

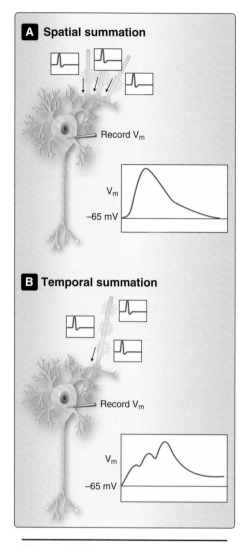

Figure 1.22
Temporospatial summation.

The synapses receiving input must be close together (Figure 1.22A) and receive input in the same time frame (Figure 1.22B). This is termed **temporospatial summation**. Figure 1.22 shows how synaptically evoked potentials received at the same time and in the same area can bring the neuron close to threshold, which results in the generation of an AP.

An AP is an all-or-none response, but a synaptically evoked potential is graduated in magnitude. Increased neurotransmitter present in the synaptic cleft results in increased receptor activation, and, in turn, more ions flow into the postsynaptic terminal.

b. **Ionotropic receptors:** Postsynaptic receptors can be either **ionotropic** or **metabotropic**. In ionotropic receptors, a neurotransmitter receptor is coupled with an ion channel. When the neurotransmitter binds to the receptor, a conformational change allows for ions to flow through the channel. The flow of ions can change the membrane potential of the postsynaptic cell, moving it closer to threshold (through the opening of Na^+ channels) or farther away from threshold (through the opening of Cl^- channels). Ionotropic receptors, thus, have a direct effect on ion movements by directly affecting an ion channel (Figure 1.23).

c. **Metabotropic receptors:** In metabotropic receptors, a neurotransmitter receptor is coupled with intracellular signaling cascades, often through G protein–coupled mechanisms (often involving enzymes). These will have an indirect effect on ion movements through the modulation of either postsynaptic channels or selective opening or closing of channels (see Figure 1.23).

D. Neurotransmitters

Neurotransmitters are molecules released by presynaptic neurons and are the means of communication at a chemical synapse. Neurotransmitters bind to neurotransmitter receptors, which can be coupled with an ion channel (ionotropic receptors) or with an intracellular signaling process (metabotropic receptors). Neurotransmitters are specific for the receptor they bind to and elicit a specific response in the postsynaptic neurons, resulting in either an excitatory or inhibitory signal (Table 1.2).

1. **Glutamate:** Glutamate is the most common excitatory neurotransmitter in the CNS. Glutamate can bind to ionotropic glutamate receptors, which include **NMDA receptors** (*N*-methyl-D-aspartate), **AMPA receptors** (α-amino-3-hydroxyl-5-methyl-4-isoxazole-propionate), and **kainate receptors**. These receptors are named for the agonists (besides glutamate) that specifically bind to them. All of these receptors cause an influx of cations (positive charge) into the postsynaptic neurons. The NMDA receptor is a bit different from the AMPA and kainate receptors in that its pore is blocked by an Mg^{2+} ion unless the postsynaptic membrane is depolarized. Once the channel is unblocked, it is permeable not only to Na^+ but to large amounts of Ca^{2+} as well. An excess of Ca^{2+} influx can result in a cascade of events that may result in cell death.

Table 1.2: Summary of Some Neurotransmitters in the CNS

	Neurotransmitter	Postsynaptic Effect
Amino acids	Acetylcholine (ACh)	Excitatory
	Glutamate	Excitatory
	γ-Aminobutyric acid (GABA)	Inhibitory
	Glycine	Inhibitory
Biogenic amines	Dopamine	Excitatory (via D1 receptors)
		Inhibitory (via D2 receptors)
	Norepinephrine	Excitatory
	Epinephrine	Excitatory
	Serotonin	Excitatory or inhibitory
	Histamine	Excitatory
Purines	Adenosine triphosphate (ATP)	Excitatory/ neuromodulatory
Neuropeptides	Substance P	Excitatory
	Metenkephalin	Inhibitory
	Opioids	Inhibitory
	Adrenocorticotropin	Excitatory

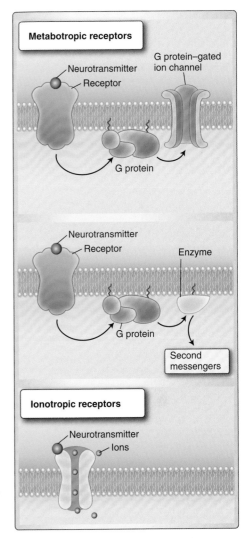

Figure 1.23
Types of receptors.

Glutamate can also bind to a family of metabotropic glutamate receptors (**mGluRs**), which initiate intracellular signaling that can modulate postsynaptic ion channels indirectly. This typically increases the excitability of postsynaptic neurons.

Glutamate is synthesized in neurons from the precursor **glutamine**. Glutamine is supplied by astrocytes, which produce glutamine from the glutamate they take up in the synaptic cleft.

2. **GABA and glycine:** γ-Aminobutyric acid (**GABA**) and **glycine** are the most important inhibitory neurotransmitters in the CNS. About half of all inhibitory synapses in the spinal cord use glycine. Glycine binds to an ionotropic receptor, which allows for Cl⁻ influx.

   Most other inhibitory synapses in the CNS use GABA. GABA can bind to ionotropic GABA receptors (GABA$_A$ and GABA$_C$), which induce Cl⁻ influx when activated. Cl⁻ influx leads to an accumulation of negative charge, which moves the membrane potential further away from reaching threshold (i.e., the neuron in "inhibited"). The metabotropic GABA receptor GABA$_B$ activates K⁺ channels and blocks Ca²⁺ channels, resulting in a net loss of positive charge, which also leads to hyperpolarization of the postsynaptic cell.

3. **Acetylcholine:** Acetylcholine (ACh) is a neurotransmitter used in both the PNS (ganglia of visceral motor system) and the CNS (forebrain). It is also the neurotransmitter used at the neuromuscular junction (see Chapter 3, "Overview of the Peripheral Nervous System").

   There are two types of receptors for ACh: (1) The **nicotinic ACh receptors** are ionotropic receptors and are coupled with a nonselective cation channel; (2) the **muscarinic ACh receptors** comprise a family of metabotropic receptors that are linked to G protein–mediated pathways.

There is no reuptake mechanism for ACh from the synaptic cleft. Its clearance depends on the enzyme **acetylcholinesterase**, which hydrolyzes the neurotransmitter and deactivates it.

4. **Biogenic amines:** Biogenic amines are a group of neurotransmitters with an amine group in their structure. They comprise the **catecholamines** dopamine, norepinephrine, and epinephrine, as well as histamine and serotonin.

   a. **Dopamine:** Dopamine is involved in many forebrain circuits associated with emotion, motivation, and reward. It acts on G protein–coupled receptors, and its main action can be either excitatory (via D_1 receptors) or inhibitory (via D_2 receptors).

   b. **Norepinephrine: Norepinephrine** (also known as *noradrenaline*) is a key neurotransmitter involved in wakefulness and attention. It acts on the metabotropic α-adrenergic and β-adrenergic receptors, both of which are excitatory. **Epinephrine** (also known as *adrenaline*) acts on the same receptors, but its concentration in the CNS is much lower than that of norepinephrine.

   c. **Histamine: Histamine** binds to an excitatory metabotropic receptor. In the CNS, it is involved in wakefulness.

   d. **Serotonin:** Serotonin can have both excitatory and inhibitory effects. It is involved in a multitude of pathways that regulate mood, emotion, and several homeostatic pathways. Most serotonin receptors are metabotropic. There is only one ionotropic receptor, which is a nonselective cation channel and is, therefore, excitatory.

5. **ATP: ATP** is best known as the **energy carrier** within cells.[2] However, it is also released by presynaptic neurons as a neurotransmitter. Because it is often released together with other neurotransmitters, it is referred to as a **cotransmitter**. ATP can be broken down in the synaptic cleft to adenosine, a purine, which binds and activates the same receptors as ATP. These purinergic receptors can be either ionotropic (P2X) or metabotropic (P2Y). The ionotropic receptors are coupled with nonspecific cation channels and are excitatory. The metabotropic receptors act on G protein–coupled signaling pathways.

ATP and purines are **neuromodulators**. Because they are co-released with other neurotransmitters, the degree of P2X or P2Y activation will modulate the response to the other neurotransmitter secreted, either enhancing that action or inhibiting it.

[2]See pp. 72 and 73 in *Lippincott Illustrated Reviews: Biochemistry*.

6. **Neuropeptides:** Neuropeptides are a group of peptides that are involved in neurotransmission. They include molecules involved in pain perception and modulation such as **substance P**, **metenkephalin**, and **opioids**. Other neuropeptides are involved in the neural response to stress such as **corticotropin-releasing hormone** and **adrenocorticotropin**.

Clinical Application 1.1: Multiple Sclerosis

Multiple sclerosis (MS) is a chronic neurological disease. The underlying pathology is loss of the myelin sheath around axons, a process called **demyelination**, and the loss of axons (**neurodegeneration**). **Severe inflammation** is observed in areas of demyelination, which is thought to be an underlying mechanism for demyelination and neurodegeneration. Demyelination can be seen as light spots on magnetic resonance imaging scans as shown in the figure.

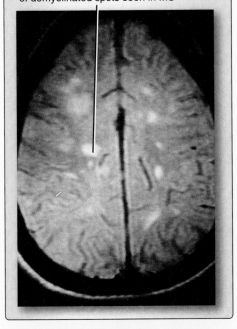

Hyperintense spots in the MRI are indicative of demyelinated spots seen in MS

Magnetic resonance imaging (MRI) scan in multiple sclerosis (MS).

Demyelination impairs function in the central nervous system (CNS). The loss of the myelin sheath leads to a **conduction block** within that axon. A myelinated axon conducts action potentials (APs) via saltatory conduction. Without the myelin sheath, the clusters of Na$^+$ (sodium) channels are too far apart and the passive current dissipates before the next cluster of Na$^+$ channels can be activated.

One way the CNS responds to the conduction block is to insert Na$^+$ channels along the demyelinated axon to allow nonsaltatory, continuous conduction. The Na$^+$ channels that are inserted, however, have a different dynamic and cause more Na$^+$ influx into the axon. The Na$^+$/Ca$^+$ exchanger can no longer maintain Na$^+$ homeostasis, proteases are activated, and the axon degenerates. In some cases, the insertion of Na$^+$ channels in the demyelinated axon is successful, continuous conductance is established, and APs can be propagated, albeit at a slower pace.

Another way the CNS aims to restore function is by **remyelinating** the axon. **Oligodendrocytes** are the myelinating cells within the CNS. In order to initiate remyelination, **polydendrocytes**, the oligodendrocyte precursor cells, are recruited to the affected area. Once they mature into oligodendrocytes, they can begin the process of remyelination. Macrophages remove myelin debris in the affected area, because it appears that myelin debris will inhibit the maturation of polydendrocytes into oligodendrocytes. Once the axon has been remyelinated, function is restored, even though the intricate relationship between the axon and its myelin sheath is not reestablished. In the healthy brain, myelin sheath thickness is tightly correlated with axon diameter and internode distance to ensure fast and efficient AP propagation. After remyelination, however, function is restored but is not as quick or as efficient.

The loss of myelin in MS will lead to a conduction block in the affected axons and with that to an acute loss of function. The loss of myelin will also affect the insulation of the axon. Under normal circumstances, current in one axon does not affect the signaling in an adjacent axon due to the insulating effects of the myelin. When the myelin sheath has been lost, "cross talk" between axons can occur, which can result in **paresthesias**, or abnormal sensations.

A permanent loss of function in MS is caused by axonal loss and neuronal death. This axonal loss is due to the loss of the protective role of the myelin sheath, the insertion of faulty Na$^+$ channels, and the failure to remyelinate.

Clinical Case:

Sarah's Epilepsy

Sarah is a 27-year-old female brought to the emergency room by someone who witnessed what they thought was a seizure: shaking of the body (flexion and extension of her arms and legs) and loss of consciousness. The description is consistent with the hallmarks of a generalized tonic–clonic seizure. Sarah's seizure lasted for 3 minutes, during which she bit her tongue and lost control of her urinary sphincters. She was drowsy for about 20 minutes after the seizure. An epileptic seizure is a transient occurrence of symptoms and/or signs due to abnormal excessive or synchronous neuronal activity in the brain. The intermittent abnormal discharges of a network of neurons is due to an imbalance

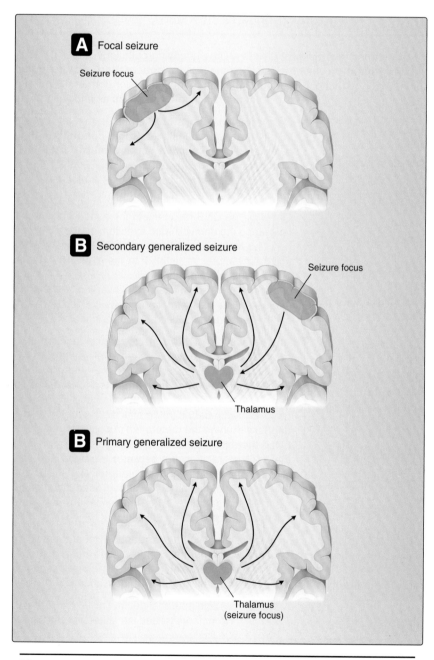

Figure 1.24
Classification of seizures. (From Golan DE, Tashjian AH, Armstrong EH, Armstrong AW. *Principles of Pharmacology*, 3rd ed. Lippincott Williams & Wilkins, A Wolters Kluwer Business, Philadelphia, PA, 2012.)

between excitation and inhibition. A tonic–clonic seizure is a type of generalized motor seizure with stiffening (tonic) followed by repetitive jerking (clonic) movements of the body.

Case Discussion

This is Sarah's fourth generalized **tonic–clonic seizure**. She also has a history of myoclonic or absence seizures. **Myoclonic seizures** are sudden brief (<100 ms) involuntary jerks of a muscle or group of muscles. **Absence seizures** are sudden-onset staring spells with interruption of ongoing activities and unresponsiveness when the patient is addressed.

Sarah was diagnosed with juvenile myoclonic epilepsy (JME). Epilepsy is a brain condition with at least two unprovoked seizures occurring more than 24 hours apart.

She takes lamotrigine, an anticonvulsant that reduces the frequency of her seizures. She has been on lamotrigine, a Na^+ channel blocker, for 5 years and was previously on valproate. Valproate has multiple mechanisms of action including blocking T-type gated calcium channels and enhancing GABA function.

Sarah's absence seizures typically last a few seconds, and on electroencephalogram (EEG) there are generalized spike and wave epileptiform discharges interictally (between her seizures).

Sarah had not taken her lamotrigine for the past 2 days because she had forgotten to get her prescription renewed. It is presumed that her most recent seizure was due to medication noncompliance (i.e., forgetting to take her lamotrigine). Sarah was advised of the importance of taking her anticonvulsant to control her epilepsy. Noncompliance is a common reason why patients whose epilepsy was previously well controlled on medications have a seizure.

How are seizures classified?

Seizures can be classified according to how they originate: they can be caused by a general abnormal discharge (**generalized**), they can originate from a specific area (**focal**), and in some cases the area of origin is unknown (Figure 1.24). Another way to classify seizures is by their underlying cause: some seizures are due to a **genetic** reason (such as a mutation in a channel), some have a **structural** etiology (such as a tumor), and some are due to **metabolic** disturbances such as might occur after an infection of the brain, or encephalitis. Some causes of seizure are yet unknown.

Generalized seizures are classified by their clinical features (tonic, clonic, tonic/clonic, atonic, myoclonic, and absence seizures) and EEG pattern.

Focal seizures originate and are typically limited to a specific area of the brain; they are subclassified according to whether there is altered awareness.

Structural–metabolic etiologies include tumor, trauma, previous strokes, and encephalitis (infection of the brain). Genetic causes may be related to gene mutations of voltage-gated or ligand-gated ion channels. For example, specific mutations of neuronal voltage-gated Na^+ channels may cause different epilepsy syndromes with variable clinical phenotypes, such as generalized epilepsy with febrile convulsions plus (GEFS+) or myoclonic epilepsy of infancy.

Chapter Summary

- The nervous system is divided into a peripheral and central nervous system (PNS and CNS). It enables us to perceive the world around us and to interact with it. In addition, the CNS is the seat of all higher cognitive functions.

- The cellular components of the central nervous system (CNS) can be roughly divided into neurons and glia. Neurons are excitable cells and are organized in networks and pathways that process all conscious and nonconscious information. Glia are the support cells in the nervous system and have multiple functions. Some are myelinating cells, such as oligodendrocytes in the CNS and Schwann cells in the PNS. Astrocytes have many roles that include the maintenance of ion and neurotransmitter homeostasis in the extracellular space as well as the shuttling of nutrients and neurotransmitter precursors to neurons. Microglia are the immune cells of the CNS. A newly identified group of cells, the polydendrocytes, are the stem cell population in the CNS and, interestingly, receive synaptic input from neurons. They appear to be the link between the neuronal and glial networks. Ependymal cells are epithelial cells that line the ventricular system and form the choroid plexus within the ventricles that secretes cerebrospinal fluid.

- The central nervous system is separated from the body environment through the blood–brain barrier. This barrier comprises the tight junction–linked epithelial cells in blood vessels and astrocyte processes. Most substances crossing the blood–brain barrier must use active transport.

- All communication in the nervous system is via electrical signals, which are mediated through ion movements. At rest, the movement of ions is at an equilibrium, expressed through the Nernst equation for a single ion and through the Goldmann equation for the sum of all ions that cross the plasma membrane. Ions move across the membrane through different types of ion channels.

Chapter Summary continued

- The currency of communication between neurons is the action potential. An action potential is generated through the opening of voltage-gated Na^+ channels. When a cell accumulates enough positive charge to reach threshold, the cell is depolarized. After a short-lived opening of Na^+ channels, K^+ leaves the cell, leading to hyperpolarization.

- Current is measured in amperes (A) and describes the movement of charge or movement of ions. The amount of work necessary to move that charge is described as the voltage and measured in volts (V). The difficulty of moving ions is referred to as resistance and measured in ohms (Ω). Conductance is the ease of moving ions and measured in siemens (S).

- Action potentials (APs) are propagated along an axon through both passive and active current. Passive current is the shuttling of charge, whereas active current is the flow of ions through ion channels. Continuous conduction means that passive current moves along an axon and opens Na^+ channels along the way (active current), effecting a continuous regeneration of the AP. Saltatory conduction happens in myelinated axons where passive current moves along the myelinated part of the axon and opens Na^+ channels at gaps in the myelin (nodes of Ranvier). The velocity of an AP depends on the velocity of active and passive current. Passive current can be accelerated through reducing resistance by increasing axon diameter and decreasing leak current through myelination. Active current can be accelerated by reducing the capacitance of the membrane, either through reducing axon diameter or through myelination.

- Neurons communicate with each other through synapses. Electrical synapses are formed by gap junctions coupling two neurons. Ions flow through these gap junctions and directly depolarize these neurons in synchrony.

- Chemical synapses are the most common type of synapse in the central nervous system. They comprise a presynaptic terminal, a synaptic cleft, a postsynaptic terminal, and an astrocyte process. When an action potential reaches an axon terminal, Ca^{2+} channels open and the influx of Ca^{2+} causes vesicles filled with neurotransmitter to fuse with the membrane releasing neurotransmitter into the synaptic cleft. There are many types of neurotransmitters, each of which binds to a specific receptor and has a specific effect.

- The neurotransmitter then binds to a neurotransmitter receptor, which can be coupled to an ion channel (ionotropic receptors) or to intracellular signaling cascades (metabotropic receptors). The resulting flow of ions creates the postsynaptic potential. When positively charged ions (i.e., Na^+ influx) flow into the postsynaptic cell, the result is an excitatory postsynaptic potential. When negatively charged ions flow (i.e., Cl^- influx) into the postsynaptic cell or cations leave the cell (i.e., K^+ efflux), the result is an inhibitory postsynaptic potential.

- When a sufficient number of excitatory postsynaptic potentials come together in time and space (temporospatial summation), the postsynaptic cell depolarizes sufficiently to reach threshold, and an action potential is generated.

Study Questions

Choose the ONE best answer.

1.1 A patient comes with a wound that requires suturing. You apply a local anesthetic, which blocks the propagation of action potentials. Action potentials are generated by which of the following mechanisms?

   A. The opening of Ca^{2+} (calcium) channels.
   B. The closing of K^+ (potassium) channels.
   C. The opening of Na^+ (sodium) channels.
   D. The opening of K^+ (potassium) channels.
   E. The closing of Ca^{2+} (calcium) channels.

> Correct answer is C. An action potential is generated through the opening of voltage-gated Na^+ channels, not any of the others. A local anesthetic, such as lidocaine, blocks Na^+ channels, and action potentials cannot be generated, which effectively prevents the propagation of the pain signal.

1.2 Which one of the following statements about glia is correct?

   A. Microglia are a stem cell population in the central nervous system.
   B. Polydendrocytes myelinate axons in the central nervous system.
   C. Astrocytes can secrete neurotransmitters.
   D. Schwann cells myelinate axons in the central nervous system.
   E. Ependymal cells are part of the blood–brain barrier.

> Correct answer is C. Polydendrocytes, not microglia, are the stem cell pool in the brain. Oligodendrocytes myelinate axons in the central nervous system, and Schwann cells myelinate axons in the peripheral nervous system. Astrocytes can secrete neurotransmitters at the synaptic cleft and thereby modulate the activity at that synapse. Ependyma is the epithelial lining of the ventricles.

1.3 Which one of the following statements best describes the cell membrane as a capacitor?

A. The cell membrane stores charges to facilitate ion movements.

B. The cell membrane accumulates charges, which hinders the movement of ions.

C. The cell membrane binds ions to its surface, allowing ions to move quickly when needed.

D. The cell membrane is selectively permeable to ions.

E. The cell membrane selectively blocks the movement of cations.

Correct answer is B. The cell membrane acts as a capacitor in that it separates and accumulates opposite charges on either side. These charges must be overcome each time Na^+ (sodium) is to enter the cell to generate (or regenerate) an action potential.

1.4 A patient is diagnosed with the peripheral nerve demyelinating disorder Guillain-Barré syndrome. He shows both sensory and motor deficits in his arms and legs. Which one of the following statements describes the underlying cause for some of his symptoms?

A. A deficit in oligodendrocyte function leads to focal demyelination of axon bundles.

B. Demyelination leads to a decrease in membrane capacitance, which delays the propagation of action potentials.

C. The most common symptom is motor weakness because Schwann cells only myelinate motor axons.

D. Axonal damage is due to microglial migration into the myelin sheath and phagocytosis of axonal segments.

E. Nerve conduction velocities are decreased because action potentials cannot be regenerated at the next cluster of Na^+ channels.

The correct answer is E. Oligodendrocytes are the myelinating cells in the central nervous system (CNS), and Schwann cells are the myelinating cells in the peripheral nervous system (PNS). Myelin decreases the membrane capacitance, and due to the insulating properties of the myelin sheath, charges are no longer accumulated and stored at the cell membrane. Demyelination increases the capacitance of the membrane, charges are accumulated, and every time a cation crosses the membrane, it needs to overcome the accumulated charge. Schwann cells myelinate both motor and sensory axons. Microglia are the macrophages in the CNS. In the PNS, phagocytosis is through blood-borne macrophages. Only in severe cases of Guillain-Barré syndrome are axons damaged. Damage to the myelin sheath is the first step in the disease process. The nerve conduction velocity is decreased due to the demyelination. Charges are lost through leak current and may not reach the next cluster of Na^+ channels at the next (now absent) node of Ranvier where the action potential would be regenerated through active current.

1.5 The blood–brain barrier isolates the neuronal environment from blood-borne pathogens and substances. This can make drug delivery to the central nervous system a challenge. What are the component parts of the blood–brain barrier?

A. The endothelium and microglia.

B. The basement membrane and the endothelium.

C. The endothelium and astrocytes.

D. The basement membrane and oligodendrocytes.

E. The endothelium and the neuropil.

The correct answer is C. Endothelial cells are linked to each other by tight junctions, and astrocyte processes ("end-feet") surround the vessel wall. This effectively separates the blood compartment from the neuropil compartment. Transport across the blood–brain barrier can be by diffusion of small lipophilic molecules, water, and gas. All other substances must use active transport.

1.6 Which of the following is a possible cause for the abnormal excessive excitation of the cortical neurons that led to Sarah's seizure in Clinical Case 1.1?

A. Impaired voltage-gated sodium channel function.

B. Impaired Na^+/K^+ pump, which normally pumps sodium into the cell.

C. Impaired ionotropic $GABA_B$ inhibitory receptor function.

D. Impaired metabotropic $GABA_A$ inhibitory receptor function.

E. Impaired metabotropic glutamate inhibitory receptor function.

The correct answer is A. Impaired voltage-gated sodium channel function due to mutations of this channel can result in various types of epilepsies. The Na^+/K^+ pump pumps sodium *out of* the cell in exchange for K^+. $GABA_B$ receptors are metabotropic ligand-gated receptors, while $GABA_A$ receptors are ionotropic ligand-gated receptors. Glutamate is an excitatory neurotransmitter.

2 Overview of the Central Nervous System

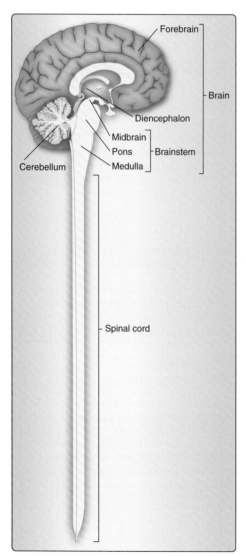

Figure 2.1
Overview of the central nervous system.

I. OVERVIEW

The human central nervous system (CNS) consists of the brain and spinal cord. The human brain weighs about 400 g at birth, and this weight triples during the first 3 years of life, primarily due to the addition of myelin and the growth of neuronal processes. The adult brain weighs approximately 1,400 g and is thus a relatively small structure, constituting about 2% of body weight. However, our distinctly human mental capacities are related not so much to the size of our brains but rather to the complexity of neuronal interconnections and the differential development of the different areas of the cerebral cortex with their unique higher cortical functions. The brain is concerned with functions as diverse as thought, language, learning and memory, imagination, creativity, attention, consciousness, emotional experience, and sleep. In addition, the brain regulates or modulates visceral, endocrine, and somatic functions.

In some ways, the spinal cord is a simpler part of the CNS than the brain in that it has a uniform organization throughout its course. However, processing within the spinal cord is complex, and it serves extremely important functions: It receives extensive sensory information about the world around us and performs the initial processing of this input. The spinal cord carries all of the motor information that supplies our voluntary muscles and thus participates directly in control of body movement. It also plays a direct role in regulating visceral functions. Importantly, it serves as a *conduit* for the longitudinal flow of information to and from the brain.

In this chapter, we provide a broad overview of the brain and spinal cord and their organization (Figure 2.1), as well as the ventricular system, meningeal coverings, and blood supply. A much more detailed discussion of each topic covered here can be found in the following chapters.

II. DEVELOPMENT OF THE NERVOUS SYSTEM

An understanding of nervous system development is important in understanding its adult geometry and organization.

Three primary germ layers develop in the early embryo: the ectoderm, mesoderm, and endoderm. The endoderm develops into the internal organs ("viscera"). The mesoderm gives rise to the somites, segmented structures that develop into bone, skeletal muscle, and dermis of the skin. The ectoderm develops into neural structures and the epidermis of the skin.

Innervation to structures derived from the somites (from mesoderm) is through the somatic part of the nervous system. Innervation to structures derived from the endoderm is through the visceral part of the nervous system.

A. Development of the neural tube

Whereas the adult nervous system is quite complex, the origin of the nervous system is from a simple ectodermal tube. Development begins around the third week of gestation when a longitudinal (rostral–caudal) band of ectoderm thickens to form the neural plate. This process is initiated by a rodlike structure, the **notochord**, which is the primary inductor in the early embryo. A midline groove soon appears on the posterior surface of the neural plate, and the neural plate begins to fold inward, as illustrated in Figure 2.2. As the groove deepens, neural folds appear on each side of the groove. These folds then begin to approach each other, and by the end of the third week of development, the neural folds begin to fuse, forming a neural tube. If a portion of the neural tube fails to develop or the neural folds do not close completely, this can cause defects in the spinal cord and in

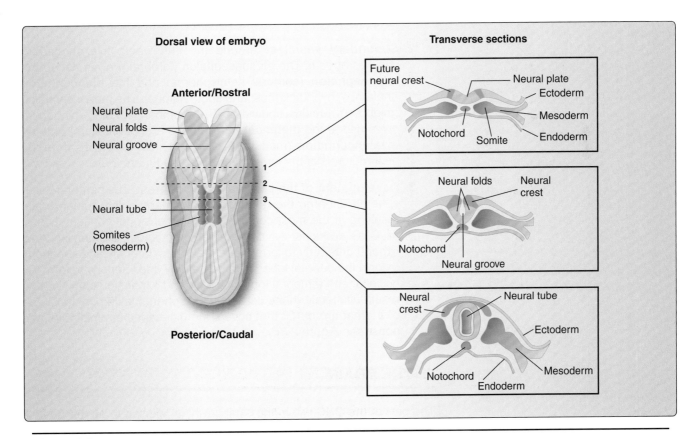

Figure 2.2
Development of the neural tube. The transverse sections show the sequence of neural tube closure.

the bones of the spine, a condition known as spina bifida. The most common location of spina bifida is the lower back, but in rare cases, it may occur in the middle back or neck and may result in a wide range of physical and cognitive disabilities.

The rostral end of this tube develops into the brain, and the remainder develops into the spinal cord (see Figure 2.2). Fusion of the neural tube occurs from the middle out toward the rostral and caudal ends, and as it occurs, cells from the top or crest of each neural fold dissociate from the neural tube. These **neural crest cells** migrate away from the neural tube and differentiate into a variety of cell types including the sensory neurons in the ganglia of the spinal nerves and some cranial nerves (V, VII, VIII, IX, and X), postganglionic neurons of the autonomic nervous system, the Schwann cells of the peripheral nervous system (PNS), and the adrenal medulla. As the neural tube closes, it separates from the ectodermal surface and thus becomes enclosed within the body.

B. Development of the brain

Development of the brain begins during the fourth week of gestation when differential growth results in enlargements (**vesicles**) and bends (**flexures**) in the neural tube.

1. **Primary vesicles:** Three primary vesicles appear at the rostral end of the neural tube: the **prosencephalon** (which becomes the forebrain), the **mesencephalon** (which becomes the midbrain), and the **rhombencephalon** (which becomes the hindbrain), the latter merging with the spinal portion of the neural tube (Figure 2.3).

2. **Secondary vesicles:** Around the fifth week, five secondary vesicles appear. The prosencephalon (forebrain) gives rise to the **telencephalon** (cerebral hemispheres) and the **diencephalon** (thalamus, hypothalamus, and subthalamus). The mesencephalon (midbrain) remains undivided. The rhombencephalon (hindbrain) gives rise to the **metencephalon** (pons and cerebellum) and the **myelencephalon** (medulla). The cerebellum is formed from the posterior part of the metencephalon (see Figure 2.3).

3. **Development of the cerebral hemispheres:** The cerebral hemispheres undergo the greatest development in the human brain, resulting in the most complex three-dimensional configuration of all CNS divisions. As development proceeds, the hemispheres expand anteriorly to form the frontal lobes, laterally and superiorly to form the parietal lobes, and posteriorly and inferiorly to form the occipital and temporal lobes. Growth and expansion continue and result, ultimately, in the cerebral hemispheres taking on the shape of a great arc or "C" that covers the diencephalon, midbrain, and pons (see Figure 2.3).

III. THE BRAIN

The part of the CNS within the skull cavity is referred to as the *brain*. It consists of the forebrain (from the prosencephalon), the midbrain (from the mesencephalon), and the hindbrain (from the rhombencephalon). The forebrain consists of the cerebral hemispheres and deep structures.

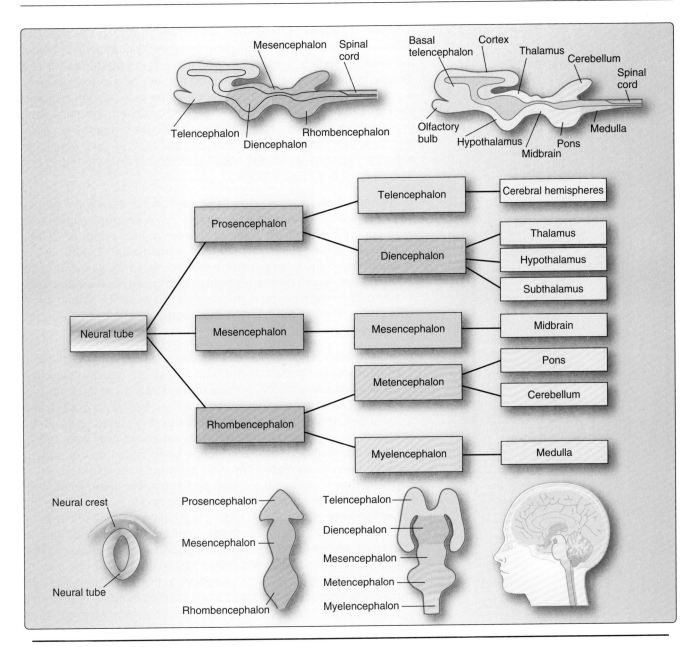

Figure 2.3
Development of the brain.

The midbrain and hindbrain are collectively referred to as the *brainstem*, with the hindbrain further divided into the *pons*, *medulla*, and *cerebellum* (see Figure 2.3).

A. Orientation in the brain

During development, the orientation of the neural tube is straightforward: There is a rostral pole (*rostrum* is Latin for "beak") and a caudal pole (*cauda* is Latin for "tail"). A ventral or anterior surface and a dorsal or posterior surface can be described. The bulging and bending that occur during brain development that ultimately allow for upright gait with two eyes looking forward change the simple arrangement of the neural tube. In the forebrain, the **ventral** surface of the brain

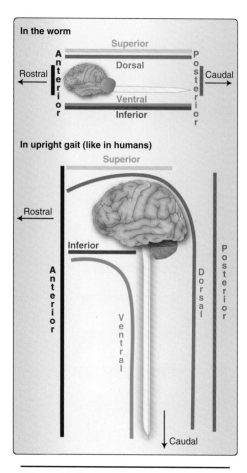

Figure 2.4
Orientation in the central nervous system.

is also the **inferior** surface, and the **dorsal** surface is the **superior** surface. In the brainstem and spinal cord, the ventral surface is also the **anterior** surface, and the dorsal surface is the **posterior** surface. **Rostral** describes anything toward the anterior pole of the forebrain. It accounts for the flexure happening at the level of the midbrain. **Caudal** describes anything toward the inferior pole of the spinal cord or toward the "tail." Throughout this book, we use the terms "anterior" and "posterior," "superior" and "inferior," and "rostral" and "caudal" to describe locations within the spinal cord and brainstem. Figure 2.4 summarizes these orientations.

1. **Planes of orientation:** The brain can be cut in different planes of orientation, as illustrated in Figure 2.5. A **coronal** section cuts through the brain from superior to inferior, like a tiara or *corona* (Latin for "crown") sitting on the head. A **horizontal** or **axial** section cuts through the brain parallel to the ground, in the same plane as the horizon would be in an upright standing person. A **sagittal** section cuts through the brain from anterior to posterior, like an arrow shooting through the brain. A **midsagittal** section separates the two hemispheres.

2. **Gray matter and white matter:** Gray matter is any collection of neuronal cell bodies. In the brain, these are found in the cortical layer (cortex) on the surface of the forebrain and cerebellum (Figure 2.6). In order to accommodate the massive number of neuronal cell bodies that make up the cortex, the surface area of the human forebrain and cerebellum has expanded over time, resulting in the grooves (sulci) and ridges (gyri) visible on the surface of the brain. Gray matter can also be found in deep structures of the forebrain, the basal ganglia, and structures of the limbic system. White matter consists of fiber tracts and is found deep to the cortical gray matter in the brain (see Figure 2.6). Gray matter is dispersed throughout the substance of the brainstem comprising intrinsic systems and cranial nerve nuclei. A **nucleus** is a collection of nerve cell bodies within the CNS. In the PNS, a collection of nerve cell bodies is called a **ganglion**. In the spinal cord, the gray matter is located centrally and is surrounded by white matter (see Figure 2.6). A **tract** is a bundle of axons traveling from one area to another within the CNS, whereas in the PNS, a bundle of axons is called a **nerve**. These axons are mostly myelinated, resulting in the "white" appearance of the white matter.

B. Forebrain

The **forebrain** consists of the **telencephalon** and the **diencephalon**, derived from the most rostral parts of the developing neural tube. The components of the forebrain are summarized in Figure 2.7. The **telencephalon** is composed of the massive cerebrum, which is divided into two cerebral hemispheres. The cerebral hemispheres consist of a covering of gray matter (the cerebral cortex); gray matter structures deep within the cerebrum, including the basal ganglia and two major limbic system structures (the hippocampus and the amygdala); and underlying white matter. The **diencephalon** is composed of the thalamus, hypothalamus, and subthalamus, which are all gray matter structures.

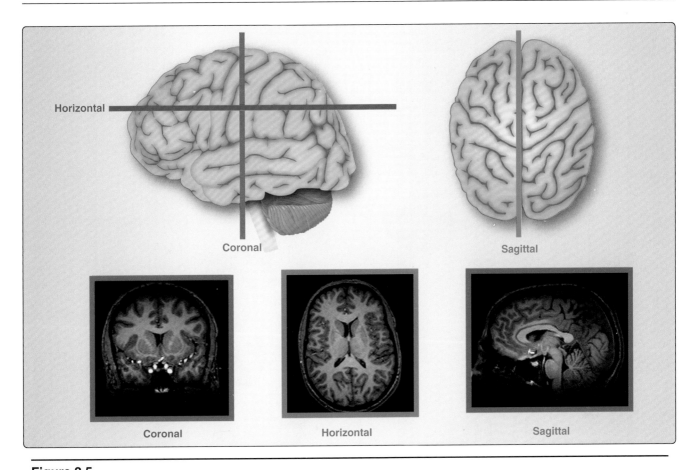

Figure 2.5
Planes of orientation.

1. **Cerebral hemispheres:** The two large cerebral hemispheres are almost mirror images of each other in terms of their gross anatomy, although there is some asymmetry of function that is discussed later (see Chapter 13, "The Cerebral Cortex," for more detail). Each hemisphere is divided into four lobes, named for the cranial bones that overlie them—the frontal, parietal, occipital, and temporal lobes. Ridges in the cortex are called **gyri** (singular, *gyrus*) and grooves are called **sulci** (singular, *sulcus*) or **fissures** (deeper grooves). The **longitudinal fissure** located along the midsagittal plane separates the two hemispheres. The **lateral** or **sylvian fissure** separates the temporal lobe from the frontal and parietal lobes. The **parietooccipital sulcus** is visible on the medial surface of the brain and separates the occipital lobe from the parietal lobe. Figure 2.8 gives an overview of the general anatomy of the cerebral hemispheres and Figure 2.9 provides a general overview of the functional areas of the cortex.

   a. **Frontal lobe:** The frontal lobe is the largest of the brain and has multiple functions. It is separated from the parietal lobe by the **central sulcus** and from the temporal lobe by the **lateral fissure**. The **precentral gyrus**, located anterior to the central sulcus, contains the primary motor areas. Areas on both the lateral and medial surfaces are essential not only

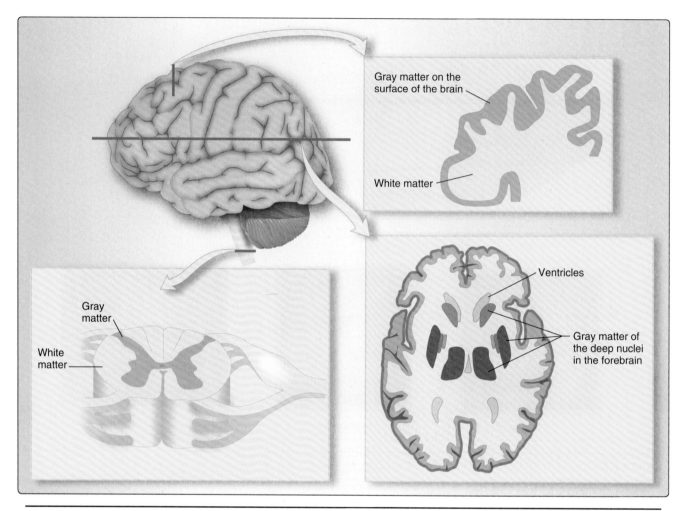

Figure 2.6
Gray matter and white matter.

for regulating voluntary motor activity or behavior but also for initiating motor behavior, that is, "deciding" which movements should be performed. Expressive or motor aspects of language are also processed on the lateral surface of the frontal lobe, primarily in the dominant (typically left) hemisphere. The remainder of the frontal lobe consists of association areas known as the **prefrontal association areas**. These are concerned with functions such as emotion, motivation, personality, initiative, judgment, ability to concentrate, and social inhibitions. An area on the medial surface, the **cingulate gyrus**, is also important for modulating emotional aspects of behavior.

b. **Parietal lobe:** The parietal lobe is important in regulating somatosensory, language, and spatial orientation functions. It is separated from the frontal lobe by the central sulcus, from the temporal lobe by the lateral fissure, and from the occipital lobe by the **parietooccipital sulcus**. The **postcentral gyrus** is the primary somatosensory area of the cortex. Initial cortical processing and perception of touch, pain, and limb position occurs

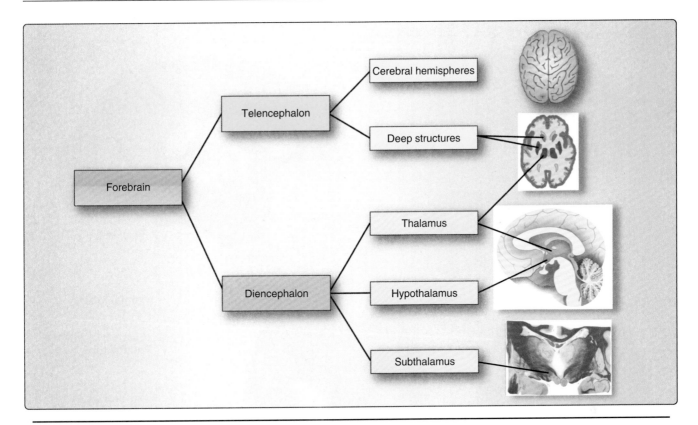

Figure 2.7
Components of the forebrain.

on both the lateral and medial aspects of the parietal lobe. Receptive or sensory aspects of language are also processed in the parietal lobe primarily in the dominant (typically left) hemisphere. The parietal lobe also mediates complex aspects of spatial orientation and perception, including self-perception and interaction with the world around us.

c. **Occipital lobe:** The occipital lobe is primarily involved in processing visual information. It is separated from the parietal lobe by the parietooccipital sulcus. The primary visual area is located on the medial surface of the occipital lobe. Visual association areas surround it and cover the lateral surface of this lobe. They mediate our ability to see and recognize objects.

d. **Temporal lobe:** The temporal lobe is important for processing auditory information, language, and certain complex functions. It is separated from the frontal and parietal lobes by the lateral fissure and from the occipital lobe by a line that can be drawn as an extension of the parietooccipital sulcus. It is important for processing auditory information. The **superior temporal gyrus** is the area where our ability to both hear and interpret what we hear is processed. In addition, an area on the lateral surface of the temporal lobe functions for perception of language. Anterior medial areas of the temporal lobe are important in complex aspects of limbic system function (see "Limbic lobe" below).

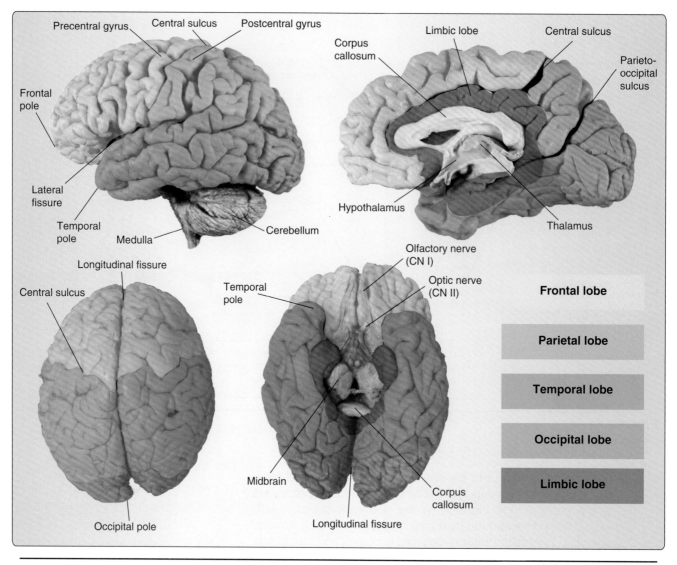

Figure 2.8
Cerebral hemispheres, major sulci and gyri, and lobes of the cerebrum. CN = cranial nerve.

e. Limbic lobe: In addition to these four lobes, a ring of cortex on the medial surface, the **cingulate** and **parahippocampal gyri**, is typically referred to as the "limbic lobe." This is not a true discrete lobe like the other four but rather covers parts of the frontal, parietal, and temporal lobes. This area of cortex overlies and is interconnected with structures of the limbic system and is important in processing complex aspects of learning, memory, and emotion.

2. Deep structures: Underneath the cortical layer containing the neuronal cell bodies, extensive white matter tracts connect the various parts of the brain with one another. In addition, large gray matter nuclei can be found deep within the forebrain. These are the basal ganglia and major structures of the limbic system including the hippocampus and amygdala.

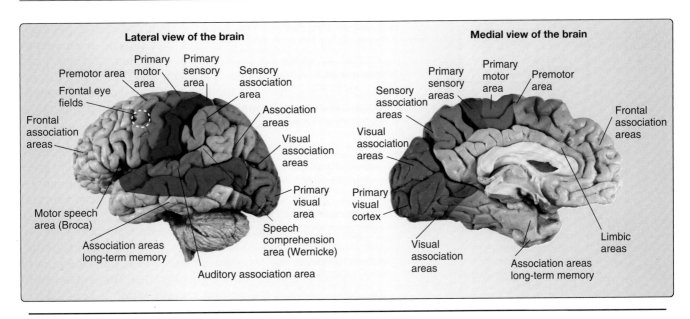

Lateral view of the brain

Premotor area
Primary motor area
Primary sensory area
Sensory association area
Frontal eye fields
Association areas
Frontal association areas
Visual association areas
Motor speech area (Broca)
Primary visual area
Association areas long-term memory
Speech comprehension area (Wernicke)
Auditory association area

Medial view of the brain

Primary sensory areas
Primary motor area
Premotor area
Sensory association areas
Frontal association areas
Visual association areas
Primary visual cortex
Visual association areas
Association areas long-term memory
Limbic areas

Figure 2.9
Functional areas of the cerebral hemispheres.

a. **Basal ganglia:** The basal ganglia are a group of interconnected, interacting nuclei within the forebrain, diencephalon, and midbrain (note that these are truly large nuclei, but the term "ganglia" is commonly used; Figure 2.10). The forebrain components of the basal ganglia lie deep within the cerebral hemispheres and include the **caudate** and **lenticular** (**putamen** and **globus pallidus**) **nuclei**. The nucleus in the diencephalon is the subthalamic nucleus and in the midbrain is the substantia nigra. Together, the basal ganglia play a critical role in the initiation and control of voluntary movements.

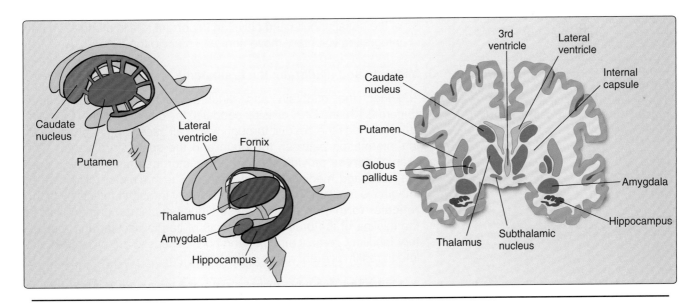

Caudate nucleus
Putamen
Lateral ventricle
Fornix
Thalamus
Amygdala
Hippocampus

3rd ventricle
Lateral ventricle
Caudate nucleus
Internal capsule
Putamen
Globus pallidus
Amygdala
Thalamus
Subthalamic nucleus
Hippocampus

Figure 2.10
Deep structures of the forebrain: basal ganglia and limbic structures.

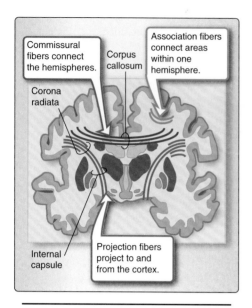

Figure 2.11
Fiber types within the central nervous system.

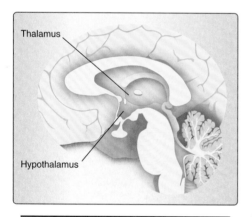

Figure 2.12
The diencephalon (thalamus and hypothalamus) on a midsagittal section through the brain.

b. Limbic structures: The limbic system consists of interconnected and interacting structures that provide a neuroanatomical substrate for drive-related, motivated, and emotional behaviors and play a role in learning and memory. The deep forebrain components of the limbic system include the **amygdala** and **hippocampus**, both of which are located in the temporal lobe (see Figure 2.10).

c. White matter: Deep to the cerebral cortex, white matter tracts are important in connecting various cortical areas to each other (Figure 2.11). **Association fibers** interconnect various cortical regions within the same hemisphere. **Commissural fibers** reciprocally connect areas of cortex in one hemisphere with corresponding areas of the opposite hemisphere. By far, the largest set of commissural fibers is the **corpus callosum**, which connects the two hemispheres. **Projection fibers** carry information to and from the cerebral cortex. The largest set of projection fibers is the **corona radiata**, which is bundled into the **internal capsule** as it passes through the cerebrum and contains all the fibers traveling between the cortex, the deep forebrain structures, and the spinal cord.

3. Diencephalon: The diencephalon consists of several sets of paired structures on either side of the third ventricle (Figure 2.12). The largest structure is the **thalamus**, consisting of two "egg-shaped" nuclear masses. The thalamus is a critical processing station for all sensory information (except olfactory) on its way to the cortex and plays key roles in processing motor information, in integrating higher-order cognitive and emotional information, and regulating cortical activity. The thalamus can be considered the "gatekeeper" for the cortex. The **hypothalamus** is structurally part of the diencephalon but functionally part of the limbic system. It plays key roles in coordinating and integrating endocrine, autonomic, and homeostatic functions. The **subthalamus** is part of the basal ganglia and plays an important role in modulating and integrating voluntary movement and muscle tone.

C. Midbrain and hindbrain: the brainstem

Together, the midbrain and hindbrain comprise the brainstem (Figure 2.13), the CNS division caudal to the diencephalon. The brainstem provides the conduit through which all ascending and descending information between the brain and the spinal cord travels. The **cranial nerves** provide sensory and motor information to and from the head and mediate special senses. The brainstem also has an integrative function. This is achieved largely through the activity of the **reticular formation nuclei**. These diffuse nuclei run primarily along the midline of the brainstem. Aspects of cardiovascular and respiratory function, cortical arousal, and consciousness are organized and integrated in different areas of the brainstem.

1. Midbrain: This most rostral area of the brainstem is characterized by the large pair of **cerebral peduncles** on its anterior surface and two pairs of nuclei, the **superior** and **inferior colliculi**, on its posterior surface (see Figure 2.13). The descending **corticospinal tract** that mediates voluntary motor activity travels through the cerebral

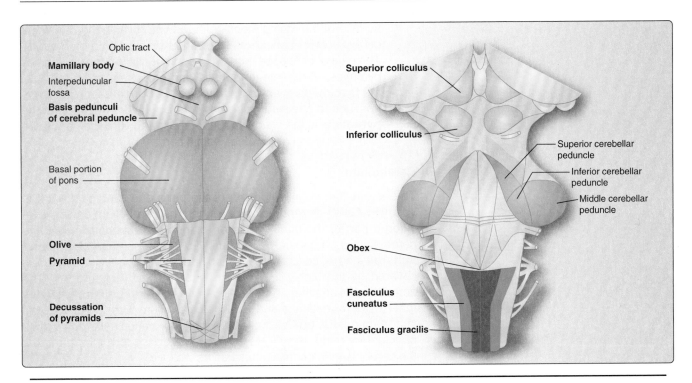

Figure 2.13
Overview of the surface anatomy of the brainstem.

peduncles. The superior colliculi in the rostral midbrain are involved in visual reflexes, whereas the inferior colliculi, in the caudal midbrain, are a major integrating center in the auditory pathway. The **cerebral aqueduct** connecting the third and fourth ventricles is found in the posterior area of the midbrain. Important internal structures of the midbrain include the red nucleus and the substantia nigra, both of which play roles in coordinating motor activity.

2. **Hindbrain:** The hindbrain comprises the **pons** and **medulla**.

   a. **The pons:** The pons is characterized by the prominent anterior or **basal pons**. The basal pons consists of bundles of descending, longitudinal **corticospinal fibers**, and transverse **pontocerebellar fibers** that carry information from the **pontine nuclei** to the opposite cerebellum through the **middle cerebellar peduncles**, which arise off the lateral surface of the basal pons. The posterior surface of the pons consists of the **fourth ventricle** and rostrally the **superior cerebellar peduncles**, containing primarily cerebellar efferents, which form most of the roof of the fourth ventricle (see Figure 2.13).

   b. **The medulla:** The medulla is the most caudal part of the brainstem and merges at its most caudal end with the spinal cord (see Figure 2.13). The rostral or "open" portion of the medulla is characterized by the **pyramids** (descending corticospinal fibers) on its anterior surface and the caudal part of the **fourth ventricle** on its posterior surface. The **inferior cerebellar peduncles**, carrying information from the spinal cord and brainstem, can also be seen on its posterolateral surface. Oval swellings on the lateral

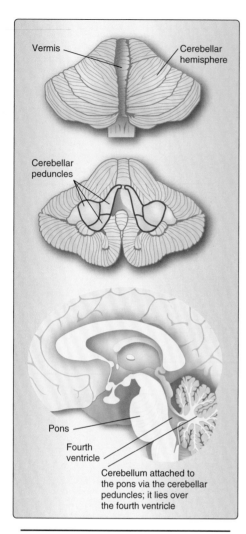

Vermis — **Cerebellar hemisphere**

Cerebellar peduncles

Pons — **Fourth ventricle** — Cerebellum attached to the pons via the cerebellar peduncles; it lies over the fourth ventricle

Figure 2.14
Overview of the surface anatomy of the cerebellum.

surface of the rostral medulla, the **olives**, overlie the prominent olivary nuclear complexes involved in modulating motor activity. The caudal or "closed" portion of the medulla is characterized by pyramids on its anterior surface and prominent sensory tracts, the **fasciculus gracilis** and **fasciculus cuneatus**, on its posterior surface. The pyramids cross, forming the **decussation of the pyramids**, in the caudal medulla. The fourth ventricle narrows to form the central canal of the spinal cord.

D. Cerebellum

The cerebellum is an outgrowth of the pons and overlies the fourth ventricle. Given its embryological origin, the cerebellum can be considered part of the pons. However, many neuroanatomists treat the cerebellum as a separate structure. The cerebellum has **two hemispheres** and a central area called the **vermis**, and its surface is covered by the highly folded **cerebellar cortex** (Figure 2.14). The cerebellum is attached to the brainstem by the **cerebellar peduncles** that carry information to and from the cerebellum. The cerebellum has important roles in processing of sensory information and coordination of voluntary motor activity. More recent evidence indicates a role for the cerebellum in cognitive function as well.

IV. THE SPINAL CORD

The spinal cord is a long, cylindrical, segmented structure that has a consistent organization throughout its course. It receives sensory input from the limbs, trunk, and many internal organs and plays an important role in the initial processing of this information. It also contains the somatic motor tracts that supply the skeletal muscles as well as visceral efferents to our viscera, smooth muscles, and glands.

A. Spinal cord organization

The spinal cord has a clear segmental organization, corresponding to the nerve roots attached to it. A continuous series of **posterior rootlets**, containing sensory axons, enter the posterior aspect of the spinal cord, and a continuous series of **anterior rootlets**, containing motor axons, emerge from the anterior aspect of the spinal cord. These sensory and motor axons merge together in the **spinal nerves**, which are part of the PNS. A bulge in the posterior root, just proximal to the point where the spinal nerve forms, is the **spinal ganglion**, containing the cell bodies of the sensory nerve fibers. Two areas within the spinal cord, the **cervical enlargement** and the **lumbosacral enlargement**, contain increased numbers of motor neurons to supply the arms and legs, respectively (Figure 2.15). At its caudal pole, the spinal cord tapers off into the **conus medullaris** and ends in the **filum terminale**, which anchors the spinal cord to the dorsum of the coccyx.

B. Spinal cord length

The segments of the spinal cord are not at the same level as the intervertebral foramina through which the spinal nerves exit. Until about the 3rd month of fetal life, the spinal cord and the vertebral column grow at about the same rate, and the spinal nerves exit through intervertebral foramina that are lateral to them and at the same level. From then on, however, the vertebral column grows faster than the spinal cord, and the posterior and anterior roots increase in length as they travel further to exit through the appropriate intervertebral foramina. The lumbosacral roots are the longest and form the **cauda equina**. In the newborn, the spinal cord ends at about the third lumbar vertebral level (L3). A small amount of growth occurs postnatally, and, in the adult, the spinal cord ends at approximately the L1–L2 level.

C. Spinal cord gray and white matter

Finally, in contrast to the cerebral and cerebellar hemispheres, where white matter forms the inner core and gray matter surrounds the white matter, in the spinal cord, gray matter forms the inner core, and white matter surrounds the gray matter.

V. THE VENTRICULAR SYSTEM

The ventricular system is a fluid-filled space within the brain, which is continuous with the central canal of the spinal cord. See Figure 2.16 for an overview. The **cerebrospinal fluid (CSF)** circulating within the ventricles is a secretion of the ependymal cells of the choroid plexus.

A. Ventricles

Two **lateral ventricles** are associated with the telencephalon, one in each cerebral hemisphere. The lateral ventricles are "C" shaped, due to the curvature of the brain during embryonic development. The **anterior horn** of the lateral ventricle is deep in the frontal and parietal lobes of the forebrain and anatomically associated with the basal ganglia, in particular the head of the caudate nucleus. The **posterior horn** extends into the occipital lobe, and the **inferior horn** is in the temporal lobe. The hippocampus is located in the floor of the inferior horn. The **septum pellucidum** separates the two lateral ventricles on the medial surface, such that they cannot communicate with each other. The lateral ventricles are connected with the third ventricle through the **interventricular foramen (foramen of Monro)**. The **third ventricle** is associated with the diencephalon, and the thalamus and hypothalamus are located on either side of the third ventricle. The third ventricle is connected to the fourth ventricle through the **cerebral aqueduct**. The **fourth ventricle** is associated with the hindbrain and lies between the cerebellum and the pons and medulla. The superior medullary velum closes off the fourth ventricle on the posterior surface between the two superior cerebellar peduncles.

B. Cerebrospinal fluid

CSF fills the ventricles and surrounds the brain and spinal cord in the subarachnoid space, the space between the arachnoid and the pia mater (see below). Its main function is to support and cushion the brain. Due to a difference in the specific gravity between the brain and the CSF, the apparent weight of the brain is lessened. In addition, the CSF has a function similar to that of the lymph system in the rest of the body. There is continuous exchange between the brain parenchyma and the CSF. This is the reason why a sampling of CSF can provide

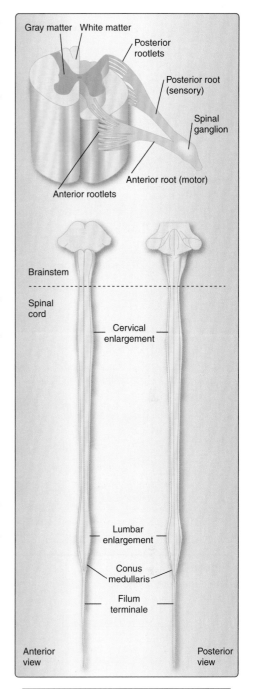

Figure 2.15
The spinal cord.

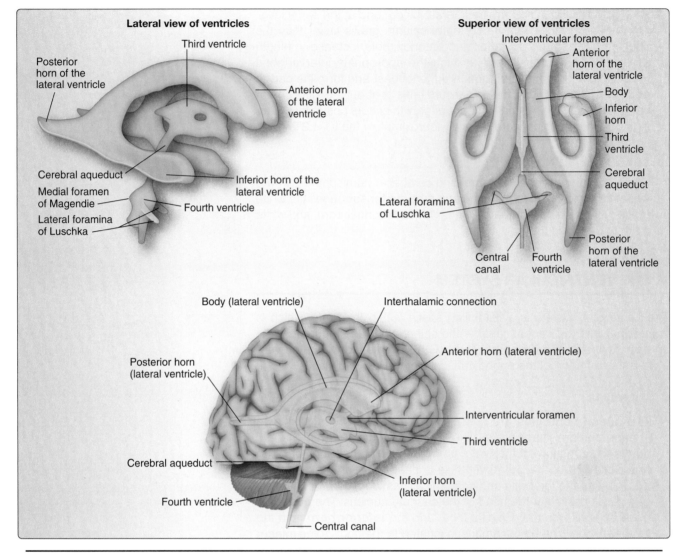

Figure 2.16
The ventricular system.

information about brain pathologies. In addition, periventricular neurons (neurons close to or around the ventricular system) can secrete neurotransmitters such as serotonin into the ventricular system, and these neurotransmitters then have widespread effects throughout the brain. Clinically, this exchange between the brain parenchyma and the CSF is used when drugs are applied intrathecally or directly into a CSF-filled space.

1. **Production of cerebrospinal fluid:** CSF is produced mainly in the **choroid plexus** located within the two lateral ventricles, the third ventricle, and the fourth ventricle. Some production of CSF also occurs through the brain parenchyma. The choroid plexus is a three-layered structure composed of the fenestrated endothelium of the choroid arteries, a pial layer, and a specialized ependymal layer (Figure 2.17). CSF is produced by passage of substances through the fenestrated endothelium and active transport through the ependymal cells. There is also a considerable amount of

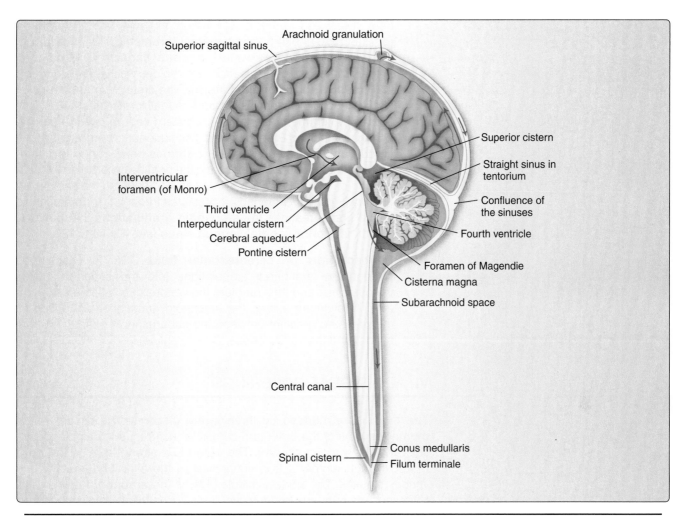

Figure 2.17
Production, circulation, and absorption of cerebrospinal fluid.

passive transport and diffusion of water to maintain osmolarity. The composition of CSF is similar to that of plasma, but there are some major differences, such as the protein content (Table 2.1). In humans, about 500 ml of CSF are produced every day.

2. **Circulation of cerebrospinal fluid:** The continuous production of new CSF is the main motor of CSF circulation. The influences of the cardiac and respiratory cycles play a minor role. CSF moves from the lateral to the third and to the fourth ventricle. The outflow

Table 2.1: Cerebrospinal Fluid and Plasma Composition

	Cerebrospinal Fluid	Serum (Arterial)
pH	7.33	7.41
Glucose	60 mg/dL	90 mg/dL
Proteins	0.035 g/dL	7.0 g/dL
Sodium (Na⁺)	135 mEq/L	135 mEq/L
Potassium (K⁺)	2.8 mEq/L	5 mEq/L

from the fourth ventricle is through the two **lateral foramina (of Luschka)** and through the **medial foramen (of Magendie)** into the subarachnoid space or into the **central canal** of the spinal cord. In the subarachnoid space, there are regions with large amounts of CSF called **subarachnoid cisterns**. The **cisterna magna**, or cerebellomedullary cistern, is located caudal to the cerebellum, just above the foramen magnum. This cistern can be used to sample CSF if a spinal tap is not possible. The superior cistern is above the cerebellum. The interpeduncular cistern is between the two cerebral peduncles of the midbrain, and the pontine cistern is at the caudal end of the pons. CSF moves posteriorly around the spinal cord and back up again anteriorly. It circulates through the subarachnoid space until it reaches the **arachnoid granulations** that protrude into the **superior sagittal venous sinus** (see Figure 2.17).

3. **Reabsorption of cerebrospinal fluid:** CSF is reabsorbed through the arachnoid granulations into the superior sagittal venous sinus and thus reenters the venous circulation (see Figure 2.17). Movement across the arachnoid granulations is driven by the pressure gradient between the subarachnoid space (150 mm saline) and the venous sinus (90 mm saline).

VI. MENINGEAL COVERINGS

The brain is surrounded by three layers of connective tissue. These **meninges** protect the brain and contain structures, such as blood vessels and the venous sinuses. The outermost strong layer is the **dura mater** ("tough mother"). It is connected to the skull and contains the venous sinuses. The spidery middle layer is the **arachnoid mater**. This layer lines the dura and bridges over the sulci on the surface of the brain. The inner layer is the **pia mater**. This fine layer of tissue is directly connected to the brain parenchyma and follows all gyri and sulci. Fine arachnoid trabeculae interconnect the arachnoid and pia. Figure 2.18 offers a conceptual overview of the meningeal layers.

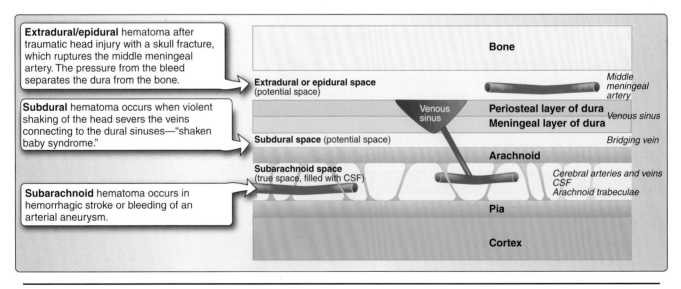

Figure 2.18
Overview of the meninges and spaces between the meningeal layers. CSF = cerebrospinal fluid.

A. Dura mater

The dura mater is the outermost and strongest layer of the meninges. It consists of two layers: a **periosteal layer** that attaches directly to the skull and a **meningeal layer**. These two layers are tightly fused, but they separate to form **venous sinuses**, into which the cerebral veins drain. **Dural reflections** (Figure 2.19) separate different compartments within the brain. The **falx cerebri**, which lies within the longitudinal fissure, partially separates the two cerebral hemispheres. It contains the **superior sagittal sinus** at its outer border and the **inferior sagittal sinus** at its free border between the hemispheres. The **tentorium cerebelli** separates the middle cranial fossa from the posterior cranial fossa. The posterior cranial fossa, or **infratentorial compartment**, contains the cerebellum and the brainstem. The **transverse sinus** runs along the outer border of the tentorium and the **straight sinus** along the attachment of the falx cerebri with the tentorium. The **falx cerebelli** is a small dural reflection that separates the two cerebellar hemispheres and contains the **occipital sinus**. The superior sagittal sinus, the transverse sinuses, the straight sinus, and the occipital sinus all meet at the posterior pole of the skull at the **confluence of sinuses**. The venous blood drains through the transverse sinus to the **sigmoid sinus** and from there to the **internal jugular vein**. The red arrows in Figure 2.19 indicate the flow of venous blood. The **diaphragma sellae** is a dural reflection that covers the pituitary fossa in the base of the skull.

1. **Innervation:** The dura mater is pain sensitive. Dura mater in the anterior and middle cranial fossae receives its innervation through meningeal branches of the **trigeminal nerve** (cranial nerve [CN] V). Dura mater in the posterior cranial fossa is innervated by meningeal branches of the vagus nerve (CN X).

2. **Blood supply:** The blood supply to the dura mater is through meningeal arteries, the most prominent of which is the middle meningeal artery. Meningeal arteries travel in the periosteal layer of the dura. A bleed in these vessels can cause the periosteal dura to dissociate from the skull creating an epidural or extradural space filled with blood (see below).

B. Arachnoid mater

The middle layer is the arachnoid mater, which is pressed against the inner surface of the dura through CSF pressure. Small strands of collagenous connective tissue, the **arachnoid trabeculae**, connect to the pia mater. **Bridging veins** pierce the arachnoid to connect to the venous sinuses within the dura. The **arachnoid granulations** are specialized parts of the arachnoid that protrude into the superior sagittal sinus and are responsible for the reabsorption of CSF. Because the arachnoid is attached to the dura, it bridges the sulci of the brain surface and the **cisterns** of the subarachnoid space (see below).

C. Pia mater

The pia mater is the innermost meningeal layer. It adheres tightly to the brain parenchyma and follows all gyri and sulci. It separates the brain from the CSF in the subarachnoid space (see Figure 2.18). As vessels penetrate the brain parenchyma from the subarachnoid space, they enter through a sleeve of pia, the **perivascular space**, which extends until the vessel becomes a capillary (Figure 2.20).

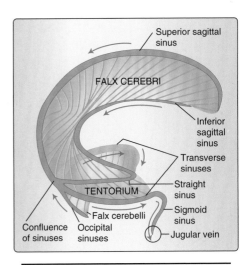

Figure 2.19
Dural reflections and sinuses.

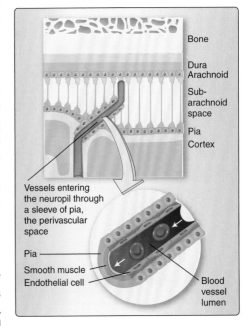

Figure 2.20
The perivascular space.

D. Spaces between the meninges

Figure 2.18 gives an overview of the true and potential spaces between the meninges.

1. **Subarachnoid space:** The only true space between the meninges is the **subarachnoid space** between the pia and the arachnoid. As noted, this space is filled with CSF and contains the arteries and veins on the surface of the brain. These vessels are suspended within the arachnoid trabeculae that connect to the pia mater. **Subarachnoid hemorrhage** occurs when an **arterial aneurysm** ruptures. Patients will present with a sudden onset of "the worst headache of my life," and their computed tomography scans usually reveal blood in the subarachnoid space. A spinal tap can also show red blood cells in the CSF. This life-threatening condition requires immediate intervention to stop the arterial bleed.

2. **Subdural space:** The **subdural space** is a potential space between the arachnoid and the dura. These two meningeal layers are usually tightly connected to each other. Bridging veins pass through the arachnoid into the dural venous sinus. When shearing forces are applied to the head, bridging veins can rupture as they pierce through the stiff dura to enter the sinus. This type of injury is seen in "shaken baby syndrome" when an infant is shaken violently. The resulting **venous hemorrhage** will seep into the newly created subdural space. A chronic subdural hematoma will develop slowly from minor trauma over weeks, whereas an acute subdural hematoma is often associated with other intracerebral injuries.

3. **Potential epidural space:** The **epidural** or **extradural space** is a potential space in the skull between the bone and the periosteal layer of the dura. This space is created through a bleed from the **meningeal arteries** that travel on the surface of the periosteal layer of dura. The most susceptible vessel is the **middle meningeal artery** in the temporal bone. An epidural hematoma develops slowly, even though it is an arterial bleed because it takes a lot of force to separate the dura from the skull. The underlying injury is typically a fracture of the temporal bone causing a rupture of the middle meningeal artery.

4. **True epidural space:** In the spinal cord, the dura has only one meningeal layer. A **true epidural space** exists between this meningeal layer of dura and the periosteum of the vertebrae. This true epidural space is filled with fatty tissue and the vertebral venous plexus.

VII. OVERVIEW OF BLOOD SUPPLY

The blood supply to the brain comes from two sources: the **internal carotid** and the **vertebral arteries**. The two vertebral arteries come together as the **basilar artery** at the level of the brainstem. This **vertebral–basilar system** supplies both the spinal cord and the brainstem. The forebrain is supplied by both the vertebral–basilar and the internal carotid systems. These two systems are joined at the base of the brain to form an arterial circle, known as the **circle of Willis**, from which major arteries supplying the brain arise (Figure 2.21).

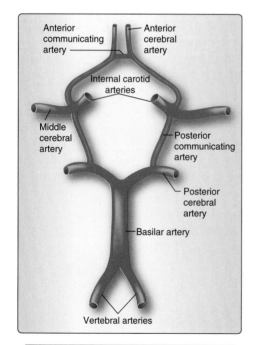

Figure 2.21
Blood supply to the brain: the cerebral arterial circle (of Willis).

The internal carotid artery enters the skull through the **carotid canal**. After branching of the **ophthalmic artery** to the orbit, the internal carotid artery joins the circle of Willis at the base of the brain. The **middle cerebral artery** is the direct extension of the internal carotid and supplies most of the lateral surface of the brain as well as deep structures, such as the basal ganglia and the internal capsule. Anteriorly, the circle of Willis gives rise to the **anterior cerebral artery**, which supplies the medial surface of the frontal and parietal lobes. The two anterior cerebral arteries are joined together by the **anterior communicating artery**. The **posterior communicating artery** joins the internal carotid system with the vertebral–basilar system. At the junction, the **posterior cerebral artery** branches off to supply the medial surfaces of the occipital and temporal lobes as well as the thalamus deep in the forebrain.

A detailed description of blood supply to specific structures is given with each chapter in this book.

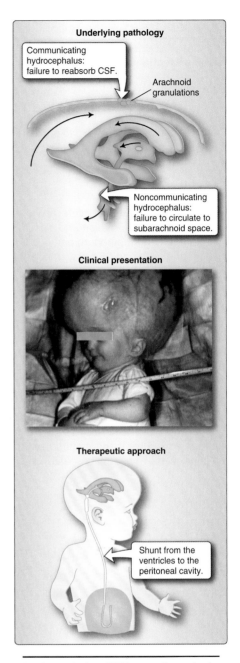

Hydrocephalus. CSF = cerebrospinal fluid.

Clinical Application 2.1: Hydrocephalus

Hydrocephalus, literally "waterhead," is a condition in which too much cerebrospinal fluid (CSF) is present in the brain. It develops when the production, circulation, or absorption of CSF is impaired. In the case of communicating, or nonobstructive, hydrocephalus, the communication between the ventricles and the subarachnoid space is intact. The underlying cause is usually a deficiency of the absorption of CSF into the sinus. This can happen if the arachnoid granulations are damaged, for instance, as a consequence of purulent bacterial meningitis. CSF would still be produced in the ventricles and circulate normally through the ventricular system and into the subarachnoid space, but absorption into the venous sinus would be impaired. A noncommunicating hydrocephalus develops when the outflow from the ventricles is obstructed, and there is no communication between the ventricles and the subarachnoid space. CSF continues to be produced within the ventricles but cannot circulate normally and be reabsorbed in the arachnoid granulations. The result is an enlargement of the ventricles. The underlying cause might be a tumor or a developmental anomaly.

Treatment of hydrocephalus involves reestablishing the normal cycle of production, circulation, and reabsorption. Often, this includes the surgical implantation of a shunt from the ventricles to the peritoneal cavity where CSF can be absorbed.

Clinical Case:

Sarah's Pregnancy Concerns

Sarah, whom we met in Chapter 1, is planning to get pregnant in the next few months. She continues to take lamotrigine and is no longer taking valproate. She has been advised that she should take folic acid to prevent neural tube defects, but she often forgets to take it. Now, she is starting to worry that the baby might be at risk.

Case Discussion

Neural tube formation occurs at week 3–4 of gestation. Neurulation is the formation and closure of the neural tube. When the anterior neuropore and posterior neuropore close (around the 24th and 26th day, respectively), neurulation is complete.

Craniorachischisis is the complete failure of neurulation and it occurs by day 20–22. Failure of anterior neuropore closure results in either anencephaly (complete failure) or encephalocele (partial failure). Anencephaly is the absence of a major portion of the brain, skull, and scalp. Encephalocele is the presence of neural tissue (brain) through a defect in the cranium. Failure of posterior neuropore closure results in spina bifida.

Spina bifida can present along a continuum (myeloschisis, myelomeningocele, meningocele, spina bifida occulta; Figure 2.22). Myeloschisis (complete failure of posterior neuropore closure) is the most severe form with no overlying membrane and a cleft spinal cord. Myelomeningocele

consists of herniation (protrusion) of the spinal cord and meninges through the vertebral defect. Meningocele is the herniation of only the meninges (not spinal cord) through the vertebral defect. Spina bifida occulta is the failure of lumbar spine fusion with no meninges or spinal cord herniation. It is the most common neural tube defect and typically asymptomatic unless it is associated with a tethered spinal cord.

What is a tethered spinal cord?

Tethered spinal cord is either a congenital (developmental) or acquired tethering of the spinal cord resulting in a short thick filum terminale below L2. Typically, in an adult, the conus medullaris (end of the spinal cord) ends at L1/2. The filum terminale anchors the spinal cord to the dural sac at S2 and finally the coccyx. Symptoms of a tethered cord may include motor/sensory impairment, sphincter dysfunction such as urinary incontinence, pain, foot deformities, and scoliosis.

What are causes of neural tube defects?

Folic acid deficiency and a positive family history are risk factors for neural tube defects. Anticonvulsant medications such as valproate and carbamazepine can reduce folic acid in the body and increase the risk of neural tube defects.

Neural tube defects can be detected by both elevated alpha fetoprotein and fetal ultrasound. Folic acid supplementation is recommended for women with epilepsy who are on anticonvulsant medications.

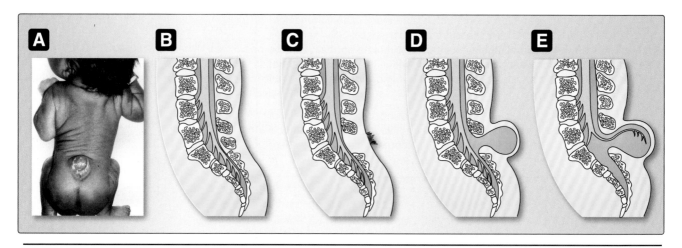

Figure 2.22
Spina bifida. **A.** Infant with spina bifida. **B.** Normal spine. **C.** Spina bifida occulta. **D.** Spina bifida with meningocele. **E.** Spina bifida with myelomeningocele. (**A** from Sadler TW. *Langman's Essential Medical Embryology*. Lippincott Williams & Wilkins, Philadelphia, PA, 2005. **B–E** from Rosdahl CB, Kowalski MT. *Textbook of Basic Nursing*, 10th ed. Lippincott Williams & Wilkins, A Wolters Kluwer Business, Philadelphia, PA, 2012.)

Chapter Summary

- The human central nervous system consists of the brain and spinal cord. The brain is concerned with unique higher cortical functions including thinking, language, learning and memory, creativity, attention, and emotional experience. The spinal cord processes and carries ascending sensory as well as descending motor information and participates directly in control of body movement.

- The nervous system develops from the neural tube. The rostral end of the neural tube develops into the brain, and the remainder develops into the spinal cord. Three primary vesicles give rise to the forebrain, midbrain, and hindbrain. Five secondary vesicles give rise to all of the structures within these three divisions.

- In the brain, gray matter (accumulation of neuronal cell bodies or nuclei) is found in the cortical layer on the surface and in deep structures of the forebrain, whereas white matter (fiber tracts or bundles of axons) is located deep to the cortex. In the spinal cord, by contrast, the gray matter is located centrally and the white matter surrounds the gray matter.

- The gray matter structures deep within the cerebral hemispheres include the basal ganglia and structures of the limbic system.

- The white matter tracts deep to the cerebral cortex are important in connecting cortical areas to each other within the same hemisphere (association fibers) or between hemispheres (commissural fibers, the largest of which is the corpus callosum) or in connecting cortical areas to brainstem and spinal cord structures (projection fibers).

- The forebrain consists of the telencephalon (two cerebral hemispheres) and diencephalon (thalamus, hypothalamus, and subthalamus). The *cerebral hemispheres* are each divided into four lobes, the frontal, parietal, temporal, and occipital.

- The *diencephalon* consists of paired structures on either side of the third ventricle. The thalamus is critical for processing and integrating sensory (except olfactory) and motor information and in regulating cortical activity. The hypothalamus is structurally part of the diencephalon but functionally part of the limbic system, with a key role in coordinating and integrating endocrine, autonomic, and homeostatic functions. The subthalamus is part of the basal ganglia and plays an important role in modulating and integrating voluntary movement and muscle tone.

- The *brainstem* comprises the midbrain and hindbrain (pons and medulla). The brainstem provides the conduit through which all ascending and descending information between the brain and the spinal cord travel. The cranial nerves provide sensory and motor information to and from the head and mediate special senses. The brainstem also has an integrative function through the reticular formation nuclei.

- The *cerebellum* overlies the fourth ventricle. It has two hemispheres and a central vermis. The cerebellum has important roles in processing of sensory information, coordination of voluntary motor activity, and aspects of cognitive function.

- The spinal cord plays a role in the initial processing of sensory information. It also contains somatic motor tracts that supply the skeletal muscles and visceral (autonomic) efferents to the viscera, smooth muscles, and glands.

- The spinal cord has a clear segmental organization, corresponding to the nerve roots attached to it. A continuous series of posterior (sensory, cell bodies in spinal ganglia) rootlets enter the posterior aspect of the spinal cord, and a continuous series of anterior (motor) rootlets emerge from the anterior aspect of the spinal cord. The sensory and motor roots merge together as the spinal nerves, which are part of the peripheral nervous system.

- The **ventricular system** is a fluid-filled space within the brain. There are two lateral ventricles, one in each cerebral hemisphere, a third ventricle associated with the diencephalon, and a fourth ventricle that lies between the cerebellum and the brainstem (pons and medulla). Cerebrospinal fluid (CSF), a secretion from the ependymal cells of the choroid plexus, fills the ventricles and surrounds the brain and spinal cord in the subarachnoid space. CSF is reabsorbed through the arachnoid granulations into the superior sagittal venous sinus.

- The brain is surrounded by three layers of connective tissue, the **meninges**. The dura mater is the outermost layer. It consists of a periosteal and a meningeal layer that are tightly fused but separate to form venous sinuses, into which the cerebral veins drain. The middle meningeal layer is the arachnoid mater, which lies against the inner surface of the dura, bridges over the sulci and cisterns, and is connected to the pia by the arachnoid trabeculae. The subarachnoid space, which is filled with CSF, is where the cerebral arteries and veins travel. The pia mater is the innermost meningeal layer. It adheres tightly to the brain parenchyma and separates the brain from the CSF in the subarachnoid space. Dural reflections (falx cerebri, tentorium cerebelli, falx cerebelli) separate different compartments within the skull. In the spinal cord, the dura has only one meningeal layer. A true epidural space exists between this meningeal layer of dura and the periosteum of the vertebrae.

- The **blood supply** to the brain comes from both the internal carotid and the vertebral arteries. The vertebral–basilar system supplies both the spinal cord and the brainstem. The forebrain is supplied by both the vertebral–basilar and the internal carotid systems. These two systems are joined at the base of the brain to form an arterial circle, known as the circle of Willis, from which major arteries supplying the brain arise.

Study Questions

Choose the ONE best answer.

2.1 Which of the following statements concerning development of the nervous system is correct?

    A. A longitudinal band of endoderm thickens to form the neural plate.

    B. Neural crest cells break away from the neural tube to form peripheral neural structures.

    C. Secondary vesicles that arise from the rhombencephalon become the pons and medulla.

    D. The cerebellum develops from the myelencephalon.

    E. The telencephalon develops into the thalamus and hypothalamus.

> The correct answer is B. Neural crest cells dissociate from the top of the neural folds and migrate away to form peripheral neural structures including sensory neurons in peripheral ganglia, postganglionic neurons of the autonomic nervous system, and glial cells. The neural plate forms from ectoderm. One secondary vesicle from the rhombencephalon, called the metencephalon, develops into the pons and cerebellum. The medulla develops from another secondary vesicle, the myelencephalon. The telencephalon forms the cerebral hemispheres, whereas the diencephalon develops into the thalamus, hypothalamus, and subthalamus.

2.2 A patient arrived in the emergency room following head trauma sustained in a car accident. A magnetic resonance image was ordered to examine brain structures from the anterior forebrain through the rostral levels of the spinal cord, and images in the coronal, sagittal, and axial planes were prepared for review by the neuroradiologist. In terms of orientation of the central nervous system:

    A. An axial section cuts through the brain from superior to inferior.

    B. A sagittal section cuts through the brain from medial to lateral.

    C. In the brainstem, the ventral surface is also the inferior surface.

    D. In the forebrain, the dorsal surface is also the superior surface.

    E. Rostral refers to anything toward the inferior pole of the spinal cord.

> The correct answer is D. In the forebrain, the dorsal surface is also the superior surface, and the ventral surface is also the inferior surface. An axial section cuts through the brain horizontally, parallel to the ground, whereas a sagittal section cuts through the brain from anterior to posterior. In the brainstem and spinal cord, the ventral surface is also the anterior surface, and the dorsal surface is also the posterior surface. Rostral refers to anything toward the anterior pole of the forebrain.

2.3 A 25-year-old man developed a fever and began having severe headaches over the entire head and back of the neck. A lumbar puncture was ordered to assess whether he had meningitis. As he was doing the lumbar puncture, the attending physician was explaining to the medical students who were observing that the puncture had to be done below the level of the caudal spinal cord where the filum terminale and cauda equina could be pushed aside without causing damage to the spinal nerves. In relation to the anatomy of the spinal cord, which statement is correct?

    A. The adult spinal cord ends at approximately the L3 vertebral level.

    B. The cauda equina forms because the spinal cord grows faster than the vertebral column.

    C. The filum terminale anchors the spinal cord to the dorsum of the coccyx.

    D. There is a continuous series of anterior rootlets that carry sensory information.

    E. White matter forms the inner core of the cord and gray matter surrounds it.

> The correct answer is C. The filum terminale, an extension of pia and supporting cells, anchors the spinal cord to the dorsum of the coccyx. The spinal cord of the adult ends at approximately L1–L2, whereas that of the newborn ends at approximately L3. The cauda equina forms when the vertebral column grows faster than the spinal cord, forcing the lumbosacral roots to travel farther before they can exit through the appropriate vertebral foramina. The posterior roots carry sensory information and the anterior roots carry motor information. In the spinal cord, gray matter forms the inner core, and white matter surrounds the gray matter.

2.4 Meningitis involves inflammation of the meninges and causes severe headache over the entire head and back of the neck. In relation to the meningeal coverings of the brain and spinal cord:

 A. Arachnoid trabeculae connect the arachnoid to the dura.
 B. A tear in the middle meningeal artery can result in a subdural bleed.
 C. Only the pia mater is pain sensitive.
 D. The dura covering both the brain and spinal cord consists of two layers.
 E. The subarachnoid space contains arteries and veins.

The correct answer is E. Arteries and veins travel in the subarachnoid space. The arachnoid trabeculae connect the arachnoid to the pia. The meningeal arteries supply the dura, and a tear in the middle meningeal artery will result in an epidural bleed. The dura mater, not the pia mater, is pain sensitive. The dura covering the brain consists of two layers, whereas that covering the spinal cord is a single layer.

2.5 A newborn baby developed hydrocephalus shortly after birth. It was determined that the reabsorption of cerebrospinal fluid (CSF) from the ventricular system was deficient, and a shunt was inserted to drain the CSF and relieve the increasing intracranial pressure. In relation to the anatomy and function of the ventricular system, which statement is correct?

 A. Cerebrospinal fluid is produced in the choroid plexus mainly in the third ventricle.
 B. The foramina of Luschka allow outflow of cerebrospinal fluid from the fourth ventricle.
 C. The two lateral ventricles communicate with each other and with the third ventricle.
 D. The posterior horn of the lateral ventricle extends into the temporal lobe.
 E. Reabsorption of cerebrospinal fluid occurs through arachnoid granulations in the inferior sagittal sinus.

The correct answer is B. Cerebrospinal fluid (CSF) can leave the ventricular system through the two lateral foramina in the medulla, the foramina of Luschka, and the medial foramen of Magendie. CSF is produced in the choroid plexus mainly in the lateral and fourth ventricles. The two lateral ventricles are separated by the septum pellucidum and cannot communicate with each other. Both communicate with the third ventricle. The posterior horn of the lateral ventricle extends into the occipital lobe, and the inferior horn extends into the temporal lobe. Reabsorption of CSF occurs through arachnoid granulations in the superior sagittal sinus.

2.6 Which of the following statements is TRUE about neurulation?

 A. Neurulation occurs in the 8th week of gestation.
 B. Neurulation is the formation and closure of the neural tube.
 C. Failure of the anterior neuropore to close results in myelomeningocele.
 D. Failure of the posterior neuropore to close results in encephalocele.
 E. Failure of the posterior neuropore to close results in anencephaly.

The correct answer is B. Neurulation is the formation and closure of the neural tube. It occurs in weeks 3–4 of gestation. The failure of the anterior neuropore to close results in anencephaly. The failure of the posterior neuropore to close results in myelomeningocele/spina bifida.

3

Overview of the Peripheral Nervous System

I. OVERVIEW

The peripheral nervous system (PNS) is composed of cranial and spinal nerves that link the brain and spinal cord with the peripheral environment and visceral tissues. Cranial nerves arise from the brain and brainstem, whereas spinal nerves arise from the spinal cord (Figure 3.1).

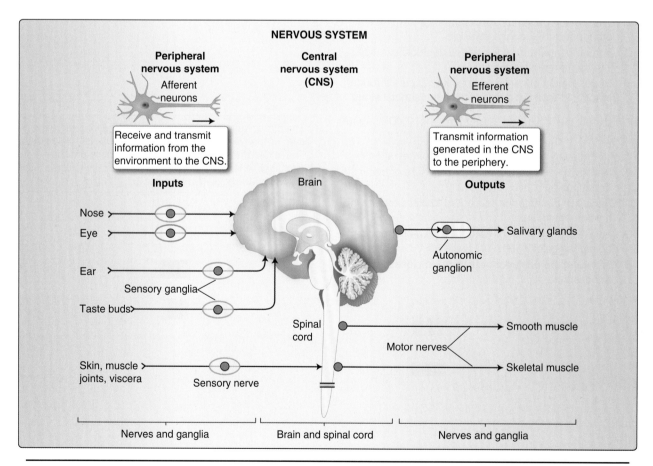

Figure 3.1
Overview of the peripheral nervous system.

Peripheral nerves carry sensory input to the central nervous system (CNS) where it is processed and carry motor output to the muscles for an appropriate motor response. Incoming information enters the sensory posterior part of the spinal cord, whereas outgoing information that will stimulate motor responses to the sensory input emerges from the anterior portion of the spinal cord (Figure 3.2).

The PNS can be further divided into somatic and visceral components. Somatic sensory or somatic afferent fibers carry information from the structures derived from somites (skin, skeletal muscle, joints). Somatic motor or somatic efferent fibers carry information to musculature derived from somites (skeletal muscle). Visceral sensory or visceral afferent fibers carry information from the viscera of the body core (thoracic, abdominal, and pelvic organs). Visceral motor or visceral efferent fibers are also referred to as the autonomic nervous system and can be divided into sympathetic and parasympathetic components (Figure 3.3). The sympathetic nerves send motor innervation to the body core (viscera) and the body periphery (e.g., blood vessels, sweat glands). The parasympathetic nerves only innervate the core (viscera).

Ganglia are aggregations of nerve cell bodies *outside* the CNS (in contrast to aggregations of nerve cell bodies *within* the CNS, which are called **nuclei**). All sensory, both somatic and visceral, nerves have their cell bodies in a spinal ganglion. In addition, visceral motor or autonomic nerves synapse in a peripheral ganglion.

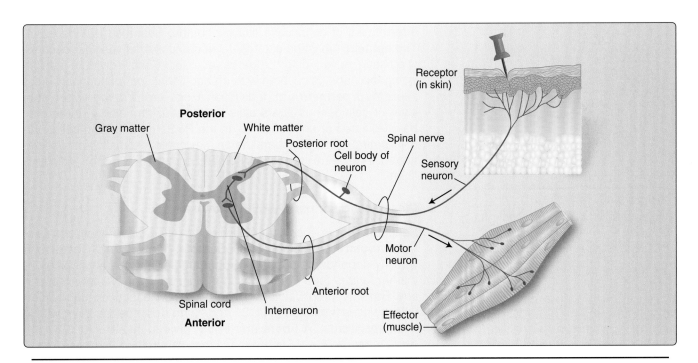

Figure 3.2
Cross section of the spinal cord with the anterior and posterior roots.

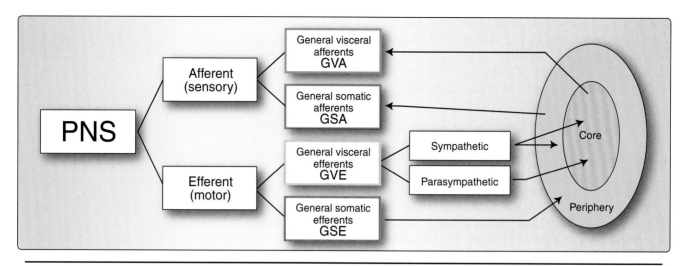

Figure 3.3
Modalities within the peripheral nervous system (PNS).

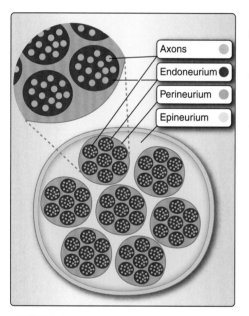

Figure 3.4
Organization of a peripheral nerve in fascicles and their protective layers of connective tissue.

II. THE PERIPHERAL NERVE

Peripheral nerves are bundles of axons or nerve fibers surrounded by several layers of connective tissue. They carry both somatic and visceral information.

A. Organization of the peripheral nerve

Peripheral nerves are arranged in bundles or **fasciculi**. They have three layers of connective tissue sheaths, illustrated in Figure 3.4. The **epineurium** is the external layer composed of vascular connective tissue surrounding the nerve fascicles. Each individual fascicle within the nerve is covered with an additional coat of connective tissue called **perineurium**. Individual axons are covered with a thin layer of collagenous fibers called **endoneurium**. Nerve fibers may be myelinated or unmyelinated. In the PNS, the Schwann cell is the myelinating cell. A peripheral nerve may contain somatic and visceral **sensory** (**afferent**) as well as somatic and visceral **motor** (**efferent**) fibers.

B. Classification of peripheral nerve fibers

Peripheral nerve fibers are classified to reflect either their **conduction velocity** or their **axonal diameter** (Table 3.1). Conduction velocity depends on both the axon diameter and the myelination of the nerve fiber (see Chapter 1, "Introduction to the Nervous System and Basic Neurophysiology"). Classification based on conduction velocity uses the letters "A," "B," and "C" to describe the fiber, with *A* being the fastest. **A fibers** are further divided into Aα, Aβ, Aδ, and Aγ. Commonly, the "A" is dropped when referring to them, for example, α-motoneuron and γ-motoneuron. **B fibers** are smaller, myelinated, and typically preganglionic visceral motor (autonomic) fibers. **C fibers** are small in diameter and unmyelinated. Postganglionic visceral motor (autonomic) fibers and some sensory fibers are classified as C fibers.

Table 3.1: Classification of Peripheral Nerve Fibers

	Group		External Diameter (µm)	Conduction Velocity (m/s)	Function
Myelinated	Aα		12–20	70–120	Motor to skeletal muscle
		Ia	12–20	70–120	Sensory from muscle spindle
	Aβ	Ib	10–15	60–80	Sensory from Golgi tendon organ and Ruffini endings
		II	5–15	30–80	Sensory from skin receptors
	Aγ		3–8	15–40	Motor to intrafusal fibers
	Aδ	III	3–8	10–30	Sensory from free nerve endings for pain and temperature and hair follicles
	B		1–3	5–15	Preganglionic autonomic fibers
Unmyelinated	C	IV	0.2–1.5	0.5–2.5	Postganglionic autonomic fibers; sensory from free nerve endings for pain and temperature; smell

Classification based on axon diameter is used only for sensory fibers and uses the Roman numerals I, II, III, and IV, with I being the largest. Sensory fibers can be further divided using "a" and "b" (e.g., Ia, Ib fibers).

III. SENSORY RECEPTORS

Sensory receptors detect information from the environment, such as light and sound, or from our own bodies, such as touch and body position. Sensory receptors act as transducers in that they transform a physical or chemical stimulus (or form of energy) into an electrical impulse (Figure 3.5). They are specialized to detect sensory information and translate stimuli into **receptor potentials**, or electrical signaling within the receptor, caused by the opening and closing of ion channels. Each sensory receptor has a receptive field, which lets us discriminate the location of the sensory stimulus.

A. Receptor potentials

Receptor potentials, also referred to as **generator potentials**, are electrical impulses transduced by the sensory receptor. The receptor potential is a graded response that depends on the magnitude of the sensory stimulus and encodes for duration and intensity. Because a receptor potential dissipates after a few millimeters, an action potential must be generated to travel the long distance between the sensory receptor and the CNS. In order to generate an action potential, the depolarization of the membrane at the sensory receptor must reach threshold. The firing frequency of the action potential in the sensory nerve is modulated by the receptor potential: The greater the stimulus, the greater the receptor potential, and the higher the frequency of action potentials produced.

B. Classification of sensory receptors

Sensory receptors can be classified by the source of the stimulus or by the mode of detection.

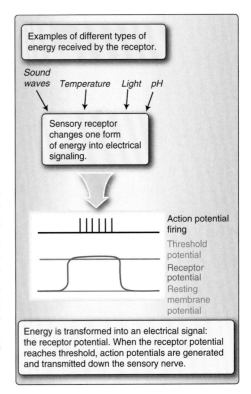

Figure 3.5
Sensory receptors generating a receptor potential.

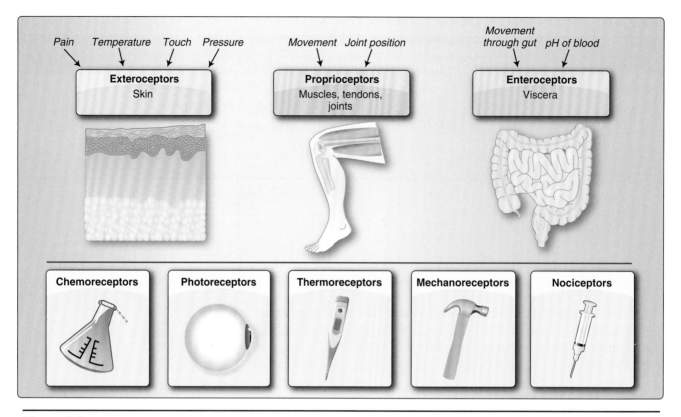

Figure 3.6
Types of peripheral receptors.

The source of the stimulus can be external or internal (Figure 3.6). Superficially located sensory endings in the skin are called **exteroceptors** and respond to pain, temperature, touch, and pressure, that is, stimuli *outside the body*. Muscles, tendons, and joints have **proprioceptors** that signal awareness of body position and movement. **Enteroceptors** monitor events *within the body* such as feeling movement through the gut.

Mode of detection can be grouped into five categories:

- **Chemoreceptors** detect molecules that bind to the receptor, for example, in the olfactory bulb.
- **Photoreceptors** detect light in the retina.
- **Thermoreceptors** detect temperature in the skin.
- **Mechanoreceptors** are stimulated by the mechanical opening of ion channels, for example, touch receptors in the skin.
- **Nociceptors** detect signals associated with tissue damage, which are interpreted as pain.

C. Receptor adaptation

Sensory receptors become less responsive to a stimulus over time, which is a process known as **receptor adaptation** (Figure 3.7).

1. Slowly adapting receptors: Slowly adapting receptors adapt very little over time and remain responsive during long stimuli. These receptors are suited to monitor unchanging stimuli such as pressure.

2. **Rapidly adapting receptors: Rapidly adapting receptors** adapt very quickly and essentially detect only the onset of a stimulus. They are suited to detect rapidly changing impulses such as vibration.

In this chapter, we discuss skin receptors and receptors in the muscles and joints. Sensory receptors related to special senses are discussed in later chapters. Nociceptors are introduced here and then elaborated on in Chapter 22, "Pain."

D. Skin receptors

We recognize five basic modalities with our skin receptors: fine touch, crude touch, vibration, temperature, and pain. The areas of skin from which sensation is perceived are referred to as **receptive fields**, and the skin is essentially a mosaic of spots dedicated to specific sensations. Receptive fields vary in size. Fingertips have a high density of small receptive fields, which allows for fine discrimination of sensory input. In contrast, receptor fields in the skin of the back are very large, and fine discrimination of sensory input is not possible.

Skin receptors (Figure 3.8) can be either encapsulated or nonencapsulated. Encapsulated receptors can have either a layered capsule or a thin capsule. These capsules are an integral part of the structure of the receptors, tightly linked to how they detect sensory information. Nonencapsulated receptors either are free nerve endings or come with accessory structures specific to how a sensory stimulus is detected.

1. **Hair follicle receptors:** These receptors are a mesh-like arrangement of axons around a hair follicle. They are rapidly adapting mechanoreceptors sensitive to touch.

2. **Merkel endings:** These nonencapsulated receptors are an expansion of a sensory fiber into a specialized cell known as a **Merkel cell**, which is located in the basal cell layer of the epidermis. Merkel

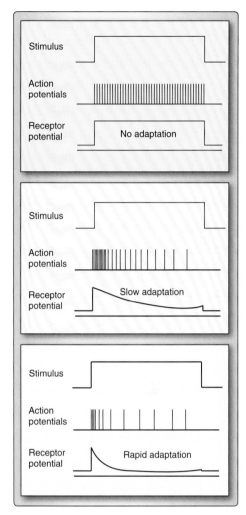

Figure 3.7
Receptor adaptation.

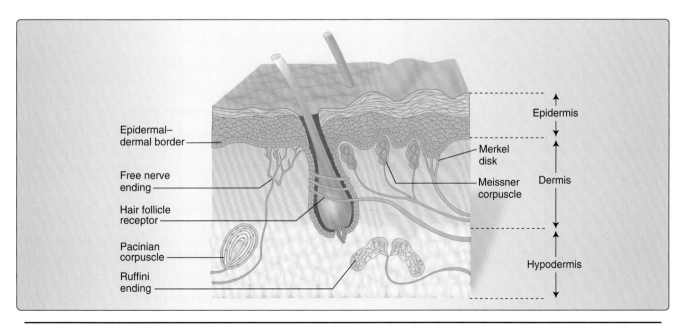

Figure 3.8
Skin receptors.

endings or disks are slowly adapting mechanoreceptors sensitive to pressure and low-frequency vibration.

3. **Meissner corpuscles:** These encapsulated receptors are composed of a stack of epithelial cells with axons winding throughout this stack. They are rapidly adapting and sensitive to light touch.

   Merkel endings and Meissner corpuscles are the receptors responsible for fine touch and fine discrimination in our fingertips.

4. **Pacinian corpuscles:** These encapsulated receptors located deeper in the skin (in the hypodermis, below the dermis) are composed of concentric layers of epithelial cells. They are very rapidly adapting and therefore can respond to quickly changing stimuli. Pacinian corpuscles detect vibration.

5. **Ruffini endings:** These receptors have a thin capsule and a mesh of longitudinally arranged collagen fibers within the capsule. Sensory fibers branch throughout the strands of collagen fibers. They are slowly adapting and detect skin stretching and pressure.

6. **Nociceptors:** These are free nerve endings that respond to tissue damage or stimuli that could result in tissue damage. They can detect mechanical (e.g., pinching the skin) and thermal (e.g., sunburn) stimuli, as well as molecules (e.g., histamine) released in tissue damage, and interpret this information as pain.

E. Proprioceptors

Proprioceptors are distributed throughout our musculoskeletal system. They detect the position of the body in space and relay this information back to the CNS. Proprioception is a keystone to the understanding of the control of movement. See Chapter 17, "The Cerebellum" for more information.

Muscles and tendons have receptors that can detect muscle length (**muscle spindles**) and muscle strength (**Golgi tendon organs**). They also contain free nerve endings that are thought to detect muscle pain and possibly extracellular fluid during muscle activity. Proprioceptors in the joints detect information about the position of the joint.

1. **Muscle spindle:** Muscle spindles detect changes in **muscle length** and are found dispersed throughout all skeletal muscles. Morphologically, they are composed of a few muscle fibers and nerve endings surrounded by a capsule (Figure 3.9).

   The muscle spindle is the proprioceptive organ of skeletal muscle. It detects and regulates muscle length through its participation in the gamma reflex loop. See Chapter 18, "The Integration of Motor Control," for details.

   a. **Intrafusal and extrafusal muscle fibers:** The muscle fibers within the muscle spindle capsule are termed **intrafusal fibers** (meaning fibers within the spindle). All other muscle fibers in a skeletal muscle are termed **extrafusal fibers** (meaning fibers outside the spindle).

      The muscle spindles are arranged in parallel with the extrafusal fibers. The intrafusal fibers are attached to the extrafusal

Clinical Application 3.1: Guillain-Barré Syndrome

Guillain-Barré syndrome is an inflammatory disorder of peripheral nerves. An autoimmune response directed at the myelin sheath of peripheral nerves after a viral infection is thought to be the underlying cause. Patients present with progressive weakness in the lower limbs that rapidly ascends to involve the upper limb and trunk musculature. Spinal reflexes are either absent or reduced. These patients are in acute need of medical attention. About 30% will need assisted ventilation due to weakness of the muscles involved in respiration. Patients stabilize after 1–2 weeks and most eventually recover completely. Therapeutically, the aim is to attenuate the autoimmune antibodies. This can be achieved through plasmapheresis to remove the autoantibodies or through intravenous administration of immunoglobulins, which bind the autoantibodies and neutralize them.

Demyelination of peripheral nerves leads to a conduction block in these nerves. Both sensory (afferent) signaling and motor (efferent) signaling are involved. This leads to the loss of sensory input as well as weakness.

Diagnostically, a nerve conduction study can be done, which monitors the conduction velocity in peripheral nerves and can show the conduction block.

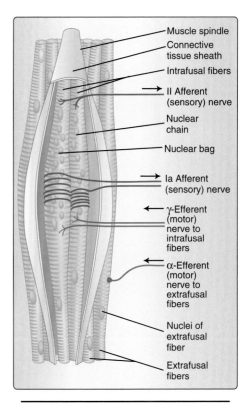

Figure 3.9
The muscle spindle.

fibers. When the muscle is stretched, the intrafusal fibers are stretched passively. This activates the sensory endings in the muscle spindle.

There are two types of intrafusal fibers: The largest intrafusal fiber has all of its nuclei bunched up in the middle of the fiber and is called a **nuclear bag fiber**. The other intrafusal fibers have their nuclei arranged in a line and are called **nuclear chain fibers**. There are also two types of sensory endings in the muscle spindle. **Type Ia sensory endings** innervate the middle portion of all intrafusal fibers, and **type II sensory endings** innervate the nuclear chain fibers. Together, these nerves detect muscle length and relay this proprioceptive information to the spinal cord. Motor innervation to the intrafusal fibers comes via γ motor neurons, which cause intrafusal muscle cells to contract and remain responsive.

b. **Muscle spindle density:** The density of muscle spindles varies among the different muscles. Those muscles that require precise movement, such as the extraocular muscles and finger muscles, have a high density of muscle spindles, whereas those responsible for gross motor movements, such as leg movements, have a low density of muscle spindles.

2. **Golgi tendon organs:** Spindle-shaped receptors called **Golgi tendon organs** are found at the tendon–muscle junction. They are slowly adapting mechanoreceptors that are stimulated by tension in the tendon. They consist of a mesh-like weave of collagenous bundles within a thin capsule (Figure 3.10). These organs are supplied by **type Ib sensory endings** that enter the capsule and break into fine nerve endings along the long axis of the capsule fibers. Here, they detect deformation of the capsule fibers resulting from tension in the tendon.

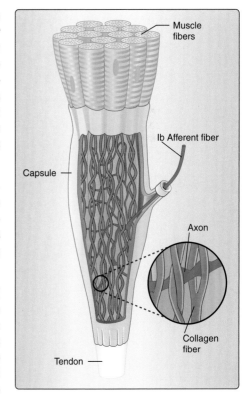

Figure 3.10
The Golgi tendon organ.

3. **Joint receptors:** Nerve endings are found within and around the capsules of joints. Ruffini endings and pacinian corpuscles in the connective tissue external to the capsule respond to the start and stop of movement as well as the position of the joint.

F. Enteroceptors

Visceral receptors are supplied by nonencapsulated nerve fibers that terminate as free nerve endings. They function mainly at a subconscious level in visceral reflexes. They can be mechanoreceptors that recognize changes in blood pressure in the aortic arch, chemoreceptors such as those in the carotid body that recognize changes in pH or blood gases, or nociceptors that signal distension of an organ (e.g., from a belly ache).

IV. EFFECTOR ENDINGS

For a response to take place as the result of sensory input, there are neuroeffector endings where axons terminate in relation to skeletal, cardiac, and smooth muscle fibers as well as cells of exocrine and endocrine glands.

A. Motor unit

The **neuromuscular junction (NMJ)**, or **motor endplate,** is a chemical synapse between motor nerve fibers and muscle fibers (Figure 3.11A). As the motor nerve axon reaches its target muscle, it branches extensively and each axonal process innervates one muscle fiber. The motor neuron

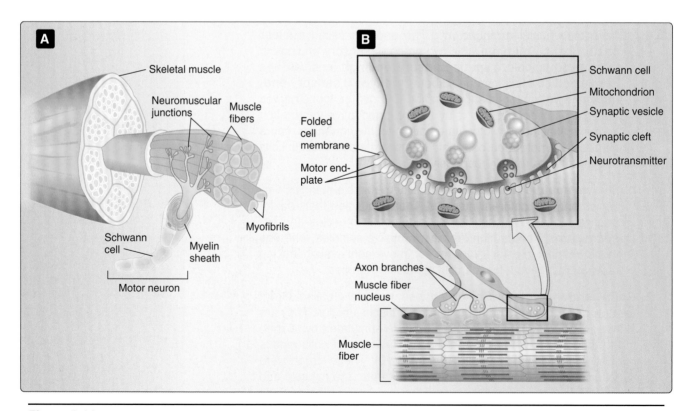

Figure 3.11
The motor unit and the neuromuscular junction.

and the muscle fibers it innervates are referred to as a **motor unit** (see Figure 3.11B). The size of a motor unit varies: The more precise the muscle movement needs to be, the smaller the motor unit. Muscles that move the eye, for example, require fine control so one nerve fiber controls very few muscle fibers. The contact between the axonal processes and the muscle fiber occurs midway along the length of a muscle fiber.

B. Neuromuscular junction

The NMJ has three components: the axonal endings from the motor neuron, the postsynaptic skeletal muscle membrane, and the associated Schwann cell (Figure 3.12).

Action potentials travel along a motor neuron and depolarize the axon terminal. This causes the influx of Ca^{2+} (calcium) through voltage-gated channels. The increase in intracellular Ca^{2+} causes the synaptic vesicles to fuse with the membrane and release the neurotransmitter **acetylcholine (ACh)** into the synaptic cleft. The ACh binds to the ACh receptor on the skeletal muscle membrane. This causes the influx of Na^+ (sodium) and the generation of an **excitatory postsynaptic potential**, or **endplate potential**. This change in the muscle membrane potential triggers the opening of voltage-gated Ca^{2+} channels in the sarcoplasmic reticulum (the endoplasmic reticulum within the muscle fiber). The influx of Ca^{2+} triggers muscle contraction.

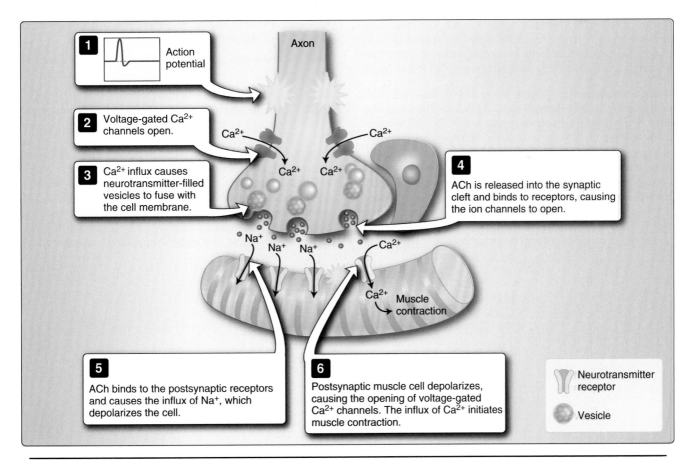

Figure 3.12
Signal transduction at the neuromuscular junction. ACh = acetylcholine.

Clinical Case:

Sven's Neuropathy

Sven is 54-year-old man who presents to the clinic with a 1-month history of intermittent numbness in his right hand when he wakes up in the morning. After a few minutes of shaking his right hand, the numbness over the palmar aspect of the right palm, thumb, and index and middle fingers resolves and he is left with intermittent paresthesia in the same region that lasts 2–5 minutes. There is no pain nor sensory symptoms in the arm. He has a 10-year history of diabetes and has noticed some intermittent paresthesia in his feet over the past few years that is not bothersome. The sensorimotor examination of his upper extremity is normal. He has a positive Phalen sign at the right wrist. In the lower extremity, he has a symmetric reduction in light touch, pinprick, and temperature sensation from midcalf to the toes (stocking distribution). He has a slightly reduced vibration threshold at the toes but normal proprioception. His tone and strength are normal. His reflexes are absent at both ankles. The rest of his neurological examination is normal.

Case Discussion

It is not uncommon to have symptoms but no signs on examination, particularly when the symptoms are intermittent; or to have no symptoms but subtle signs on examination. In Sven's case, his clinical history was suggestive of carpal tunnel syndrome and the only positive sign was the Phalen sign.

Phalen maneuver is a special maneuver in which the wrist is maximally flexed (Figure 3.13A). A positive Phalen sign is when the patient reports paresthesia within 30–120 seconds on the flexion; it is typically associated with compression of the median nerve in the carpal tunnel (carpal tunnel syndrome). Sven also has sensory findings in his lower extremities that are consistent with sensory neuropathy (see Figure 3.13B). He has absent ankle reflexes but normal deep tendon reflexes elsewhere. All of these signs and symptoms are localized to the distal parts of peripheral nerves and are indicative of what is referred to as **distal axonal neuropathy**.

Diabetes can cause various types of neuropathy depending on the type of peripheral nerve that is primarily affected: Distal sensory/sensorimotor polyneuropathies indicate damage to mixed peripheral nerves, autonomic neuropathies indicate damage to sympathetic or parasympathetic fibers, cranial neuropathies indicate damage to one or more of the 12 cranial nerves, and finally, mononeuropathies indicate damage to just one peripheral nerve. In addition, diabetes can affect the nerve roots and plexuses.

The distal polyneuropathies, such as in the case of Sven's neuropathy, are dependent on length: The longer nerves in the legs are affected first, and the shorter nerves in the arms show symptoms as the disease progresses. The distal neuropathies in diabetes can affect all types of peripheral nerve fibers.

What are the symptoms of a small fiber and large fiber neuropathy?

Small fiber neuropathies are typically painful, characterized by burning and painful paresthesia—Aδ fibers and C fibers that carry pain perception are small-diameter fibers.

Large fiber neuropathies are characterized by numbness and paresthesia. Patients typically have difficulty walking (unsteady gait) due to loss of the proprioceptive input from their feet, which is critical to guide their gait. The affected fibers are Aα, Aβ, and Aγ—all large-diameter fibers.

What are the autonomic symptoms of a diabetic autonomic neuropathy?

Autonomic symptoms refer to deficits from either parasympathetic or sympathetic fibers (see Chapter 4). The peripheral nerve fibers are typically smaller-diameter B and C fibers. Symptoms manifest in the function of thoracic and abdominal viscera, blood vessels, and glands: Resting tachycardia (fast heart beat), orthostatic hypotension (drop in blood pressure when standing that causes light-headedness or fainting), postprandial bloating or early satiety (due to slowing of gastrointestinal mobility), impaired bladder sensation, erectile dysfunction, and reduced sweating with flushing or dry skin are some symptoms of a diabetic autonomic neuropathy.

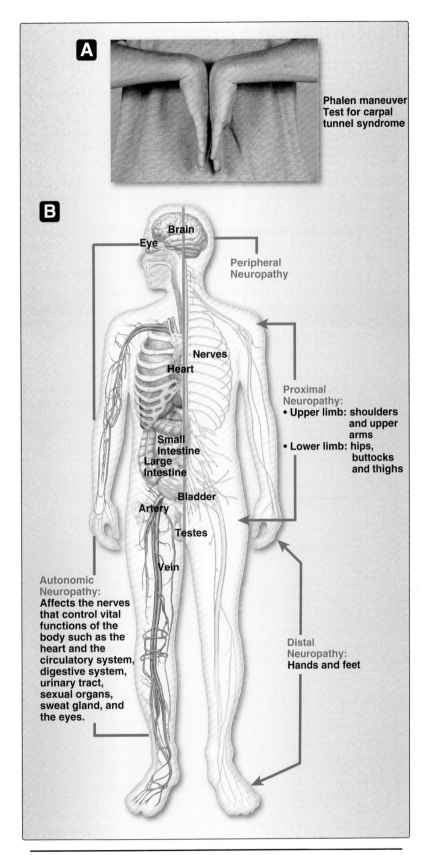

Figure 3.13
A. Phalen maneuver. **B.** Summary of clinical presentation of diabetic neuropathy. (**A** from Anderson MK, Parr GP. *Foundations of Athletic Training: Prevention, Assessment, and Management*, 5th ed. Lippincott Williams & Wilkins, A Wolters Kluwer Business, Philadelphia, PA, 2013. **B** from *Anatomical Chart Company: Understanding Diabetic Neuropathy Anatomical Chart*. Lippincott Williams & Wilkins, Philadelphia, PA, 2008.)

Chapter Summary

- The peripheral nervous system is composed of cranial and spinal nerves that link the brain and spinal cord with the peripheral environment and visceral tissues. This is achieved through afferent input from peripheral receptors. This information is interpreted by the central nervous system, which elicits an appropriate response through the efferent output to muscles and viscera.

- Peripheral nerves are bundles of axons or nerve fibers surrounded by several layers of connective tissue. They carry both somatic and visceral information. Peripheral nerves are arranged in bundles or fasciculi and can be myelinated or unmyelinated. They are classified by their conduction velocity or their axonal diameter (sensory fibers only).

- Sensory receptors detect information from the environment and translate it into receptor potentials. Sensory receptors are classified by the source of the stimulus or by the mode of detection, be it external (exteroceptors) or internal (enteroceptors). Over time, sensory receptors adapt to the stimulus and become less receptive to it. Skin has a variety of encapsulated and unencapsulated receptors that appreciate fine touch, crude touch, vibration, temperature, and pain.

- Proprioceptors are distributed throughout our musculoskeletal system and detect the body's position in space through muscle spindles (length) and Golgi tendon organs (strength). Muscle fibers may be within a capsule (intrafusal) or outside the capsule (extrafusal). Golgi tendon organs are found at tendon–muscle junctions, and joint receptors are found within and around the capsules of joints.

- Visceral receptors are nonencapsulated nerve fibers terminating as free nerves endings that monitor visceral reflexes at a subconscious level.

- Neuroeffector endings enable a response to take place in response to sensory input. This requires a neuromuscular junction at which signal transduction takes place. Depolarization occurs at the axon terminal. Excitatory postsynaptic potentials are generated resulting in influx of Ca^{2+} ions, and muscle contraction results.

Study Questions

Choose the ONE best answer.

3.1 Which statement about peripheral nerve characteristics is correct?

A. They are bundled into fasciculi (fascicles).
B. Their outermost covering is called endoneurium.
C. All peripheral nerves are myelinated.
D. Conduction velocity of a peripheral nerve is determined only by its diameter.
E. Autonomic nerve fibers are not carried in peripheral nerves.

The correct answer is A. Epineurium (*epi* means "upon" or "outer") is the outer covering. Nerves may be myelinated or unmyelinated, thereby contributing to differences in their conduction velocity. Two factors contribute to conduction velocity: nerve fiber diameter and myelination. Autonomic nerve fibers are carried in peripheral nerves.

3.2 In testing a patient for the integrity of sensory function, a physician struck a tuning fork, placed it on bony landmarks of the patient's upper limb, and asked the patient if he could feel the sensation. What type of receptor mediates the patient's ability to detect the tuning fork?

A. Meissner corpuscles.
B. Ruffini endings.
C. Nociceptors.
D. Pacinian corpuscles.
E. Proprioceptors.

The correct answer is D. Pacinian corpuscles are rapidly adapting receptors that can respond to quickly changing stimuli and, therefore, can detect vibration. Meissner corpuscles are rapidly adapting and sensitive to light touch. Ruffini endings are slowly adapting and detect skin stretching and pressure. Nociceptors are free nerve endings that respond to tissue damage or stimuli that could result in tissue damage and interpret this information as pain. Proprioceptors detect the position of the body in space and relay this information back to the central nervous system.

3.3 An elderly patient presents complaining of tingling or numbness in her hands. The physician conducts a sensory exam to test for the integrity of sensory systems. The patient is asked to close her eyes, and the doctor then examines her arms and legs by touching them gently with the sharp end and then the blunt end of a pin, asking the patient if the touch feels sharp or dull. Sensory information about pain is carried by which of the following types of fibers?

A. Aα
B. Aβ
C. Aγ
D. Aδ
E. B

The correct answer is D. Aδ fibers carry sensory information about pain and temperature from free nerve endings. Aα fibers are motor neurons. Aβ fibers carry sensory information from Golgi tendon organs and Ruffini endings. Aγ fibers are motor to intrafusal fibers of the muscle spindle. B fibers are preganglionic autonomic.

3.4 Diseases such as myasthenia gravis can affect the neuromuscular junction, resulting in weakness in muscle contraction. In order for muscles to contract, action potentials must travel along a motor neuron and depolarize the axon terminal. Which of the following statements relating to action potentials is correct?

A. There is an influx of Na^+ ions into the intracellular space.
B. Ca^{2+} ions generate the excitatory postsynaptic potential.
C. An influx of Na^+ triggers muscle contraction.
D. Acetylcholine (ACh) binds to the ACh receptor on the skeletal muscle membrane.
E. Action potentials cause hyperpolarization at the axon terminal.

The correct answer is D. Depolarization at the axon terminal causes the release of acetylcholine (ACh), binding of ACh to its receptor, and the generation of an endplate potential. It is the influx of Ca^{2+} ions that causes the synaptic vesicles to fuse with the skeletal muscle membrane. It is the Na^+ ions that generate the excitatory postsynaptic potential or endplate potential. It is the influx of Ca^{2+} ions that triggers muscle contraction. Action potentials act to depolarize the axon terminal.

3.5 When muscles are contracted or relaxed, or when a physician checks reflexes, changes in muscle length are detected by specialized receptors in the muscle. The specialized receptor that detects muscle length is the:

A. Golgi tendon organ.
B. Muscle spindle.
C. Pacinian corpuscle.
D. Meissner corpuscle.
E. Motor endplate.

The correct answer is B. The muscle spindle detects the amount and rate of stretch in muscles (muscle length). Golgi tendon organs are found at the tendon–muscle junction and detect muscle strength. They are slowly adapting mechanoreceptors that are stimulated by tension in the tendon. Pacinian corpuscles are encapsulated receptors deep in the skin that detect vibration. Meissner corpuscles are rapidly adapting encapsulated receptors that are sensitive to light touch. The motor endplate is a chemical synapse between motor nerve fibers and muscle fibers.

3.6 In Sven's case, what is the cause of the intermittent numbness over the palmar aspect of the right palm, thumb, and index and middle finger?

A. Median nerve neuropathy.
B. Ulnar nerve neuropathy.
C. Radial nerve neuropathy.
D. C5 radiculopathy.
E. C6 radiculopathy.

The correct answer is A (median nerve neuropathy). Sven likely has carpal tunnel syndrome (CTS) due to compression of the median nerve in the carpal tunnel. The median nerve distribution includes the palmar aspect of the hand, thumb, index, middle, and lateral half of the 4th digit. Patients with diabetes are at higher risk of developing carpal tunnel syndrome. Shaking of the hand helps to relieve the compression of the median nerve in the carpal tunnel and is a classic feature of CTS. The ulnar nerve innervates the dorsal and palmar aspect of the medial hand, medial half of the 4th digit and the 5th digit. The radial nerve innervates the dorsal aspect (not palmar aspect) of the hand. The 5th cervical root innervates the lateral aspect of the arm, not the hand; and the 6th cervical root innervates the lateral aspect of the forearm/hand, thumb, and index finger, but not the middle finger.

4

Overview of the Visceral Nervous System

I. OVERVIEW

The visceral nervous system maintains homeostasis within our body. It monitors and controls the function of our internal organs (viscera, core), blood vessels, and structures in the skin (periphery). The visceral nervous system, like the somatic nervous system, has an afferent (sensory) and efferent (motor) component. General visceral afferent (GVA) fibers carry information from the body core, from our internal organs, but not from the periphery. General visceral efferent (GVE) fibers control smooth muscle, cardiac muscle, and secretory glands without our apparent conscious control. The efferent component has two divisions: the **parasympathetic system** and the **sympathetic system**. Together, these two efferent systems are also referred to as the **autonomic nervous system**. Parasympathetic neurons originate from nuclei in the brainstem and from the sacral spinal cord (S2–S4), and sympathetic neurons originate from the lateral horn in the spinal cord (T1–L2).

Sympathetic fibers travel to both the core and the periphery, whereas parasympathetic fibers only innervate the core (Figure 4.1). Throughout the digestive system, these efferent branches influence the **enteric nervous system**, which is responsible for gut motility.

II. VISCERAL SENSORY SYSTEM

Visceral sensory afferents bring information from the body's internal organs (core) to the central nervous system (CNS). The visceral receptors are **nociceptors** (pain), **mechanoreceptors** (fullness), or **specialized receptors** to detect the internal chemical or physical environment (e.g., acid–base balance, blood pressure). Afferents from the visceral organs and blood vessels are critical in the initiation of visceral reflexes, and most visceral afferents do not reach the level of consciousness. Those that do reach consciousness (e.g., related to hunger, nausea, fullness of the bladder) result in vague and poorly localized sensations probably due to low receptor density.

The unipolar cell bodies of visceral afferent neurons are located in the ganglia of spinal nerves from thoracic to sacral levels (T1–L2; S2–S4) and

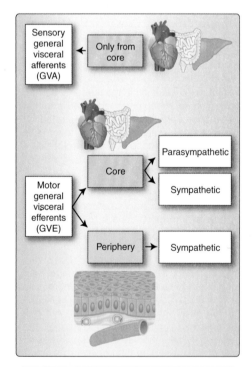

Figure 4.1
Distribution of visceral afferent and visceral efferent innervation in the body.

in the ganglia of cranial nerves (CNs) IX and X. Visceral *afferent* fibers travel with visceral *efferent* or motor fibers (sympathetic and parasympathetic). Afferents carrying pain travel mainly with sympathetic efferents, whereas information from both the mechanoreceptors and specialized receptors related to physiological functions travel mainly with parasympathetic efferents.

A. Visceral afferents related to pain

The cell bodies of the visceral afferents related to pain are localized primarily in the spinal ganglia from T1–L2 and S2–S4 and travel with sympathetic efferent fibers. Afferent fibers enter the posterior horn of the spinal cord where the majority cross the midline to travel with the **anterolateral system** on the contralateral side (Figure 4.2). These fibers terminate in the ventral posterolateral (VPL) nucleus of the thalamus, from which they project to the **insular cortex**, where the sensation of **visceral pain** is interpreted. Like other visceral sensations, visceral pain is poorly localized and often experienced as referred pain to superficial somatic structures (dermatomes) (see Chapter 22, Pain).

A subset of fibers carrying pain information ascends *bilaterally* to the nuclei of the reticular formation (see Chapter 12, Brainstem Systems and Review). These **spinoreticular fibers** have widespread influences on neurotransmitter systems and general arousal.

B. Visceral afferents related to physiological functions

Visceral afferents concerned with physiological functions, mainly visceral reflexes, can arise from the **pelvic nerves** (S2–S4) or from **CNs**. The cell bodies of sacral fibers carrying *GVAs from the pelvic viscera* are located in the spinal ganglia (S2–S4) and, interestingly, enter the spinal cord along parasympathetic efferents through the **anterior root** in contrast to all somatosensory fibers (see Figure 4.2). These fibers contain information on **bladder** and **rectum fullness** as well as **genital sensations** and participate in **visceral reflex pathways** (see below). Visceral fibers carrying *pain from some pelvic viscera* (e.g., sigmoid colon, prostate gland, cervix of the uterus), also accompany sacral parasympathetic nerves, and enter the spinal cord through the anterior root. This is an exception to the general rule that pain information mainly accompanies sympathetic fibers.

Fibers carrying visceral afferent information from sacral areas ascend *contralaterally* through the **anterolateral system** and *bilaterally* through the **spinoreticular system**.

Visceral afferents from the **thoracic** and most **abdominal viscera** (as far as the left colic flexure) are carried by the **vagus nerve (CN X)**, which also provides the visceral efferent (motor, parasympathetic) innervation of these organs. The associated **mechanoreceptors** are located in the smooth muscles of the thoracic and abdominal organs and detect **fullness** and **cramps. CN IX**, the **glossopharyngeal nerve**, carries **visceral afferen**t information from **chemoreceptors** and **baroreceptors** in the carotid sinus, which detect changes in blood gases and blood pressure. CN IX also carries visceral afferent information from the pharynx and is the afferent limb of the **gag reflex**. (See Chapter 12, "Brainstem Systems and Review," for more detail.)

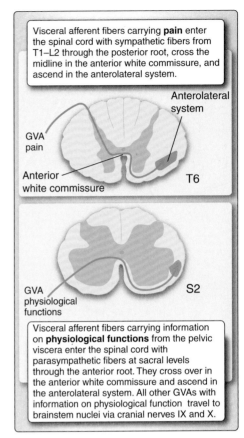

Visceral afferent fibers carrying **pain** enter the spinal cord with sympathetic fibers from T1–L2 through the posterior root, cross the midline in the anterior white commissure, and ascend in the anterolateral system.

Anterolateral system

GVA pain

Anterior white commissure

T6

GVA physiological functions

S2

Visceral afferent fibers carrying information on **physiological functions** from the pelvic viscera enter the spinal cord with parasympathetic fibers at sacral levels through the anterior root. They cross over in the anterior white commissure and ascend in the anterolateral system. All other GVAs with information on physiological function travel to brainstem nuclei via cranial nerves IX and X.

Figure 4.2
General visceral afferent (GVA) fibers enter the spinal cord.

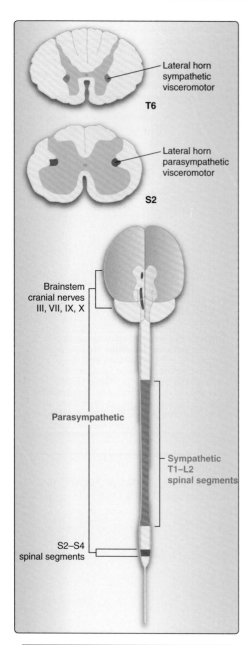

Figure 4.3
Sympathetic and parasympathetic
neurons in the CNS.

III. VISCERAL MOTOR SYSTEM

The visceral motor, or autonomic, nervous system is under central control. Signals originate in the **hypothalamus** and travel directly to autonomic nuclei in the brainstem and spinal cord. Nuclei in the **brainstem reticular formation** and direct projections from limbic structures, such as the **amygdala**, can also influence preganglionic autonomic neurons. This is how mental, emotional, visceral, and environmental influences can alter visceral motor function. For example, this is why we sweat when we are nervous, have butterflies in our stomach when we are excited, or get red in the face when embarrassed.

The parasympathetic and sympathetic systems, the two components of the visceral motor system, are antagonistic to each other. Preganglionic cell bodies of the sympathetic system can be found in the lateral horn of the spinal cord from levels T1–L2. The preganglionic parasympathetic cell bodies are found both rostral and caudal to these sympathetic cell bodies ("para" to the sympathetics) in the brainstem and in the sacral spinal cord at S2–S4 (Figure 4.3).

A. Structure of the visceral motor system

These visceral motor pathways consist of a two-neuron chain in the periphery: a preganglionic neuron leaves the CNS and synapses in a peripheral ganglion with the postganglionic neuron. This architecture is critical for the function of the visceral motor system. One preganglionic neuron can synapse with several postganglionic neurons in the ganglion. This divergence of input allows a small number of central neurons to influence a large number of peripheral neurons. Moreover, as the visceral afferents (sensory fibers) travel through the visceral ganglia, a certain degree of sensory–motor integration takes place in these ganglia. This allows visceral motor fibers to respond directly to visceral sensory input without influence from the CNS, which gives the visceral nervous system additional autonomy and the ability to respond rapidly.

B. Neurotransmitter activation of the visceral motor system

Sympathetic ganglia are located close to the CNS, whereas parasympathetic ganglia are typically located close to or within the organs they will innervate. Both parasympathetic and sympathetic nervous systems use the same neurotransmitter, acetylcholine (ACh), to activate their postganglionic neurons. Release of ACh from postganglionic parasympathetic neurons to the core increases gut motility and decreases cardiac muscle contraction. Sympathetic neurons that project to the periphery to innervate sweat glands and the erector pili muscles also use ACh as their neurotransmitter. By contrast, postganglionic sympathetic fibers to internal organs and most blood vessels use the neurotransmitter norepinephrine, which decreases gut motility and increases cardiac muscle contraction (Figure 4.4).

One subset of sympathetic neurons projects not to a peripheral ganglion, but to the adrenal medulla, without synapsing in a peripheral ganglion. These neurons use the neurotransmitter ACh and cause the secretion of hormones from the chromaffin cells in the adrenal

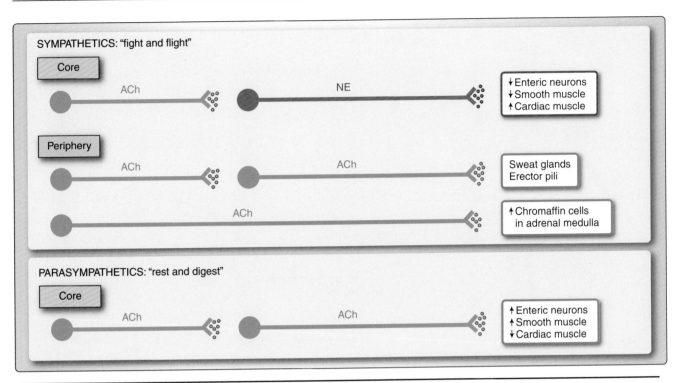

Figure 4.4
Neurotransmitters used by the sympathetic and parasympathetic systems. ACh = acetylcholine; NE = norepinephrine.

medulla. Chromaffin cells are of the same embryological origin as the peripheral ganglia (from the neural crest) and secrete norepinephrine, epinephrine, and endorphins into the bloodstream, causing a widespread sympathetic ("fight-or-flight") response.

The **enteric nervous system** comprises a complex plexus of neurons within the gastrointestinal system that is autonomous but under the influence of the parasympathetic and sympathetic nervous systems.

C. Parasympathetic nervous system

The parasympathetic system acts to maintain the status quo of the body. Also known as the **"rest-and-digest" system**, it targets thoracic, abdominal, and pelvic viscera. When activated, the parasympathetic system decreases cardiac output and blood pressure, speeds up peristalsis in the gastrointestinal tract, increases salivation, and causes pupillary contraction.

The preganglionic neurons in the parasympathetic system are located in the cranial and sacral regions of the CNS (**craniosacral outflow**) (Figure 4.5). Parasympathetic innervation to the ciliary ganglion of the eye is from the Edinger-Westphal nucleus (CN III, oculomotor). The preganglionic parasympathetic neurons to the ganglia of the salivary glands in the head are found in the brainstem: the superior salivatory nucleus (CN VII, facial) and the inferior salivatory nucleus (CN IX, glossopharyngeal). The dorsal motor nucleus of vagus contains the cell bodies of parasympathetic neurons that travel with CN X to parasympathetic ganglia of thoracic and abdominal viscera, and to enteric

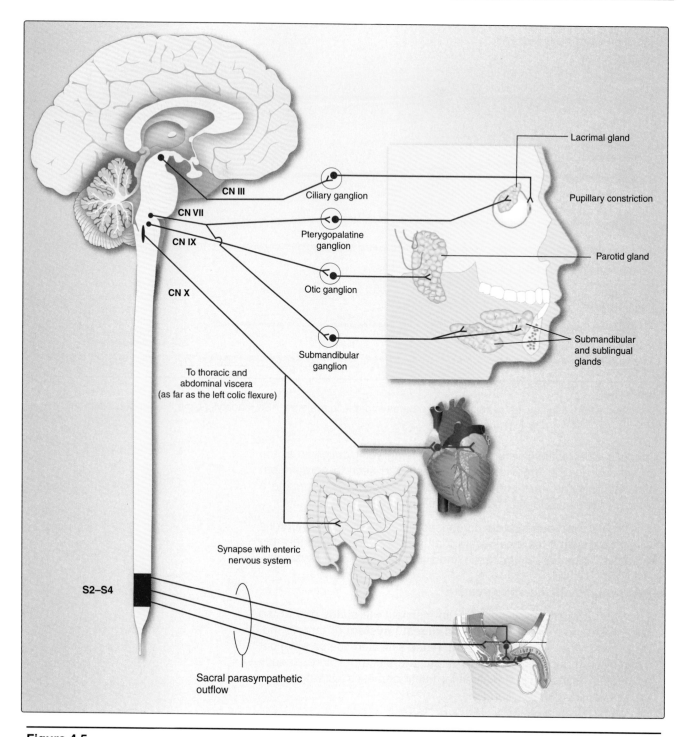

Figure 4.5
The distribution of parasympathetic fibers in the body. CN = cranial nerve.

ganglia (up to the left colic flexure). The nucleus ambiguus contains parasympathetic cell bodies for the innervation of cardiac muscle; these fibers also travel with CN X. It should be noted that nucleus ambiguus contains mostly motor neurons to skeletal muscles of the pharynx and larynx (see Chapter 10, "Sensory and Motor Innervation of the Head"). The cell bodies of the preganglionic parasympathetic fibers that innervate the descending colon and the pelvic viscera are located in the lateral horn of the spinal cord (S2–S4). Their postganglionic neurons are in specific ganglia located close to or within the walls of the target organ. An overview of the peripheral ganglia can be seen in Figure 4.5.

The parasympathetic nature of the sacral autonomic outflow is currently under review; evidence around their anatomy, molecular profile, and function on the organs of the pelvis suggests a sympathetic identity. These new findings will help to establish a new approach to understanding the diverse roles of the autonomic nerves to the pelvis.

D. Sympathetic nervous system

This system is commonly called the **"fight-or-flight" system**. It has widespread action resulting in a rise in blood pressure, an increase in blood sugar levels, redirection of blood flow to the skeletal muscles and away from visceral and cutaneous circulation, dilation of the pupils, and decreased salivation.

The sympathetic outflow originates in the spinal cord from the **lateral horn** in the thoracic and lumbar spinal cord (T1-L2) (see Figure 4.3). The axons of the preganglionic neurons leave the spinal cord through the anterior root. Just after the anterior and posterior roots merge to form a spinal nerve, a **white communicating ramus** branches off and connects the preganglionic fibers to the sympathetic trunk (Figure 4.6). The sympathetic trunk is a chain of ganglia, which contains the cell bodies of the postganglionic fibers. Sympathetic preganglionic fibers destined for the **head region** will ascend in the sympathetic trunk and synapse in the **superior cervical ganglion**. From here, the postganglionic fibers will travel into the head region along the surface of the internal carotid artery. The axons to **abdominal** and **pelvic viscera** will travel through the sympathetic chain *without synapsing* and proceed via **splanchnic nerves** to specific prevertebral ganglia to synapse on their postganglionic target neurons. Sympathetics destined for the trunk and limbs synapse in the ganglia of the sympathetic chain and send **postganglionic fibers** via **gray rami communicantes** (gray because they are unmyelinated) to be distributed with the spinal nerves to their target tissues.

Interestingly, innervation of the pelvis is a combination of the sympathetic patterns seen for both core and peripheral structures; postganglionic fibers may secrete either norepinephrine or ACh.

E. Enteric nervous system

The number of neurons in the walls of the gastrointestinal system is as great as the number of neurons in the entire spinal cord. They are arranged in clusters within the walls of the gut. These clusters comprise the **myenteric plexus (of Auerbach)** between the longitudinal and circular muscle layers, and the **submucosal plexus (of Meissner)** between the circular muscle layer and the muscular layer of the mucosa (Figure 4.7). The plexuses consist of small ganglia, which are connected by small unmyelinated nerve fibers. The myenteric

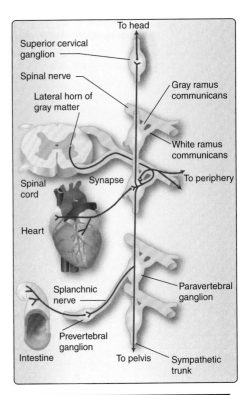

Figure 4.6
Sympathetic fibers can synapse either in the sympathetic trunk or in distal ganglia.

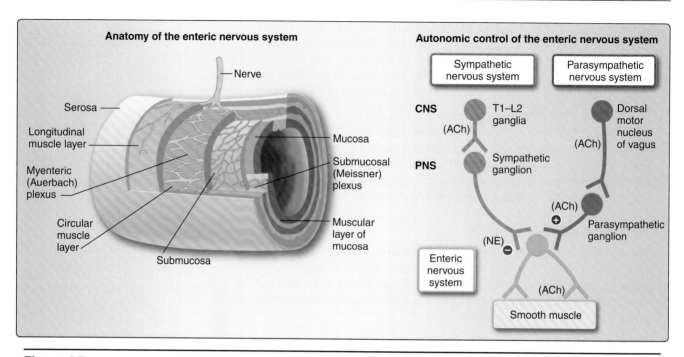

Figure 4.7
The enteric nervous system. ACh = acetylcholine; NE = norepinephrine.

plexus is the main controller of gut motility, whereas the submucosal plexus is the main controller for the secretion and absorption of fluids. Both excitatory and inhibitory neurons from these plexuses end on smooth muscle and gland cells of the gut. Neurons in the enteric system function independently but are influenced by the activity of parasympathetic and sympathetic systems.

F. Input to the visceral nervous system

Descending pathways from the **hypothalamus** regulate the visceral motor system. The hypothalamus sends projections to the parasympathetic nuclei in the brainstem (of CNs III, VII, IX, and X) as well as to sympathetic neurons in the lateral horn in spinal cord segments T1–L2.

The main tracts that influence the visceral motor neurons are the **hypothalamospinal** and **hypothalamomedullary tracts**. These fibers originate mainly from the **paraventricular nucleus**. They descend through the **periaqueductal gray (PAG)** and nearby reticular formation of the midbrain and rostral pons and then move anterolaterally in the medulla. The hypothalamomedullary fibers terminate on all brainstem nuclei containing the cell bodies of GVE fibers (Table 4.1 and Figure 4.8). The **hypothalamospinal fibers** continue through the **anterolateral medulla** into the lateral column of the spinal cord, ending in the **lateral horn**, and providing regulation of sympathetic outflow. These two tracts are an essential link between the hypothalamus and the autonomic nuclei of the brainstem and spinal cord.

Table 4.1: Overview of the Parasympathetic System

Nucleus	Nerve	Peripheral Ganglion	Target	Function
Cranial				
Edinger-Westphal	CN III Oculomotor	Ciliary ganglion	Sphincter pupillae	Pupillary constriction
			Ciliaris muscles	Accommodation
Superior salivary nucleus	CN VII Facial	Pterygopalatine ganglion	Lacrimal gland	Secretion of tears
		Submandibular ganglion	Salivary glands	Stimulation of salivation
Inferior salivary nucleus	CN IX Glossopharyngeal	Otic ganglion	Parotid gland	Stimulation of salivation
Nucleus ambiguus	CN X Vagus	Thoracic visceral plexus	Heart	Decrease heart rate
Dorsal motor nucleus of vagus	CN X Vagus	Thoracic ganglia	Thoracic and abdominal viscera	Constrict airways
		Abdominal ganglia		Stimulate digestion

CN = cranial nerve.

IV. VISCERAL REFLEXES

Visceral afferents are involved in the **regulation** of visceral function through **visceral reflex systems** that maintain homeostasis.

These visceral reflex pathways use both somatic and visceral nerves. Some of the regulation is at a nonconscious level, and for others, there are voluntary (conscious) aspects as well. Blood pressure, for instance, is regulated at a nonconscious level. Other visceral functions such as emptying the bladder, however, combine visceral and voluntary somatic function. We discuss the control of blood pressure as an example of a nonconscious visceral reflex. The control of bladder function is clinically extremely important when assessing spinal cord injuries and is an example of how the somatic and visceral systems work together. Finally, a section on sexual function shows how the somatic and visceral parts of the nervous system work together with the cortex, especially limbic areas.

A. Control of blood pressure

Blood pressure is continuously monitored through input from **baroreceptors** in the carotid sinus and aortic arch. Information from the baroreceptors travels via CN IX (carotid sinus) and X (aortic arch) to the **nucleus solitarius**. In turn, the nucleus solitarius sends information to the dorsal motor nucleus of the vagus, to medial cell bodies in the nucleus ambiguus that are associated with the heart, and to rostral areas of the anterolateral medulla. A rise in blood pressure initiates a **"vasodepressor"** response. Projections from the nucleus solitarius activate cells in the **dorsal motor nucleus of the vagus** and the **nucleus ambiguus**, which in turn send input via the **vagus nerve** to **parasympathetic ganglia in the heart**. This causes a decrease in the force of heart muscle contraction and a concomitant decrease in blood pressure. By contrast, a fall in blood pressure initiates a **"vasopressor"** response. Projections from the nucleus solitarius activate

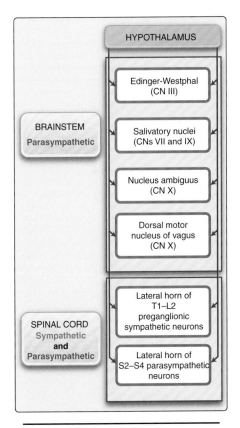

Figure 4.8
Hypothalamic input to the visceral nervous system.

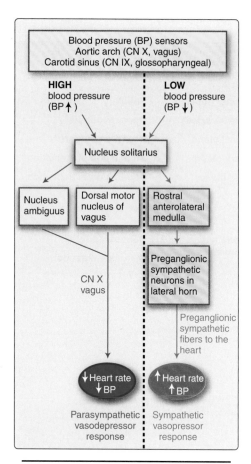

Figure 4.9
The control of blood pressure.
BP = blood pressure; CN = cranial nerve.

cells in the rostral anterolateral medulla, which project to the **preganglionic sympathetic neurons** in the **lateral horn** of the spinal cord. From there, fibers project to the **sympathetic ganglia** and travel as **postganglionic sympathetic fibers** to the heart, where they cause an increase in the force of heart muscle contraction and a rise in blood pressure (Figure 4.9). This **baroreceptor reflex** keeps blood pressure at a healthy, normal level.

B. Control of bladder function

Bladder function is controlled by the integration of sensory, motor, and visceral functions. **Somatic afferents** sense pain and temperature from the bladder, whereas **visceral afferents** sense bladder fullness, although some overlap of function likely exists here. The internal urethral sphincter (found only in males) is controlled by **sympathetic efferents**. The detrusor muscle is controlled by **parasympathetic efferents**. The external urethral sphincter is controlled by **somatic motor efferents**.

All of these functions must be coordinated and monitored by central nuclei localized both in the pons and in the forebrain, as shown in Figure 4.10.

1. **Pain and temperature:** Somatic afferent neurons encoding for pain and temperature from the mucosa of the fundus of the bladder travel with sympathetic fibers and reach the spinal cord at the T12–L1 level. Afferents from the neck of the bladder travel with parasympathetic fibers and reach the spinal cord at the S2–S3 level. All of these afferents then cross the midline and travel with the spinothalamic tract to the thalamus and then to the cortex.

2. **Fullness of the bladder:** Fullness of the bladder is sensed by mechanoreceptors in the bladder wall. This information is relayed to the sacral parasympathetic neurons at S2–S4 as well as to the cortex via the spinothalamic tract. Afferents from the trigone of the bladder, indicating that micturition (urination) is imminent, travel to the cortex via the posterior column–medial lemniscus system.

3. **Innervation of the urethral sphincters:** The **internal urethral sphincter** is located at the neck of the bladder in males (females do not have an internal urethral sphincter). The **sympathetic visceromotor** neurons that innervate the internal sphincter have their cell bodies in the lateral horn between T12 and L2. They exit the spinal cord with the lumbar splanchnic nerves, synapse in the inferior mesenteric ganglion, and innervate the internal urethral sphincter. The internal urethral sphincter is not under voluntary control and contracts reflexively during ejaculation to prevent retrograde flow into the bladder.

   The **external urethral sphincter** is a striated, skeletal muscle located in the **deep perineal pouch**. In males, there is only one external urethral sphincter, but in females, this function is taken on by the sum of three separate muscles (external urethral sphincter, sphincter urethrovaginalis, and compressor urethrae). The **somatic motor neurons** that innervate the external urethral

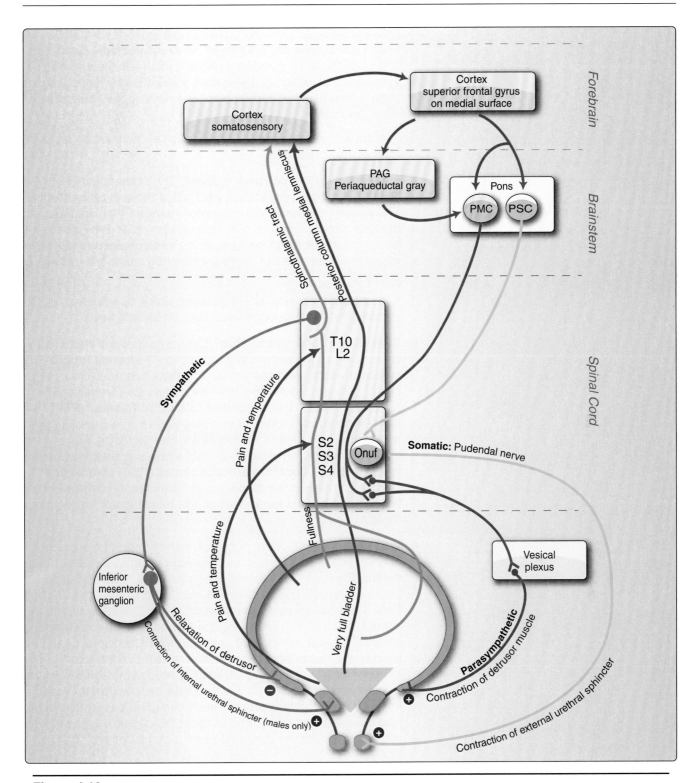

Figure 4.10

The somatic and visceral innervation of the bladder. PAG = periaqueductal gray; PMC = pontine micturition center; PSC = pontine storage center.

sphincter(s) have their cell bodies in the anterior horn at S2–S4. These cell bodies are referred to as **Onuf nucleus**. From here, fibers travel with the **pudendal nerve** to innervate the external urethral sphincter, which is under tonic stimulation by these somatic motor efferents. Only inhibition of Onuf nucleus from the pons will relax this muscle.

The somatic motor neurons allow *voluntary* contraction or relaxation of the sphincters.

4. **Innervation of the detrusor muscle:** The detrusor muscle of the bladder is **smooth muscle** located within the wall of the bladder. The **parasympathetic visceromotor neurons** that innervate the detrusor muscle originate in the lateral horn at S2–S4. From there, parasympathetic fibers synapse in the **vesical plexus** before terminating on the detrusor muscle where they excite that muscle to contract.

It is unclear if there is any significant inhibitory influence on the relaxation of the detrusor muscle via sympathetic fibers.

5. **Central control of micturition:** Central control and coordinating centers for micturition are located in both the **frontal lobe of the cerebral cortex** and the **pons.** The **PAG** has a role as a relay center between inputs from both the cortex and the spinal cord to the pons. In the pons, a **pontine micturition center (PMC)** will activate neurons that facilitate voiding (detrusor muscle via sacral parasympathetics) and inhibit those that would facilitate urine storage (urethral sphincters via Onuf nucleus and thoracolumbar sympathetics). A **pontine storage center (PSC)** coordinates the storage of urine in the bladder and facilitates the contraction of the urethral sphincters as well as the relaxation of the detrusor muscle.

6. **Voiding:** In infants, there is no voluntary control of bladder function. Voiding is coordinated via a **spinopontospinal reflex mechanism** (Figure 4.11). Visceral afferents in the bladder wall sense bladder fullness. This information is relayed to the PMC via the PAG. The PMC inhibits sympathetics at the T10–L2 level and the somatic Onuf nucleus, causing the relaxation of the internal and external urethral sphincters, respectively. In addition, excitatory input to the sacral parasympathetics results in contraction of the detrusor muscle and, with that, voiding. Around the age of 3 years, voluntary control of bladder function becomes possible through the involvement of cortical areas. These cortical areas in the frontal lobe assess whether or not it is acceptable to void and will then influence the pontine micturition and storage centers (PMC and PSC).

7. **Bladder control after spinal cord injury:** In spinal cord injury rostral to the lumbosacral levels, voluntary influences from the cortex and the pons to the spinal cord are eliminated. At first, the bladder is **areflexic** with complete urinary retention, which requires catheterization. The next step is **automatic micturition** via the spinal reflex pathway (see Figure 4.11), where mechanoreceptors

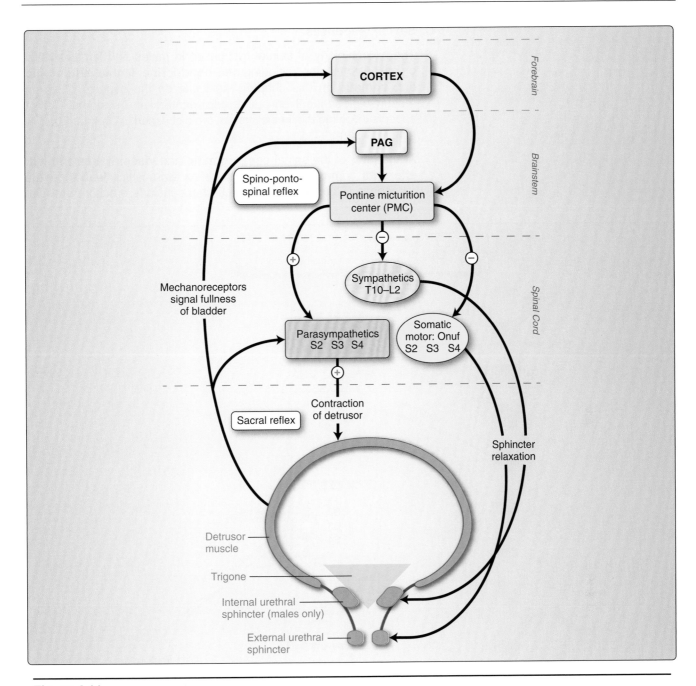

Figure 4.11
The control of micturition. PAG = periaqueductal gray.

sensing the fullness of the bladder directly activate sacral parasympathetics causing detrusor contraction. Because the influence of the PMC is gone, the external urethral sphincter is not relaxed during contraction of the detrusor. This leads to incomplete emptying of the bladder due to **detrusor sphincter dyssynergia**. Patients will require daily catheterization to ensure complete emptying of the bladder.

C. Sexual responses

The neurobiology of sexual responses in males and females relies on multiple systems, including **neuroendocrine**, **limbic**, **autonomic**, and **somatic**. In this chapter, we focus on the aspects of sexual response dependent on spinal cord centers and systems. Central aspects involving the limbic system are discussed in Chapter 20, "The Limbic System."

At the level of the spinal cord, **somatic** and **visceral afferents** and **efferents** come together to form reflex arcs, which lead to sexual responses (Figure 4.12). In males, these include erection, emission

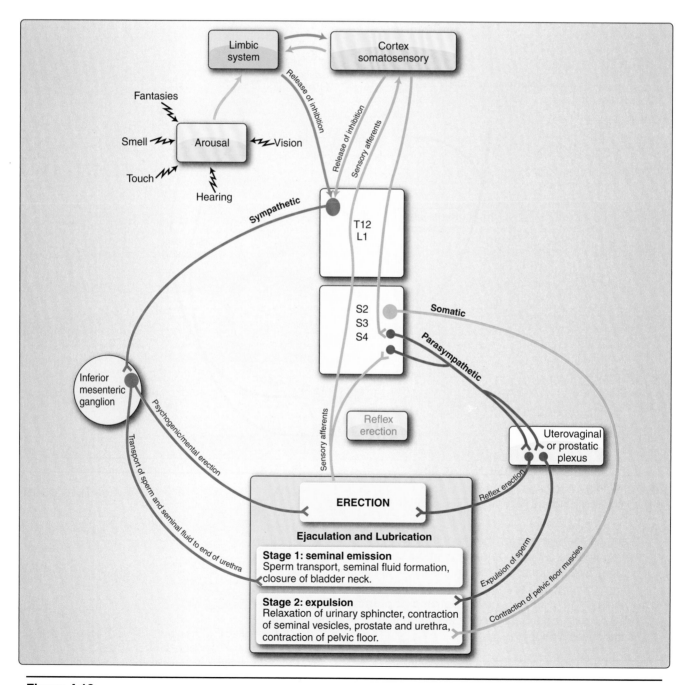

Figure 4.12

The neuronal pathways involved in sexual responses.

and ejaculation, and components of orgasm. In females, these include vaginal lubrication, engorgement of erectile tissue, and components of orgasm.

Sexual response is dependent on both the somatic and the visceral nervous systems. The **pudendal nerve** (S2–S4) carries the somatic innervation to the perineum, including the genitalia. Sensory information from the erogenous areas of the perineum is carried by the pudendal nerve, as is voluntary motor information to the pelvic floor muscles. Their rhythmic contraction is part of orgasm in both males and females. The **visceral nervous system** innervates the blood vessels and glands and is responsible for erection and vaginal lubrication as well as emission and ejaculation in males and components of orgasm in both sexes.

1. **Engorgement:** Penile erection is controlled by one of two mechanisms. A psychogenic or mental erection is mediated by sympathetic fibers (T12–L1/L2). These fibers are usually directly inhibited by higher centers, but arousal can release this inhibition and lead to activation of these fibers, which in turn leads to erection. The second mechanism for erection is dependent on sensory input through the pudendal nerve (S2–S4) to sacral parasympathetic neurons (S2–S4). This can lead to a reflex erection independent from descending influences. Parasympathetic fibers can also be activated via descending influences from cortical areas, which have been activated through other erotic sensory influences.

   In females, **vaginal lubrication** and engorgement of erectile tissue are mediated by autonomic fibers.

2. **Ejaculation:** In males, **ejaculation** is preceded by a phase of **seminal emission**, which includes sperm transport, the formation of seminal fluid, and the closure of the bladder neck to prevent retrograde ejaculation. This phase is controlled by **thoracic sympathetic fibers** (T12–L1). The next stage is the **expulsion of sperm**. This is mediated by **sacral parasympathetic fibers** (S2–S4) and includes the contraction of the seminal vesicles, prostate, and urethra. The rhythmic **contraction of the pelvic floor** is mediated by the **pudendal nerve** (S2–S4, somatic).

3. **Orgasm: Orgasm** is a complex physiological response that differs in the way it is triggered and experienced in males and females. It is an experience that integrates cortical arousal levels with sensory stimuli and motor expressions such as ejaculation, the rhythmic contraction of the pelvic floor musculature, or the contraction of the vagina and uterus, among others. A multitude of erotic stimuli are integrated in cortical and subcortical regions. Hormone and neurotransmitter release play a role in the intensity of the orgasm experience.

Clinical Case:

Autonomic Dysreflexia

A 52-year-old man, who had sustained a complete spinal cord injury 20 years ago, presents to the emergency room with high blood pressure and severe headache. Upon examination, his blood pressure is 160/90 mm Hg, with a pulse rate of 50 beats/minute. His temperature is 38.5°C. He is conscious and oriented, with no corroborating signs of meningitis or migraine. His paraplegia has a sensory level at

T2 and he has an indwelling urinary catheter. Simple analgesia for headache does not improve his symptoms. Urinalysis shows signs of a urinary tract infection with both blood and leukocytes present. His previous medical history shows that since his spinal cord injury, his blood pressure has been around 90/60 mm Hg.

The history of spinal cord injury, headache, hypertension with bradycardia, and urinary tract infection suggest **autonomic dysreflexia**. He is treated for the hypertension and given antibiotics for the urinary tract infection. Within a few hours, the hypertension improves and the headache disappears.

Case Discussion

This is a typical case of autonomic dysreflexia precipitated by a urinary tract infection in a patient with a high spinal cord injury (Figure 4.13). The infection appears to activate a sympathetic response below the level of the injury, resulting in hypertension. The increased blood pressure is sensed by baroreceptors in the carotid sinus and aortic arch, which send projections to the nucleus solitarius. From here, a parasympathetic vasodepressor response is initiated, which slows the heart rate, but cannot influence the blood pressure, which is still under increased sympathetic tone. This is because there are no descending modulatory influences on sympathetic neurons in the lateral horn of the spinal cord. The sympathetic neurons therefore remain active and blood pressure remains elevated, while parasympathetic influences slow the heart rate, thus resulting in the seemingly paradoxical presentation of hypertension with bradycardia. Excessive parasympathetic output causes vasodilation above the level of the injury, which accounts for the headache. Patients may also present with nasal congestion, sweating, and flushing.

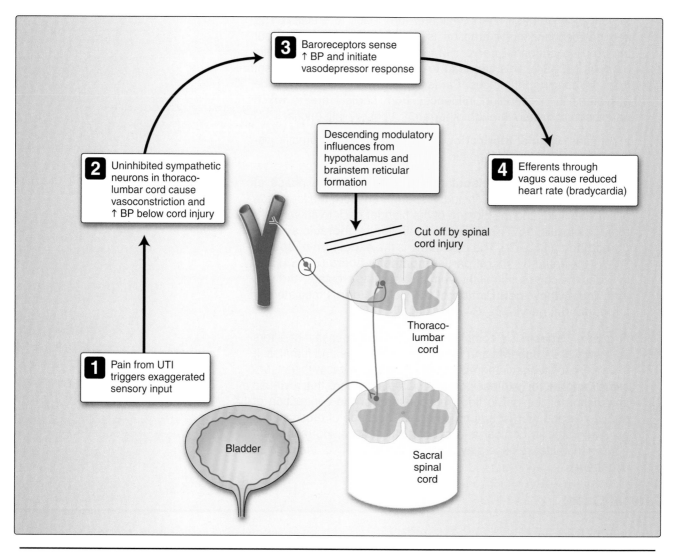

Figure 4.13
Autonomic dysregulation affecting the cardiovascular system following spinal cord injury.

Chapter Summary

- The visceral nervous system is composed of both a motor and a sensory component. **Visceral afferent information** is carried to the spinal cord and brainstem on both sympathetic and parasympathetic efferent fibers. Pain information travels mainly with sympathetic fibers, and information related to physiological functions and visceral reflexes travels mainly with parasympathetic fibers. Both types of information ascend via the contralateral anterolateral system.

- Importantly, the visceral afferents also participate in nonconscious reflex systems with visceral efferents to maintain homeostasis.

- Visceral efferents or visceral motor fibers can be grouped in two systems that are, by and large, antagonistic to each other: the parasympathetic system (rest and digest) and the sympathetic system (fight or flight). The visceral motor system is also referred to as the autonomic nervous system. The parasympathetic fibers originate from cranial nerve nuclei and from sacral levels (S2–S4) of the spinal cord. They project only to structures in the head and the core. Sympathetic fibers originate from spinal cord levels T1–L2. Sympathetic fibers project to the body core (to the viscera) as well as to the periphery where they innervate sweat glands and blood vessels. Together, the sympathetic and parasympathetic systems help maintain homeostasis and allow for visceral functions to be adjusted to current needs. The enteric nervous system comprises the myenteric and submucosal plexuses in the gut wall. These plexuses are under the influence of the visceral motor system. The myenteric plexus controls gut motility, whereas the submucosal plexus controls the secretion and absorption of fluid.

- Descending fibers to the **visceral motor system** originate from the **hypothalamus**. These fibers descend through the periaqueductal gray and the anterolateral brainstem to influence the **parasympathetic nuclei of the brainstem** (related to cranial nerves III, VII, IX, and X) and the **sacral spinal cord** (S2–S4) as well as the **preganglionic sympathetic neurons** in the lateral horn of the spinal cord (T1–L2).

- Complex systems such as **bladder control** and **sexual responses** are controlled by the interplay of visceral and somatic neuronal groups in the spinal cord. These reflexes rely on the coordination of sympathetic and somatic neurons. For bladder function in infants, a spinopontospinal reflex will initiate voiding when the bladder is full. Voluntary control of bladder function is achieved through the influence of descending systems from the brainstem and the cortex. For sexual function, a reflex erection can be achieved through sensory afferents directly contacting sympathetic neurons in the spinal cord. The more complex responses of sexual behavior integrate cortical information, which is then relayed to the spinal cord where the activation of different neuronal groups will result in a sequence of sexual responses.

Study Questions

Choose the ONE best answer.

4.1 A young man is brought to the emergency department following a motor vehicle accident. Initial examination shows motor weakness ("spinal shock") in the myotomes for the legs. A computed tomography scan reveals crushed vertebrae that have lesioned his spinal cord at T8. What are the immediate consequences for his bladder function?

    A. He will feel when his bladder is full and can then self-catheterize.

    B. He will not be able to store any urine in his bladder.

    C. He will have reflex bladder function.

    D. He will have complete urinary retention.

    E. He will be able to control emptying of the bladder.

The correct answer is D. At first, the bladder is areflexic with complete urinary retention, which requires catheterization. The next step is automatic micturition via the spinal reflex pathway, through the activation of sacral parasympathetics, causing detrusor contraction. Because the influence of the pons is gone and the sphincter and detrusor are no longer working together, there will be incomplete emptying of the bladder. He will require daily catheterization to ensure complete emptying of the bladder.

4.2 A patient complains about diffuse pain in her pelvis. These sensations are carried through visceral afferent fibers. The visceral sensory system is characterized by which one of the following?

A. The cell bodies of general visceral afferents from pelvic viscera are located in the lumbar spinal ganglia.

B. Visceral afferents from the pelvic viscera travel with CN X, vagus.

C. Fibers carrying visceral afferent information from sacral areas ascend in the contralateral anterolateral system.

D. Information from the sacral region to the reticular formation ascends ipsilaterally.

E. Visceral afferent fibers synapse in the spinal ganglion.

The correct answer is C. The cell bodies are in sacral spinal ganglia (S2–S4); these do not contain synapses. CN X, vagus, does not innervate structures in the pelvis. Information from the sacral region to the reticular formation of the brainstem ascends bilaterally.

4.3 The visceral motor system has a sympathetic and a parasympathetic branch. The anatomy of the pathways for these systems is important when localizing a lesion, which involves these systems. Which of the following statements about the *parasympathetic* nervous system is correct?

A. Nerve cell bodies are located in the T1–L2 spinal cord regions.

B. When activated, the parasympathetics act to increase cardiac output.

C. Preganglionic parasympathetic neurons that innervate the abdominal viscera are located in the anterior horn of the spinal cord.

D. The preganglionic parasympathetic neurons innervating salivary glands in the head are found in the brainstem.

E. The preganglionic parasympathetic neurons use noradrenaline as their neurotransmitter.

The correct answer is D. Nerve cell bodies are located in the brainstem, the parasympathetics act to decrease cardiac output, and preganglionic parasympathetic neurons to the abdominal viscera are located in the brainstem (dorsal motor nucleus of vagus). Preganglionic sympathetic neurons are found in the lateral horn from T1–L2. Acetylcholine is the neurotransmitter used by both sympathetic and parasympathetic preganglionic neurons, but only sympathetic neurons use norepinephrine in the postganglionic neurons.

The Spinal Cord

5

I. OVERVIEW

In the previous chapters, the general organization and major features of the spinal cord were introduced. In this chapter, the discussion includes the surface anatomy, internal structures, spinal meninges, spinal nerves, blood supply, and some of the major clinically relevant intrinsic systems of the cord.

The spinal cord is a direct continuation of the caudal brainstem and extends from the foramen magnum of the skull to the conus medullaris. It is surrounded by meningeal layers, cerebrospinal fluid (CSF), and the bones of the vertebral column.

The spinal cord has both **conduit function** for tracts going to and from higher centers and **intrinsic functions**, such as reflexes. The spinal cord tracts are located in the white matter, which surrounds the gray matter containing nerve cell bodies. The spinal cord tracts carry sensory information to higher centers, as well as motor information from the cortex to the motor neurons of the spinal cord. Motor neurons in the cortex are referred to as **upper motor neurons (UMNs)**, and motor neurons in the spinal cord, which innervate muscles, are referred to as **lower motor neurons (LMNs)**.

In cross section, the distribution of gray matter is butterfly shaped. Each side contains an **anterior (ventral) horn** and a **posterior (dorsal) horn** (Figure 5.1). A **lateral horn** is visible at sympathetic spinal cord levels T1–L2 and S2–S4.

The spinal cord is divided into **31 segments**, and a pair of spinal nerves is associated with each spinal segment. There are 8 cervical segments and nerves (C1–C8), 12 thoracic (T1–T12), 5 lumbar (L1–L5), 5 sacral (S1–S5), and 1 coccygeal (Figure 5.2). Each spinal nerve contains both sensory and motor information.

Sensory information enters the spinal cord through the **posterior roots**. Sensory cell bodies are located in the spinal ganglion of each spinal nerve.

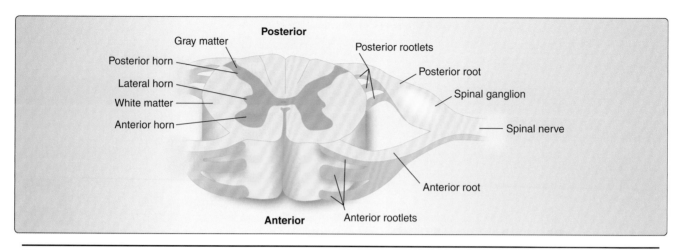

Figure 5.1
Cross section of spinal cord.

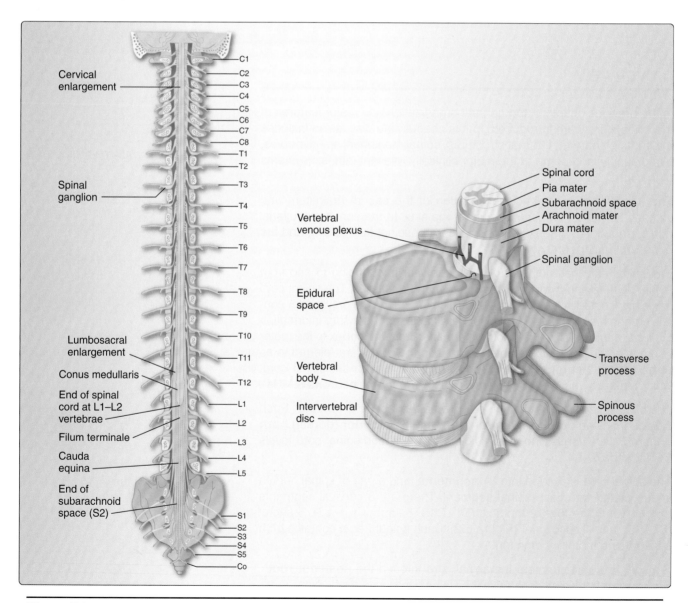

Figure 5.2
Overview of the spinal cord. Co = coccygeal.

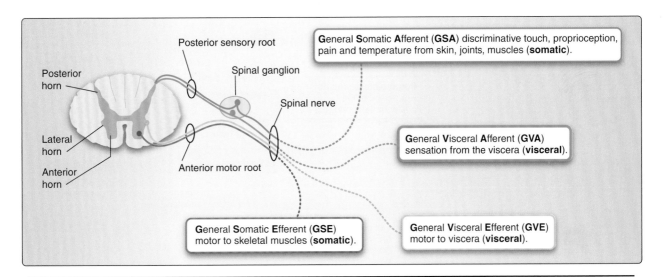

Figure 5.3
Spinal nerve.

Motor information leaves the spinal cord through the **anterior roots**, and LMNs are located in the anterior horn at each spinal level (Figure 5.3). Each segment of the spinal cord innervates a specific area of the skin, referred to as a **dermatome**, and a specific muscle group, referred to as a **myotome**. Primary efferent autonomic fibers have their cell bodies in the lateral horn and leave the spinal cord through the anterior root. The afferent autonomic fibers travel with the somatic afferents through the posterior root.

II. SURFACE ANATOMY OF THE SPINAL CORD

In newborns, the spinal cord extends to vertebral level L3. Because the vertebral canal grows to a greater extent than the spinal cord, in adults, the spinal cord extends to the L1–L2 level in the vertebral column. Spinal nerves exit the spinal cord at each vertebral level. The posterior and anterior roots that travel through the lumbar cistern from the end of the spinal cord at L1–L2 to their respective vertebral levels are referred to as **cauda equina** (Latin for "horse tail").

The spinal cord is cylindrical in shape and slightly flattened from anterior to posterior. Two major **enlargements** can be found in the **cervical** and **lumbar** regions, where neurons that supply the plexuses for the upper and lower limbs are located. The spinal cord ends at the **conus medullaris** and is attached to the dorsum of the first coccygeal segment by the **filum terminale** (see Figure 5.2).

The spinal cord is marked on its external surface by longitudinal **fissures** and **sulci** (Figure 5.4). On the anterior surface is a prominent **anterior median fissure** that is apparent the entire length of the spinal cord. This is where the anterior spinal artery can be found in the subarachnoid space. Deep to this fissure, the **anterior white commissure** is located (visible on cross section) where some sensory and motor fibers cross the midline. The **anterolateral sulcus** is the site of exit for the anterior motor rootlets.

On the posterior surface of the spinal cord, a **posterior median sulcus** and septum can be identified. It separates the posterior surface of the spinal cord into two halves. The **posterolateral sulcus** marks the entry of the posterior sensory rootlets of the spinal cord.

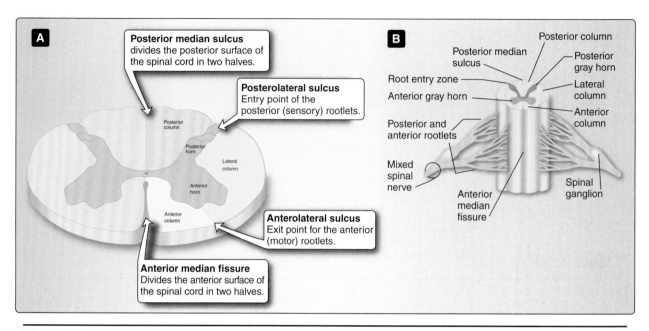

Figure 5.4
Fissures and sulci.

Between the longitudinally arranged fissures and sulci are elevations called **funiculi** (singular, *funiculus*). This is where the ascending and descending fiber tracts, or **columns**, of the spinal cord are located. The three columns are posterior, lateral, and anterior.

The **posterior column** is located bilaterally between the posterior median sulcus and the entry point of the posterior roots (posterolateral sulcus). In the cervical and upper thoracic regions, the posterior column is divided into large fasciculi (tracts) carrying sensory information up the spinal cord to the medulla (Figure 5.5). The **fasciculus gracilis** carries information from the lower trunk and lower limbs, and the **fasciculus cuneatus** carries information from the upper trunk and upper limbs. This information includes discriminative (fine) touch, vibration, and proprioception from joints and muscles.

The **lateral column** is located between the entry point of the posterior rootlets (posterolateral sulcus) and the exit point of the anterior rootlets (anterolateral sulcus). This is where major ascending and descending tracts are located. The **lateral corticospinal tract** carries motor information from the contralateral cortex to the anterior horn cells. The **anterolateral tract** carries pain and temperature information from the contralateral side of the body.

The **anterior column** is located between the anterior median fissure and the exit point of the anterior rootlets (anterolateral sulcus). This column contains ascending and descending tracts such as the **anterior corticospinal tract**. (See Figures 5.4 and 5.5 for a depiction of surface anatomy as it relates to the underlying tracts.)

A. Spinal meninges

The meningeal coverings of the spinal cord are an extension of the meningeal coverings of the brain in the skull. There are three layers: an outermost **dura mater**, the **arachnoid mater**, and the **pia mater**, which adheres tightly to the surface of the spinal cord (Figure 5.6; also see Chapter 2, "Overview of the Central Nervous System").

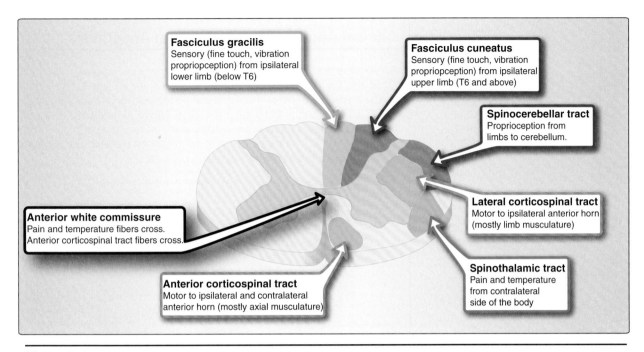

Fasciculus gracilis
Sensory (fine touch, vibration propriopception) from ipsilateral lower limb (below T6)

Fasciculus cuneatus
Sensory (fine touch, vibration propriopception) from ipsilateral upper limb (T6 and above)

Spinocerebellar tract
Proprioception from limbs to cerebellum.

Anterior white commissure
Pain and temperature fibers cross. Anterior corticospinal tract fibers cross.

Lateral corticospinal tract
Motor to ipsilateral anterior horn (mostly limb musculature)

Anterior corticospinal tract
Motor to ipsilateral and contralateral anterior horn (mostly axial musculature)

Spinothalamic tract
Pain and temperature from contralateral side of the body

Figure 5.5
Fasciculi and major tracts.

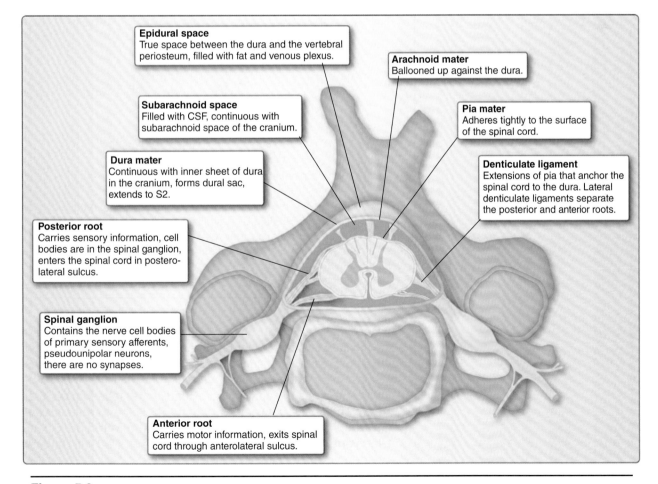

Epidural space
True space between the dura and the vertebral periosteum, filled with fat and venous plexus.

Arachnoid mater
Ballooned up against the dura.

Subarachnoid space
Filled with CSF, continuous with subarachnoid space of the cranium.

Pia mater
Adheres tightly to the surface of the spinal cord.

Dura mater
Continuous with inner sheet of dura in the cranium, forms dural sac, extends to S2.

Denticulate ligament
Extensions of pia that anchor the spinal cord to the dura. Lateral denticulate ligaments separate the posterior and anterior roots.

Posterior root
Carries sensory information, cell bodies are in the spinal ganglion, enters the spinal cord in postero-lateral sulcus.

Spinal ganglion
Contains the nerve cell bodies of primary sensory afferents, pseudounipolar neurons, there are no synapses.

Anterior root
Carries motor information, exits spinal cord through anterolateral sulcus.

Figure 5.6
Overview of spinal meninges and meningeal spaces. CSF = cerebrospinal fluid.

Clinical Application 5.1: Epidural Anesthesia

An effective, localized anesthesia can be achieved by applying local anesthetic into the spinal epidural space, as shown in the figure. The epidural space is located between the vertebral periosteum and the dural sac. It is filled with fatty tissue and a venous plexus. Nerve roots exiting or entering the dural sac travel laterally through the epidural space. Application of local anesthetic will act on those nerve roots mostly by blocking Na^+ channels, which will block the generation of action potentials.

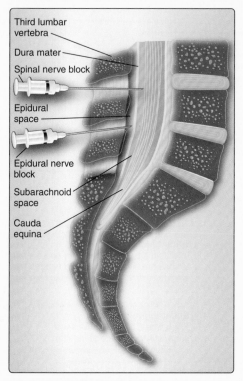

Spinal anesthesia.

In spinal anesthesia, the needle is inserted into the subarachnoid space filled with cerebrospinal fluid. The application of local anesthetic will block the spinal nerve rootlets as they enter the spinal cord. The amount of anesthetic needed for a spinal block is less than the amount needed in an epidural. A combination of spinal–epidural anesthesia is commonly used, especially in obstetrics.

1. **Dura mater:** The **dura mater** of the spinal cord has a single layer continuous with the inner layer of the cranial dura. The dura forms the dural sac, which surrounds the entire spinal cord. Whereas the spinal cord extends only to vertebral level L1–L2, the dural sac that surrounds the spinal cord extends to S2 (see Figures 5.2 and 5.6). The space between L1–L2 and S2 is the **lumbar cistern**. A **true epidural space** separates the dura from the periosteum of the vertebral column. This epidural space is filled with fat and the vertebral venous plexus.

2. **Arachnoid mater:** In life, the **arachnoid mater** is ballooned out against the dura. The **subarachnoid space** between the arachnoid and the pia mater is filled with CSF, and the spinal blood vessels are suspended in arachnoid trabeculae in this space.

3. **Pia mater:** The **pia mater** adheres tightly to the surface of the spinal cord. The pia gives off the paired **denticulate ligaments** laterally, which pierce the arachnoid and attach to the dura. The denticulate ligament separates the anterior and posterior rootlets. This pial outgrowth attaches the spinal cord laterally to the dura, suspending it in the dural sac. At the caudal end of the spinal cord, the conus medullaris, the **filum terminale** (another pial outgrowth), attaches the spinal cord to the coccyx. These pial attachments anchor the spinal cord and give it stability within the dural sac.

B. Spinal nerves

There are 31 pairs of spinal nerves. Each spinal nerve (see Figures 5.2 and 5.3) is composed of **posterior roots** (sensory) and **anterior roots** (motor). The roots come together in the **intervertebral foramen** of adjoining vertebrae where the prominent **spinal ganglion** is located. When the mixed spinal nerve emerges from the intervertebral foramen, it divides into **anterior** and **posterior rami** that supply the anterior and posterior aspects of the body, respectively.

As the posterior root approaches the spinal cord, it separates into rootlets that enter the spinal cord in the posterolateral sulcus. The anterior root is formed by a series of rootlets that emerge from the anterolateral sulcus. They come together as the anterior (motor) root that joins with the posterior (sensory) root to form a mixed spinal nerve. Each spinal nerve can potentially carry somatic sensory and motor fibers as well as visceral sensory and motor fibers (see Figures 5.3 and 5.4).

1. **Dermatomes:** Each segment of the spinal cord, or each spinal nerve, innervates a specific area of skin, referred to as a **dermatome** (Figure 5.7). Even though there is some overlap, specific areas for each spinal segment can be identified on a so-called dermatome map (Figure 5.8).

2. **Myotomes:** A **myotome** is the sum of all muscle fibers supplied by a single spinal nerve (see Figure 5.7). The anterior spinal roots provide motor control to the muscles. A skeletal muscle can be innervated by the motor fibers from several spinal cord segments (and with that, spinal nerves). The peripheral nerves in a limb are the product of recombining fibers from various spinal nerves into a plexus. This is due to the development of limb musculature: each muscle is derived from more than one somite. Each somite is supplied by nerve fibers from its associated specific spinal cord segment.

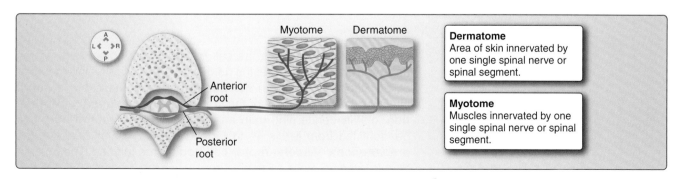

Figure 5.7
Myotome and dermatome.

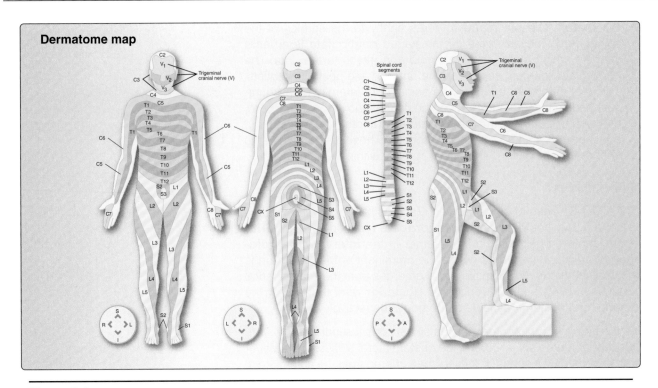

Figure 5.8
Dermatome map.

When testing muscle function, muscles derived from several somites are tested, and therefore, the movements assessed will give information about the integrity of several spinal cord segments. Table 5.1 summarizes the movements assessed in a neurological exam and the corresponding spinal cord segment. One particular movement is usually supplied by a predominant segment or root, and by testing the movement, a lesion in the spinal cord root or a specific segment can be determined.

III. INTERNAL STRUCTURE OF THE SPINAL CORD

In the spinal cord, the gray matter is localized centrally and is surrounded by white matter. Gray matter contains the nerve cell bodies: motor neurons and interneuron circuits in the anterior horn and sensory neurons in the posterior horn. White matter is divided into three columns: the **anterior**, **lateral**, and **posterior columns**.

A. Gray matter

The gray matter contains the cell bodies of the spinal cord. In the **anterior horn**, these are the **LMNs** and modulatory interneurons called **Renshaw cells**. In the **posterior horn**, these are a set of interneurons responsible for the first integration of **sensory information**, mainly concerning **pain** and **temperature**. A **lateral horn** can be identified from levels T1 to L2 and S2 to S4. This is where the **preganglionic visceral motor cell bodies** are located. The anterior horn is larger in those segments from which the brachial (C5–T1) and

Table 5.1: Movements Tested in Neurological Exam and Their Predominant Spinal Segment

Spinal Segment	Movement
C2	Neck flexion/extension
C3	Neck lateral flexion
C4	Initiation of shoulder elevation
C5	Shoulder abduction
C6	Elbow flexion/wrist extension
C7	Elbow extension/wrist flexion
C8	Finger flexion
T1	Finger abduction
L1/2	Hip flexion
L3	Knee extension
L4/5	Ankle dorsiflexion
L5	Great toe extension
S1	Knee flexion
S2	Ankle plantar flexion/toe flexion
S3/4	Anal wink

lumbosacral (L2–S4) plexuses arise, due to the increased population of motor neurons to supply the upper and lower limbs.

1. **Subdivisions of the gray matter:** The gray matter can be subdivided into 10 distinct layers, **Rexed laminae** I through X (Table 5.2). Layers I–VI comprise the posterior horn, layers VII–IX

Table 5.2: Important Structures of Rexed Laminae

Laminae	Location	Important Structures and Their Functions
I	Posterior horn, below the tract of Lissauer	Posterior (dorsal) root fibers mediating pain, temperature, and touch; posteromarginal nucleus
II	Posterior horn	Substantia gelatinosa neurons mediating pain transmissions
III and IV	Posterior horn	Proper sensory nucleus receiving inputs from substantia gelatinosa and contributing to anterolateral system mediating pain, temperature, and touch
V	Posterior horn	Neurons receiving afferent input from viscera, skin, and muscle
VI	Present only in cervical and lumbar segments	Afferents from muscle spindles and joint afferents
VII	Most anterior part of posterior horn and lateral horn	Clarke nucleus extending from C8–L2 receives muscle and tendon afferents; axons from this nucleus form spinocerebellar tract; intermediolateral cell column containing sympathetic and parasympathetic preganglionic neurons from T1-L2 and S2-S4 segements, respectively; Renshaw cells
VIII and IX	Anterior horn	α and γ motor neurons innervating skeletal muscles; neurons in medial aspect receive inputs from vestibulospinal and reticulospinal tracts and innervate axial musculature for posture and balance; neurons in lateral aspect receive inputs from corticospinal and rubrospinal tracts and innervate distal musculature
X	Gray matter surrounding central canal	Gray matter surrounding central canal

Modified from Siegel A. *Essential Neuroscience*, rev. 1st ed. Baltimore, MD: Lippincott Williams & Wilkins, 2006:140, with permission.

comprise the anterior horn, and layer X surrounds the central canal. The Rexed laminae are particularly useful when studying the posterior horn and the different levels of pain modulation occurring there. Another way of navigating the gray matter is through specific subnuclei. The substantia gelatinosa, the proper sensory nucleus (nucleus proprius), and Clarke nucleus (dorsal nucleus of Clarke) are located in the posterior horn. An intermediolateral nucleus (IML) contains the visceral motor cell bodies in the lateral horn, and specific motor nuclei are located in the anterior horn (Figure 5.9).

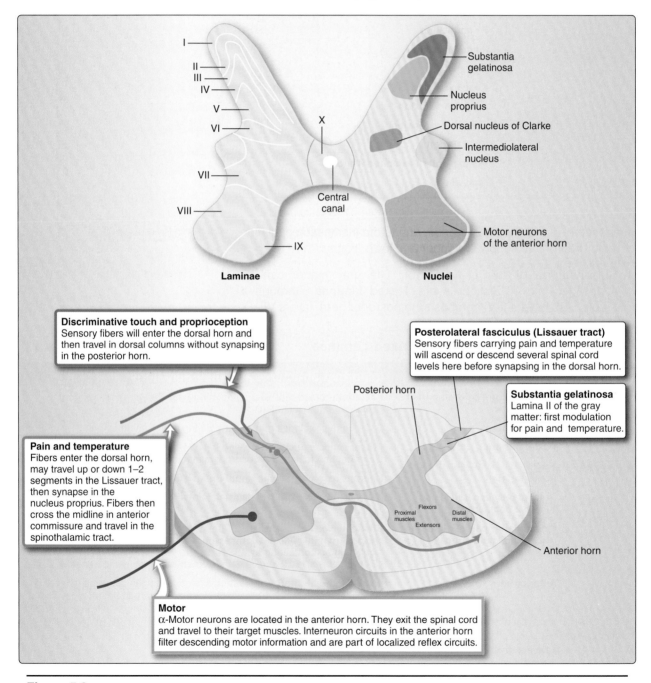

Figure 5.9
Gray matter: major nuclei in the anterior horn and posterior horn and fiber pathways.

2. **Anterior horn:** The **motor neurons** in the anterior horn project to the muscles supplied by the spinal cord level from which they exit. Alpha motor neurons are regulated by **Renshaw cells.** In the cervical and lumbar enlargements, the anterior horn is very large, because the upper and lower limbs are supplied from this region. In thoracic levels, the anterior horn is relatively small, because it supplies only axial musculature. The neurons in the anterior horn are arranged in a **somatotopic distribution,** meaning that they correspond both anatomically and functionally with the structures they serve. The neurons supplying the flexor muscles are more posterior, and those supplying the extensor muscles are more anterior. Distal muscle groups have their neurons more lateral in the anterior horn, and proximal and axial (trunk) muscle group motor neurons are medial (see Figure 5.9).

3. **Lateral horn:** A lateral horn lies at spinal cord levels T1–L2 and S2–S4; it contains the motor efferent cell bodies of the **visceral motor system** as shown in Figure 5.10. These cell bodies are collectively referred to as the intermediolateral nucleus.

4. **Posterior horn:** All **sensory information** from the periphery enters the spinal cord through the posterior horn at the various spinal levels. Fibers carrying **discriminative touch** and **proprioception** do not synapse in the posterior horn but ascend in the ipsilateral posterior columns. Fibers carrying **pain** and **temperature** enter the spinal cord in the posterior horn where they then ascend or descend several spinal levels in the posterolateral funiculus or Lissauer tract. Next, they synapse in lamina I and in the nucleus proprius (laminae III and IV). Cells from the nucleus proprius extend processes into the substantia gelatinosa (lamina II). A great deal of processing and modulation of the incoming pain signals occur in the substantia gelatinosa before the impulse travels to the brainstem and higher cortical centers. A detailed review of the physiology of pain is discussed in Chapter 22, "Pain."

5. **Clarke nucleus:** A specialized nucleus, called the **Clarke nucleus,** extends from C8 to L3 (Figure 5.11). It is an important relay station for nonconscious **proprioception** going to the cerebellum. Ia, Ib, and II fibers containing information on muscle tone and joint position send afferents to the Clarke nucleus. From there, postsynaptic fibers travel to the cerebellum via the posterior **spinocerebellar tract.** Proprioceptive fibers from below L3 travel in the fasciculus gracilis to L3 where they synapse in Clarke nucleus. Proprioceptive fibers above C8 travel in the fasciculus cuneatus and synapse in the accessory cuneate nucleus (the equivalent of Clarke nucleus, in the brainstem). These fibers then travel in the cuneocerebellar tract to the cerebellum.

B. White matter

Whereas gray matter is made up of cell bodies, white matter is composed of neuronal axons. The white matter in the spinal cord contains all ascending information to higher centers in the brainstem and forebrain as well as all descending information to spinal cord neurons. It appears white in color from the lipid-rich myelinated axons it

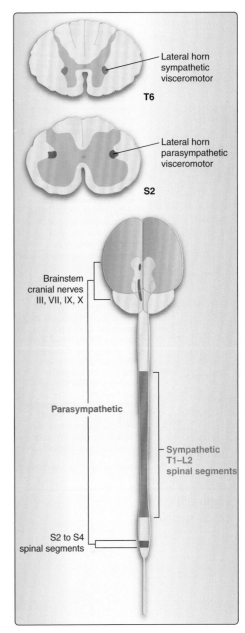

Figure 5.10
Visceral nervous system in the spinal cord.

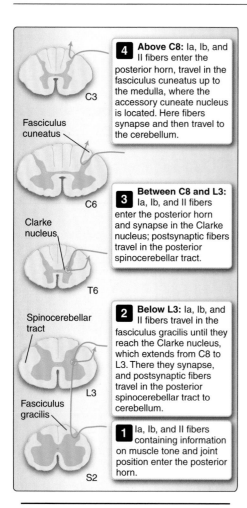

4 **Above C8:** Ia, Ib, and II fibers enter the posterior horn, travel in the fasciculus cuneatus up to the medulla, where the accessory cuneate nucleus is located. Here fibers synapse and then travel to the cerebellum.

3 **Between C8 and L3:** Ia, Ib, and II fibers enter the posterior horn and synapse in the Clarke nucleus; postsynaptic fibers travel in the posterior spinocerebellar tract.

2 **Below L3:** Ia, Ib, and II fibers travel in the fasciculus gracilis until they reach the Clarke nucleus, which extends from C8 to L3. There they synapse, and postsynaptic fibers travel in the posterior spinocerebellar tract to cerebellum.

1 Ia, Ib, and II fibers containing information on muscle tone and joint position enter the posterior horn.

Figure 5.11
Clarke nucleus.

contains. The ascending information is mainly sensory from the body. The descending information is motor to the LMNs. In addition, there are descending sensory fibers that modulate sensory input (mainly pain). White matter in the spinal cord is divided into three columns or **fasciculi** (Latin for "bundles"). In cervical regions, there is more white matter than in more caudal regions. Rostral, or cervical, segments contain the accumulation of ascending fibers, as well as all descending fibers on their way to lower segments.

1. **Anterior column:** The anterior column is located between the entry point of the anterior root and the midline. The main tract in the anterior column is the **anterior corticospinal tract**.

   a. **Anterior corticospinal tract:** The anterior corticospinal tract lies medially (Figure 5.12). It carries motor information from the cortex to LMNs concerned with proximal trunk musculature. These UMNs do not cross in the brainstem, as do the UMNs in the lateral corticospinal tract. Rather, the anterior corticospinal tract neurons cross over at the level of the spinal cord at which they innervate the LMNs. Most of the innervation to the trunk is bilateral to allow for maintaining posture during upright gait.

   b. **Other tracts:** Other descending motor tracts, such as the vestibulospinal and reticulospinal tracts, also travel in the anterior column (not shown).

2. **Lateral column:** The lateral column is located between the entry points of the posterior and anterior roots. This is where the main motor tract to the spinal cord, the **lateral corticospinal tract**, descends from the forebrain, having crossed in the brainstem, to reach the LMNs at each spinal cord level.

   a. **Lateral corticospinal tract:** The corticospinal tract is located medially in the lateral column, adjacent to the gray matter. Its fibers are arranged somatotopically, so fibers to the upper body are located most medially. These are the first fibers to synapse with the LMNs in the anterior horn. Fibers to the lower part of the body are located most laterally and are the last fibers to reach their destination on the LMNs of lumbar and sacral levels (see Figure 5.12).

   b. **Spinocerebellar tracts:** Lateral to the corticospinal tract lie the **spinocerebellar tracts**, as seen in Figure 5.12. These tracts carry proprioceptive information to the ipsilateral cerebellum.

   c. **Spinothalamic tracts:** The spinothalamic tract is located in the anterior part of the lateral column and is part of the anterolateral system. (See Chapter 7, "Ascending Sensory Tracts.") It carries pain and temperature from the contralateral side of the body (see the next section).

   d. **Other tracts:** Additional tracts, such as the ascending spino-olivary and the descending rubrospinal tracts, can also be found in the lateral column. These tracts are important in the coordination of movement between the LMN circuits and the cerebellum and cortex (not shown).

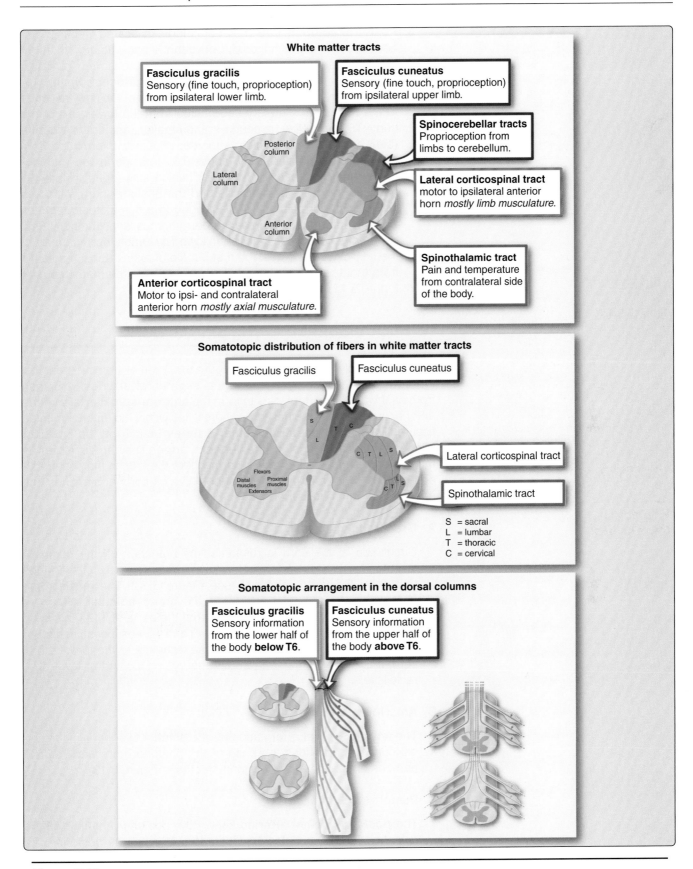

Figure 5.12
White matter: tracts and somatotopy.

3. **Posterior column:** The posterior column is located on the posterior aspect of the spinal cord, between the entry point of the posterior root and the midline. This is where all sensory information concerning **discriminative touch** and **proprioception** (awareness of body position) from the ipsilateral side of the body ascends to the brainstem. Sensory information from the lower part of the body, T6 and below, ascends in the smaller **fasciculus gracilis** (Latin for "graceful"), which is located medially. Information from the upper part of the body, above T6, ascends in the **fasciculus cuneatus** (Latin for "wedge"), which is wedged in laterally between the fasciculus gracilis and the posterior horn. All fibers in the posterior columns are arranged in a somatotopic manner. Sensory information from sacral dermatomes is most medial (first to enter the fasciculus gracilis), followed by lumbar dermatomes. In the fasciculus cuneatus, the first bundle of axons (most medial) is from thoracic dermatomes followed by cervical dermatomes (see Figure 5.12).

IV. BLOOD SUPPLY TO THE SPINAL CORD

The blood supply to the spinal cord (Figure 5.13) comes from two sources: the **vertebral–basilar system** and the **segmental arteries**. A branch from each vertebral artery joins to form the **anterior spinal artery** on the anterior surface of the spinal cord, where it lies within the anterior median fissure. The paired **posterior spinal arteries** branch off the posterior inferior cerebellar arteries or the vertebral artery. They are localized on the posterior surface of the spinal cord, where they run in the posterolateral sulci. Anterior and posterior arteries give off coronal branches that anastomose with each other.

A. Great radicular artery

Circulation to the cord is reinforced by branches off the segmental arteries of the aorta, which in turn give rise to radicular arteries that enter the vertebral canal (see Figure 5.13). The most prominent of these segmental arteries is the **great radicular artery** (of Adamkiewicz), which usually arises from the aorta at T12. This is an important artery in spinal cord surgeries and injuries, because disruption of this important vessel can cause ischemia to the lumbosacral cord. It is also vulnerable in aortic aneurysms located at lower thoracic levels.

B. Anterior spinal artery

The **anterior spinal artery** supplies the anterior two-thirds of the spinal cord. This includes the base of the posterior horn and variable portions of the lateral corticospinal tract (see Figure 5.13).

C. Posterior spinal arteries

The **posterior spinal arteries** supply the posterior one-third of the spinal cord. This includes the posterior columns, the posterior root entry zone and Lissauer tract, the substantia gelatinosa, and variable portions of the lateral corticospinal tract (see Figure 5.13).

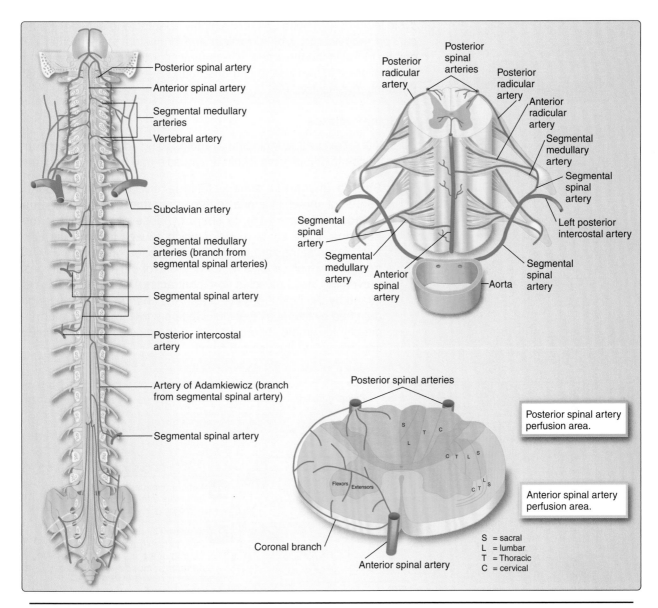

Figure 5.13
Blood supply to the spinal cord.

V. SPINAL CORD SYSTEMS

The LMN is the final common pathway for signal transduction to skeletal muscles. The input to the LMN will determine its firing rate, which will in turn determine muscle contraction. The UMNs synapsing with the LMNs come from the cortex (corticospinal path) to initiate voluntary movements and from the brainstem to control balance and posture. Another set of influences on the LMN comes from inputs directly within the spinal cord. These spinal cord systems comprise the spinal reflexes and more complex interactions for the control of bladder function and sexual responses. (See Chapter 4, "Overview of the Visceral Nervous System.")

Another system within the spinal cord comprises the **central pattern generators (CPGs)** that for example are responsible for parts of the gait cycle in humans. These CPGs are highly adaptable to sensory input and can be modulated by both cortical and brainstem tracts. Other CPGs can be found in the brainstem where they coordinate complex motor patterns involved in swallowing and breathing. (See Chapter 12, "Brainstem Systems and Review.")

A. Spinal reflexes

Spinal reflexes allow for the rapid and ongoing adjustment of posture in response to stimuli from within the muscle and from the environment.

1. **Stretch (myotatic) reflex:** LMNs receive direct input about the degree of stretch in a muscle through **Ia fibers** from within **muscle spindles** (Figure 5.14). Ia fibers synapse directly with **α motor neurons** of the same muscle, which causes that muscle to contract. At the same time, the Ia fiber synapses with **inhibitory interneurons**, which inhibit α motor neurons of the antagonist muscles, preventing those muscles from contracting. Muscle length is determined by the input from the UMN. Any change in muscle length is detected by the muscle spindle. Ia fibers then relay this information directly to α motor neurons so that the muscle can be readjusted immediately to the desired length.

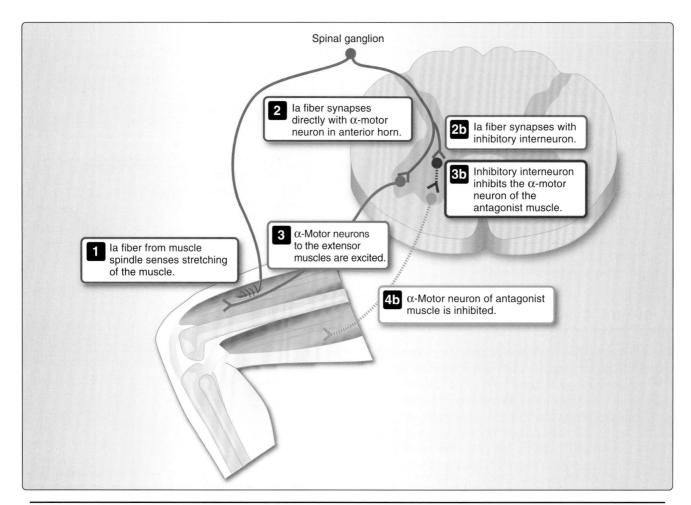

Figure 5.14
Stretch reflex commonly tested by tapping the tendon to elicit the deep tendon reflexes.

Small γ **motor neurons** innervate the muscle spindle and determine its level of excitability. This is how the muscle spindle gets the necessary information about the desired muscle length. Any deviation will result in reflex adjustment.

Clinically, this reflex is tested by tapping the tendon of a muscle. This stretches the muscle, stimulating the muscle spindle, which initiates the reflex arc. The same muscle will contract, whereas the antagonist muscle is inhibited.

2. **Withdrawal reflex:** The **withdrawal reflex**, also named **flexor reflex**, is an archaic reflex at the level of the spinal cord designed to get a body area away from a noxious stimulus very quickly.

A noxious stimulus (i.e., pain) is detected by **Aδ** and **C fibers** in the skin. These fibers synapse directly with the α motor neurons of the flexor muscles for the limb in which the noxious stimulus was detected. At the same time, the extensor muscles are inhibited through an interneuron, as shown in Figure 5.15. The net result of this reflex arc is the very quick withdrawal (flexion) of a limb, taking it away from a noxious stimulus.

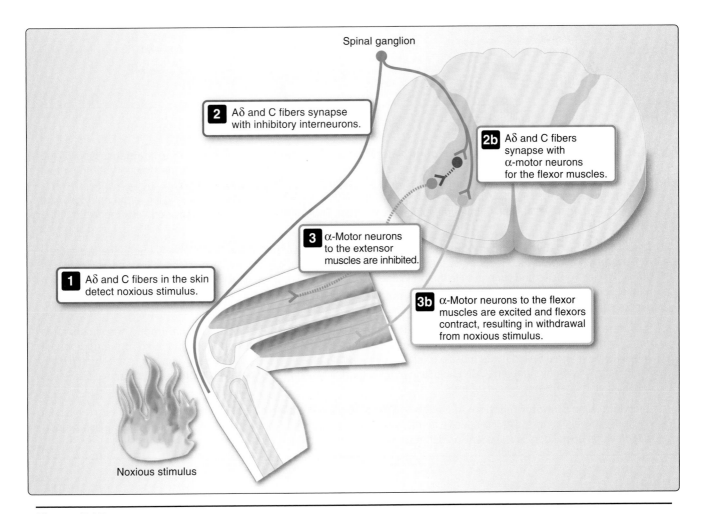

Spinal ganglion

2 Aδ and C fibers synapse with inhibitory interneurons.

2b Aδ and C fibers synapse with α-motor neurons for the flexor muscles.

3 α-Motor neurons to the extensor muscles are inhibited.

1 Aδ and C fibers in the skin detect noxious stimulus.

3b α-Motor neurons to the flexor muscles are excited and flexors contract, resulting in withdrawal from noxious stimulus.

Noxious stimulus

Figure 5.15
Withdrawal reflex.

Clinical Application 5.2: Hemisection Through the Spinal Cord: Brown-Séquard Syndrome

Hemisection through the spinal cord can occur due to injury and results in a typical array of symptoms due to the tracts that have been transected on one side.

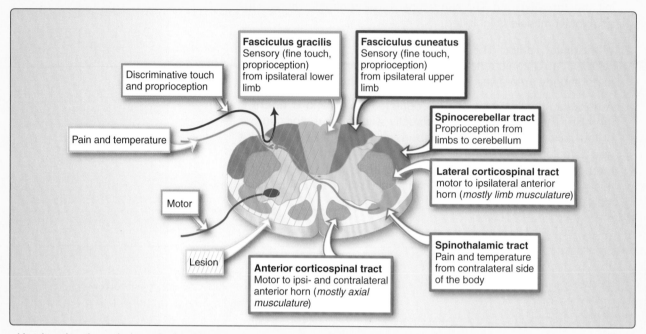

Hemisection through the spinal cord.

1. **Loss of pain and temperature on the contralateral side of the body**: This loss is due to the transection of the **spinothalamic tract**. The loss usually occurs a few segments below or above the level of the lesion. When pain and temperature fibers enter the spinal cord, they travel a few segments in the Lissauer tract before they synapse, which is why the loss of function is a few levels below or above the lesion. The fibers then cross the midline in the anterior white commissure and travel in the contralateral spinothalamic tract. Thus, the loss of sensation is on the side opposite to the side of the lesion.

2. **Loss of discriminative touch and proprioception on the ipsilateral side of the body**: This is due to the transection of the **posterior columns**. Fibers enter the posterior horn and without synapsing travel in the ipsilateral posterior columns (fasciculi gracilis and cuneatus) to the brainstem. Nonconscious proprioception (e.g., joint position) is lost due to a transection of the **spinocerebellar tract**. This is where fibers from the ipsilateral side of the body travel after they have synapsed in the Clarke nucleus.

3. **Ipsilateral loss of motor function**: This is due to the transection of the **lateral corticospinal tract**. These fibers arise from the contralateral cerebral cortex, cross over in the medulla, and then travel in the spinal cord to the appropriate spinal levels where they innervate muscle groups on the ipsilateral side of the body. Loss of upper motor neuron (UMN) input to the lower motor neuron (LMN) results in paresis (partial weakness) or paralysis (complete weakness) as well as spasticity and brisk deep tendon reflexes.

4. **Ipsilateral loss of motor function**: At the level of the lesion within the spinal cord, damage to anterior horn cells can occur. This would affect the LMNs specific to that level. A lesion of these LMNs can result in weakness to muscles innervated by that spinal cord level as well as decrease or loss of deep tendon reflexes at that spinal cord level.

Clinical Cases:

A Day in the Spine Clinic

The following five people have an appointment at the spine clinic for assessment. They all have different neurological symptoms and signs that affect different areas of the spinal cord and at different spinal cord levels. For each mini-case below, identify the type of spinal cord lesion and the spinal cord level (i.e., T6, C7) if possible.

Case Histories

Case 1. Upper back pain

A 36-year-old female with a history of neurofibromatosis type 2 presents with a 3-month history of upper back pain with a tight band sensation across her upper chest and dysesthesia in her hands. There is no weakness or bladder/bowel dysfunction. She has reduced pinprick and temperature sensation in her index fingers and the medial aspects of her upper limbs and thorax to the nipple line. Sensation to pinprick and temperature elsewhere is normal. Vibration, proprioception, gait, strength, tone, and reflexes are normal.

Case 2. Unsteady gait and clumsiness

A 41-year-old female has a 3-week history of progressively unsteady gait with clumsiness and numbness in her arms and legs. She has fallen a few times in the shower and during the night when walking to the bathroom in the dark. She is a vegetarian. When she flexes her neck, she has a sensation like electricity running down her spine (L'Hermitte sign). She has a loss of vibration sense in her toes, ankles, knees, and hips and reduction in proprioception to her ankles. She also has a loss of vibration sense to her metacarpal bones (knuckles) and reduction in proprioception at the distal interphalangeal (DIP) joints in her fingers. Pinprick and temperature sensation are normal. Her strength, tone, and reflexes are normal.

Case 3. Symptoms after a fall

A 58-year-old male fell and landed on his back. Immediately after the fall, he was unable to feel or move his legs. Now, 3 months later, he presents at the clinic with the following signs: loss of pinprick, temperature, and vibration sensation to the umbilicus; absent proprioception to both hips; spasticity in both legs; 0/5 strength in both legs; 3+ deep tendon reflexes (DTR); and bilateral upgoing toes. In addition, there is bladder and bowel dysfunction.

Case 4. Unsteady gait with right leg spasticity

A 29-year-old male has a history of unsteady gait with numbness over the entire left leg and weakness of the right leg. He is dragging his right leg and it feels stiff upon examination. There is no bladder/bowel dysfunction. He has reduced pinprick and temperature sensation of the entire left leg up to the inguinal region, including the perineum on the left. There is reduced vibration sensation and proprioception up to the ankle on the right. Sensation elsewhere is normal, including the trunk. In the right leg, he has 4-/5 strength of his hip/knee/

toe flexors and dorsiflexion of his ankle and 4/5 strength of his knee extensors and plantar flexion and toe flexion. His strength elsewhere is normal. He has spasticity in his right leg with normal tone in the left leg and both arms. He has 3+ DTR at the right ankle and knee and normal DTR in the left leg and both arms. His right great toe is upgoing, and his left toe is downgoing.

Case 5. Paraplegia after aortic aneurysm surgery

A 67-year-old male had abdominal aortic aneurysm surgery. He awoke from surgery with midthoracic back pain, flaccid paraplegia, areflexia, and urinary retention. There was reduction in pinprick and temperature sensation from T6 down. Vibration and proprioception were normal. The spine clinic has been asked to consult on this case and develop a diagnosis.

Case Discussions

See Figure 5.16.

Case 1 (upper back pain) is indicative of a central cord lesion from C7 to T4. The loss of pinprick and temperature sensation included the medial half of the hand (C7–C8) as well as the medial forearm/arms (C8–T2) to the level of the nipples (T4). It was a small lesion affecting both spinothalamic tract fibers that cross the ventral commissure of the central spinal cord from C7 to T4—this pattern of lesion is referred to as a suspended sensory level. The anterior horn, corticospinal tracts, and posterior column were spared; therefore, there were no motor signs and no loss of vibration/proprioception. An MRI of the cervical spine showed an expanding intramedullary tumor.

Case 2 (unsteady gait and clumsiness) is typical for a posterior cord syndrome in the cervical spine. The exact level could not be determined on clinical examination, but the sensory symptoms (loss of proprioception and joint sensation) in the upper limb indicate a lesion to the cervical spine. An MRI of the cervical spine revealed a long segment T2 hyperintensity in the posterior spinal cord from C3 to the conus. Pinprick and temperature sensation as well as strength were normal; this indicates that the spinothalamic tracts and corticospinal tracts/anterior horn cells were spared. The L'Hermitte sign was due to abnormal hyperexcitability from stretching of the posterior column when the neck is flexed. Gait is critically dependent on proprioceptive and visual input. With a lesion affecting the proprioceptive input, there is additional difficulty walking in the dark due to the additional reduction in visual guidance to aid with sensory feedback. Her vitamin B12 level was very low, which can also affect the health of sensory neurons.

Case 3 (symptoms after a fall) shows a complete transverse cord at T10. The umbilicus is within the T10 dermatome. The spinothalamic tract (loss of pinprick and temperature sensation), posterior column (loss of vibration and proprioception), and corticospinal tract (weakness, spasticity, and increased/brisk DTR) are all involved. An MRI

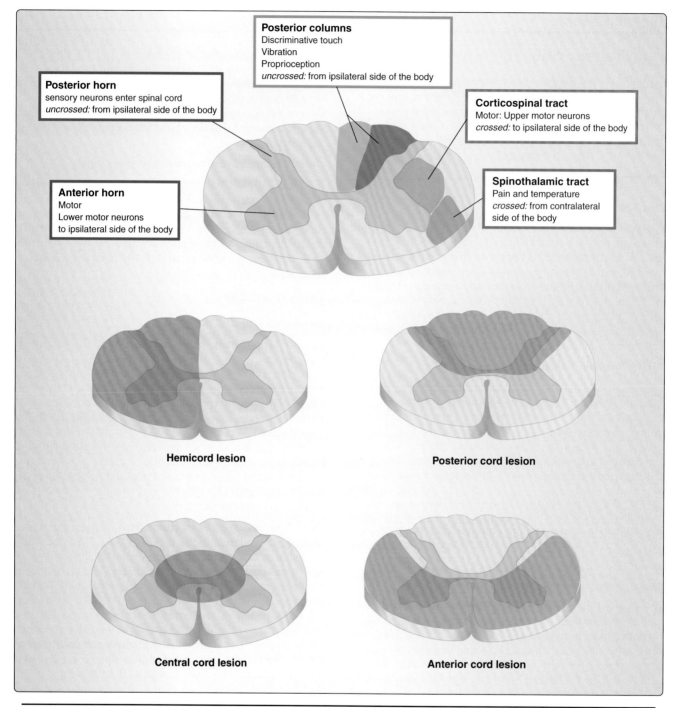

Posterior columns
Discriminative touch
Vibration
Proprioception
uncrossed: from ipsilateral side of the body

Posterior horn
sensory neurons enter spinal cord
uncrossed: from ipsilateral side of the body

Corticospinal tract
Motor: Upper motor neurons
crossed: to ipsilateral side of the body

Anterior horn
Motor
Lower motor neurons
to ipsilateral side of the body

Spinothalamic tract
Pain and temperature
crossed: from contralateral
side of the body

Hemicord lesion

Posterior cord lesion

Central cord lesion

Anterior cord lesion

Figure 5.16
Lesion patterns in the spinal cord corresponding to the clinical cases.

of his thoracic spine showed vertebral fracture with extrinsic compression of his spinal cord at T10.

Case 4 (unsteady gait with right leg spasticity) is typical of a hemicord (Brown-Séquard lesion) at right L1. There is loss of pinprick and temperature sensation ipsilaterally (spinothalamic tract) to the inguinal region (L1) and impairment

of proprioception and vibration (posterior column) as well as UMN features (corticospinal tract) on the opposite side. He was found to have an intradural extramedullary tumor compressing the spinal cord on the right. Hemicord (Brown-Séquard) syndrome causes ipsilateral corticospinal tract and posterior column dysfunction and contralateral spinothalamic dysfunction (see Clinical Application Box 5.2).

Case 5 (paraplegia after aortic aneurysm surgery) is indicative of an anterior cord syndrome at T6 due to ischemia of the anterior spinal artery. This results in loss of pinprick/temperature (spinothalamic tract) and flaccid paraplegia with absent deep tendon reflexes (DTRs) in the legs (anterior horn cells). Anterior cord syndrome can also affect the corticospinal tract (UMN fibers). In the acute phase, UMN lesions cause flaccid paraplegia with reduced reflexes; however, hours to days later, spasticity and brisk reflexes with upgoing toes develop.

Further Discussion

By definition, spinal cord pathology should not cause cranial symptoms. If there is facial weakness, speech/swallowing difficulties, or diplopia, then the lesion cannot be *only* in the spinal cord.

Central cord syndromes may be due to an intramedullary tumor such as ependymoma or astrocytoma. A late consequence of trauma to the spinal cord can be the development of a syrinx—a widening of the central canal within the spinal cord. All of these pathologies cause damage to the center of the cord.

Posterior cord syndrome may be due to vitamin B12 deficiency. Syphilis, a bacterial infection, can affect both the posterior columns and corticospinal tracts. Posterior spinal arteries (PSAs) supply the posterior 1/3 of spinal cord (posterior roots, posterior horns, and posterior column) on each side. A lesion to one of the PSAs will cause ischemia to the ipsilateral posterior 1/3 of the spinal cord. Other causes of posterior cord syndrome include multiple sclerosis, trauma, and extrinsic compression of the spinal cord due to tumors.

Complete **transverse cord lesions** may occur due to inflammatory conditions such as multiple sclerosis, infection, trauma, tumor, or extrinsic compression from disc herniation.

Hemicord (Brown-Séquard) syndrome may be due to inflammation, infections, trauma, or tumors.

Spinal cord tumors can be extradural (outside the dura) such as a meningioma, intradural extramedullary (between the spinal cord and dura) such as schwannoma, and intramedullary (within the spinal cord) such as ependymoma.

Anterior cord syndrome occurs from ischemia to the anterior 2/3 of the spinal cord. The anterior spinal artery supplies the anterior 2/3 of the cervical and upper/midthoracic spinal cord. The T4–T8 region of the spinal cord is susceptible to ischemia (watershed region). Hypotension during aortic surgery is a possible mechanism of the ischemia to the anterior 2/3 of the spinal cord from T6 down.

If there was a central cord lesion at C7 to T1 level that expanded to involve the anterior horn cells (AHC), what type of motor symptoms would you expect to find?

A lesion of the AHC would produce lower motor neuron signs such as weakness, reduced (flaccid) tone, reduced or absent reflexes, and loss of muscle bulk. Muscles innervated by the lower motor neurons originating from the C7 to T1 level of the anterior horn cell would be weak. In addition, damage to the anterior horn cells at these levels would result in a decrease in the triceps reflex (C7/8) and normal brachioradialis and biceps reflex (C5/6) if there was no involvement of the corticospinal tract.

What large anterior radicular artery supplies the lower thoracic and upper lumbar spinal cord?

The great radicular artery is a large anterior radicular artery that arises from the left side of the aorta between T8 and L1 segment. It anastomoses with the anterior spinal cord and supplies the lower thoracic and upper lumbar region including the conus medullaris (see Figure 5.14).

Chapter Summary

- The spinal cord extends from the foramen magnum to vertebral level L1–L2. It is surrounded by three meningeal layers: the **pia mater**, the **arachnoid mater**, and the **dura mater**. The subarachnoid space is filled with cerebrospinal fluid and is continuous with the subarachnoid space in the cranium. A **true epidural space** filled with fat and a venous plexus exists between the dura mater and the periosteum of the vertebrae.

- **Thirty-one pairs of spinal nerves** arise from the spinal cord. Each spinal nerve carries somatic afferent (**sensory**) and somatic efferent (**motor**) information. Some spinal nerves also carry **visceral afferent** and **efferent** information. Sensory information enters the spinal cord through the **posterior root**. Motor information exits the spinal cord through the **anterior root**. An area of skin supplied by one spinal nerve is a **dermatome**, and **myotome** refers to a muscle group supplied by an individual spinal nerve.

- The blood supply to the spinal cord arises from the **vertebral–basilar system** and the **segmental arteries of the aorta**. The anterior spinal artery supplies the anterior two-thirds of the spinal cord, and the paired posterior spinal arteries supply the posterior one-third. The spinal arteries anastomose to form a **corona** around the spinal cord.

- The internal structure of the spinal cord can be divided into gray matter and white matter. **Gray matter** is butterfly shaped and has a posterior horn and an anterior horn. The **posterior horn** receives **sensory input,** and the **anterior horn** contains the **lower motor neurons. White matter** can be divided into **three columns**: posterior, lateral, and anterior. This is where the ascending and descending fiber tracts are located.

> ## Chapter Summary continued
>
> - Ascending tracts include all sensory tracts traveling from the spinal cord to higher centers: the **posterior columns** for discriminative touch and proprioception and the **spinothalamic tract** for pain and temperature. The **spinocerebellar tracts** provide information about proprioception to the cerebellum. Descending fiber tracts include the **corticospinal tract** with the primary motor input to the lower motor neuron in the anterior horn as well as additional motor tracts descending from brainstem centers.
>
> - Several reflex systems exist at the level of the spinal cord. The **stretch reflex** is a pathway for postural stability in which sensory input from the muscle spindle has a direct influence on the α motor neuron, keeping the muscle at a set length. The **withdrawal** or **flexor reflex** is designed to allow rapid withdrawal from a noxious stimulus. Pain fibers (Aδ and C fibers) from the skin directly influence the α motor neurons for the flexor muscles of that particular area, allowing for rapid flexion of the limb to remove it from the source of pain.

Study Questions

Choose the ONE best answer.

5.1 A patient presented with symptoms that made the doctor suspect that he had bacterial meningitis. Because analysis of the cerebrospinal fluid can provide evidence to confirm this diagnosis, the patient was sent for a lumbar puncture. At what vertebral level is a lumbar puncture typically done?

   A. T1–T2
   B. S1–S2
   C. T11–T12
   D. L1–L2
   E. L3–L4

The correct answer is E. Because the vertebral column grows to be longer than the spinal cord, in the adult, the cord generally ends at vertebral level L1–L2. A lumbar puncture is therefore done below this level, either between L3 and L4, or between L4 and L5 to withdraw cerebrospinal fluid (CSF) from the lumbar cistern. Importantly, in the newborn, the spinal cord typically ends at the L3 vertebral level, and a lumbar puncture must be done below this level. T1–T2 and T11–T12 vertebral levels are too high, as insertion of a needle at these levels could damage the spinal cord. Similarly, the lumbar cistern extends from the end of the spinal cord at vertebral level L1–L2 to the end of the dural sheath at vertebral level S2. Although the S1–S2 vertebral level is at the end of the lumbar cistern, it may be too low to obtain sufficient CSF.

5.2 Understanding the circulation of the cerebrospinal fluid is critical for the ability to diagnose disorders where circulation is abnormal and for the ability to take samples of CSF for analysis. Which statement concerning spinal meninges and cerebrospinal fluid is correct?

   A. CSF travels in the subarachnoid space.
   B. CSF is typically sampled from the epidural space.
   C. The dural sac that surrounds the spinal cord ends at the L1–L2 vertebral level.
   D. The pia mater adheres tightly to the arachnoid mater.
   E. Extensions from the arachnoid help anchor the spinal cord within the dural sac.

The correct answer is A. Cerebrospinal fluid (CSF) travels in the space between the pia mater and the arachnoid mater, known as the subarachnoid space. There is no CSF in the epidural space. Although the spinal cord in the adult ends at the L1–L2 vertebral level, the dural sac extends to the S2 vertebral level. The pia mater adheres tightly to the surface of the CNS and is separated from the arachnoid by the subarachnoid space, which is filled with CSF. Extensions off the pia (denticulate ligaments, filum terminale) anchor the spinal cord within the dural sac.

5.3 A lesion of the medial aspect of the lateral white column of the spinal cord is most likely to cause deficits in:

    A. Discriminative touch and proprioception on the ipsilateral side.

    B. Pain and temperature sensations on the contralateral side.

    C. Pain and temperature sensations on the ipsilateral side.

    D. Voluntary movement on the ipsilateral side.

    E. Coordinated movements of the lower limbs.

> The correct answer is D. The lateral corticospinal tract, which mediates voluntary motor activity, travels in the lateral column of the spinal cord. It carries motor information from the cerebral cortex to the anterior horn cells in the spinal cord. Fibers carrying information on discriminative touch and proprioception ascend in the ipsilateral posterior white column. Fibers carrying information on pain and temperature ascend in the contralateral lateral white column. The spinocerebellar tract, which carries proprioceptive information to the ipsilateral cerebellum, travels in the lateral area of the lateral white column.

5.4 On examination, the patient showed a loss of discriminative touch and proprioception in the right lower limb but intact pain and temperature sensations and motor function. The left lower limb and both upper limbs were entirely normal. Based on these findings, which statement below is correct?

    A. Information concerning discriminative touch and proprioception from the right lower limb travels from the spinal cord to the brainstem in the left lateral white column.

    B. Information concerning discriminative touch and proprioception from the right lower limbs ascends in the right fasciculus gracilis.

    C. Information concerning pain and temperature sensations from the right lower limb travels to the brainstem in the right spinothalamic tract.

    D. The blood supply to the tracts carrying discriminative touch and proprioception comes primarily from the anterior spinal artery.

    E. Fibers carrying information on discriminative touch and proprioception cross the midline in the spinal cord.

> The correct answer is B. Discriminative touch and proprioception ascend ipsilaterally in the posterior white columns, not the lateral white column. Information from the lower limb travels in the fasciculus gracilis and that from the upper limb travels in the fasciculus cuneatus. Information concerning pain and temperature travels in the lateral white columns. The blood supply to the posterior white columns comes from the posterior spinal arteries, whereas that to the anterior white columns comes from the anterior spinal artery. Fibers carrying information on discriminative touch and proprioception cross the midline in the caudal medulla.

5.5 The anterior spinal artery is formed from branches of the:

    A. Basilar artery.
    B. Posterior inferior cerebellar arteries.
    C. Segmental arteries of the aorta.
    D. Vertebral arteries.
    E. Great radicular artery.

> The correct answer is D. Single branches from each vertebral artery join together to form the anterior spinal artery lying deep within the anterior median fissure of the spinal cord. The basilar artery is formed by joining of the two vertebral arteries. The posterior inferior cerebellar arteries arise as branches off the vertebral arteries. Segmental arteries of the aorta provide additional circulation to the spinal cord to reinforce that from the anterior and posterior spinal arteries. The great radicular artery (of Adamkiewicz) is one of the larger segmental arteries.

6 Overview and Organization of the Brainstem

I. OVERVIEW

This chapter is a conceptual overview of the brainstem. Due to its complexity, the brainstem cannot be covered in this one chapter. The framework provided here facilitates the understanding of brainstem nuclei, tracts, and systems, which will be discussed in more detail in upcoming chapters.

We will relate surface anatomical landmarks to internal structure to provide a framework for orientation in the brainstem. In addition, this chapter introduces cranial nerves and their positions at each level of the brainstem. An overview of the blood supply provides the basis for understanding clinical syndromes.

The brainstem is a small but complex part of the central nervous system (CNS). Tracts traveling up and down between the spinal cord and the cortex, as well as tracts interconnecting both the cortex and the spinal cord with the cerebellum, all travel through the brainstem. In addition, the brainstem contains cranial nerve nuclei and their afferent/efferent fibers as well as intrinsic systems and relay nuclei.

The brainstem is located in the **posterior cranial fossa**. It merges with the spinal cord caudally at the level of the foramen magnum and with the diencephalon rostrally. Three distinct areas can be identified in the brainstem: the **medulla oblongata** (or **medulla**) caudally, the **pons**, and the **midbrain** rostrally (Figure 6.1). The **cerebellum** is developmentally part of the pons, but its functions are so distinct that it is often considered as its own entity. We treat the cerebellum as such in this book (see Chapter 17, "The Cerebellum"). The cerebellum is connected to the brainstem via three cerebellar **peduncles** (inferior, middle, and superior), composed of large white matter tracts, which carry information between the brainstem and the cerebellum.

The brainstem contains the nuclei associated with CNs III to XII (with the exception of XI) (see discussion in Section III). These cranial nerves supply the head and neck and also contain parasympathetic fibers for the thorax and abdomen. The special senses (olfaction [CN I], vision [CN II], taste [CNs VII and IX], and hearing/balance [CN VIII]) are discussed in chapters dedicated to those systems.

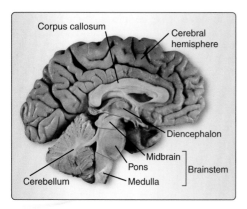

Figure 6.1
Midsagittal view of the brainstem, showing the midbrain, pons, and medulla.

All **ascending** and **descending pathways** travel through the brainstem. Many synapse here in **relay nuclei**, and others arise directly from or synapse in other brainstem nuclei.

Small lesions in the brainstem can result in substantial deficits, often involving multiple motor, sensory, and neuroregulatory modalities.

II. SURFACE ANATOMY AND RELATIONSHIP TO INTERNAL STRUCTURES

The medulla, pons, and midbrain all have characteristic surface features that typically reflect major internal or underlying structures. Knowledge of the surface features can be helpful in understanding and remembering the internal structures of the brainstem.

In transverse section, the internal structure of the brainstem can be divided into four areas: from posterior to anterior, these are the **tectum** or roof over the ventricular system; the **ventricular system** itself; the core of the brainstem, referred to as the **tegmentum**; and the **basal portion** situated most anteriorly (Figure 6.2). The superior and inferior colliculi, along with the superior and inferior medullary vela, comprise the tectum of the brainstem that roofs the cerebral aqueduct and fourth ventricle. The tegmentum is the largest, most central portion of the brainstem. It contains the reticular formation, cranial nerve nuclei, ascending pathways from the spinal cord, and some descending pathways. The majority of the descending motor pathways travel through the basal portion of the brainstem. The **reticular formation** consists of a loose network of nuclei and tracts, which span the tegmentum of the brainstem, clustered primarily along the midline.

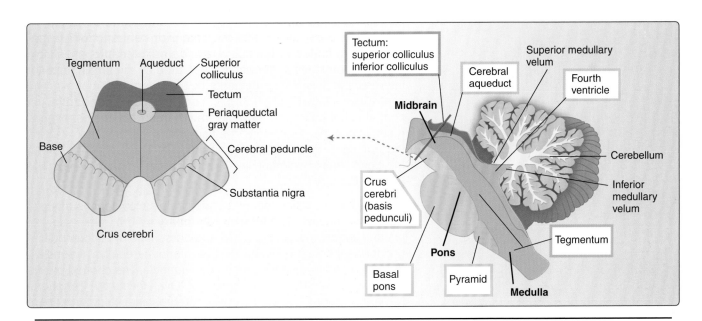

Figure 6.2
Components of the tectum (*dark blue*), tegmentum (*green*), base (*orange*), and the ventricular system (*light blue*).

A. Medulla oblongata

The medulla is the rostral continuation of the spinal cord beginning at the level of the foramen magnum. It rests on the basilar portion of the occipital bone. Its rostral end merges with the pons at a clearly visible sulcus. The sulci of the spinal cord are continuous with the sulci on the surface of the medulla (anterior median, anterolateral, posterolateral, and posterior median sulci [Figure 6.3]).

1. **Anterior surface—motor information:** Lateral to the anterior median fissure, paired longitudinal elevations, the **pyramids**, are visible (see Figure 6.3). These contain motor fibers of the descending **corticospinal** and **corticobulbar tracts**. Corticospinal fibers travel from the motor cortex to the lower motor neurons in the anterior horn of the spinal cord. The majority of these fibers cross the midline in the caudal medulla. Note that the median fissure on the anterior surface of the medulla is interrupted by these crossing or decussating corticospinal fibers (the **decussation of the pyramids**). In contrast, corticobulbar fibers travel primarily bilaterally to the lower motor neurons of cranial nerve nuclei in the brainstem. Lateral to the pyramids, at the level of the rostral medulla, lie oval bulges called the olives (see Figure 6.3). These indicate the position of the underlying **inferior olivary nuclear complex**. Between the pyramids and the olives, in the anterolateral sulcus, a series of rootlets emerge to form the **hypoglossal nerve (CN XII)**. Posterior to the olive, in the posterolateral sulcus (not labeled, see Figure 6.5), another series of rootlets emerge to form the **glossopharyngeal (CN IX)**, **vagus (CN X)**, and **accessory (CN XI) nerves** (Figure 6.4; also see Figure 6.3).

2. **Posterior surface—sensory information:** On the posterior surface, the medulla is divided into a closed and an open portion (Figure 6.5; also see Figure 6.3). The **closed, or caudal, medulla** contains the rostral extension of the **central canal** of the spinal cord. On its posterior surface, the **posterior columns**, which contain the **fasciculus gracilis** and **fasciculus cuneatus**, are visible as paired longitudinal elevations separated by the small posterior intermediate sulcus. The posterior columns form the roof over the central canal (see Figure 6.3). In the **open, or rostral, medulla**, the central canal widens to form the **fourth ventricle**, which extends rostrally into the pons. Here, the floor of the fourth ventricle forms the posterior surface of the open medulla. The junction between the closed and open medulla is marked by the V-shaped **obex** (see Figures 6.3 and 6.5).

The inferior cerebellar peduncles arise from the posterior surface of the rostral medulla. The **inferior medullary velum**, a thin sheet of pia and ependyma, adheres to the undersurface of the cerebellum and forms a bridge between the two inferior cerebellar peduncles, forming a roof and sealing the fourth ventricle. The choroid plexus of the fourth ventricle (see Chapter 3, "Overview of the Peripheral Nervous System"), which secretes cerebrospinal fluid, forms from the ependyma of the inferior medullary velum. The fasciculi gracilis and cuneatus continue from the spinal cord to end in their respective nuclei, the **nucleus gracilis** and **nucleus cuneatus**. These nuclei are visible on the surface as the **gracile tubercle** and **cuneate tubercle** (see Figures 6.3 and 6.5).

Moving posterolaterally toward the rootlets of CNs IX, X, and XI, there is a slight longitudinal elevation (**tuberculum cinereum**) that overlies the **spinal trigeminal nucleus and tract** (see Figures 6.3 and 6.5).

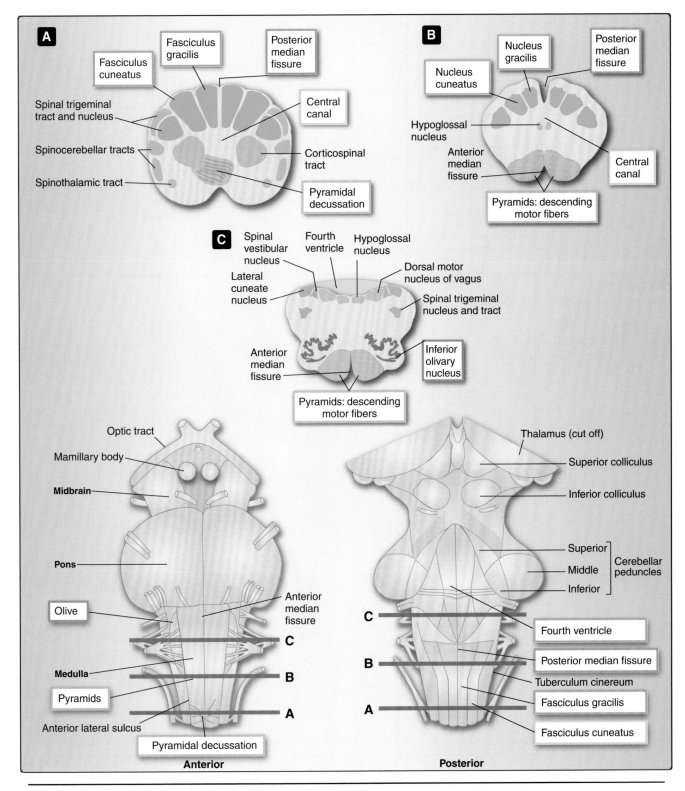

Figure 6.3
Relationship of external to internal structures in the medulla.

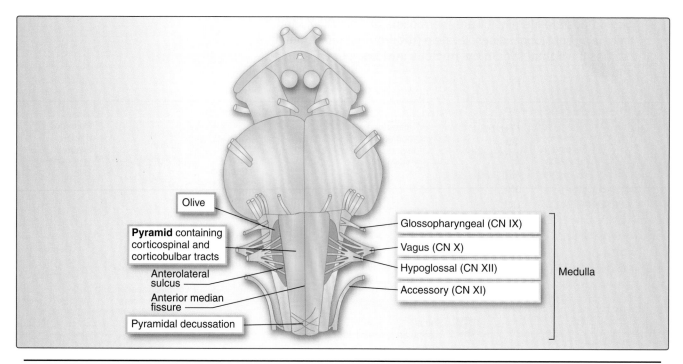

Figure 6.4
Brainstem from anterior highlighting surface structures of the medulla. CN = cranial nerve.

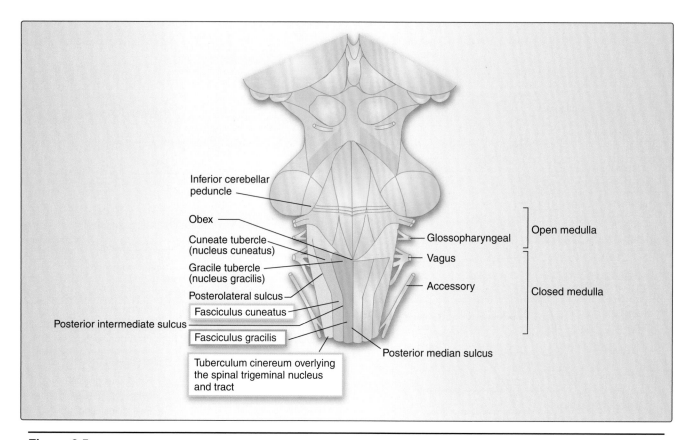

Figure 6.5
Brainstem from posterior highlighting surface structures of the medulla.

B. Pons

The pons is located between the medulla and the midbrain. It is characterized by a large bulge on the anterior surface and the three cerebellar peduncles on the posterior surface (Figure 6.6).

1. **Anterior surface (basal pons):** The distinctive bulge on the anterior or **basal** surface of the pons contains pontine nuclei, transverse fibers, and descending motor tracts. The transverse fibers originate from the pontine nuclei and cross to the contralateral cerebellar hemispheres (see Figure 6.6). The descending motor tracts travel to the brainstem (corticobulbar) and spinal cord (corticospinal) motor nuclei (see Figure 6.6). The **trigeminal nerve (CN V)** emerges at the lateral edge of the basal pons, at the junction of the basal pons with the middle cerebellar peduncle. **CN VI (abducens)** emerges from the anterior surface close to the midline at the junction of the basal pons with the medulla. **CNs VII** and **VIII** emerge from the brainstem lateral to CN VI, in the groove between

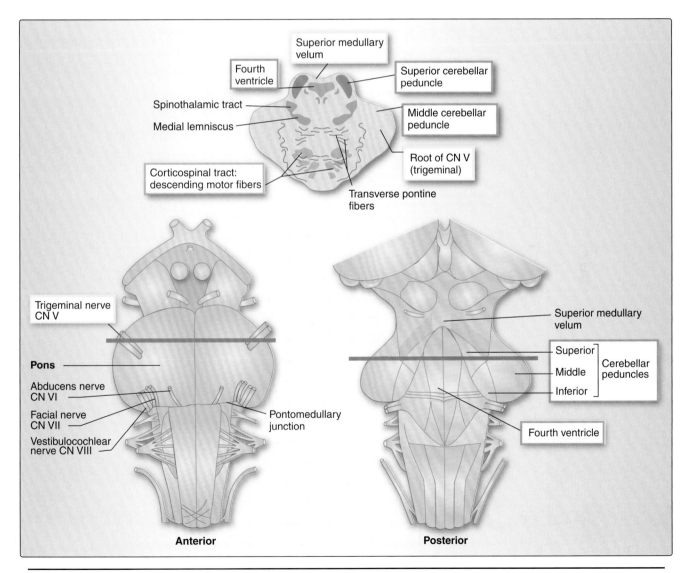

Figure 6.6
Relationship of external to internal structures in the pons. CN = cranial nerve.

the pons and medulla (**pontomedullary junction**) and close to the inferior border of the middle cerebellar peduncle (see Figure 6.6).

2. **Posterior surface (roof):** The posterior surface of the pons is formed by the floor of the **fourth ventricle**. The roof overlying the rostral part of the fourth ventricle is formed by the cerebellum and the superior cerebellar peduncles. The **superior medullary velum**, a thin sheet of white matter, forms a bridge between the two **superior cerebellar peduncles** and completes the roof (see Figure 6.6).

C. Midbrain

The midbrain is located rostral to the pons and caudal to the diencephalon.

1. **Anterior surface:** The anterior surface of the midbrain is characterized by paired prominent bundles of fibers called the **cerebral peduncles**. The most anterior portion is called the **basis pedunculi (crus cerebri)** and contains descending motor information—the corticobulbar and corticospinal tracts. The substantia nigra is located posterior to the basis pedunculi (Figure 6.7). **CN III (oculomotor)**

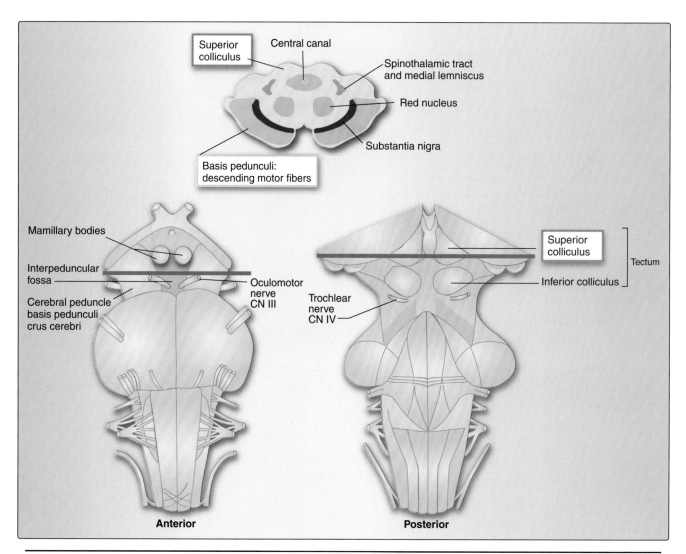

Figure 6.7
Relationship of external to internal structures in the midbrain. CN = cranial nerve.

emerges in the **interpeduncular fossa** between the cerebral peduncles (see Figure 6.7).

2. **Posterior surface:** The posterior surface is characterized by four rounded elevations—the paired **inferior** and **superior colliculi**. These form the **tectum** (roof) of the midbrain (see Figures 6.2 and 6.7). **CN IV (trochlear)** emerges just caudal to the inferior colliculus (see Figure 6.7). It is the only cranial nerve to emerge from the posterior surface of the brainstem.

III. INTRODUCTION TO THE CRANIAL NERVES

There are **12 pairs** of cranial nerves, 9 of which have their nuclei within the brainstem (Figure 6.8). Individual cranial nerves may be purely sensory, purely motor, or "mixed," containing both sensory and motor components. Like spinal nerves, cranial nerves can carry **somatic motor and sensory** information as well as **visceral motor and sensory** information. In addition, cranial nerves are associated with **special senses** such as olfaction, vision, taste, hearing, and balance.

The organization of the cranial nerves is analogous to that of the spinal nerves. Cranial nerves with somatic and visceral sensory components have associated ganglia containing the sensory cell bodies, and processing occurs in nuclei associated with each specific cranial nerve. Some nuclei are associated with more than one cranial nerve. The somatic and visceral motor cell bodies are located in specialized nuclei in the brainstem.

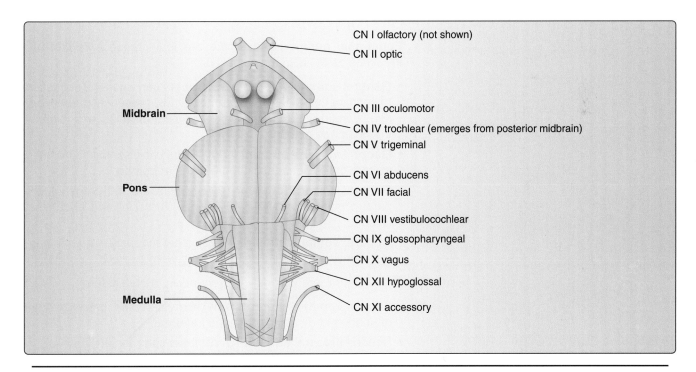

Figure 6.8
Cranial nerves on the anterior surface of the brain. CN = cranial nerve.

The arrangement of the cranial nerve nuclei in the brainstem can be explained by the embryological development of the brainstem.

A. Development of the cranial nerve nuclei

As the **neural tube** develops, changes take place that differentiate the orientation of spinal cord functional elements from brainstem functional elements. The neural tube has an **alar plate**, positioned dorsally (will be posterior), containing the sensory components, and a **basal plate**, positioned ventrally (will be anterior), containing the motor components of the nerves. These two plates are separated by a groove called the **sulcus limitans** (Figure 6.9). This arrangement is consistent throughout the spinal cord, where sensory information is located in the posterior gray horn and motor information is located in the anterior gray horn. Of note, visceral sensory and visceral motor information are both located closest to the sulcus limitans.

Changes occur in this arrangement as the brainstem begins to develop (see Figure 6.9). In the caudal area of the brainstem, specifically in the area that will become the rostral or open medulla, the central canal widens to form the fourth ventricle. As this occurs, the alar plates move out laterally. The sulcus limitans still separates the alar and basal plates. Sensory components of the cranial nerves are now located lateral (rather than posterior) to the sulcus limitans, whereas motor components are now located medial (rather than anterior) to the sulcus limitans (see Figure 6.9). In the rostral pons, the fourth ventricle narrows down as the cerebral aqueduct forms. At this point, the alar plate rotates back to a more posterior position relative to the basal plate (see Figure 6.9).

The cranial nerves can be grouped according to their functional components (Figure 6.10).

1. **Functional components:** There are four components that are the same as those in the spinal cord:

   - **General somatic afferent (GSA)** fibers carry general sensation (discriminative touch, pain, temperature, pressure, proprioception, vibration) from receptors in the skin, muscles, and joints of the head and neck.
   - **General visceral afferent (GVA)** fibers carry sensation from the viscera of the head, neck, thorax, and abdomen.
   - **General visceral efferent (GVE)** fibers are the preganglionic parasympathetic neurons for cranial, thoracic, and abdominal viscera.
   - **General somatic efferent (GSE)** fibers innervate skeletal muscles derived from somites, hence the word *somatic*.

   There are three additional components in the brainstem:

   - **Special somatic afferent (SSA)** fibers carry the special senses of hearing and balance from the inner ear.
   - **Special visceral afferent (SVA)** fibers carry taste.
   - **Special visceral efferent (SVE)**, also called **branchial efferent (BE)**, fibers innervate skeletal muscle derived from the branchial (or pharygeal) arches, that is, muscles of the jaw, face, larynx, and pharynx.

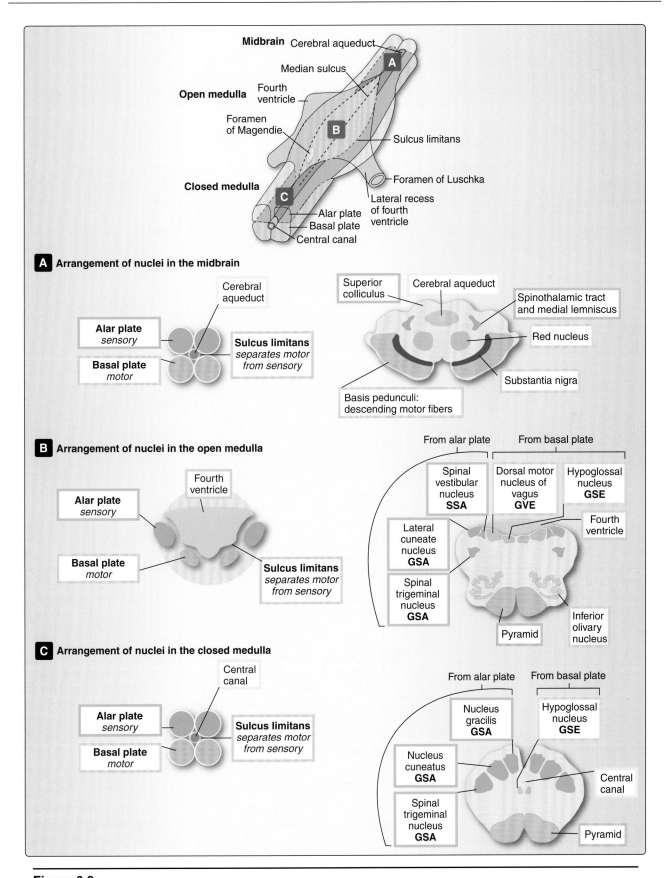

Figure 6.9
Arrangement of nuclei derived from the alar and basal plates in the midbrain and open and closed medulla. SSA = special somatic afferent; GVE = general visceral efferent; GSE = general somatic efferent; GSA = general somatic afferent.

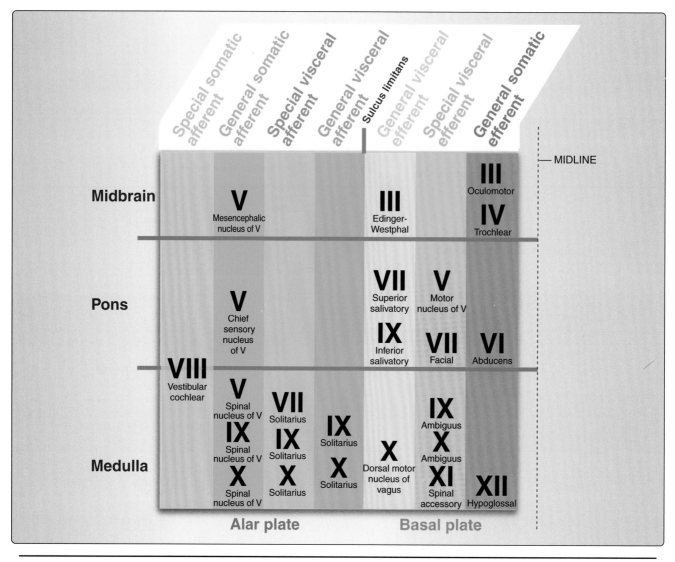

Figure 6.10
The seven cell columns derived from the alar and basal plates and the cranial nerve nuclei derived from them as seen in open medulla: (B) in Figure 6.9.

2. **Columnar organization:** During development, these seven components are organized into **columns** (see Figure 6.10). Not all columns are present at every level of the brainstem. The position of nuclei derived from these columns changes slightly from their original developmental locations, due to lateral movement of some nuclei as additional structures develop in the brainstem (Figure 6.11; also see Figure 6.10).

B. **Overview of the cranial nerves**

The **12 pairs** of cranial nerves are arranged in groups along the longitudinal axis of the brainstem (see Figure 6.8). Knowing the brainstem level for the cranial nerve nuclei and where the fibers enter or leave the brainstem is extremely helpful in localizing lesions in the brainstem. If a lesion involves a cranial nerve, that tells you the level of the

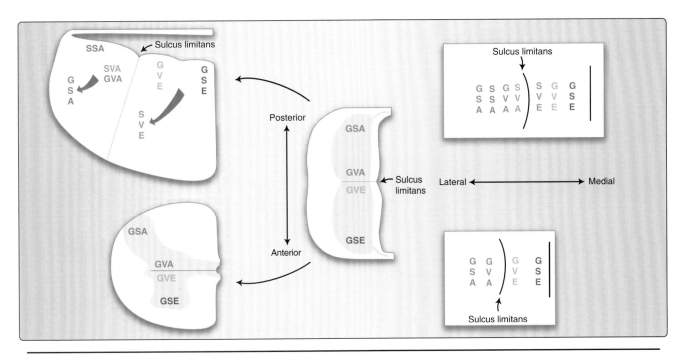

Figure 6.11
Migration of cell columns during development. SSA = special somatic afferent; SVA = special visceral afferent; GVA = general visceral afferent; GSA = general somatic afferent; GVE = general visceral efferent; GSE = general somatic efferent; SVE = special visceral efferent. (Modified from Haines DE. *Neuroanatomy: An Atlas of Structures, Sections, and Systems,* 7th ed. Baltimore, MD: Lippincott Williams & Wilkins, 2007.)

lesion. Figure 6.12 gives an overview of all brainstem nuclei associated with the cranial nerves.

Medulla: CNs IX, X, XI, and XII emerge from the medulla, and all but CN XI have their nuclei within the medulla. The trigeminal tract and nucleus and the inferior (spinal) vestibular nuclei lie within the medulla.

Pons: CNs VI, VII, and VIII emerge from the brainstem at the junction between the medulla and the pons. The nuclei of CNs VI, VII, and VIII (cochlear and remaining vestibular nuclei) are also located in the pons. The chief sensory nucleus and motor nucleus of CN V lie within the pons, and CN V emerges from the lateral basal pons.

Midbrain: CNs III and IV have their nuclei within the midbrain, from which they emerge. The mesencephalic nucleus of CN V is also located in the midbrain.

1. **CN I (olfactory):** CN I has historically been classified as a nerve, but it is actually a CNS tract, the **olfactory tract**. It carries information from the olfactory bulb to the olfactory areas of the cerebral cortex (see Chapter 21, "Smell and Taste").

2. **CN II (optic):** CN II carries visual information from the retina of each eye to the optic chiasm. After the chiasm, these fibers are referred to as optic tracts. Calling CN II a "nerve" is really a misnomer, because it is a CNS tract. We will, however, keep with the convention (see Chapter 15, "The Visual System").

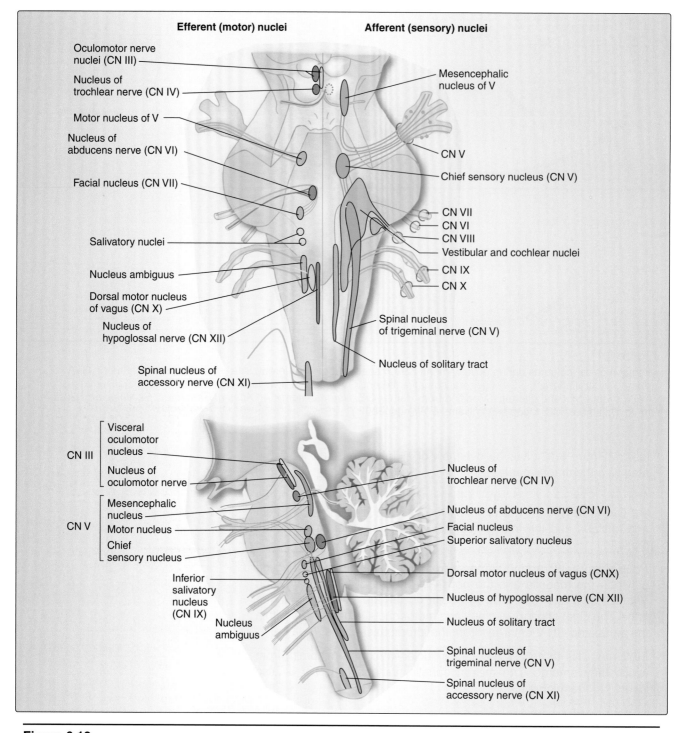

Figure 6.12
Overview of the location of cranial nerve nuclei in the brainstem (posterior and lateral views). CN = cranial nerve.

3. **CN III (oculomotor):** CN III has two components (Table 6.1; also see Figure 6.10). A **GSE** nucleus (oculomotor nuclear complex) contains the nerve cell bodies of the somatic motor fibers that supply extraocular muscles of the eye including the superior, medial, and inferior recti and the inferior oblique in addition to the

Table 6.1: Cranial Nerve Nuclei and their Functions

Cranial Nerve	Nucleus	Modality	Function
CN III—oculomotor	Oculomotor nucleus	GSE	Motor to levator palpebrae superioris, superior rectus, inferior rectus, medial rectus, and inferior oblique
	Edinger-Westphal	GVE	Parasympathetic: motor to sphincter pupillae, ciliary muscles for accommodation
CN IV—trochlear	Trochlear nucleus	GSE	Motor to superior oblique muscle
CN V—trigeminal	Mesencephalic nucleus of CN V	GSA	Proprioception from muscles of mastication
	Chief sensory nucleus of CN V	GSA	Discriminative touch and vibration from the head
	Spinal nucleus of CN V	GSA	Pain and temperature from the head
	Motor nucleus of CN V	SVE	Motor to muscles of mastication
CN VI—abducens	Abducens nucleus	GSE	Motor to lateral rectus
CN VII—facial	Facial nucleus	SVE	Motor to muscles of facial expression
	Chief sensory nucleus of CN V	GSA	Sensory from external acoustic meatus and skin posterior to the ear
	Superior salivatory nucleus	GVE	Parasympathetic motor to lacrimal, sublingual, and submandibular glands
	Nucleus solitarius	SVA	Taste from anterior two-thirds of the tongue
CN VIII—vestibulocochlear	Vestibular nuclear complex	SSA	Balance
	Cochlear nuclear complex	SSA	Hearing
CN IX—glossopharyngeus	Spinal nucleus of CN V	GSA	General sensation from posterior third of the tongue, tonsil, skin of external ear, internal surface of tympanic membrane, pharynx
	Nucleus solitarius	GVA	Chemoreceptors and baroreceptors in the carotid body, visceral afferent information from the tongue and pharynx (gag reflex)
	Nucleus solitarius	SVA	Taste from the posterior third of the tongue
	Nucleus ambiguus	SVE	Motor to stylopharyngeus
	Inferior salivatory nucleus	GVE	Parasympathetic to parotid gland
CN X—vagus	Spinal nucleus of CN V	GSA	Sensory from posterior meninges, external acoustic meatus, and skin posterior to the ear
	Nucleus solitarius	GVA	Sensory from the larynx, trachea, esophagus, thoracic and abdominal viscera, stretch receptors in aortic arch, chemoreceptors in aortic bodies adjacent to the arch
	Nucleus solitarius	SVA	Sensory from taste buds in the epiglottis
	Nucleus ambiguus	SVE	Motor to pharyngeal muscles and intrinsic muscles of the larynx
	Dorsal motor nucleus of vagus	GVE	Parasympathetic to smooth muscles and glands of the pharynx, larynx, and thoracic and abdominal viscera
	Nucleus ambiguus	GVE	Cardiac muscle
CN XI—accessory	Spinal accessory nucleus	SVE	Sternocleidomastoid and trapezius
CN XII—hypoglossal	Hypoglossal nucleus	GSE	Hyoglossus, genioglossus, styloglossus, and all intrinsic muscles of the tongue

GSA = general somatic afferent; GVA = general visceral afferent; SSA = special somatic afferent; SVA = special visceral afferent; GSE = general somatic efferent; GVE = general visceral efferent (parasympathetic); SVE = special visceral efferent.

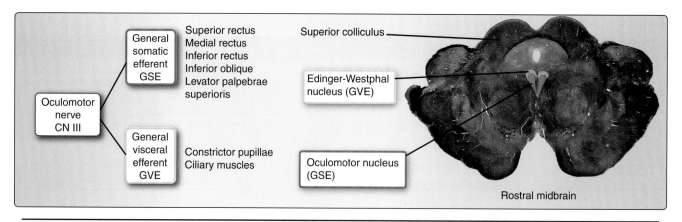

Figure 6.13

Cross-section through the midbrain showing the nuclei associated with cranial nerve (CN) III.

levator palpebrae superioris. The nucleus is located in the tegmentum of the rostral midbrain at the level of the superior colliculus (Figure 6.13; also see Figure 6.12). CN III also has a **GVE** component, the **visceral oculomotor nucleus** (**Edinger-Westphal**) that is located posterior to the oculomotor nuclear complex (see Figure 6.12). These parasympathetic fibers are involved in the pupillary light reflex and the accommodation reflex (see Chapter 9, "Control of Eye Movements").

4. **CN IV (trochlear):** The trochlear nerve is the smallest of the cranial nerves. Its nucleus is located in the tegmentum of the midbrain. Its fibers cross before exiting caudal to the inferior colliculus. It carries **GSE** fibers to innervate the contralateral superior oblique muscle of the eye (Figure 6.14; also see Figure 6.12 and Table 6.1). It is the only nerve to emerge from the posterior surface of the brainstem (see Figure 6.12).

5. **CN V (trigeminal):** The trigeminal (meaning "three twins") nerve is the principal **GSA** nerve for the head as well as supplying motor control (**SVE**) to the muscles of mastication and some other small muscles (see Figure 6.10 and Table 6.1). It has three major divisions: **ophthalmic (V$_1$)**, **maxillary (V$_2$)**, and **mandibular (V$_3$)**. Afferent fibers of the three divisions join together at the **trigeminal ganglion** that houses the sensory nerve cell bodies. Central processes from the trigeminal ganglion form the sensory root of the trigeminal nerve and enter the pons at its midlateral point (see Figure 6.12). The sensory nucleus is extremely long, extending from the midbrain to the caudal medulla. Axons carrying touch information synapse in the **trigeminal chief sensory nucleus** and axons carrying pain and temperature synapse in the **spinal nucleus of the trigeminal nerve** (Figure 6.15; also see Figure 6.14). Proprioception is processed by the **mesencephalic nucleus**, a thin column of cells extending from the pons into the rostral midbrain. It is unique in that it is the only nucleus in the CNS that contains first-order sensory neurons. The **motor nucleus of CN V** is located in the pons and contains the SVE neurons to the muscles of mastication (see Figures 6.12 and 6.15).

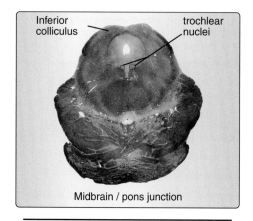

Figure 6.14

Cross-section through the midbrain showing the root of cranial nerve (CN) IV.

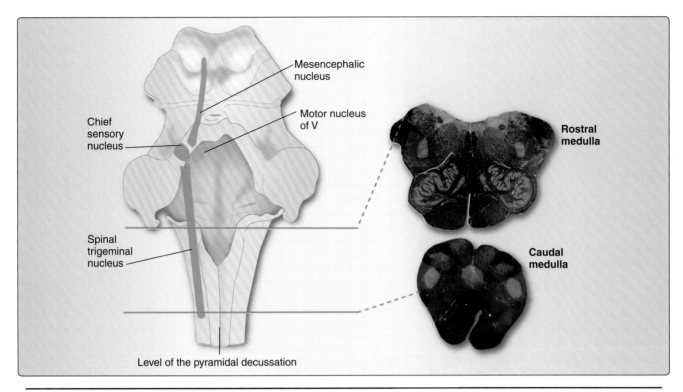

Figure 6.15
Overview of the brainstem nuclei associated with cranial nerve V.

6. **CN VI (abducens):** The abducens nerve is a motor nerve supplying **GSE** input to the lateral rectus muscle, which abducts the eye (moves the eye laterally) (see Figure 6.10 and Table 6.1). Its nucleus is located in the pons close to the midline and its axons emerge from the brainstem at the junction between the medulla and the pons (Figure 6.16; also see Figure 6.12).

7. **CN VII (facial):** The facial nerve has two primary components and emerges at the pontomedullary junction.

   a. **Facial nerve proper:** One component is the facial nerve proper that supplies motor fibers (SVE) to the muscles of facial expression (see Figure 6.10 and Table 6.1). These fibers arise from the facial nucleus in the lateral part of the caudal pons.

   b. **Nervus intermedius:** The other component, called the **nervus intermedius**, lies lateral to the facial nerve proper (Figure 6.17). It contains a **parasympathetic component (GVE)** and two sensory components: **Special sensory fibers (SVA)** carry **taste** from the anterior two-thirds of the tongue (see Figure 6.10; see also Table 6.1). The SVA fibers (taste) synapse in the rostral part of the **nucleus solitarius** (gustatory nucleus), as shown in Figure 6.17. A minor **GSA** component carries some sensory information from the skin behind and within the ear, which projects to the spinal trigeminal nucleus and tract. The GVE (**parasympathetic**) fibers are **secretomotor** to the lacrimal, sublingual, and submandibular glands as well as to the mucosa of the nose, paranasal sinuses, and the hard and soft palates. The preganglionic nerve cell bodies are located in the **superior salivatory nucleus** in the pontine tegmentum (see Figures 6.10, 6.12, and 6.17; see also Table 6.1).

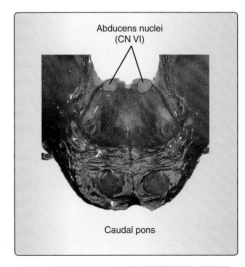

Abducens nuclei (CN VI)

Caudal pons

Figure 6.16
Cross-section through the pons showing abducens nucleus (cranial nerve [CN] VI).

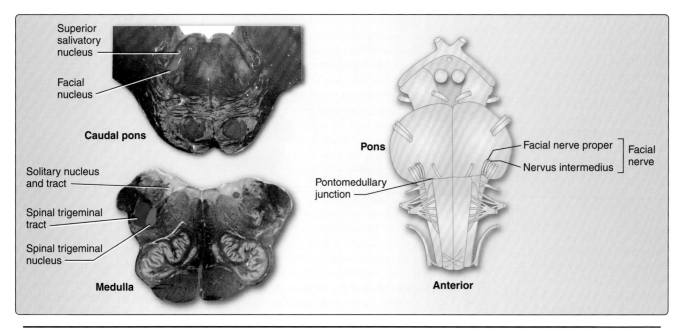

Figure 6.17
Cross-section through the caudal pons and rostral medulla showing the nuclei associated with cranial nerve (CN) VII and diagram of CN VII emerging at the pontocerebellar angle.

8. **CN VIII (vestibulocochlear):** This nerve carries two types of **SSA** fibers—**vestibular** for **balance** and **cochlear** for **hearing** (see Figures 6.10 and 6.12; see also Table 6.1). The sensory receptors are located in the walls of the membranous labyrinth in the petrous part of the temporal bone. The sensory nerve cell bodies lie in the vestibular and spiral ganglia. The central processes from these ganglia form CN VIII that enters the brainstem lateral to CN VII at the pontomedullary junction. The central processes terminate in the vestibular and cochlear nuclei in the brainstem (Figure 6.18) (see Chapter 11, "Hearing and Balance," for more information).

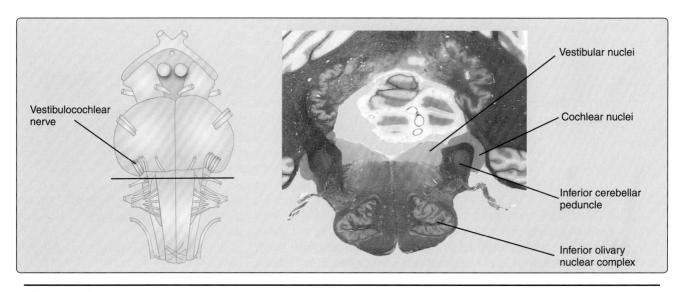

Figure 6.18
Cross-section through the rostral medulla showing the nuclei associated with cranial nerve (CN) VIII and a diagram of CN VIII emerging at the pontocerebellar angle.

9. **CN IX (glossopharyngeal):** As its name implies, this nerve supplies the tongue (glossus) and pharynx with a total of five modalities (see Figures 6.10 and 6.12; see also Table 6.1). A minor **GSA** component carries general sensation from the pharynx and the skin behind and within the ear. These sensory cell bodies are located in the superior glossopharyngeal ganglion and processed further in the **spinal trigeminal nucleus. SVA** fibers carry **taste** from the posterior third of the tongue. These fibers project to the rostral, gustatory part of the **nucleus solitarius. GVA** fibers carry information from the carotid body and carotid sinus and project to the caudal part of the **nucleus solitarius. GVE**, or parasympathetic, fibers from the **inferior salivatory nucleus** supply the parotid gland. **SVE** fibers have their cell bodies in the **nucleus ambiguus** and supply the small stylopharyngeus muscle (Figure 6.19). CN IX emerges from the medulla as the most rostral rootlets between the olive and the inferior cerebellar peduncle (see Figure 6.8).

10. **CN X (vagus):** Vagus comes from the Latin meaning, "wandering," and that is just what this nerve does. It has a **parasympathetic motor or GVE component** to the viscera of the neck, thorax, and abdomen, which originates in the dorsal motor nucleus of the vagus, the main parasympathetic nucleus in the brainstem. The vagus nerve is the main parasympathetic nerve in the body. It is also the largest visceral afferent **(GVA)** nerve of the body, receiving information from the viscera of the neck, thorax, and abdomen, and these fibers project to the nucleus solitarius. There is also a small **SVA—taste** component, carrying special sensory information from the taste buds in the epiglottis to the

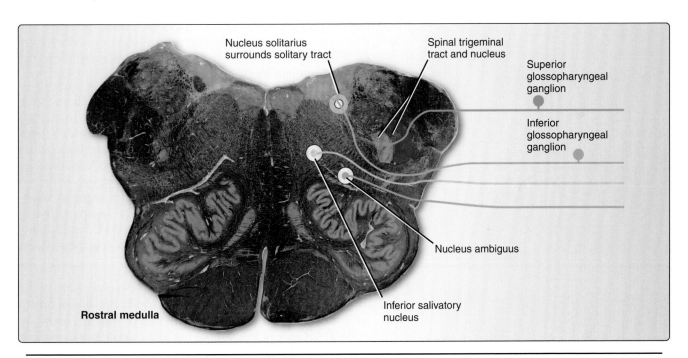

Figure 6.19
Cross-section through the rostral medulla showing the nuclei associated with cranial nerve IX.

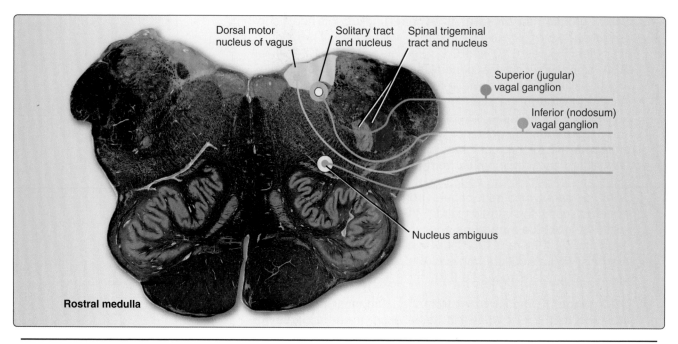

Figure 6.20
Cross-section through the rostral medulla showing the nuclei associated with cranial nerve X.

nucleus solitarius (see Figures 6.10 and 6.11; see also Table 6.1). **GSA** information comes from the posterior meninges, the skin behind and within the ear (a minor component), the pharynx, and the larynx. These fibers project to the spinal trigeminal nucleus and tract. The **SVE** fibers arise from the **nucleus ambiguus** in the medulla and are motor to muscles of the palate, pharynx, larynx, and tongue—all structures that are derived from pharyngeal arches (Figure 6.20). There is also a GVE component from the nucelus ambiguus to cardiac muscle.

11. **CN XI (accessory):** This nerve carries motor (**SVE**) fibers to the sternocleidomastoid and trapezius muscles. The axons arise from the **spinal accessory nucleus** in the spinal cord cervical segments C1 to C5/C6. The rootlets converge and ascend through the foramen magnum of the skull to join with CNs IX and X, where they exit the skull through the jugular foramen (see Figures 6.10 and 6.12; see also Table 6.1). There is some controversy around classification of these fibers (whether SVE or GSE). The most recent evidence points to their development from the SVE cell column. There is no functional or clinical consequence to how these fibers are classified.

12. **CN XII (hypoglossus):** CN XII is the principal motor nerve of the tongue. It carries **GSE** fibers to all of the intrinsic and all but one of the extrinsic muscles of the tongue. The motor neurons lie in the hypoglossal nucleus in the caudal medulla, close to the midline. The rootlets of the nerve emerge from the anterolateral aspect of the medulla between the pyramid and the olive and converge to form the hypoglossal nerve (see Figures 6.8, 6.10, and 6.12; see also Table 6.1).

IV. OVERVIEW OF BLOOD SUPPLY TO THE BRAINSTEM

The CNS needs a constant and rich supply of oxygen and nutrients. As neither glucose, the principle source of energy to the brain, nor oxygen is stored in significant amounts, disruption of the blood supply to the brain and brainstem, even for a few minutes, can cause serious and often irreversible damage. Accurate knowledge and understanding of the normal blood supply to the CNS are critical for the appreciation of abnormal or pathological states and the localization of lesions following strokes so that prompt and appropriate treatment can occur.

A. Arterial systems of the CNS

The blood supply to the CNS comes from two arterial systems: the anterior (internal carotid) system, which arises from the internal carotid arteries, and the posterior (vertebral–basilar) system, which arises from the vertebral arteries (Figure 6.21). The circle of Willis interconnects the anterior and posterior systems. The anterior system, which supplies most of the cerebral hemispheres, and the circle of Willis are discussed in Chapter 13, "The Cerebral Cortex." Here, we discuss the posterior system, which supplies the brainstem.

B. The vertebral–basilar system

The vertebral–basilar (or vertebrobasilar) system can be studied on the anterior surface of the brainstem. For consideration of the blood supply, the brainstem in cross section can be divided into a paramedian area, a lateral area, and a posterior or posterolateral area (Figure 6.22).

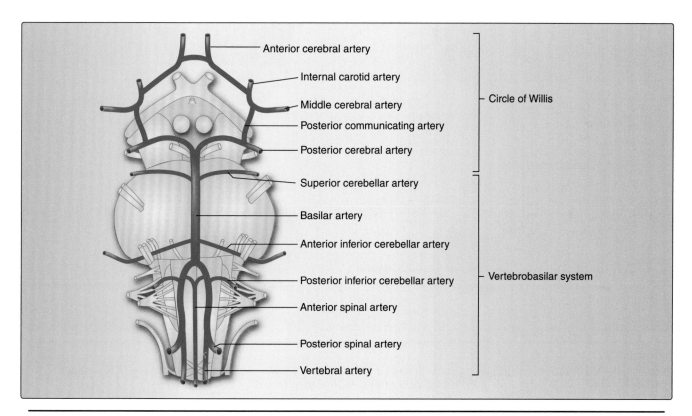

Anterior cerebral artery
Internal carotid artery
Middle cerebral artery
Posterior communicating artery
Posterior cerebral artery
Superior cerebellar artery
Basilar artery
Anterior inferior cerebellar artery
Posterior inferior cerebellar artery
Anterior spinal artery
Posterior spinal artery
Vertebral artery

Circle of Willis

Vertebrobasilar system

Figure 6.21
Overview of vessels supplying the brainstem on the anterior surface.

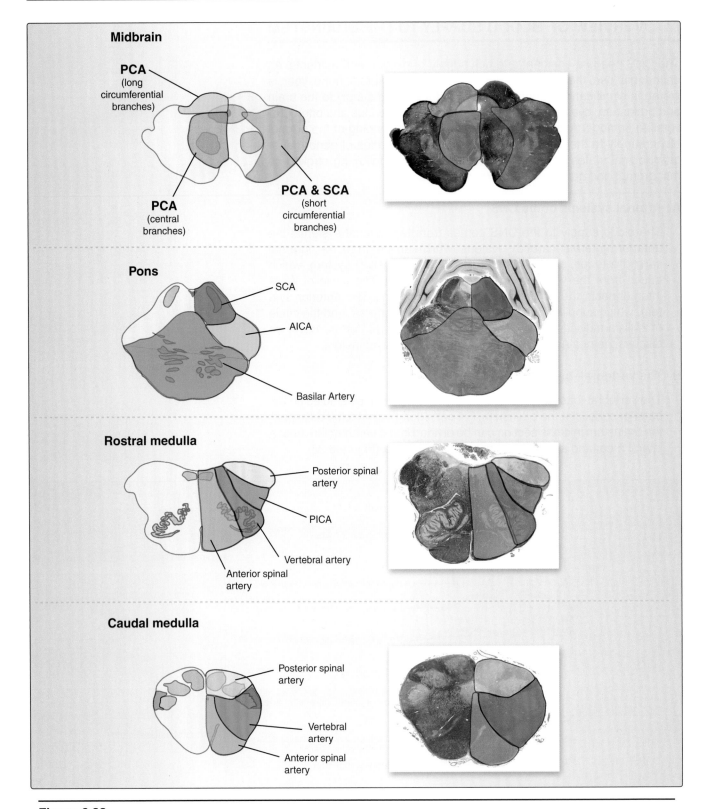

Figure 6.22
Perfusion areas of the vessels supplying the brainstem in cross-section. PCA = posterior cerebral artery; SCA = superior cerebellar artery; PICA = posterior inferior cerebellar artery; AICA = anterior inferior cerebellar artery.

1. **The vertebral arteries:** Two **vertebral arteries** are the first branches off the subclavian arteries. The vertebral arteries enter the foramina in the transverse processes of the cervical vertebrae at approximately C6. They then continue to ascend through the cervical vertebrae and pierce the dura as they enter the foramen magnum. At approximately the junction between the medulla and the pons, the two vertebral arteries join to form a single basilar artery. The vertebral arteries give rise to the **posterior spinal arteries**, the **anterior spinal artery**, and the **posterior inferior cerebellar arteries (PICAs)**.

2. **The posterior spinal arteries:** The two posterior spinal arteries run caudally along the posterolateral aspect of the spinal cord and supply the posterior third of each half of the spinal cord. In addition, the posterior spinal arteries supply the posterior columns in the caudal medulla (see Figures 6.21 and 6.22).

3. **The anterior spinal artery:** Each vertebral artery gives off a small anterior branch. The two branches then join to form a single anterior spinal artery that descends along the anterior midline of the spinal cord and supplies the anterior two-thirds of the spinal cord. In addition, the anterior spinal artery supplies the paramedian area of the medulla (see Figures 6.21 and 6.22).

4. **The posterior inferior cerebellar arteries:** The PICAs are the largest branches of the vertebral arteries. They typically arise proximal to the origin of the basilar artery and supply most of the posterior inferior surface of the cerebellar hemispheres. Importantly, as the PICAs curve around the medulla on their way to the cerebellum, they give off multiple small branches that supply the posterolateral area of the medulla. The vertebral arteries themselves supply the paramedian area of the caudal medulla, together with the anterior spinal artery, and both the paramedian and the lateral areas of the rostral medulla (see Figures 6.21 and 6.22).

5. **The basilar artery:** The **basilar artery** ascends along the midline of the pons and, at the level of the caudal midbrain, bifurcates to form the two **posterior cerebral arteries**.

   The **basilar artery** and its branches supply the entire pons. Paramedian branches supply the paramedian area, and short and long circumferential branches supply the lateral and posterolateral areas. The basilar artery also gives off the anterior inferior cerebellar arteries (AICAs) and the superior cerebellar arteries (see Figures 6.21 and 6.22).

6. **The anterior inferior cerebellar arteries:** The **AICAs** arise close to the start of the basilar artery and supply the more anterior areas of the inferior surface of the cerebellar hemispheres. In addition, on their way to the cerebellum, the AICAs supply the middle cerebellar peduncles as well as a small part of the posterior pontine tegmentum (see Figures 6.21 and 6.22).

7. **The superior cerebellar arteries:** The **superior cerebellar arteries** arise from the basilar artery just before it bifurcates to form the posterior cerebral arteries. They, too, curve around the brainstem, supply the superior surfaces of the cerebellum and rostral pons (superior cerebellar peduncles), and may provide some

supply to the roof (inferior colliculi) of the caudal midbrain (see Figures 6.21 and 6.22).

8. **The posterior cerebral arteries:** The **posterior cerebral arteries** curve around the caudal midbrain to supply the medial and inferior surfaces of the temporal and occipital lobes of the cerebral hemispheres. As they curve around the brainstem, these arteries provide blood supply to the entire midbrain (the paramedian, lateral, and posterolateral areas) (see Figures 6.21 and 6.22).

V. CRANIAL NERVE SCREENING EXAM

A neurological examination includes a screening exam for the cranial nerves. Any abnormalities found in the screening exam will then result in a more in depth investigation. These cranial nerves, their function, and the reflexes they are associated with will be discussed in more detail in the upcoming chapters.

A summary of a cranial nerve (CN) examination can be found in Table 6.2.

A. Cranial nerve I—olfactory

Test the sense of smell by presenting the patient with nonirritating odors such as coffee or vanilla.

B. Cranial nerve II—optic

Perform a visual acuity test, visual field test, and test afferent component of the pupillary light reflex

C. Cranial nerves III, IV, and VI—oculomotor, trochlear, and abducens

These cranial nerves innervate the extraocular muscles that move the eye. An H-test (see Chapter 9 "Control of Eye Movements" for details) tests the function of the extraocular muscles and the cranial nerves that innervate them. The GVE (parasympathetic) component of CN III is tested through the pupillary light reflex.

Table 6.2: Summary of a Cranial Nerve Screening Exam

Cranial Nerve	Clinical Assessment
I	Smell
II	Visual acuity, visual fields
II, III	Pupillary light reflex
III, IV, VI	Extraocular movements
V	Facial sensation, jaw movements
VII	Facial movements
VIII	Hearing
IX, X	Swallowing, rise of palate, gag reflex
V, VII, X, XII	Voice and speech
XI	Shoulder and neck movements
XII	Tongue symmetry and position

Data from Bickley LS. *Bates' Guide to Physical Examination and History Taking*, 11th ed. Wolters Kluwer Health, Philadelphia, 2013.

Cranial nerve V—Trigeminal. CN V has a motor component and a sensory component. The motor component is tested by assessing strength of the muscles of mastication. The sensory component is assessed by checking for sensation from the forehead, cheeks, and jaw; this tests all three peripheral branches of the trigeminal nerve.

D. Cranial nerve VII—facial

In a screening exam, the innervation of the muscles of facial expression is examined. Motor function in both the upper and lower face is assessed, while ensuring symmetry is maintained. For example, ask the patient to wrinkle his/her forehead, close his/her eyes tightly, and show his/her teeth or smile.

E. Cranial nerve VIII—vestibulocochlear

In a screening exam, only the acoustic component is typically assessed; the vestibular function is rarely included. Assess hearing by whispering close to each ear.

F. Cranial nerves IX and X—glossopharyngeal and vagus

These nerves are assessed together when asking the patient to swallow. A hoarse voice would indicate a lesion to CN X—the recurrent laryngeal nerves innervate the vocal cords. Asymmetry of the soft palate with deviation of the uvula can indicate a lesion to CN X. The gag reflex tests both CN IX and X, which are the afferent and efferent limbs of the reflex, respectively. It is elicited through light stimulation of the back of the throat on each side.

G. Cranial nerve XI—accessory

Test the strength of the trapezius and sternocleidomastoid muscles by asking the patient to shrug his/her shoulder and turn the head against resistance.

H. Cranial nerve XII—hypoglossal

Inspection of the patient's tongue will show any atrophy of the intrinsic muscles of the tongue. Ask the patient to protrude the tongue and check for asymmetry and deviation of the tongue from the midline.

Chapter Summary

- The brainstem is a small but complex part of the central nervous system. It contains tracts traveling up and down between the spinal cord and the cortex and tracts interconnecting both the cortex and the spinal cord with the cerebellum, cranial nerve nuclei and tracts, intrinsic systems, and relay nuclei.
- Three distinct areas can be identified: the **medulla oblongata** (or **medulla**) caudally, the **pons**, and the **midbrain** rostrally. The **cerebellum** is developmentally part of the pons but is considered as its own entity.
- In transverse section, the internal structure of the brainstem can be divided into four areas: from posterior to anterior, these are the **tectum** or roof over the ventricular system, the **ventricular system** itself, the **tegmentum** or core of the brainstem, and the **basal portion** situated most anteriorly. Whereas these four areas occur throughout the brainstem, the medulla,

Chapter Summary continued

pons, and midbrain all have characteristic surface features that typically reflect major internal or underlying structures. In brief, the major gross landmarks and internal structures at each level of the brainstem are as follows:

- ○ **Medulla**: The caudal medulla contains the pyramids anteriorly (overlying the descending corticospinal and corticobulbar tracts) and the posterior columns posteriorly (overlying the ascending fasciculi gracilis and cuneatus). The rostral medulla contains the pyramids anteriorly, the olives laterally (overlying the inferior olivary nuclear complex), and the caudal part of the fourth ventricle and the inferior cerebellar peduncles posteriorly.

- ○ **Pons**: The anterior, basal pons contains the transverse pontine fibers and the descending corticospinal and corticobulbar tracts. The fourth ventricle, middle cerebellar peduncles, and superior cerebellar peduncles comprise the posterior pons.

- ○ **Midbrain**: The cerebral peduncles form the anterior part of the midbrain. These contain the descending corticospinal and corticobulbar tracts. The inferior and superior colliculi mark the posterior surface of the caudal and rostral midbrain, respectively, and the cerebral aqueduct lies deep to the colliculi, connecting the third and fourth ventricles.

- There are **12 pairs of cranial nerves** arranged in groups along the longitudinal axis of the brainstem. Cranial nerves can carry **somatic motor and sensory** information as well as **visceral motor and sensory** information. An additional type of motor fiber targets those muscles derived from the pharyngeal arches during development. In addition, cranial nerves are associated with **special senses** such as olfaction, vision, taste, hearing, and balance. Although they have both sensory and motor components, individual cranial nerves may be purely sensory (I, II, VIII), purely motor (III, IV, VI, XI, XII), or "mixed," containing both sensory and motor components (V, VII, IX, X).

- The blood supply to the brainstem comes from the **vertebral–basilar system**, which is formed by the joining of the two **vertebral arteries** to form the **basilar artery**. The anterior spinal artery is formed by the joining of branches from the vertebral arteries. The basilar artery ascends and bifurcates into the **posterior cerebral arteries** at the level of the midbrain. Three branches off the basilar artery supply the posterior aspect of the brainstem: the **superior cerebellar arteries**, the **anterior inferior cerebellar arteries**, and the **posterior inferior cerebellar arteries**.

Study Questions

Choose the ONE best answer.

6.1 Brainstem nuclei that receive general somatic afferent information include the:

- A. Facial nucleus.
- B. Mesencephalic nucleus of cranial nerve V.
- C. Nucleus solitarius.
- D. Oculomotor nucleus.
- E. Vestibular nuclei.

The correct answer is B. The mesencephalic nucleus of V is one of the three components of the trigeminal sensory nucleus. It receives general somatic afferent information related to proprioception from the muscles of mastication. The facial nucleus is a motor nucleus that controls muscles of facial expression. The nucleus solitarius is a sensory nucleus that receives special visceral afferent information about taste and general visceral afferent information from the viscera of the head, neck, thorax, and abdomen. The oculomotor nucleus is a motor nucleus that controls four extraocular muscles and the levator palpebrae superioris. It also has a parasympathetic component that is motor to the sphincter pupillae and the ciliary muscles. The vestibular nuclei are sensory nuclei that receive special sensory afferent information about balance.

6.2 A patient presented with paralysis on the left side of his body. Magnetic resonance imaging showed a bleed in the area of the right pyramid at the level of the caudal medulla. The paralysis most likely resulted from a lesion of:

- A. Right-sided branches of the anterior spinal artery.
- B. Right paramedian branches of the basilar artery.
- C. Paramedian branches of the right posterior cerebral artery.
- D. The right posterior inferior cerebellar artery.
- E. The right posterior spinal artery.

The correct answer is A. The anterior spinal artery arises when branches off the two vertebral arteries join. It descends along the midline on the anterior surface of the medulla and spinal cord and supplies the anterior two-thirds of the cord. In addition, it also supplies the paramedian area of the caudal medulla where the pyramids are located. Right paramedian branches of the basilar artery supply the right paramedian area of the basal pons. Paramedian branches of the right posterior cerebral artery supply the right cerebral peduncle. The right posterior inferior cerebellar artery supplies the posterolateral area of the right rostral medulla. The right posterior spinal artery supplies the posterolateral areas of the right caudal medulla.

6.3 Which statement about the oculomotor nerve is correct?

  A. It exits the brainstem in the rostral pons.
  B. It has both somatic motor and parasympathetic components.
  C. It has its cell bodies in the lateral part of the midbrain.
  D. It innervates the superior oblique muscle of the eye.
  E. It receives its blood supply from branches of the basilar artery.

The correct answer is B. The oculomotor nerve has both somatic motor and parasympathetic components. It exits from the anterior surface of the midbrain in the interpeduncular fossa. Its nucleus is at the midline of the rostral midbrain (motor nuclei are medial, whereas sensory nuclei are lateral to the sulcus limitans). It innervates four extraocular muscles and the levator palpebrae superioris. It also has a parasympathetic component that mediates pupillary constriction and accommodation. Cranial nerve IV, the trochlear nerve, innervates the superior oblique muscle. The nuclei of CN III receive their blood supply from long circumferential branches of the posterior cerebral artery.

6.4 A defining anatomical feature of the rostral medulla is the:

  A. Cerebral aqueduct.
  B. Decussation of the pyramids.
  C. Middle cerebellar peduncle.
  D. Olive.
  E. Posterior column tracts.

The correct answer is D. The olive is a feature on the lateral surface of the rostral medulla, overlying the inferior olivary nuclear complex. The cerebral aqueduct is in the midbrain and connects the third and fourth ventricles. The decussation of the pyramids occurs at the level of the caudal medulla and is visible on the anterior surface as the medulla merges into the spinal cord. The middle cerebellar peduncles arise from the lateral surface of the basal pons at the midpons level. The posterior column tracts (fasciculus gracilis, fasciculus cuneatus) ascend on the posterior aspect of the spinal cord and medulla and synapse in the nucleus gracilis and nucleus cuneatus, respectively, in the caudal medulla.

6.5 A defining anatomical feature of the caudal midbrain is the:

  A. Chief sensory nucleus of cranial nerve V.
  B. Inferior colliculus.
  C. Superior cerebellar peduncle.
  D. Superior colliculus.
  E. Vertebral artery.

The correct answer is B. The two inferior colliculi form the tectum or roof of the caudal midbrain, whereas the superior colliculi form the tectum, or roof, of the rostral midbrain. The chief sensory nucleus of cranial nerve V is located in the pons, in the lateral area of the tegmentum. The two superior cerebellar peduncles arise in the rostral pons and form part of the roof over the fourth ventricle. The vertebral arteries supply the medulla.

7 Ascending Sensory Tracts

I. OVERVIEW

This chapter presents detailed information on three major ascending systems or tracts: the posterior column–medial lemniscus system (discriminative touch, vibration, pressure, proprioception), the anterolateral system (pain, temperature, nondiscriminative touch), and the spinocerebellar tracts (mainly proprioceptive information to the cerebellum).

Sensory information is detected by specific receptors that are sensitive to stimuli arising from both outside and within our bodies. **Exteroceptors** sense stimuli from the external world and respond to pain, temperature, touch, vibration, and pressure. **Proprioceptors** sense stimuli from within the body (e.g., muscles, tendons, joints), signal awareness of body position and movement in space, and allow us to plan movement accordingly. **Enteroceptors** monitor events within the body (e.g., viscera, gut, and other internal organs) and allow us to feel the body's inner working. Figure 7.1 presents an overview of the flow of sensory information from the periphery to the spinal cord.

All sensory information must ultimately be conveyed to the central nervous system via tracts ascending through the spinal cord and brainstem. Some of this information goes to brainstem centers, some goes to the cerebellum for the planning and refinement of movement, and some travels to the cortex, which mediates conscious awareness of stimuli.

Sensory neurons have their cell bodies in the **spinal ganglia** along the spinal cord or, for the head and neck, the sensory ganglia associated with their respective cranial nerves. From the periphery, sensory information enters the spinal cord through the **posterior root** or enters the brainstem to synapse in the appropriate sensory nuclei (see Chapter 6, "Overview and Organization of the Brainstem").

Sensory information can either be conscious and make us aware of the sensation or be on a nonconscious level and serve to adjust the movement or function of our body (Figure 7.2).

Sensory information is bundled according to modality, and different modalities will travel in different tracts.

One modality is dedicated to the detection of **mechanical stimuli** and is associated with information about **discriminative touch, pressure, vibration**, and **proprioception**. This information travels through the **posterior column–medial lemniscus system**. A second modality is

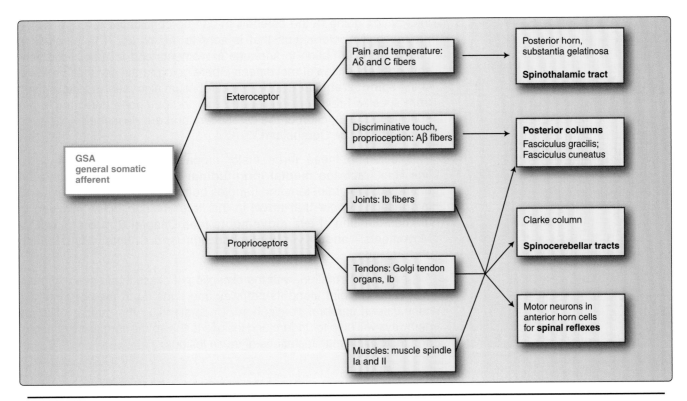

Figure 7.1
Flow of sensory information from the peripheral receptors to the spinal cord.

dedicated to the detection of **nociceptive stimuli** (mechanical, chemical, and thermal) and is associated with information about **pain** and **temperature**. This information travels through the **anterolateral system**. A third modality (in a sense, a "submodality") encompasses information about **proprioception** that is conveyed to the cerebellum. The

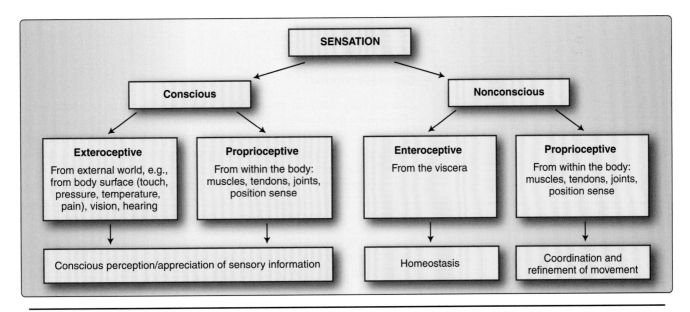

Figure 7.2
Overview of conscious and nonconscious sensory modalities.

information going to the cerebellum can be considered a "copy" of the proprioceptive information that is sent to the cortex. The cerebellum needs this information to fine-tune movements and to predict the sensory consequences of movement. Fibers going to the cerebellum travel in the **spinocerebellar tracts**. The cerebellum also receives information from specific brainstem nuclei and the cortex. These latter tracts are specific to the function of the cerebellum and are described in detail in Chapter 17, "The Cerebellum."

In addition to these three major modalities, there is a long midline fiber tract, the **medial longitudinal fasciculus**, that spans the entirety of the brainstem and handles both ascending and descending information. Fibers that travel in this tract include those involved in the coordination of **eye movements** (see Chapter 9, "Control of Eye Movements") and in **vestibular** function (see Chapter 11, "Hearing and Balance").

All sensory information, with the exception of olfaction, crosses the midline at some point along its pathway and ends up in the cortex of the contralateral cerebral hemisphere. Understanding the course of these pathways will help to understand lesions to the spinal cord and brainstem and the clinical manifestations of these lesions.

II. POSTERIOR COLUMN–MEDIAL LEMNISCUS SYSTEM

The posterior column–medial lemniscus system carries **general somatic afferent** (general sensory) information about **discriminative touch**, **pressure**, **vibration**, and **proprioception**. This pathway allows for high-conduction velocity through the fast-conducting large-diameter fibers (Aβ) and a limited number of synaptic relays. A high degree of resolution and an efficient transfer of information are achieved through additional mechanisms discussed later in this chapter. Sensory information traveling in this pathway will reach the cortex and result in conscious awareness or perception of that information, quickly and with high resolution.

A. General anatomy

The primary afferent fibers have their peripheral processes associated with the various types of peripheral sensory receptors (see Chapter 3, "Overview of the Peripheral Nervous System"). The cell bodies of these **pseudounipolar neurons** are located in the **spinal ganglia** of associated spinal nerves along the spinal cord. The signal from the periphery travels along the **dendritic axon**, bypasses the cell body in the ganglion, and continues along the central process (or axon) to the **posterior horn** of the spinal cord.

In the spinal cord, axon collaterals synapse with motor neurons in the anterior horn to form **spinal reflex arcs** (see Chapter 5, "The Spinal Cord"). Although some collaterals synapse in the posterior horn, allowing additional processing of information, the majority of fibers ascends in the **posterior columns** (Figure 7.3) and does not synapse until the caudal medulla. The posterior columns are composed of two

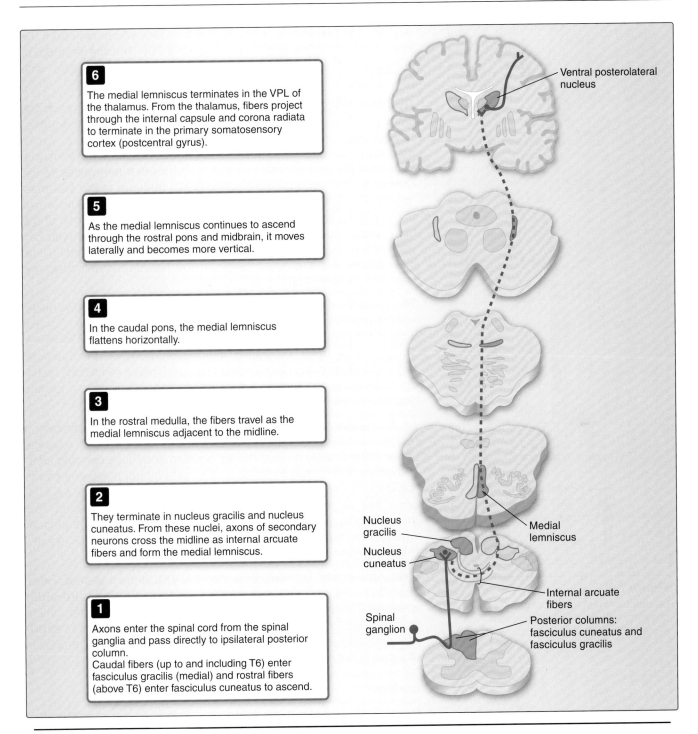

6

The medial lemniscus terminates in the VPL of the thalamus. From the thalamus, fibers project through the internal capsule and corona radiata to terminate in the primary somatosensory cortex (postcentral gyrus).

5

As the medial lemniscus continues to ascend through the rostral pons and midbrain, it moves laterally and becomes more vertical.

4

In the caudal pons, the medial lemniscus flattens horizontally.

3

In the rostral medulla, the fibers travel as the medial lemniscus adjacent to the midline.

2

They terminate in nucleus gracilis and nucleus cuneatus. From these nuclei, axons of secondary neurons cross the midline as internal arcuate fibers and form the medial lemniscus.

1

Axons enter the spinal cord from the spinal ganglia and pass directly to ipsilateral posterior column.
Caudal fibers (up to and including T6) enter fasciculus gracilis (medial) and rostral fibers (above T6) enter fasciculus cuneatus to ascend.

Ventral posterolateral nucleus

Nucleus gracilis

Nucleus cuneatus

Medial lemniscus

Spinal ganglion

Internal arcuate fibers

Posterior columns: fasciculus cuneatus and fasciculus gracilis

Figure 7.3
Longitudinal diagram of the posterior column medial–lemniscus system, which carries discriminative touch, vibration, pressure, and conscious proprioception to the primary somatosensory cortex.

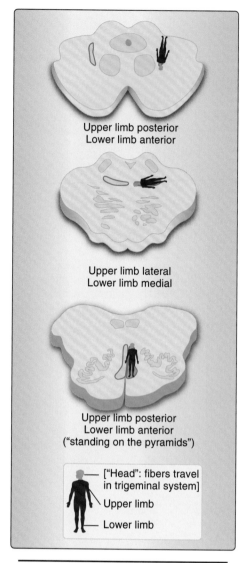

Figure 7.4
Somatotopic distribution of fibers in the posterior column–medial lemniscus system throughout the brainstem.

fasciculi: the **fasciculus gracilis**, which carries information from the lower limb and trunk (up to and including T6), and the **fasciculus cuneatus**, which carries information from the trunk and upper limb (above T6).

This **somatotopic organization** continues as the fibers reach the medulla, where they synapse on their respective second-order neurons located in the nucleus gracilis and nucleus cuneatus. Axons from these second-order neurons cross the midline as the internal arcuate fibers and ascend bundled as the **medial lemniscus** to the contralateral **ventral posterolateral (VPL) nucleus** of the thalamus. Information from the lower half of the body travels in the anterior part of the medial lemniscus, and information from the upper half of the body travels in the posterior part of the medial lemniscus (you can think of this as a "headless person standing on the pyramids" [Note: information from the head travels in the nearby trigeminothalamic tract; see Chapter 10]).

The posterior column nuclei are more than just a set of relay nuclei from which the fibers cross over to the contralateral side. Cortical input to these nuclei is critical in streamlining the information received. Through selectively facilitating information from common inputs, there is a significant level of processing and "noise reduction" resulting in efficient and targeted transfer of information, critical for fine discrimination.

As the **medial lemniscus** moves rostrally through the brainstem, it rotates laterally. The result is that fibers carrying information from the upper limbs occupy a more medial position and those carrying information from the lower limbs occupy a more lateral position (i.e., the "headless" people fall over and wind up "neck to neck"). In the midbrain, the medial lemniscus moves more laterally: Information from the lower limb is now most posterior, and information from the upper limb is now most anterior ("headless" people now "upside down") (Figure 7.4).

Finally, the medial lemniscus reaches the VPL in the thalamus. Here, fibers from the second-order neurons synapse with the **third-order neurons**. These third-order neurons relay the information through the posterior limb of the **internal capsule** and **corona radiata** to the **primary somatosensory cortex**, which allows for conscious perception of the sensory information.

B. Lesions

Interestingly, lesions to the posterior columns do not have the devastating effect that one would expect. Whereas there is a loss of proprioception and a loss of the ability to distinguish the finer aspects of tactile stimuli, there is only a small effect on the performance of tasks requiring tactile information processing. This can be explained, at least partially, through an apparent redundancy in the system: A small percentage of fibers encoding for discriminative touch travels with the anterolateral system, which contains mainly fibers carrying pain and temperature.

III. THE ANTEROLATERAL SYSTEM

The anterolateral system comprises a set of fibers encoding for **pain** and **temperature** and **nondiscriminative touch**. These fibers can be divided into several different tracts. The majority of the fibers travels to the thalamus (**spinothalamic tract**) and plays a key role in mediating conscious perception or awareness of pain and temperature. The present discussion will focus on the spinothalamic tract. Other smaller tracts are involved in modulating these sensations and terminate in various targets in the brainstem and diencephalon. These targets include the midbrain (spinomesencephalic), reticular formation (spinoreticular), brainstem nuclei (spinobulbar), and hypothalamus (spinohypothalamic). Together, these tracts modulate pain and initiate responses to pain sensation. The function of these additional tracts is discussed in detail in Chapter 22, "Pain." See Figure 7.5 for an overview of the tracts of the anterolateral system and their targets.

All of the nerve fibers coming in from the periphery are slowly conducting fibers: either **lightly myelinated Aδ fibers** or **unmyelinated C fibers.** All of these fibers have **free nerve endings** in the periphery and do not have specialized sensory transduction organelles associated with them, as do the fibers in the posterior column–medial lemniscus system.

A. Aδ mechanosensitive receptors and fibers

Pain and temperature are sensed by **Aδ mechanosensitive receptors** and carried by **Aδ fibers. Aδ thermoreceptors** can be divided into those that are **heat activated** (at 35°C–45°C) and those that are **cold activated** (at 17°C–35°C). Aδ fibers that encode for pain encode primarily for well-localized, sharp primary (or "pricking") pain that is conducted relatively rapidly at 20 m/s. **Nondiscriminative touch** also

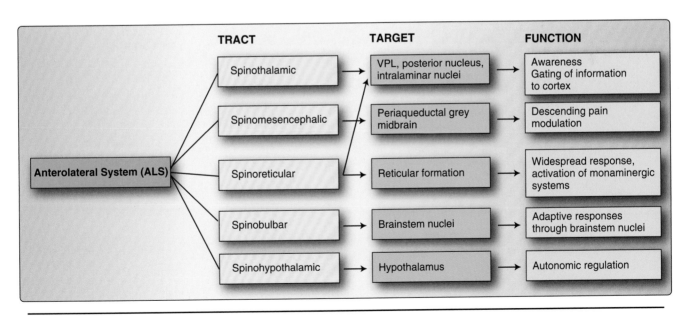

Figure 7.5
Overview of the anterolateral system with its various tracts and targets. VPL = ventral posterolateral nucleus.

results from the stimulation of **Aδ mechanosensitive fibers**. Any intense mechanical stimulus that does not result in tissue damage will activate these fibers.

B. C fibers

C fibers are **polymodal**, having many functions associated with them, and have a slower conduction at 2 m/s. They are the **chemonociceptors** that are activated by substances released during tissue damage (bradykinin, histamine, changes in pH) and are responsible for the sensation of dull, poorly localized pain. Histamine selective C fibers are responsible for the sensation of **itch**, and C fibers in muscles convey the burning sensation in muscles that occurs with extreme exercise. C fibers are also sensitive to **thermal** and **mechanical stimulation**, similar to Aδ fibers. It is due to their slower conduction velocity that they elicit pain that is dull and poorly localized (often referred to as **secondary pain**) (Figure 7.6).

C. General anatomy

The neuronal cell bodies of all fibers sensitive to pain and temperature are located in the spinal ganglia. The central processes enter the spinal cord through the **posterior horn** and ascend or descend a few spinal cord levels in **Lissauer tract** before entering the gray matter. They synapse primarily in superficial laminae (I and II) and in the nucleus proprius (laminae III and IV). Cells

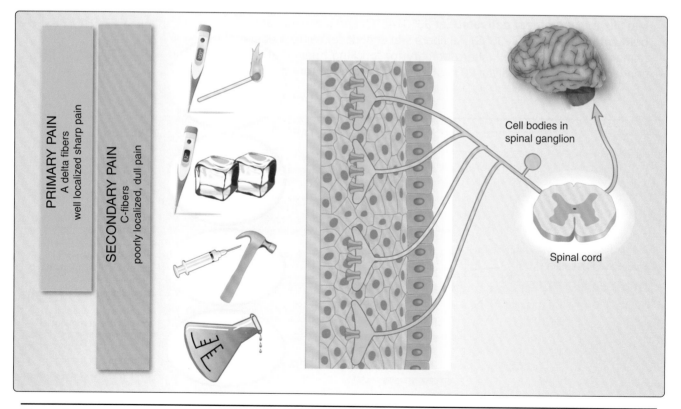

Figure 7.6
Overview of nociceptive fibers.

from the nucleus proprius extend processes into the substantia gelatinosa (lamina II). A great deal of pain modulation occurs in the substantia gelatinosa before the information travels to higher cortical centers. This is discussed in more detail in Chapter 22, "Pain," which is dedicated entirely to pain. Fibers arising from nuclei in the posterior horn (laminae I, III, and IV) cross the midline within the spinal cord in the **anterior white commissure** and ascend as the **anterolateral system**. In the rostral medulla, the anterolateral system lies between the inferior olivary nucleus and the nucleus of the spinal trigeminal tract. In the pons and midbrain, the anterolateral system lies lateral to the medial lemniscus (Figure 7.7). The fibers in the anterolateral system are arranged **somatotopically**, with fibers carrying information from the lower limb located more laterally and those carrying information from the upper limb more medially. The anterolateral system becomes smaller as it ascends through the brainstem, because it gives off fibers that terminate in brainstem structures that are critical to pain modulation (see Chapter 22, "Pain"). The spinothalamic tract of the anterolateral system terminates in the VPL. From the thalamus, fibers project through the posterior limb of the **internal capsule** and **corona radiata** to the **primary somatosensory cortex**.

D. Lesions

Because fibers carrying pain and temperature travel up or down a few levels before synapsing in the posterior horn and crossing the midline within the spinal cord, a lesion of the spinothalamic tract causes a loss of pain and temperature sensation on the side of the body **contralateral** to the injury, beginning a few levels below or above the level of the injury (Figure 7.8).

Clinical Application 7.1: Brown-Séquard Syndrome

Hemisection of the spinal cord due to traumatic injury leads to a characteristic pattern of symptoms referred to as Brown-Séquard syndrome (see Figure 7.8). The patient will experience a loss of discriminative touch, vibration, and proprioception on the side *ipsilateral* to the spinal cord lesion. This is due to a disruption of the posterior columns, which are uncrossed in the spinal cord. In addition, there will be a loss of pain and temperature sensation on the side *contralateral* to the lesion, due to a disruption of the spinothalamic tract fibers. The fibers of the anterolateral system cross to the contralateral side in the spinal cord.

The transection of the corticospinal tract fibers will also lead to hemiparesis on the side ipsilateral to the lesion. The corticospinal tract fibers arise from the contralateral cerebral hemisphere, cross in the medulla, and in the spinal cord, travel ipsilateral to the motor neurons they will innervate (see Chapter 5, Box 5.2, and Chapter 8, Descending Motor Tracts).

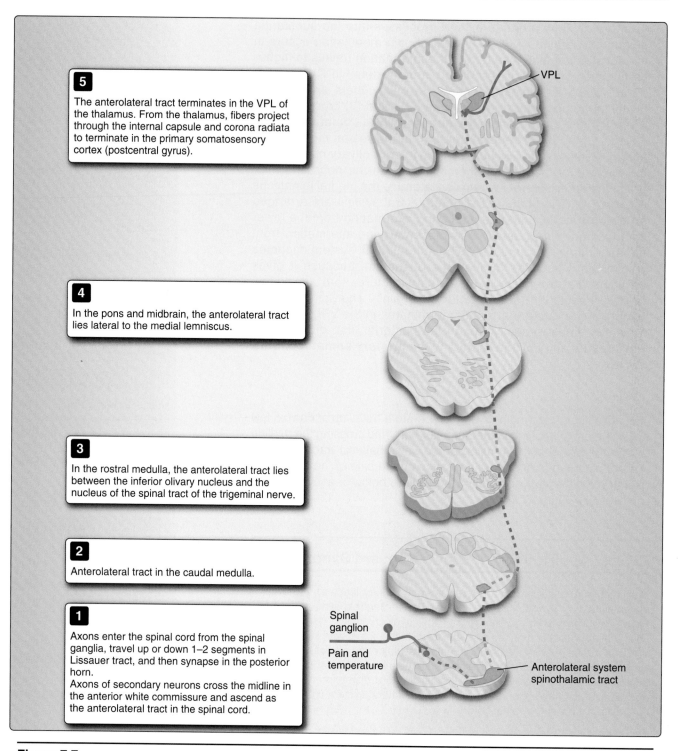

5

The anterolateral tract terminates in the VPL of the thalamus. From the thalamus, fibers project through the internal capsule and corona radiata to terminate in the primary somatosensory cortex (postcentral gyrus).

VPL

4

In the pons and midbrain, the anterolateral tract lies lateral to the medial lemniscus.

3

In the rostral medulla, the anterolateral tract lies between the inferior olivary nucleus and the nucleus of the spinal tract of the trigeminal nerve.

2

Anterolateral tract in the caudal medulla.

1

Axons enter the spinal cord from the spinal ganglia, travel up or down 1–2 segments in Lissauer tract, and then synapse in the posterior horn.
Axons of secondary neurons cross the midline in the anterior white commissure and ascend as the anterolateral tract in the spinal cord.

Spinal ganglion

Pain and temperature

Anterolateral system spinothalamic tract

Figure 7.7
Longitudinal diagram of the spinothalamic tract, which carries pain and temperature as well as nondiscriminative touch, to the primary somatosensory cortex. VPL = ventral posterolateral nucleus.

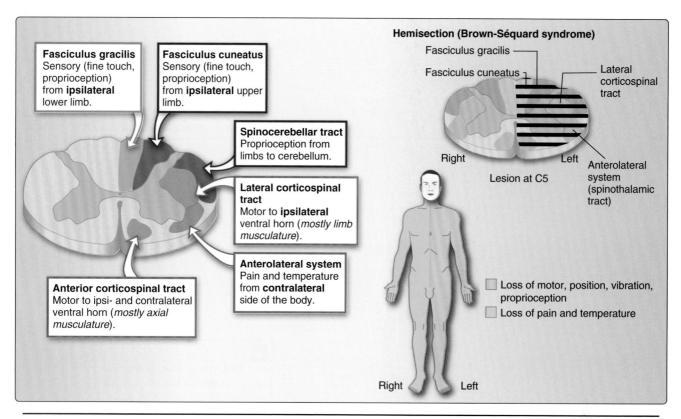

Figure 7.8
Sensory loss in Brown-Séquard syndrome.

IV. THE SPINOCEREBELLAR TRACTS

The spinocerebellar tracts relay information about limb position and movement to the cerebellum. To do this effectively, **proprioceptive** information from muscles, tendons, and joints and **exteroceptive** information from skin receptors are integrated to allow for a complete picture of the body in motion. Movement of the body is sensed not only through proprioceptors but through movement of the overlying skin as well, which is why all of this information needs to travel to the cerebellum.

There are several different pathways to the cerebellum. The major ones are the **posterior spinocerebellar tract**, which carries information from the lower limb, and the **cuneocerebellar tract**, which is the upper limb equivalent of the posterior spinocerebellar tract. Two other pathways play a smaller role in the control of movement: The **anterior spinocerebellar tract** integrates proprioceptive information from the lower limb with descending input, and the **rostral spinocerebellar tract** integrates information from the upper limb with descending input.

Together, these four tracts convey a complete picture of the body in motion to the cerebellum, where that information is then used to fine-tune and adjust movement as well as to facilitate motor learning.

A. The posterior spinocerebellar tract

The posterior spinocerebellar tract (Figure 7.9) carries proprioceptive and tactile information from the lower limb. Type Ia and II fibers from the muscle spindles, Ib fibers from Golgi tendon organs and joint

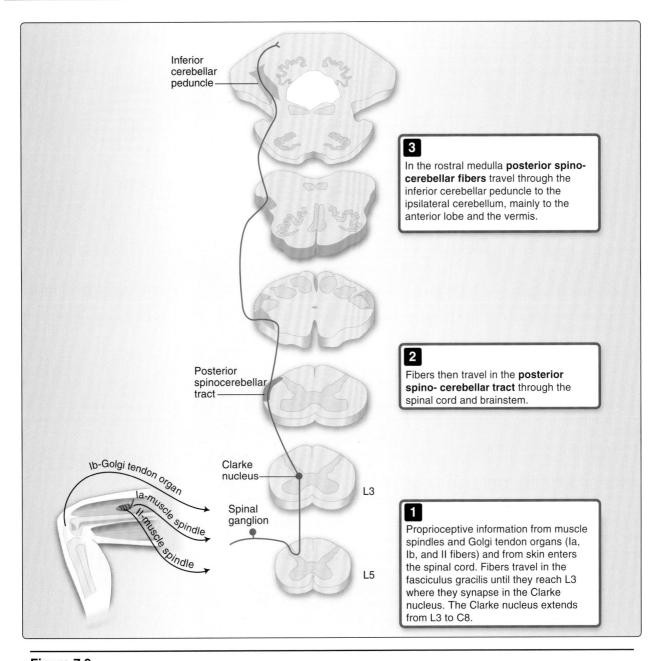

Inferior
cerebellar
peduncle

3

In the rostral medulla **posterior spino-cerebellar fibers** travel through the inferior cerebellar peduncle to the ipsilateral cerebellum, mainly to the anterior lobe and the vermis.

Posterior
spinocerebellar
tract

2

Fibers then travel in the **posterior spino- cerebellar tract** through the spinal cord and brainstem.

Ib-Golgi tendon organ

Ia-muscle spindle

II-muscle spindle

Clarke
nucleus

Spinal
ganglion

L3

L5

1

Proprioceptive information from muscle spindles and Golgi tendon organs (Ia, Ib, and II fibers) and from skin enters the spinal cord. Fibers travel in the fasciculus gracilis until they reach L3 where they synapse in the Clarke nucleus. The Clarke nucleus extends from L3 to C8.

Figure 7.9
Longitudinal diagram of the posterior spinocerebellar tract, which carries nonconscious proprioception to the cerebellum.

receptors (all proprioceptors), and type II and III fibers from cutaneous receptors (exteroceptors) enter the spinal cord at their respective levels through the posterior root and synapse in Clarke nucleus, which extends from C8 to L3.

1. **Clarke nucleus:** Fibers from below L3 travel in the fasciculus gracilis to L3, where they then synapse in Clarke nucleus. From Clarke nucleus, fibers travel in the ipsilateral posterior spinocerebellar tract to the brainstem, where they enter the cerebellum through the inferior cerebellar peduncle and project mainly to the ipsilateral anterior lobe of the cerebellum with projections to the vermis and paravermis of the ipsilateral posterior lobe. See Chapter 5, "The Spinal Cord," and Chapter 17, "The Cerebellum," for more information.

2. **Lower limb position and movement:** The sum of information carried in the posterior spinocerebellar tract and, in particular, information processed in Clarke nucleus gives the cerebellum feedback and information on the **position** and the **movement** of the lower limb. Neurons in Clarke nucleus have widespread connectivity and can thus integrate information about the entire limb. Together, information about muscle length, muscle force, and proprioception is processed and compared to reference points of a "steady-state" or "in-motion" value. This is integrated with the information from exteroceptors in the skin, which results in complete sensory data on the lower limb motion in a particular environment. Surprisingly, recent studies have shown that descending inputs from the *corticospinal tract* also modulate incoming proprioceptive information through local spinal circuits. Integration of information about the entire limb in Clarke nucleus streamlines the information going to the cerebellum by filtering out superfluous input. Thus, we now know that a large part of the processing of proprioceptive information related to motor planning and evaluation that was thought to occur in the cerebellum actually occurs at the level of Clarke nucleus.

B. The cuneocerebellar tract

The **cuneocerebellar tract** (Figure 7.10) has functions analogous to the posterior spinocerebellar tract but carries information about the **upper limb** to the cerebellum. **Proprioceptive** and **exteroceptive** information enters the spinal cord through the posterior horn from C8

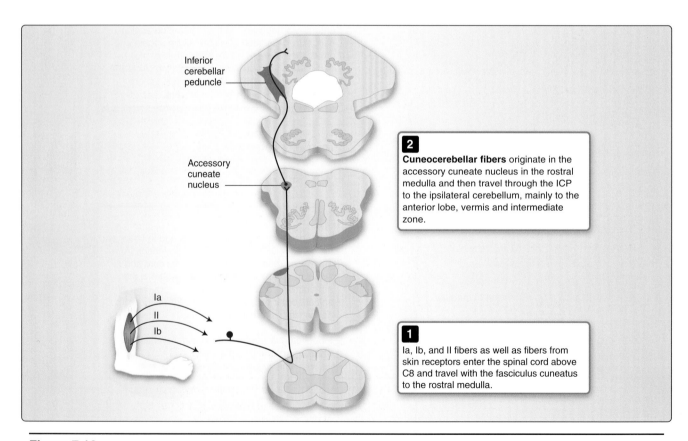

Inferior cerebellar peduncle

Accessory cuneate nucleus

Ia
II
Ib

2
Cuneocerebellar fibers originate in the accessory cuneate nucleus in the rostral medulla and then travel through the ICP to the ipsilateral cerebellum, mainly to the anterior lobe, vermis and intermediate zone.

1
Ia, Ib, and II fibers as well as fibers from skin receptors enter the spinal cord above C8 and travel with the fasciculus cuneatus to the rostral medulla.

Figure 7.10
Longitudinal diagram of the cuneocerebellar tract, which carries nonconscious proprioception to the cerebellum. ICP = inferior cerebellar peduncle.

to C1. Unlike the fibers that end up in the posterior spinocerebellar tract, however, they do not synapse in the spinal cord, because the Clarke column does not extend to the cervical levels. Instead, these fibers travel with the **fasciculus cuneatus** to the medulla where they synapse in the **accessory cuneate nucleus**. From there, the fibers travel in the **ipsilateral cuneocerebellar tract** to the inferior cerebellar peduncle and, like fibers in the posterior spinocerebellar tract, terminate mainly in the ipsilateral anterior lobe of the cerebellum with projections to the vermis and paravermis of the ipsilateral posterior lobe.

The processing in the accessory cuneate nucleus is analogous to the processing in the Clarke column. The cerebellum therefore receives information about the upper limb in motion in the environment, enabling it to fine-tune and adjust the movement immediately.

C. The anterior spinocerebellar tract

The **anterior spinocerebellar tract** (Figure 7.11) has an anatomical organization and function somewhat different from those of the posterior spinocerebellar and cuneocerebellar tracts. It arises from a set of neurons in the **anterior horn**, located on its anterolateral border,

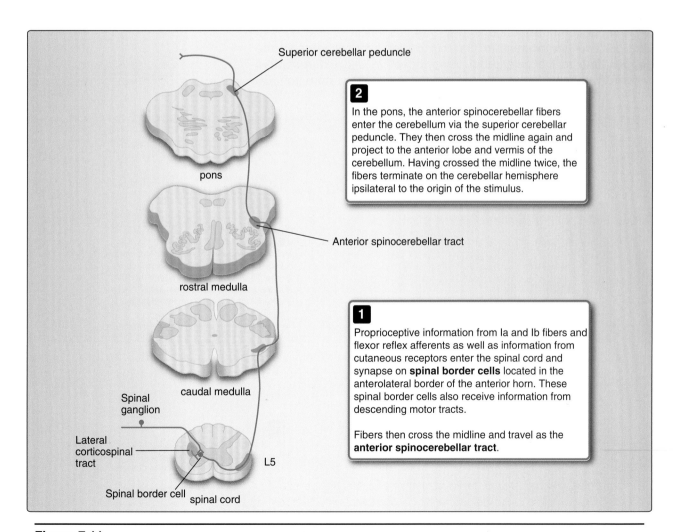

Superior cerebellar peduncle

pons

2

In the pons, the anterior spinocerebellar fibers enter the cerebellum via the superior cerebellar peduncle. They then cross the midline again and project to the anterior lobe and vermis of the cerebellum. Having crossed the midline twice, the fibers terminate on the cerebellar hemisphere ipsilateral to the origin of the stimulus.

Anterior spinocerebellar tract

rostral medulla

1

Proprioceptive information from Ia and Ib fibers and flexor reflex afferents as well as information from cutaneous receptors enter the spinal cord and synapse on **spinal border cells** located in the anterolateral border of the anterior horn. These spinal border cells also receive information from descending motor tracts.

Fibers then cross the midline and travel as the **anterior spinocerebellar tract**.

caudal medulla

Spinal ganglion

Lateral corticospinal tract

L5

Spinal border cell   spinal cord

Figure 7.11
Longitudinal diagram of the anterior spinocerebellar tract, which integrates proprioception with descending motor modulation and feeds this information back to the cerebellum.

called **spinal border cells**, extending through the **lumbar segments** of the spinal cord. The axons of these neurons *cross the midline* and ascend as the anterior spinocerebellar tract to the cerebellum, where the majority of the fibers *cross again* terminating on the side of the cerebellum ipsilateral to the original peripheral input (again, primarily in the vermis and paravermis of the **anterior** and **posterior lobes**).

The spinal border cells receive input from lower limb muscles (Ia from muscle spindles and Ib from Golgi tendon organs), from **modulating descending tracts** to the lower motor neurons, and from **flexor reflex arcs** in the spinal cord. These descending pathways adjust the output from the lower motor neurons independently from the influences of the corticospinal tract. It is interesting that these cells are located in the anterior horn, which contains the lower motor neurons, and that they receive both sensory and motor input. They use this information to compare the situation in the periphery (through the proprioceptors) with the modulatory input coming to the lower motor neurons. They can then feed this integrated information back to the cerebellum and adjust the descending motor information in real time. The sum of the input to the spinal border cells gives information to the cerebellum about the **postural stability** of the lower limb, which is of particular importance in the coordination and stability of upright gait.

D. The rostral spinocerebellar tract

The **rostral spinocerebellar tract** (Figure 7.12) is the upper limb equivalent of the anterior spinocerebellar tract. Sensory afferents synapse with neurons in the anterior horn, and the second-order fibers travel **uncrossed** (ipsilaterally) to the cerebellum through the superior cerebellar peduncle. The inputs and functions of this tract are the same as those of the anterior spinocerebellar tract—proprioceptive information from the upper limb and descending modulating tracts converge on the anterior horn neurons, and the information is then fed back to the cerebellum.

E. Lesions of the spinocerebellar tracts

Lesions of the spinocerebellar tracts lead to **ataxias**, or loss of muscle coordination, due to a loss of proprioceptive input to the cerebellum. The coordination of movement and the adjustment of movement due to input from ongoing stimuli are no longer possible. Clinically, however, the spinocerebellar tracts are rarely, if ever, damaged in isolation. For example, in a patient with a hemisection of the spinal cord (Brown-Séquard syndrome; see Clinical Application Box 7.1), one would expect ataxia (loss of muscle coordination) due to a lesion of the spinocerebellar tracts. Ataxia does indeed occur but is typically masked by the weakness or hemiplegia resulting from damage to the descending lateral corticospinal tract. Even in the family of inherited diseases known as the spinocerebellar atrophies, damage to tracts other than the spinocerebellar tracts appears to occur. In Friedreich ataxia, the most common of the spinocerebellar disorders, the patient does experience a lack of coordination, which is consistent with damage to the spinocerebellar tracts and neurons in the Clarke column that is known to occur. In addition, however, degenerative changes in the posterior columns, corticospinal tracts, and spinal ganglion cells are observed, resulting in disturbed tactile sensation and proprioception as well as muscle weakness and even paralysis.

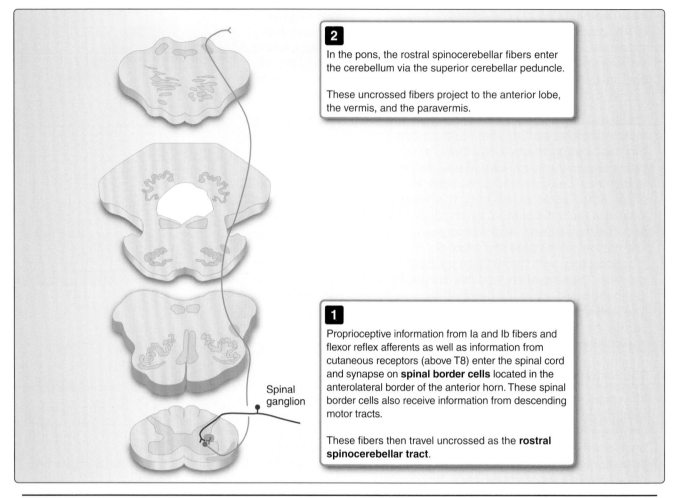

2

In the pons, the rostral spinocerebellar fibers enter the cerebellum via the superior cerebellar peduncle.

These uncrossed fibers project to the anterior lobe, the vermis, and the paravermis.

1

Proprioceptive information from Ia and Ib fibers and flexor reflex afferents as well as information from cutaneous receptors (above T8) enter the spinal cord and synapse on **spinal border cells** located in the anterolateral border of the anterior horn. These spinal border cells also receive information from descending motor tracts.

These fibers then travel uncrossed as the **rostral spinocerebellar tract**.

Spinal ganglion

Figure 7.12

Longitudinal diagram of the rostral spinocerebellar tract, which integrates proprioception with descending motor modulation and feeds this information back to the cerebellum.

Clinical Case:

John's Unsteady Gait

John is a 55-year-old male presenting with progressively unsteady gait with paresthesia (abnormal sensations without external stimuli, e.g., tingling, numbness), dysesthesia (abnormal sensation to touch), and pain radiating from his back into his legs (radicular pain). On examination, he has normal sensation in his face and arms. He has reduced vibration and impaired proprioception up to his hips. Sensation to pinprick and temperature are normal. His reflexes are absent in his legs and reduced in his arms. His strength and tone are normal. He is unsteady and nearly falls with his eyes closed. He looks at his feet when walking.

Case Discussion

The lesion is located in the posterior column and posterior root entry zone. The deficits in vibration and proprioception upon clinical examination are indicative of a posterior column lesion. The dysesthesia with radicular pain from the

back into the legs is indicative of a posterior root entry zone lesion of the nerve roots in the lumbosacral region. When there is a proprioceptive deficit, balance can still be maintained by visual input. When eyes are closed, visual input is removed, which results in instability. The clinical test for this is the Romberg test. The patient is asked to stand with feet together and eyes closed. A positive Romberg sign refers to excessive swaying or falling when the eyes are closed.

What infectious disease causes impairment of the posterior columns?

Neurosyphilis causes multiple deficits. Here, we focus on tabes dorsalis with sensory ataxia, loss of sensation to vibration and proprioception, and loss of deep tendon reflexes (Figure 7.13). Neurosyphilis typically does not appear for years after the initial infection. The posterior roots are particularly involved resulting in degeneration of the posterior columns (fasciculus gracilis and fasciculus cuneatus). Cerebrospinal fluid from a lumbar puncture would show lymphocytic infiltration. Treatment consists of intravenous antibiotics.

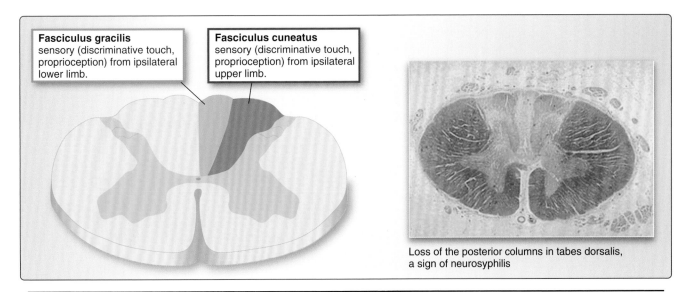

Figure 7.13
Tabes dorsalis.

Chapter Summary

- The sensory systems ascending through the spinal cord and brainstem can be divided into three major systems: the posterior column–medial lemniscus system, the anterolateral system, and the spinocerebellar tracts.

- The **posterior column–medial lemniscus** system carries discriminative touch, vibration, pressure, and proprioception. The sensory fibers in this tract are associated with specific sensory endings that are transducers for the mechanical stimuli. These fibers travel uncrossed in the posterior columns of the spinal cord, synapse in their respective nuclei in the caudal medulla, cross to the contralateral side in the internal arcuate fibers, and ascend through the brainstem as the medial lemniscus (Figure 7.14). After terminating in the ventral posterolateral nucleus of the thalamus, information is relayed to the primary somatosensory cortex, where conscious appreciation of this sensory information occurs.

- The **anterolateral system** carries pain, temperature, and nondiscriminative touch. The sensory fibers in this tract all have free nerve endings in the periphery. These fibers enter the posterior horn and ascend or descend several spinal cord levels before synapsing. Processing and modulation of pain occur in the posterior horn, with a major role for the substantia gelatinosa. Postsynaptic fibers then cross to the contralateral side in the anterior white commissure and ascend through the spinal cord and brainstem as the anterolateral system. Subsets of fibers synapse with various structures in the brainstem and diencephalon for pain modulation, whereas the spinothalamic fibers synapse in the ventral posterolateral nucleus of the thalamus. From the thalamus, information is relayed to the cortex, where conscious awareness of this sensory information occurs (see Figure 7.14).

- The **spinocerebellar tracts** carry both proprioceptive and exteroceptive information from the periphery and ascend through the spinal cord and brainstem to the cerebellum (Figure 7.15). This information is integrated with both ascending and descending inputs, and modulation of the proprioceptive afferents occurs. This gives the cerebellum information about joint position and muscle activity (posterior spinocerebellar and cuneocerebellar tracts) as well as integrated information about the influence of descending tracts on joint position and muscle activity (anterior and rostral spinocerebellar tracts) and allows the cerebellum to adjust and fine-tune motor output.

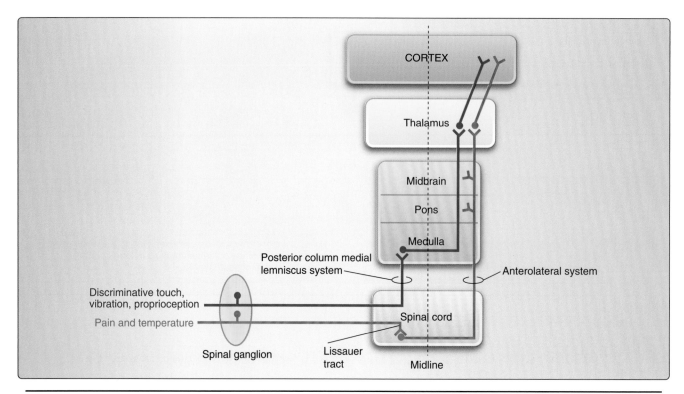

Figure 7.14
Review of conscious sensory input and tracts to the cortex.

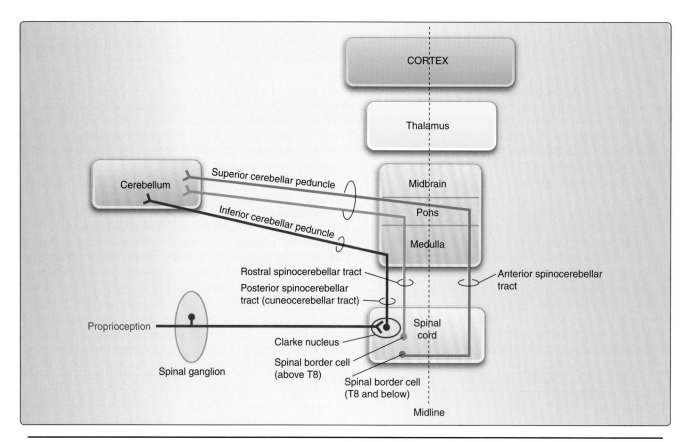

Figure 7.15
Review of nonconscious sensory input and tracts to the cerebellum.

Study Questions

Choose the ONE best answer.

7.1 A patient presents with selective loss of pain and temperature sensations to both hands. Other sensory modalities and voluntary motor activity are intact. What is the most likely cause of this problem?

    A. A lesion to the left and right anterolateral white columns at the level of the cervical spinal cord.

    B. A lesion to the left and right anterolateral white columns at the level of the caudal medulla.

    C. A lesion around the central canal at the cervical level of the spinal cord, extending into the white matter anteriorly.

    D. A lesion to both left and right branches of the anterior spinal artery at the level of the cervical spinal cord.

    E. A lesion to both left and right branches of the anterior spinal artery at the level of the caudal medulla.

Correct answer is C. This is a case of syringomyelia, a central cavitation of the spinal cord beginning around the central canal. It can result from inflammation and other causes. The cavitation can then extend further into the central gray matter, affecting the anterior gray horns, and/or can extend into the white matter. In this case, the cavitation has extended into the white matter anterior to the central canal, disrupting the crossing fibers of the left and right anterolateral or spinothalamic tracts. At the cervical level of the cord, specifically C7–C8, this would result in loss of pain and temperature sensations to the hands. The other choices all would involve a fairly large bilateral lesion, which would present with additional symptoms. For a review of the blood supply to the spinal cord and brainstem, see Chapters 5 and 6.

7.2 Which statement is true of sensory information on discriminative touch, vibration, and proprioception that enters the spinal cord?

    A. It ascends ipsilaterally until reaching the pons.

    B. It crosses the midline in the caudal medulla.

    C. It has second-order neurons in the Clarke column.

    D. It synapses in the posterior horn before ascending.

    E. It ultimately synapses in the ipsilateral primary and association sensory cortices.

Correct answer is B. Cell bodies of the first-order neurons are in the spinal ganglia. Information enters the spinal cord and ascends without synapsing to terminate in the nuclei gracilis and cuneatus in the caudal medulla. From there, second-order neurons cross the midline as the internal arcuate fibers and ascend to the thalamus. From there, fibers travel to the primary somatosensory cortex. The Clarke column contains second-order neurons whose processes ascend as the posterior spinocerebellar tract. There are no major synapses in the posterior horn for fibers carrying touch, proprioception, and vibration.

7.3 A 76-year-old woman complains that her right hand feels a bit numb and "clumsy." When she reaches for a glass of water, for example, she sometimes knocks it over. Which one of the following arteries is most likely to be occluded by an embolus in this scenario?

    A. Branches of the right posterior spinal artery at the level of the cervical spinal cord.

    B. Branches of the right anterior spinal artery at the level of the cervical spinal cord.

    C. Branches of the right posterior inferior cerebellar artery at the level of the rostral medulla.

    D. Branches of the left vertebral artery at the level of the caudal medulla.

    E. Right paramedian branches of the basilar artery at the level of the caudal pons.

Correct answer is A. Branches of the right posterior spinal artery supply the posterior column tracts (i.e., fasciculus gracilis and fasciculus cuneatus), which carry fine touch, proprioception, and vibration sense. Information for the hand related to these modalities is carried by the fasciculus cuneatus. A lesion of the branches of the posterior spinal artery supplying the fasciculus cuneatus at the level of the cervical spinal cord could result in loss of discriminative touch and proprioception, which, in turn, could result in the feelings of numbness and clumsiness. Branches of the right anterior spinal artery at the level of the cervical spinal cord supply the right anterior two-thirds of the cord and would not affect the posterior column tracts. Branches of the right posterior inferior cerebellar artery at the level of the rostral medulla supply the posterolateral area of the medulla. Because the medial lemniscus, which is carrying fine touch and proprioception, travels along the midline of the medulla, it would be unaffected by a lesion of the right posterior inferior cerebellar artery. Because branches of the left vertebral artery at the level of the caudal medulla supply the lateral area of the medulla, this lesion would not affect the medial lemniscus. Right paramedian branches of the basilar artery at the level of the caudal pons do supply the midline area where the medial lemniscus now lies. However, information from the right hand would now be traveling on the left side of the brainstem, and thus, a lesion of the right side would not cause a problem with the right hand. For a review of the blood supply to the spinal cord and brainstem, see Chapters 5 and 6.

7.4 Which of the following statements about the spinocerebellar tracts is correct?

A. Proprioceptive and exteroceptive information from the lower limb is carried to the cerebellum in the cuneocerebellar tract.

B. The rostral spinocerebellar tract integrates information from the lower limb with descending input.

C. The Clarke nucleus plays an important role in integrating proprioceptive information from the upper and lower limbs.

D. The anterior spinocerebellar tracts project directly to the side of the cerebellum ipsilateral to the original peripheral input.

E. Spinal border cells convey information to the cerebellum about the postural stability of the lower limb.

Correct answer is E. Spinal border cells are unique cells in the anterior horn of the spinal cord. They receive input from lower limb muscles as well as input from modulating descending tracts to the lower motor neurons and from flexor reflex arcs in the spinal cord and can feed this integrated information back to the cerebellum. The sum of the input to the spinal border cells gives information to the cerebellum about the postural stability of the lower limb. Proprioceptive and exteroceptive information from the lower limb is carried to the cerebellum in the posterior spinocerebellar tract. The rostral spinocerebellar tract integrates information from the upper limb with descending input. The Clarke nucleus receives proprioceptive and exteroceptive information about the lower limb and provides feedback to the cerebellum on the position and the movement of the lower limb. The anterior spinocerebellar tracts originate from the spinal border cells. Axons of these neurons *cross the midline* and ascend to the cerebellum, where the majority of the fibers *cross again*, terminating on the side of the cerebellum ipsilateral to the original peripheral input.

7.5 A patient presents with a large glioma that has invaded the rostral medulla on the right side. Which one of the following sets of symptoms would result from this lesion?

A. Loss of touch, vibration, and proprioception, as well as pain and temperature on the right side of the body.

B. Loss of touch, vibration, and proprioception as well as pain and temperature on the left side of the body.

C. Loss of touch, vibration, and proprioception on the right side of the body and loss of pain and temperature on the left side of the body.

D. Loss of touch, vibration, and proprioception on the left side of the body and loss of pain and temperature on the right side of the body.

E. Loss of discriminative touch, vibration, and proprioception on the left side of the body but intact pain and temperature sensations.

Correct answer is B. Information on touch, vibration, and proprioception from the left side of the body enters the spinal cord and ascends in the posterior columns, synapsing in the left nuclei gracilis and cuneatus. This information then crosses the midline in the caudal medulla and travels in the right medial lemniscus. Information on pain and temperature from the left side of the body synapses in the left posterior horn shortly after entering the spinal cord. Second-order fibers then cross the midline and travel in the anterolateral aspect of the spinal cord and brainstem. Thus, information about discriminative touch, vibration, and proprioception as well as information about pain and temperature are both traveling on the right at the level of the rostral medulla, and a lesion on the right side of the rostral medulla will interrupt all of this ascending information that originated on the left side of the body. The other choices are all incorrect because they contain incorrect information about the side of the body or the modality affected by the lesion.

Descending Motor Tracts

8

I. OVERVIEW

Motor activity is controlled by two separate and equally important systems: the upper motor neuron (UMN) system and the lower motor neuron (LMN) system. This division into UMN and LMN systems reflects the anatomy and function of these systems and is clinically useful for observing and describing lesions.

The term UMN refers to the sum of supraspinal (above or *rostral* to the spinal cord) descending influences on the LMNs. Descending motor tracts can arise from the **cortex (corticospinal and corticobulbar)** and from the **brainstem (vestibulospinal, reticulospinal, rubrospinal, and tectospinal)** (Figure 8.1). Descending tracts from the cortex and brainstem act in parallel and thus function as part of a large network rather than as multiple independent inputs to the LMNs.

LMNs are located in the anterior horn of the spinal cord and the motor nuclei of cranial nerves in the brainstem. They are organized in circuits in order to sort and regulate the multitude of descending influences so that activity of truncal, proximal, and distal muscle groups is coordinated and results in fluid movement (see Figure 8.1).

Table 8.1 provides an overview of the descending motor tracts, their targets, and their main functions.

Movement is a composite of **voluntary** and **involuntary** motor activity. **Voluntary movement** is mostly directed at the distal muscle groups, which are involved in functions such as gait in the lower limb and the intricate movements of the hand in the upper limb. Movement of the extremities must be fine-tuned, and each side of the body must be able to act completely independently. LMN innervation, therefore, must be precise and without overlap or redundancy. This is mediated mainly through the **lateral corticospinal tract.**

Involuntary movement involves mainly proximal and truncal muscle groups and is critical for postural stability. **Postural adjustments** *related to movement of the extremities* are under the influence of the **anterior corticospinal tract** (Figure 8.2).

Postural stability is mediated via **phylogenetically older pathways**, meaning that from an evolutionary standpoint, they bear resemblance to the nervous system structures of much older, more primitive organisms. These include the **vestibulospinal, reticulospinal, rubrospinal, and**

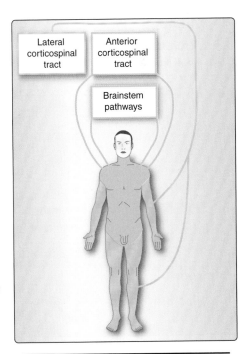

Figure 8.1
Descending motor tracts from the cortex and the brainstem.

Table 8.1: Descending Motor Tracts and Their Functions

Tract	Origin	Target	Crosses Midline	Function
From Cortex				
Lateral corticospinal tract	Cortex, primary motor, premotor, supplementary motor, and sensory	LMN in the spinal cord	Yes, in caudal medulla and pyramidal decussation	Skilled voluntary movements of the distal extremities
Anterior corticospinal tract	Cortex, primary motor, premotor, supplementary motor, and sensory	LMN in the spinal cord	Yes, at spinal segments, often bilateral	Posture adjustments to compensate for voluntary movements of the distal extremities
Corticobulbar tract	Cortex, mainly primary motor	LMN in brainstem nuclei	Mostly bilateral innervation, some exceptions	Motor control of the head and neck
From Brainstem				
Medial vestibulospinal tract	Medial vestibular nuclei	LMN in the spinal cord	Bilateral to neck muscles	Postural adjustment in response to changes in balance
Lateral vestibulospinal tract	Lateral vestibular nuclei	LMN in the spinal cord	Ipsilateral	Facilitates extensors and inhibits flexors Postural adjustment in response to changes in balance
Reticulospinal tract	Reticular formation	LMN in the spinal cord	Mostly ipsilateral and some bilateral projections	Changes in muscle tone associated with voluntary movement and postural stability
Rubrospinal tract	Red nucleus	LMN in the spinal cord	Yes, in midbrain	Flexor movements in the upper limb
Tectospinal tract	Superior colliculus	LMN in cervical spinal cord	Bilateral	Orientation of the head and neck in response to eye movements

LMN = lower motor neuron.

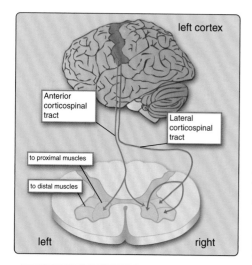

Figure 8.2
Descending motor tracts from the cortex and the spinal cord targets they innervate.

tectospinal tracts, which arise from the brainstem (Figure 8.3). These brainstem pathways receive input from the corticospinal pathways, so that all postural adjustments can occur simultaneously with and are coordinated with voluntary movements.

Truncal stability *with or without voluntary movement* needs to be coordinated along the entire trunk. Thus, innervation of the trunk musculature is often **bilateral**, and redundancies extend over many spinal cord segments.

An understanding of these descending pathways, their targets, and their function is critical in analyzing lesions following injury to these pathways as well as determining a prognosis for recovery of function.

II. DESCENDING PATHWAYS FROM THE CORTEX

The main function of the pathways descending from the cortex is to influence LMNs in the anterior horn of the spinal cord or in the motor nuclei of cranial nerves in the brainstem. These descending pathways include the

lateral and **anterior corticospinal** tracts, also known as the **pyramidal tracts**, which target LMNs in the spinal cord, and the **corticobulbar** tract, which targets motor nuclei of cranial nerves in the brainstem. These descending cortical pathways are **crossed** for the innervation of the **limbs** and **bilateral** for the innervation of the **trunk**.

Beyond their main function of influencing LMNs, the descending tracts have a number of additional functions that are important in the control of movement. They provide gating for **spinal reflexes**; they are responsible for the descending **influences on afferent (ascending, sensory) systems**; and they perform a **trophic**, or nutritive, function for the neuron groups they contact.

A. Lateral corticospinal tract

The cell bodies of the lateral corticospinal tract are located quite broadly in the cortex. Interestingly, only one-third of these fibers originates in the primary motor cortex. The remaining fibers come from the frontal (premotor and supplementary motor) and parietal (primary sensory) areas (see Chapter 13, "The Cerebral Cortex"). Importantly, the primary motor cortex receives direct and indirect input from association areas. Thus, motor output from the cerebral cortex is influenced by sensory input, allowing movements to be guided by touch and other senses as well as higher cortical function.

The axons of these UMNs descend through the **corona radiata** and converge in the **posterior limb of the internal capsule**. From here, the fibers travel as distinct bundles through the **cerebral peduncles** and as dispersed bundles through the **basal pons**. Caudal to the pons, the fibers converge again to form the **pyramids** on the **anterior** surface of the **medulla**. In the caudal medulla, 85%–90% of the fibers cross the midline in the **pyramidal decussation** and continue to descend in the lateral column of the spinal cord to the LMNs in the anterior horn. Figure 8.4 depicts the course of the lateral corticospinal tract.

1. **Function:** The **target neurons** for the **lateral corticospinal tract** are located in the lateral portion of the **anterior horn**. They are the LMNs responsible for the innervation of the **distal muscle groups** (Figure 8.5). This innervation is a direct contact between an axon originating in the cortex and a motor neuron directed to a specific muscle, which allows for precise skilled movements.

   The lateral corticospinal tract is also needed for cortical control of the movement of an **entire limb**. This is achieved through contact with LMN circuits or **central pattern generators**, which are a network of interneurons that integrate motor and sensory input and innervate LMN groups that will, in turn, innervate flexor and extensor muscles over several segments. These are important networks for automated movement cycles such as the gait cycle.

   Together, the inputs of the corticospinal tract to LMN circuits allow for precise skilled movements and synergistic movement of an entire limb, both of which are necessary for complex, skilled motor activities.

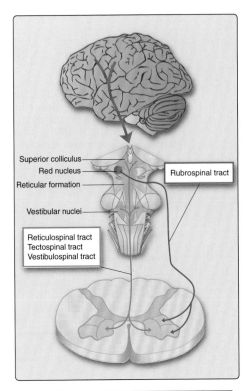

Figure 8.3
Descending motor tracts from the brainstem and the spinal cord targets they innervate.

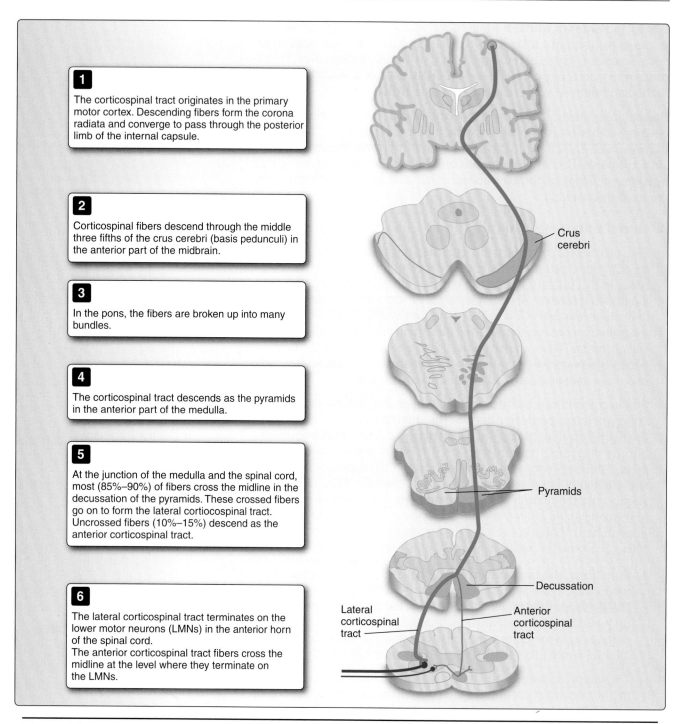

1

The corticospinal tract originates in the primary motor cortex. Descending fibers form the corona radiata and converge to pass through the posterior limb of the internal capsule.

2

Corticospinal fibers descend through the middle three fifths of the crus cerebri (basis pedunculi) in the anterior part of the midbrain.

3

In the pons, the fibers are broken up into many bundles.

4

The corticospinal tract descends as the pyramids in the anterior part of the medulla.

5

At the junction of the medulla and the spinal cord, most (85%–90%) of fibers cross the midline in the decussation of the pyramids. These crossed fibers go on to form the lateral cortiocospinal tract. Uncrossed fibers (10%–15%) descend as the anterior corticospinal tract.

6

The lateral corticospinal tract terminates on the lower motor neurons (LMNs) in the anterior horn of the spinal cord.
The anterior corticospinal tract fibers cross the midline at the level where they terminate on the LMNs.

Crus cerebri

Pyramids

Decussation

Lateral corticospinal tract

Anterior corticospinal tract

Figure 8.4
Corticospinal tracts.

The same LMNs contacted by the lateral corticospinal tract are also innervated by descending tracts from the brainstem. Whereas the lateral corticospinal tract activates these neuron groups for precise, skilled, voluntary movements of the limbs, the brainstem tracts activate them predominantly for muscle tone and postural adjustments.

2. **Somatotopic arrangement:** Throughout its course, the corticospinal tract features a **somatotopic arrangement** of fibers from medial to lateral: arm, trunk, and leg. In the cortex, this somatotopy results in a cortical map of the body indicating the location and amount of cortex associated with each body part, known as a **motor homunculus** (see Chapter 13, "The Cerebral Cortex," for more detail).

B. Anterior corticospinal tract

The **anterior corticospinal tract** originates in the same cortical areas as the lateral corticospinal tract and descends together with it. However, anterior corticospinal fibers do not cross, or decussate, to the contralateral side in the pyramidal decussation (see Figures 8.4 and 8.5). Rather, these fibers continue to descend on the **ipsilateral**, or *same*, side of the spinal cord, and the majority of fibers then cross over at the segmental level at which they will terminate. Anterior corticospinal fibers terminate on LMNs controlling trunk and proximal musculature, and innervation from the anterior corticospinal tract is often **bilateral**, that is, to *both sides* of the body. Below L2, the only innervations to LMNs are from the lateral corticospinal tract, and the myotomes are all related to the lower limb (no more trunk).

The anterior corticospinal tract is important for **ipsilateral postural adjustments** when voluntary movements are performed with the contralateral extremity. For example, to perform a precise movement with the hand, such as screwing in a light bulb, motor input comes from the contralateral cortex through the lateral corticospinal pathway. At the same time, bilateral cortical input activates truncal and proximal muscle groups to stabilize the body during this activity (Figure 8.6).

C. Corticobulbar tract

The corticobulbar tract arises primarily from areas of motor cortex related to the head and face and descends through the **corona radiata**. Fibers converge primarily in the **genu** (but may also occupy the most anterior part of the posterior limb) of the **internal capsule** from which they then descend together with corticospinal fibers. Figure 8.7 shows the location of descending fibers in the internal capsule.

Corticobulbar fibers terminate on the **motor nuclei of the brainstem**. Most of the innervation to the cranial nerve motor nuclei is **bilateral**, with a few exceptions that are discussed in subsequent chapters that deal with the specific cranial nerves. Figure 8.8 gives an overview of the descending corticobulbar tract and its targets throughout the brainstem.

An additional component of the corticobulbar tract is a subset of cortical fibers that innervate brainstem nuclei that will exert descending influences on spinal cord neurons. These **descending cortical fibers** terminate in the **reticular formation**, the **superior colliculus**, and the **red nucleus**. These connections are thought to coordinate the cortical and brainstem motor systems and are discussed further in the section below.

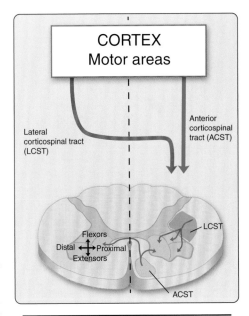

Figure 8.5
Spinal cord projections of the lateral and anterior corticospinal tracts.

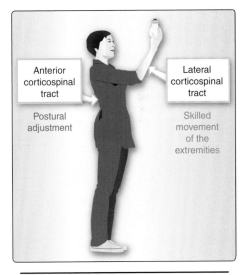

Figure 8.6
Postural adjustments in response to voluntary movements through the lateral corticospinal tract are mediated by the anterior corticospinal tract.

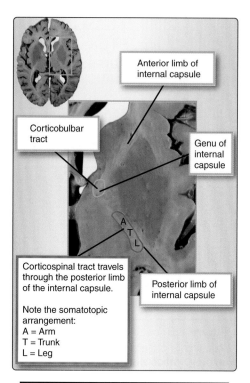

Figure 8.7
Horizontal section through the forebrain illustrating the internal capsule. The corticospinal tracts travel through the posterior limb of the internal capsule, whereas the corticobulbar tracts travel through the genu of the internal capsule.

III. DESCENDING PATHWAYS FROM THE BRAINSTEM

The **descending motor pathways** from the brainstem include the **vestibulospinal**, **reticulospinal**, **rubrospinal**, and **tectospinal** tracts. They are **phylogenetically** older tracts and are involved mainly in the control of **proximal** and **truncal musculature**. These pathways coordinate **body movement** and assure **postural stability** during voluntary movements and changes in body position so that a purposeful and coordinated motor behavior can result.

A. Vestibulospinal tracts

The **vestibulospinal system** originates in the **vestibular nuclei** located in the caudal pons/rostral medulla and comprises **medial** and **lateral vestibulospinal tracts**. The **medial vestibulospinal tract**, also called the **descending medial longitudinal fasciculus**, arises mainly in the **medial vestibular nucleus**. It descends **bilaterally** through the brainstem and travels in the anterior white columns of the spinal cord where it terminates on LMN circuit neurons of cervical and upper thoracic levels. The medial vestibulospinal tract influences motor neurons controlling the neck muscles, is responsible for stabilizing the head as we move our bodies or as our head moves in space, and also plays a role in coordinating head movements with eye movements. The **lateral vestibulospinal tract** arises in the **lateral vestibular nucleus**. It descends **ipsilaterally** in the **anteromedial** area of the brainstem and travels in the anterior white column of the spinal cord (Figure 8.9).

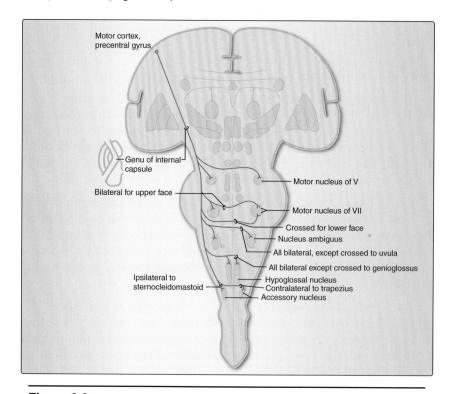

Figure 8.8
Corticobulbar tract: overview of the projections from the cortex to the motor nuclei of the cranial nerves. (Modified from Haines DE. *Neuroanatomy: An Atlas of Structures, Sections, and Systems*, 7th ed. Baltimore MD: Lippincott Williams & Wilkins, 2007.)

The fibers of the lateral vestibulospinal tract terminate at all levels of the ipsilateral spinal cord to facilitate the activity of the extensor muscles and inhibit the activity of the flexor muscles. Through these actions, this tract serves to mediate postural adjustments to compensate for movements and changes in position of the body and to coordinate the orientation of the head and body in space. For example, if you whirl around quickly and start to feel dizzy, activation of the lateral vestibulospinal tract will help you adjust your posture so you do not fall. Relevant to this, the vestibular nuclei receive and interpret information about balance from the **inner ear**. Direct output to spinal cord neurons ensures that **postural adjustments** can be made immediately when a change in position of the head is detected by the inner ear. The lateral vestibulospinal tract also influences muscle tone (see Chapter 11, "Hearing and Balance").

B. Reticulospinal tract

The **reticulospinal tract** is the most primitive descending motor system and arises from cell bodies in the reticular formation. Briefly, the reticular formation comprises a "meshwork" of neurons located mainly near the midline throughout the core or central portion of the brainstem. Whereas cells at many levels of the reticular formation may contribute to the reticulospinal tract, most of the fibers involved in somatomotor function originate in the **pontine and rostral medullary reticular formation**. They descend mainly **ipsilaterally** in the **anteromedial** area of the **brainstem** and the anterior white column of the spinal cord and terminate on LMN circuits throughout the length of the spinal cord (Figure 8.10). The reticulospinal tract coordinates muscle group activation for primitive motor behaviors such as the **orientation of the body** toward or away from a stimulus and motor behaviors that do not require dexterity. It also integrates distal muscle actions with proximal muscle actions and initiates changes in **muscle tone** related to voluntary movements of the limbs.

1. **Control of breathing:** One important function of the motor nuclei in the reticular formation is the **control of breathing**. Tonically active neurons in the medulla project to neurons in the spinal cord that will in turn activate the skeletal muscles involved in respiration. Voluntary control of these neuron groups involved in respiration occurs through cortical projections. This system is discussed in more detail in Chapter 12, "Brainstem Systems and Review."

2. **Emotional motor system:** The reticular formation also contains a network of neurons that use **monoaminergic neurotransmitters**, such as serotonin, noradrenaline, and dopamine. These neurons receive input from **limbic system** structures that have widespread connections, including descending projections to spinal cord motor neurons (see Chapter 20, "The Limbic System," for more information).

Together, these descending monoaminergic fibers comprise an **emotional motor system** that mediates the expression of emotion. For example, an individual with depression will exhibit posture and body language different from those of someone who is happy. The emotional motor pathways can determine the **excitability of**

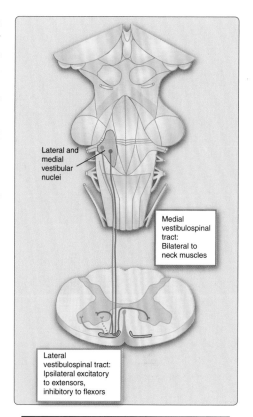

Lateral and medial vestibular nuclei

Medial vestibulospinal tract: Bilateral to neck muscles

Lateral vestibulospinal tract: Ipsilateral excitatory to extensors, inhibitory to flexors

Figure 8.9
Vestibulospinal tract.

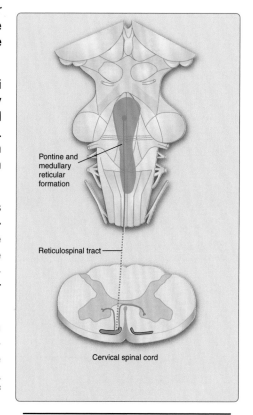

Pontine and medullary reticular formation

Reticulospinal tract

Cervical spinal cord

Figure 8.10
Reticulospinal tract.

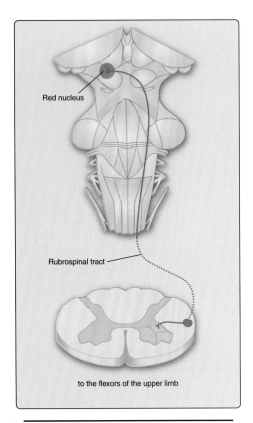

Figure 8.11
Rubrospinal tract.

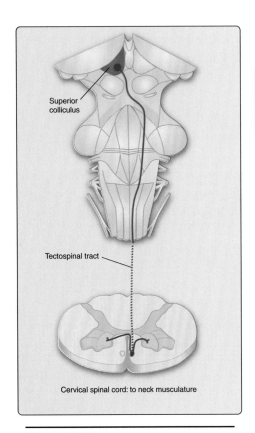

Figure 8.12
Tectospinal tract.

spinal motor neurons and are important contributors to emotional body language. This change in excitability is what makes somebody "jumpy" when stressed or excited. The threshold for activation has been shifted through the influence of the emotional motor system. This is discussed in more detail in Chapter 18, "The Integration of Motor Control."

C. Rubrospinal tract

The **rubrospinal tract** originates in the **red nucleus** in the midbrain. Fibers decussate in the midbrain, travel through the brainstem tegmentum, and descend on the **contralateral side** in the lateral column of the spinal cord. Fibers terminate on LMNs that innervate the **flexors of the upper limb** (Figure 8.11).

In humans, the function of the rubrospinal tract is outweighed by the corticospinal pathway. The primary function of the red nucleus has shifted from a structure responsible for the motor control of the upper limbs to an important integrator and relay nucleus in cerebellar circuits, as discussed in further detail in Chapter 17, "The Cerebellum." With loss of the corticospinal tract, rubrospinal influence on upper limb flexors can become apparent (see Clinical Application Box 8.1).

D. Tectospinal tract

The **tectospinal tract** originates in the **superior colliculus** of the midbrain. Fibers travel contralaterally through the brainstem and in the anterior white columns of the spinal cord and project to the **cervical spinal cord** where they innervate motor neurons responsible for neck movements (Figure 8.12). This tract is responsible for orienting the head and neck during eye movements.

Clinical Application 8.1: Posturing in the Comatose Patient as a Diagnostic Tool

In a comatose patient, abnormal posturing is common because of the activity of descending motor tracts. The type of posturing can give an indication of the location of the lesion or the extent of the brain damage.

Posturing with both upper and lower limbs extended is referred to as **extensor posturing**. In this case, the lesion is at the level of the midbrain or rostral pons (see **A** in the figure). Importantly, the lesion includes the red nucleus and with it, the rubrospinal tract. In this case, all descending cortical systems (corticospinal to the spinal cord, corticorubral and corticoreticular to the brainstem) are interrupted. Thus, lower motor neurons receive no input from the corticospinal tract or the rubrospinal tract, whereas excitatory and inhibitory components of the reticular formation (reticulospinal tract) are intact but receive no cortical modulation or input. The extensor rigidity is likely the result of excessive input to spinal cord interneurons or circuit neurons via the reticulospinal tract.

Posturing with the upper limbs flexed and the lower limbs extended is referred to as **flexor posturing**. The lesion in this case is rostral to or above the red nucleus (see **B** in the figure). All descending input to both the spinal

Clinical Application 8.1: Posturing in the Comatose Patient as a Diagnostic Tool continued

cord and the brainstem is thus interrupted, but both the rubrospinal tract and the reticulospinal tract remain intact. The lower extremities show increased extensor tone due to the input from the reticulospinal tract, and the upper extremities are under the influence of the rubrospinal tract, which gives input to the neuron groups involved in flexor activity.

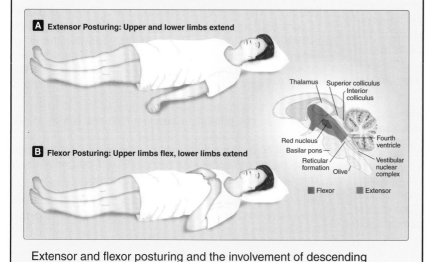

A Extensor Posturing: Upper and lower limbs extend

B Flexor Posturing: Upper limbs flex, lower limbs extend

Thalamus
Superior colliculus
Interior colliculus
Red nucleus
Basilar pons
Reticular formation
Olive
Fourth ventricle
Vestibular nuclear complex

■ Flexor   ■ Extensor

Extensor and flexor posturing and the involvement of descending fibers from the red nucleus.

Clinical Case:

Nicola's Gait Problems

Nicola is a 60-year-old female with a 5-month history of progressive gait difficulties with impaired dexterity in her right hand. She first noticed dragging of her left foot with tripping and falls. One month later, she developed a right foot drop. In the past couple of months, she has difficulty with fine finger movements and noticed that her hand muscles are smaller on the right. Her husband has noticed some slurring of her speech. There is no difficulty swallowing or breathing. She has no sensory symptoms and no pain. Her vision and hearing are normal. On examination, she has slow soft speech. She has full range of extraocular movements with normal facial sensation and strength. She has a brisk jaw jerk. She has tongue fasciculations and her tongue movements are slow. Her neck flexion is 4/5. The rest of her cranial nerve examination is normal. She has atrophy of the intrinsic muscles of her right hand with 3/5 finger abduction, 4/5 finger flexion and extension, and 4/5 strength elsewhere on the right arm. Her strength is normal in the left arm with no atrophy. She has a spasticity (spastic catch) in both arms. Her hip flexion and knee flexion strength is 4/5 on both sides. Her knee extension strength is 5/5. Ankle dorsiflexion and toe extension is 3/5 on both sides. She has a spasticity in the arms and legs. Her deep tendon reflexes are 3+ and both toes are upgoing. She has fasciculation in both gastrocnemius muscles.

Case Discussion

Nicola's most likely diagnosis is amyotrophic lateral sclerosis (ALS) since she has progressive, purely motor symptoms consisting of both upper motor neuron (pyramidal weakness, spasticity, and brisk reflexes) and lower motor neuron (fasciculations, atrophy) symptoms in cranial (tongue), cervical (arms), and lumbar (legs) regions.

Amyotrophic lateral sclerosis is a motor neuron disease that affects both the upper motor neurons (UMNs) and the lower motor neurons (LMNs). About 5%–10% of cases are familial. At autopsy, there is loss of pyramidal cells in the motor cortex and atrophy of the brainstem, spinal cord, and anterior nerve roots. ALS begins as a localized degeneration of the motor neurons and then spreads through neuronal contacts throughout the entire somatic motor system. The LMNs affected are either in the anterior horn of the spinal cord or in the motor nuclei of the cranial nerves in the brainstem. The loss of these neurons leads to atrophy of muscles supplied by these neurons that are reliant on the trophic or nutritive factors supplied by the motor neurons (*amyotrophic*, "muscle atrophy"). Damage to anterior horn cells leads to lower motor neuron symptoms in that myotomal region. For example, damage to the anterior horn cells at C8/T1 level would result in weakness in muscles innervated by C8/T1 region. The UMNs that are lost include both projections from the cortex (corticospinal and corticobulbar tracts) and projections from the brainstem. The degeneration of the motor tracts in the lateral column of the spinal cord and the resulting glial scars lead to sclerosis of this portion of the spinal cord (*lateral sclerosis*, "hardening of the lateral column"). Interestingly, both oculomotor and visceral motor neurons are spared in this disease process (Figure 8.13).

The clinical manifestations of the disease can be divided into the symptoms related to the loss of LMNs and those related to the loss of UMNs, and during the course of the disease, both LMNs and UMNs will be affected.

LMN symptoms typically begin as a weakness in one limb, usually in the distal muscle groups. There is characteristic cramping of the muscles early in the morning. The muscle weakness can then move from one side to the same muscle

groups on the other side of the body. The muscles atrophy, and muscle fasciculations are observed. When the LMNs in the brainstem are affected, the clinical symptoms include difficulty chewing (motor nucleus of cranial nerve [CN] V), swallowing (nucleus ambiguus), and moving the face (CN VII) and tongue (CN XII).

The UMN symptoms include spasticity (velocity-independent increase in tonic stretch reflex), clonus, and increased deep tendon reflexes (**hyperreflexia**). Babinski (upgoing big toe) is seen when the UMN pathway is damaged and is indicative of a lesion to the corticospinal tract. The degeneration of corticobulbar fibers can lead to difficulty speaking (**dysarthria**) due to the lack of coordination between the different cranial nerve nuclei involved in speech. The disruption of input to the reticular formation can lead to inappropriate motor expression of emotions such as excessive crying or laughing.

Mortality is not primarily dependent on the length of disease or the progression of disease but solely on the nuclei affected. When the motor system responsible for innervation of the main respiratory muscles is affected, the patient dies. This can occur early or later on in the disease.

Because the disease is limited to the somatic motor system, there is no loss of sensory function or in bowel or bladder function even in late stages.

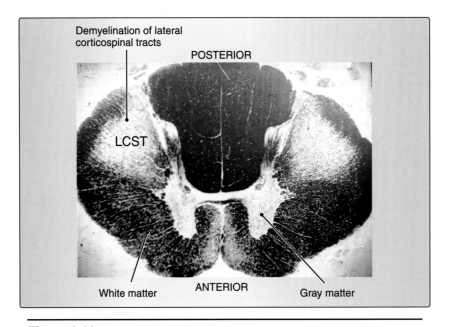

Figure 8.13
Amyotrophic lateral sclerosis. Cross section of spinal cord: myelin is stained black. The lateral corticospinal tracts (LCST in the section) are pale due to loss of myelin.

Chapter Summary

- Descending systems that are part of the motor system have two primary origins: the **cortex** and the **brainstem**. Descending tracts from the cortex influence **voluntary movements** and **adjust posture** in relation to voluntary movement. The **lateral corticospinal tract** originates in the **primary motor cortex** and other cortical areas and descends through the **posterior limb of the internal capsule** and then in the **anterior portion of the brainstem**. Fibers cross over in the caudal medulla in the **pyramidal decussation** and then travel in the **lateral column of the spinal cord**. These fibers influence the neurons that will innervate the **distal muscle groups** of the extremities and execute **precise voluntary movements**. The postural adjustments related to these voluntary movements of the extremities are also under cortical control and occur via the **anterior corticospinal tract**. These fibers do not cross in the pyramidal decussation. They descend in the **anterior column** of the spinal cord white matter and project to both the **ipsilateral** and **contralateral sides** of the spinal cord where they influence interneurons that coordinate postural stability as well as the tone and movement of proximal and truncal muscle groups. The projections to the contralateral side **decussate** in the **anterior white commissure** of the spinal cord.

- Descending motor fibers from the brainstem are phylogenetically older pathways that coordinate **truncal** and **proximal movements**. These fibers have the following points of origin and functions.

  - The **vestibulospinal tract** originates in the vestibular nuclei and adjusts body posture in response to changes in head position detected by the inner ear.

  - The **reticulospinal tract** is a collection of fibers with a wide range of functions. It influences muscle tone and posture of truncal musculature, it comprises the neurons that stimulate respiratory muscles, and it contains emotional motor system fibers that can set the excitation threshold of spinal cord motor neurons through monoaminergic pathways.

  - The **rubrospinal tract** influences the flexors of the upper limb. In humans, this function has been largely taken over by the corticospinal tract.

  - The **tectospinal tract** originates in the superior colliculus and orients the head and neck in response to eye movements.

Study Questions

Choose the ONE best answer.

8.1 A child goes on the whirligig ride at an amusement park. This ride spins him first in one direction and then in the other. When he steps off the ride, he staggers to regain his balance. The ride has activated descending tracts that are critical in making postural adjustments to changes in head position. These tracts are the:

    A. Lateral corticospinal tracts.
    B. Reticulospinal tracts.
    C. Corticobulbar tracts.
    D. Lateral vestibulospinal tracts.
    E. Tectospinal tracts.

Correct answer is D. The fibers of the lateral vestibulospinal tract descend uncrossed (ipsilaterally) and terminate at all levels of the spinal cord to facilitate the activity of the extensor muscles and inhibit the activity of the flexor muscles. This tract mediates postural adjustments to compensate for movements and changes in position and to coordinate the orientation of the head and body in space. The lateral corticospinal tracts carry descending voluntary motor information. The reticulospinal tracts coordinate muscle group activation allowing orientation of the body toward or away from a stimulus. They also play a role in muscle tone. The corticobulbar tracts descend initially along with the corticospinal tracts and terminate on lower motor neurons of the cranial nerve motor nuclei. The tectospinal tracts help orient our head towards an object of interest within our visual field.

8.2 A patient was admitted to the hospital following a car accident in which he was thrown hard against the windshield and hit his head. He was conscious but confused and disoriented when admitted. A computed tomography scan revealed an accumulation of blood consistent with a subdural hematoma, but no evidence of a skull fracture. Pressure from this accumulating blood was causing a leftward shift of the brain, resulting in compression of the left cerebral peduncle against the tentorium. Further examination is most likely to reveal:

A. Weakness and increased muscle tone and reflexes on the right side of the body.

B. Weakness and decreased muscle tone and reflexes on the left side of the body.

C. Decreased pain and temperature sensations on the right side of the body.

D. Decreased discriminative touch, proprioception, and vibration sense on the right side of the body.

E. Decreased pain and temperature sensations on the right side of the face.

8.3 A patient presents with marked weakness of the right arm and leg, increased muscle tone of both right limbs, and increased deep tendon reflexes on the right. In addition, examination reveals intact vibration sense on both sides of the body but no response to pinprick on the left side of the body. The most likely cause of these symptoms is:

A. Occlusion of the left posterior inferior cerebellar artery at the level of the rostral medulla.

B. Occlusion of the right branches of the anterior spinal artery at approximately cervical level 5.

C. Occlusion of the right branches of the anterior spinal artery at the level of the caudal medulla.

D. Occlusion of the left paramedian branches of the basilar artery at the level of the caudal pons.

E. Occlusion of the left short circumferential branches of the posterior cerebral artery at the level of the rostral midbrain.

Correct answer is A. Descending corticospinal fibers originate from motor and sensory areas of the cortex and descend through the internal capsule, the cerebral peduncles, the basal pons, and the pyramids. They then cross at the junction between the spinal cord and medulla to descend in the lateral area of the spinal cord and innervate lower motor neurons in the anterior gray horn of the spinal cord. Fibers traveling in the left cerebral peduncle will innervate the right side of the body, and compression of this peduncle will, thus, cause upper motor neuron (UMN) signs on the right (i.e., weakness with increased tone and increased reflexes). The left cerebral peduncle will control the right side of the body, and there will be UMN, not lower motor neuron (decreased tone and reflexes) signs. Pain and temperature sensations, as well as discriminative touch, proprioception, and vibration sense, will be normal on both sides of the body, because these modalities will be spared with a lesion of the cerebral peduncle. Similarly, pain and temperature sensations on the right side of the face will be normal, because they will not be affected by a lesion in the cerebral peduncle.

Correct answer is B. Right-sided branches of the anterior spinal artery supply the right anterior two-thirds of the spinal cord. Occlusion of these vessels at a cervical level will affect blood supply to several structures in the right half of the spinal cord, including the right corticospinal tract (descending corticospinal fibers from the left cortex have already crossed in the pyramidal decussation), resulting in upper motor neuron (UMN) signs (weakness, increased tone, and increased reflexes) on the right, and the right anterolateral column, resulting in loss of pain and temperature sensation to the left side of the body below the level of the lesion. Other descending motor systems may also be affected, but deficits from these will not be as noticeable given the UMN signs. Occlusion of the left posterior inferior cerebellar artery at the rostral medulla will result in lateral medullary syndrome. There will be a number of deficits but no UMN signs and no loss of pain and temperature sensations to the body. Occlusion of the right branches of the anterior spinal artery at the level of the caudal medulla will affect the right pyramid, resulting in UMN signs on the left. There may be some disruption of discriminative touch, proprioception, and vibration sense, perhaps bilaterally, because this is where ascending posterior column fibers cross. Pain and temperature should be intact as the anterolateral area is not affected by this lesion. Occlusion of the left paramedian branches of the basilar artery at the level of the caudal pons will result in UMN signs on the right. Discriminative touch, vibration, and proprioception on the right may be affected. Pain and temperature sensations should be intact as the anterolateral tracts are located more laterally. Occlusion of the left circumferential branches of the posterior cerebral artery at the level of the rostral midbrain will result in UMN signs on the right. However, pain and temperature sensations will be intact.

8.4 Which of the following statements about the corticobulbar tracts is correct?

A. They arise ipsilaterally from both sensory and motor areas of the cortex.

B. They travel primarily in the posterior limb of the internal capsule.

C. They provide bilateral innervation to most of the motor nuclei they serve.

D. They contact circuit neurons to coordinate motor activity of the entire limb that they innervate.

E. They innervate both somatic and visceral motor nuclei in the brainstem.

Correct answer is C. Corticobulbar fibers provide innervation to cranial nerve somatic motor nuclei. Innervation is primarily bilateral with just a few exceptions where the pattern of innervation is more complex. Corticobulbar fibers arise primarily from motor areas of the cortex related to the head and face. Corticobulbar fibers travel primarily in the genu of the internal capsule. The corticospinal fibers contact circuit neurons to coordinate motor activity of an entire limb. Visceral motor nuclei receive their descending input from the hypothalamus.

8.5 Although they are the most primitive descending motor system, the reticulospinal tracts have a fairly wide range of important functions. Which statement about the reticulospinal tracts is correct?

A. Most of the reticulospinal fibers involved in somatomotor function originate in the midbrain reticular formation.

B. Reticulospinal fibers play a role in muscle tone related to voluntary movements of the limbs.

C. Reticulospinal fibers play a role in the control of breathing through the direct innervation of skeletal muscles involved in respiration.

D. Reticulospinal fibers play an important role in innervating neck muscles and in orienting the head and neck during eye movements.

E. Neurotransmitters released by reticular formation neurons inhibit the activity of spinal motor neurons involved in emotional body language.

Correct answer is B. Reticulospinal fibers play a role in integrating distal and proximal muscle actions and, thus, in muscle tone related to voluntary movements of the limbs. Most of the reticulospinal fibers involved in somatomotor function originate in the pontine and rostral medullary reticular formation. Reticulospinal fibers play a role in the control of breathing through direct projections to spinal cord neurons, which, in turn, activate skeletal muscles involved in respiration. Tectospinal fibers play an important role in innervating neck muscles and in orienting the head and neck during eye movements. Neurotransmitters released by reticular formation neurons play a role in the excitability of spinal motor neurons involved in emotional body language. Their role is modulatory, not specifically inhibitory.

Control of Eye Movements

9

I. OVERVIEW

In previous chapters, we introduced both the general organization of the central nervous system (CNS) and the more detailed organization of the brainstem including the major ascending and descending tracts that travel through the brainstem. This chapter focuses on the brainstem structures and cortical areas involved in the control of eye movements. These include the vestibular system, as well as three pairs of cranial nerves that innervate our extraocular muscles, tracts interconnecting the nuclei of these nerves, and cortical areas that send input to the brainstem.

The precise coordination of eye movements is important so that both eyes can focus together on a specific target. The two eyes must move synergistically so that we can orient ourselves in our environment and track objects in our visual fields. Any disruption to the synergistic movement of our eyes will result in changes in our visual perception (most notably double vision) caused by the fact that the eyes are not perfectly aligned.

Disorders associated with alterations in eye movements are often the first sign of pathology in the brainstem. An understanding of eye movements and the wiring underlying these movements is an important diagnostic tool when assessing brainstem disorders.

In this chapter, we look at the anatomy of the extraocular muscles and how they move the eyeball in the orbit. We describe the wiring of the cranial nerves that innervate these muscles and see how synergistic movements are possible. We then look at the different types of eye movements and the link between the vestibular system and the control of eye movements. Finally, we look at the clinical assessment of eye movements and commonly associated clinical syndromes.

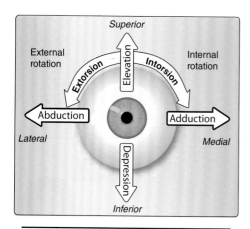

Figure 9.1
Movements of the eyeball.

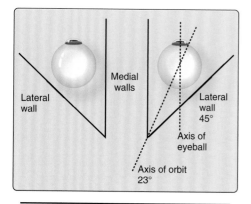

Figure 9.2
Axis of the eyeball and axis of the orbit as seen from above.

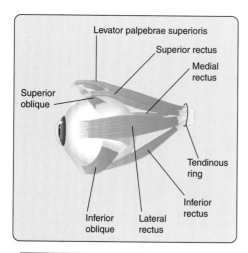

Figure 9.3
Overview of the extraocular muscles in the orbit.

II. STRUCTURES INVOLVED IN EYE MOVEMENTS

Coordinated movements of the two eyes are critical to keep the two eyes precisely aligned and maintain focus on an object of interest in the visual field. Six extraocular muscles (detailed below) in each orbit control these eye movements (Figure 9.1). These muscles are innervated by three pairs of cranial nerves. Coordinating centers in the forebrain and brainstem as well as fiber tracts that interconnect the cranial nerve nuclei provide the ultimate control for the synergistic movements required for this precise alignment of both eyes.

A. Extraocular muscles

Six extraocular muscles move the eye in the orbit, primarily along a horizontal or a vertical axis. Along **the horizontal axis**, the eye can be pulled toward the midline (**adduction**) or pulled away from the midline (**abduction**). Along the **vertical axis**, the eye can be pulled upward (**elevation**) or downward (**depression**). In addition, the eye can rotate inward (**intorsion**) or outward (**extorsion**). These primary eye movements are depicted in Figure 9.1. All extraocular muscles have a primary movement and participate in a secondary movement.

In order to understand the actions of these muscles, it is important to understand the orientations of the eyes relative to the orbits (Figure 9.2). Whereas the medial walls of each orbit are oriented in the sagittal plane, the lateral walls are oriented at an angle 45 degrees lateral to the sagittal plane. Thus, with the eyes looking straight ahead, the axis of the orbit is deviated about 23 degrees laterally from the axis of the eyeball. This difference between the axes is particularly important in understanding the action of the extraocular muscles in elevating or depressing the eyeball, as discussed below.

1. **Rectus muscles:** There are four rectus muscles (medial, lateral, superior, inferior). The four **rectus muscles** originate from a **tendinous ring** at the back of the orbit and attach to the eyeball anteriorly (Figure 9.3). The medial and lateral recti move the eye along the horizontal plane and their actions are relatively straightforward. The **medial rectus** adducts the eye, and the **lateral rectus** abducts the eye (Figure 9.4A).

   In contrast, because the orbital axis and the visual axis are not aligned, the actions of the superior and inferior recti and the two oblique muscles (see below), all of which move the eyes along the vertical plane, are more complex. If we consider just the primary actions of the superior and inferior recti, the superior rectus elevates the eye and the inferior rectus depresses the eye. However, because of the orientation and attachment of the rectus muscles (see Figure 9.4A,B), the superior rectus also causes intorsion of the eyeball, whereas the inferior rectus causes extorsion of the eyeball.

2. **Oblique muscles:** The **inferior oblique muscle** originates on the medial side of the floor of the orbit and attaches to the lateral aspect of the eyeball anteriorly, passing under the lateral rectus (see Figure 9.3). Contraction of this muscle causes **elevation** and **extorsion** of the eyeball, as shown in Figure 9.5. The **superior oblique muscle** originates at the back of the orbit, lateral to the tendinous ring. It then

goes through a pulley (**trochlea**) at the anterior medial pole of the orbit and attaches to the eye superiorly (Figure 9.6). The function or action of the superior oblique muscle depends on the position of the eye in the orbit. When the eye is turned toward the midline, or **adducted**, the angle of the muscle attaching to the eyeball and the pulley is such that contraction of the muscle will cause **depression** of the eyeball. When the eye is **abducted**, or turned away from the midline, the angle of superior oblique is such that contraction of the muscle will cause **intorsion** of the eye, as seen in Figure 9.5.

B. Cranial nerves that control eye movements

Three pairs of cranial nerves control the movements of the extraocular muscles. The **oculomotor nerve**, cranial nerve (CN) III, innervates the lateral rectus, superior rectus, inferior rectus, and inferior oblique muscles. The **abducens nerve**, CN VI, innervates the lateral rectus muscle. The superior oblique muscle is innervated by the **trochlear nerve**, CN IV.

A mnemonic for remembering the innervation of the extraocular muscles is (**LR$_6$ SO$_4$)3**. That is, the **l**ateral **r**ectus is innervated by CN VI (**6**), the **s**uperior **o**blique is innervated by CN IV (**4**), and all the rest are innervated by CN III (**3**).

1. **The oculomotor nerve:** CN III has two components: The **general somatic efferent (GSE)** component innervates the extraocular muscles—the **superior, medial**, and **inferior rectus** muscles and the **inferior oblique** muscle, all of which move the eyeball, as well as the **levator palpebrae superioris**, which elevates the eyelid. The **general visceral efferent**, or **parasympathetic**, component innervates the intraocular muscles—the **constrictor pupillae** and **ciliary muscles**, which control the size of the pupil and the rounding of the lens, respectively (Figure 9.7).

   The somatic motor nucleus of CN III, known as the **oculomotor nuclear complex**, lies close to the **midline**, in the tegmentum of

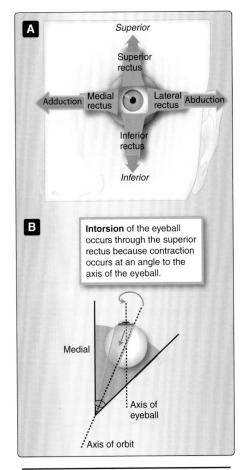

Intorsion of the eyeball occurs through the superior rectus because contraction occurs at an angle to the axis of the eyeball.

Figure 9.4
A and B. Actions of the rectus muscles for moving the eyeball.

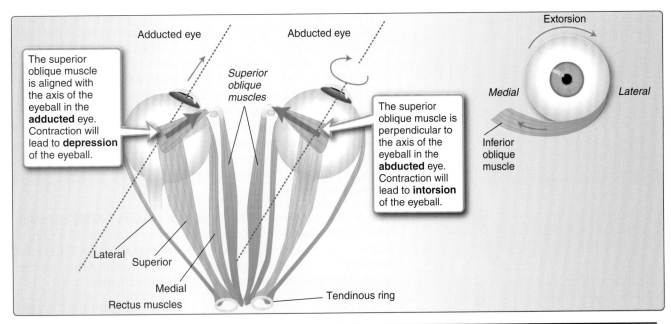

Figure 9.5
Actions of the oblique muscles for moving the eyeball.

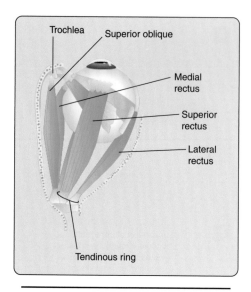

Figure 9.6
Pulley system for the superior oblique
muscle.

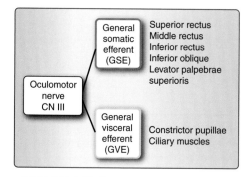

Figure 9.7
Components of the oculomotor nerve.
CN = cranial nerve.

the **rostral midbrain**, at the level of the superior colliculus. It is anterior to the cerebral aqueduct and has the medial longitudinal fasciculus (MLF) as its inferior and lateral boundary (Figure 9.8).

The lower motor neurons (LMNs) leaving the oculomotor nuclear complex course anteriorly in the tegmentum of the midbrain, through the medial aspect of the red nucleus and the cerebral peduncle, to emerge in the interpeduncular fossa at the junction of the midbrain and the pons. They then enter the orbit where they supply the inferior oblique and inferior and medial rectus muscles as well as the superior rectus and levator palpebrae superioris.

The **visceral motor (parasympathetic)** nucleus of CN III, known as the Edinger-Westphal nucleus, lies posterior to the main oculomotor nucleus in the **periaqueductal gray (PAG)**. The visceral motor (preganglionic) fibers travel with the oculomotor nerve to the **ciliary ganglion** where their cell bodies are located. Postganglionic axons leave the ganglion as short ciliary nerves, enter the eyeball close to the optic nerve, and innervate the ciliary body and the constrictor pupillae muscle.

Interestingly, the parasympathetic fibers are the most external fibers in CN III and, therefore, are the most susceptible to compression. Dilation of the pupil is often the first sign of CN III compression.

Sympathetic fibers originating from the superior cervical ganglion travel through the ciliary ganglion and reach their target muscle, the dilator pupillae muscle in the eye, through the long ciliary nerves.

2. **The trochlear nerve:** CN IV, the smallest cranial nerve, has only **GSE** fibers and supplies only one muscle—the **superior oblique**. The trochlear nucleus is located in the **tegmentum of the midbrain** at the level of the inferior colliculus (Figure 9.9). From the nucleus, the **axons course posteriorly** around the cerebral aqueduct and the PAG, cross the midline, and emerge from the **posterior surface** of the midbrain just caudal to the inferior colliculus. Once their fibers emerge from the midbrain, the LMNs of the trochlear nerve curve anteriorly around the cerebral peduncle and course between the posterior cerebral and superior cerebellar arteries, lateral to CN III. They then enter the cavernous sinus where they lie lateral to the internal carotid artery and, from there, enter the orbit through the superior orbital fissure, above the tendinous ring. These axons then course medially close to the orbital roof to reach the superior oblique muscle. Note that because they cross the midline before emerging from the midbrain, the axons from the trochlear nerve nuclei supply the contralateral superior oblique muscles. However, a lesion of the crossed emerging fibers is more common than a lesion to the nucleus.

3. **The abducens nerve:** CN VI has one modality—**GSE** to the **lateral rectus muscle**. The abducens nucleus, on the other hand, has two components: it contains the cell bodies of the GSE motor neurons and it houses interneurons that function as a **relay center** for the **coordination of horizontal eye movements**.

The abducens nucleus is located in the caudal **pons** at the level of the facial colliculus close to the midline (Figure 9.10). The axons

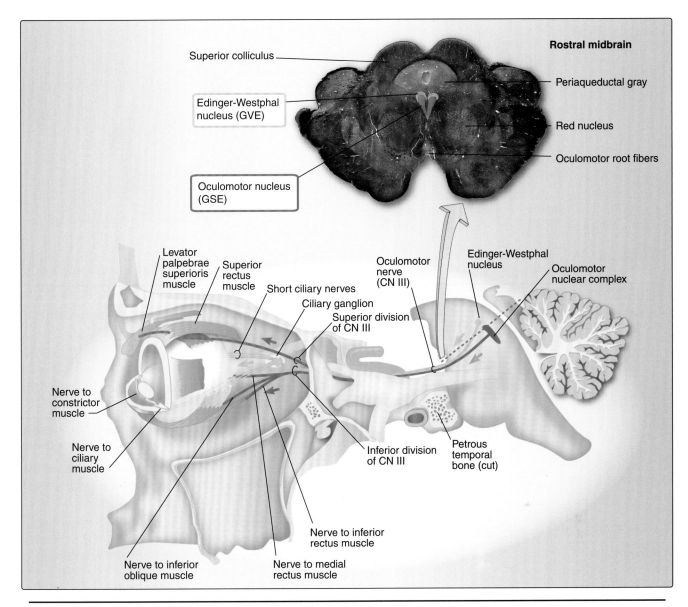

Figure 9.8
Overview of the oculomotor nerve (cranial nerve [CN] III). GSE = general somatic efferent; GVE = general visceral efferent. (Modified from Wilson-Pauwels L, et al. *Cranial Nerves: Function and Dysfunction*, 3rd ed. PMPH, CT, 2010.)

of the LMNs emerge on the anterior surface of the pons at the pontomedullary junction. The nerve runs through the posterior cranial fossa and follows a subdural course before it enters the cavernous sinus. It then enters the orbit at the medial end of the superior orbital fissure and travels laterally to enter the deep surface of the lateral rectus muscle.

C. Medial longitudinal fasciculus

The **MLF** is a fiber tract that interconnects the third, fourth, and sixth cranial nerve nuclei to each other and to the vestibular nuclei, thus allowing synergistic or coordinated movements of the two eyes

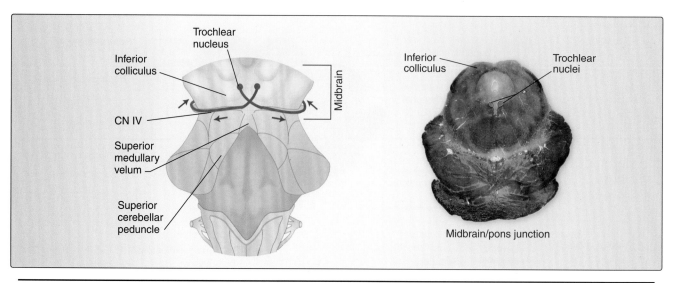

Figure 9.9
Overview of the trochlear nerve (cranial nerve [CN] IV). (Modified from Wilson-Pauwels L, et al. *Cranial Nerves: Function and Dysfunction*, 3rd ed. PMPH, CT, 2010.)

and adjustments of eye position in response to movements of the head (Figure 9.11).

The MLF originates from the **vestibular nuclei** in the rostral medulla/caudal pons and has both **descending** and **ascending** components.

The descending MLF is the **medial vestibulospinal tract,** important in **controlling the head and neck position** (see Chapter 8, "Descending Motor Tracts").

The **ascending** component of the MLF arises from the medial **vestibular nuclei**, with some input from the superior vestibular nuclei, and is critical for the coordination and synchronization of all major classes of **eye movements**. Because of its origin in the vestibular nuclei, the MLF mediates eye movements generated in response to vestibular stimuli and movements of the head. For example, the **vestibuloocular reflex** (see *Gaze* below) adjusts eye movements to head movements. Projections from the vestibular nuclei innervate the *contralateral* sixth or abducens nucleus. As noted, this nucleus contains the cell bodies of somatic motor neurons that innervate the ipsilateral lateral rectus muscle as well as interneurons. It is the axons of these interneurons that cross the midline immediately after leaving the nucleus and ascend as the contralateral MLF, making connections with the trochlear and oculomotor nuclei. By linking the abducens, trochlear, and oculomotor nuclei to each other, the MLF mediates linked movements of the two eyes so that **conjugate gaze** in all directions can occur. The ascending MLF comprises a small pair of **heavily myelinated** tracts near the midline, just anterior to the fourth ventricle in the medulla and pons and anterior to the cerebral aqueduct in the midbrain.

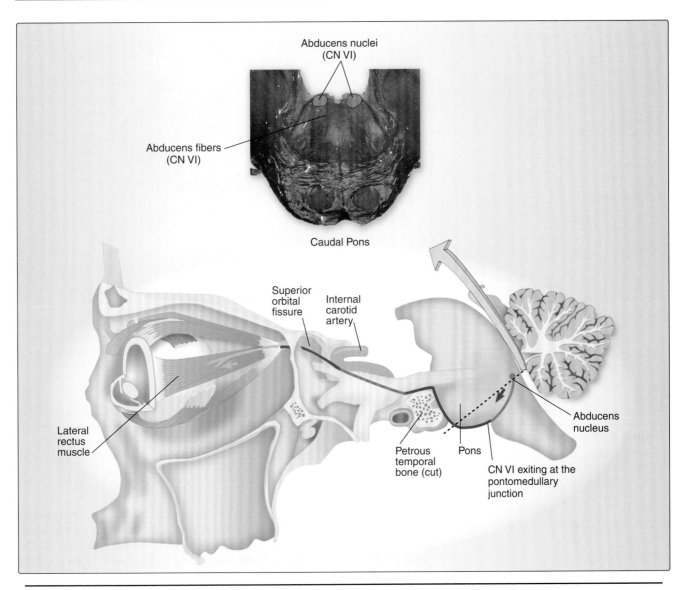

Figure 9.10
Overview of the abducens nerve (cranial nerve [CN] VI). (Modified from Wilson-Pauwels L, et al. *Cranial Nerves: Function and Dysfunction*, 3rd ed. PMPH, CT, 2010.)

III. GAZE

Gaze is the coordinated, **synergistic movement of both eyes** to a target in the environment. Gaze directs the projection of a target onto the area of highest visual acuity in the retina, the **fovea**. Gaze is coordinated by higher centers in the cortex and the brainstem, which project to the motor neurons of CNs III, IV, and VI, whose fibers innervate the extraocular muscles. Importantly, these cranial nerve nuclei receive input only from these gaze systems, but none from the corticobulbar tract. Gaze is tightly connected to both the visual and the motor systems and, thus, plays a pivotal role in the visual orientation necessary for purposeful motor activity. It uses visual stimuli, such as objects detected in a visual field, and motor commands, in order to direct both eyes to an object of interest.

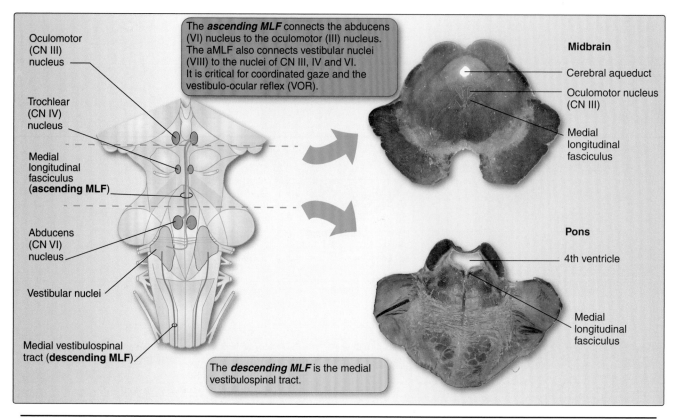

Figure 9.11
The ascending and descending medial longitudinal fasciculi (MLF). Vestibular input to the aMLF is both crossed and uncrossed (not shown) Micrographs show the ascending MLF. CN = cranial nerve.

Gaze, or eye movements involved in gaze, can occur in a **horizontal plane** or in a **vertical plane**. In addition, eyes can **rotate** to compensate for rotational movement of the head. For all of these movements, the eyes must work together so that focus on the fovea of both eyes is on the same object. This synergistic movement of the eyes is controlled by gaze centers in the brainstem and the cortex. The **horizontal gaze** center in the brainstem is located in the **paramedian pontine reticular formation (PPRF),** and the **vertical gaze** center is located in the rostral midbrain reticular formation (rMRF) and pretectal area. A lesion to the pons will lead to a deficit in horizontal gaze due to damage of the PPRF. A lesion of the midbrain will lead to a deficit in vertical gaze due to damage to the rMRF and pretectal area.

The horizontal and vertical gaze centers act together to allow the two eyes to see images in the same visual field and for a smooth projection of these images onto the retina. Any discrepancy in the projections onto the retina will result in double vision.

A. Horizontal gaze

1. **Wiring of horizontal gaze:** In order to move both eyes in the same direction, the lateral rectus muscle on one side and the medial rectus muscle on the other side must work together. This means that the motor neurons innervating the **lateral rectus**

(CN VI, abducens) on one side and the motor neurons innervating the **medial rectus (CN III, oculomotor)** on the other side must be activated at the same time (Figure 9.12). Although there are different types of horizontal movement (see *Saccades* below), and the wiring will differ for each of these movements, all input for horizontal eye movements will eventually end in the **abducens nucleus**. For example, if gaze is to be directed to the left side, the abducens nucleus on the left side will be stimulated. Within this nucleus, cell bodies of somatic motor neurons will send projections to the lateral rectus on the ipsilateral or left side. Axons from interneurons immediately cross the midline and ascend as the contralateral MLF to the oculomotor nucleus on the contralateral or right side. Motor neurons in the right oculomotor nucleus then send axons to the right medial rectus muscle. Together, this results in contraction of the lateral rectus on the left side, causing the left eye to abduct. Simultaneously, the medial rectus on the right side contracts, causing the right eye to adduct, and both eyes move to the left.

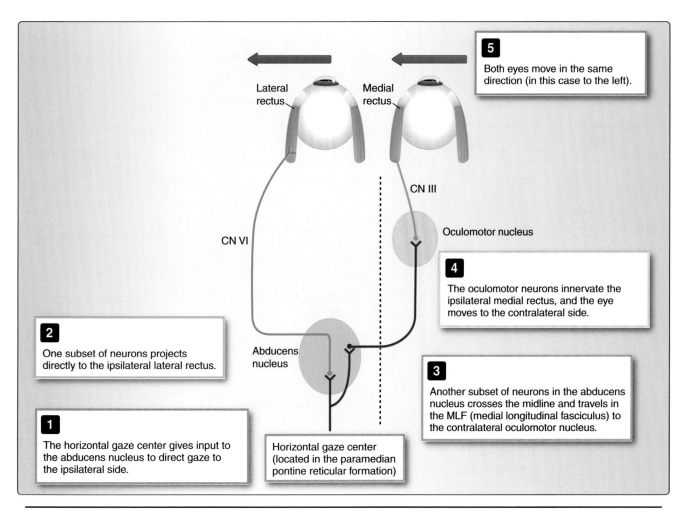

5
Both eyes move in the same direction (in this case to the left).

Lateral rectus

Medial rectus

CN III

CN VI

Oculomotor nucleus

4
The oculomotor neurons innervate the ipsilateral medial rectus, and the eye moves to the contralateral side.

2
One subset of neurons projects directly to the ipsilateral lateral rectus.

Abducens nucleus

3
Another subset of neurons in the abducens nucleus crosses the midline and travels in the MLF (medial longitudinal fasciculus) to the contralateral oculomotor nucleus.

1
The horizontal gaze center gives input to the abducens nucleus to direct gaze to the ipsilateral side.

Horizontal gaze center (located in the paramedian pontine reticular formation)

Figure 9.12
Wiring diagram for horizontal gaze. CN = cranial nerve.

2. Saccades: Saccades are **rapid eye movements** that redirect gaze to an object of interest and result in the projection of that object onto the fovea. Saccades **orient our gaze** in the visual environment (visual field). In a saccade, the eyes jump from one object or point in space to another (or one fixation point to the next), stopping briefly at each point to allow detailed inspection by the fovea. During a saccade, the visual system suppresses the incoming visual input, making us unaware of these saccadic movements.

Saccadic eye movements are the result of cortical and subcortical input to the horizontal gaze center, or **PPRF**, as shown in Figure 9.13. The cortical areas involved in the generation of saccadic eye movements are located in both the parietal and the frontal lobes. The best-described cortical areas controlling saccadic eye movements are the **frontal eye fields (FEFs)**. Whereas previous studies in experimental animals identified an area in the middle frontal gyrus as the key cortical area, more recent functional

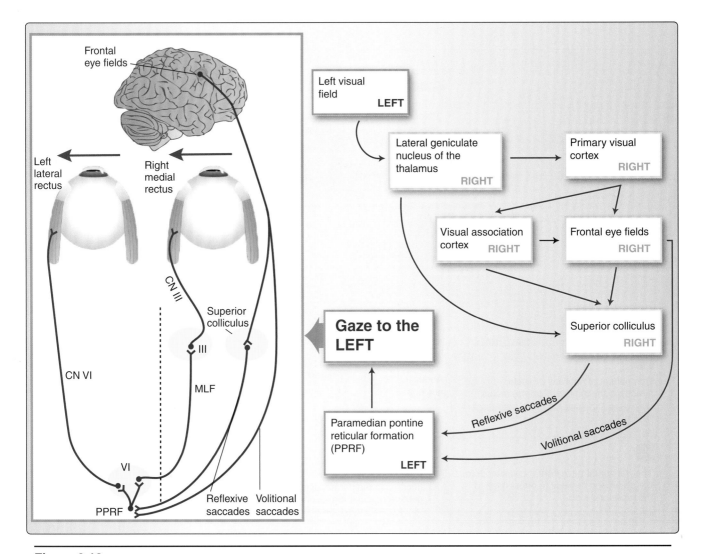

Figure 9.13
Wiring diagram for saccadic eye movement. CN = cranial nerve; MLF = medial longitudinal fasciculus.

imaging studies in humans suggest that the key area may actually be more posterior, on the border of the precentral gyrus, just anterior to the area for primary hand representation. From there, fibers project to the ipsilateral **superior colliculus** and to the contralateral **PPRF**. Inputs from the PPRF to the adjacent abducens nucleus initiate horizontal gaze ipsilaterally (contralateral to the FEF that activated the horizontal eye movements).

Exploration of the visual environment requires two types of saccades: **reflexive** and **volitional saccades**. A reflexive saccade is a visually guided saccade. It occurs in response to external cues, and it is used to orient ourselves in our visual environment. Reflexive saccades are also called **prosaccades**, because they direct the eyes toward visual cues or points in space. A volitional or intentional saccade, on the other hand, is independent of a visual stimulus. It is more cognitively complex and requires higher-order processes such as spatial memory and analysis of contextual cues.

a. **Reflexive saccades:** The **superior colliculus** provides the main input for reflexive saccades (Figure 9.14). The superior

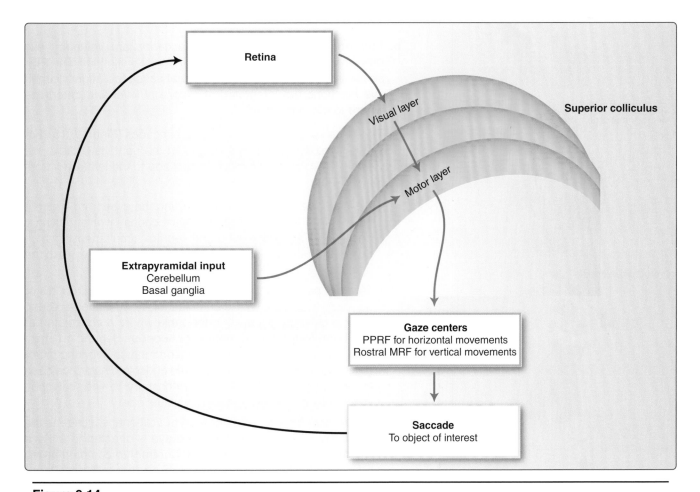

Figure 9.14
The role of the superior colliculus in reflexive saccades. PPRF = paramedian pontine reticular formation; MRF = midbrain reticular formation.

colliculus is a layered structure on the posterior surface of the rostral midbrain. The most **superficial layer** receives visual input from the retina via the **lateral geniculate nucleus**. The **deeper layers** have **motor neurons** that project to the **PPRF** and initiate saccades. The sensory information and the motor output are linked via synaptic contact between the superficial visual layer and the deep motor layer, resulting in a correlation between the sensory map of the visual environment and a motor map for eye movement.

For example, an object may be detected in the left visual field. This activates the **right lateral geniculate nucleus** and the right superior colliculus. Neurons in a specific area of the visual map in the superficial layers of the superior colliculus are activated, and these, in turn, activate their corresponding motor neurons in the deep layer. From there, the input goes to the contralateral (left) PPRF, which initiates a horizontal gaze to the left, to the exact location from which the visual stimulus came. At the same time that input is sent to the PPRF, neurons in the superior colliculus also project to motor neurons in the cervical spinal cord that innervate neck muscles to facilitate head movements toward the object in the visual field (tectospinal tract; see Chapter 8, "Descending Motor Tracts").

b. **Volitional saccades:** Volitional saccades are voluntary eye movements that are independent of a visual stimulus. The **FEF** sends input to the PPRF for volitional saccades. Higher cognitive functions are integrated at a cortical level and determine where to move the eyes.

There are several types of volitional saccades (Figure 9.15). For example, to avoid seeing an object in the right visual field, the viewer can voluntarily perform a saccade to the left (away from the stimulus).

* *Antisaccade*: This process of averting the eyes from a stimulus is called an **antisaccade**. Antisaccades involve making a decision to avert the eyes and require simultaneous inhibition of the reflexive saccade that would direct the eyes to the stimulus.

* *Memory saccades*: A viewer can also voluntarily direct his or her gaze to a place where an object or a visual cue had been (e.g., a child gazing back to the table where a chocolate bar had been lying). These **memory saccades** depend on intact working memory as well as spatial orientation. Working memory is required to remember that the chocolate had once been there and spatial orientation is necessary to find that spot, which may no longer contain any visual cue.

* *Predictive saccade*: Another type of volitional saccade is the **predictive saccade**, in which the eye is directed to a place where a target is expected to be. Contrary to common belief, it is not possible to "keep your eye on the ball" because a thrown ball moves far too quickly. Rather, we predict the trajectory of the ball and anticipate it landing in a certain place. Our eyes perform a saccade to this predicted place so that we

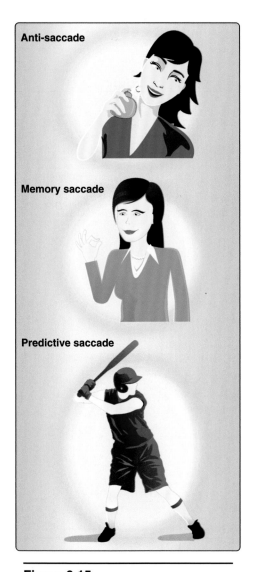

Figure 9.15
Types of volitional saccades.

can catch the ball or hit it with a baseball bat. (Curve balls are so effective because they have a trajectory different from the one we expect. Our eyes perform a volitional predictive saccade to the anticipated target and then a reflexive saccade to where the ball actually ends up. The ball appears to "curve.")

3. **Smooth pursuit:** Smooth pursuit is used when **tracking a slowly moving object**. The aim is to keep that object on the fovea, the area of highest visual acuity, in both eyes. Smooth pursuit stabilizes and holds the image on the retina during movement of the object itself or the person. In order to accomplish this, there must be integration of **cortical information** (primary visual and visual association cortices, "see the object" and FEFs, "move eyes"), **cerebellar information** ("How am I moving in the environment?"), and **vestibular information** ("Where am I in the environment?"), as well as an interconnection between CN VI from one side and the contralateral CN III (via the MLF) to achieve synergistic eye movements (Figure 9.16).

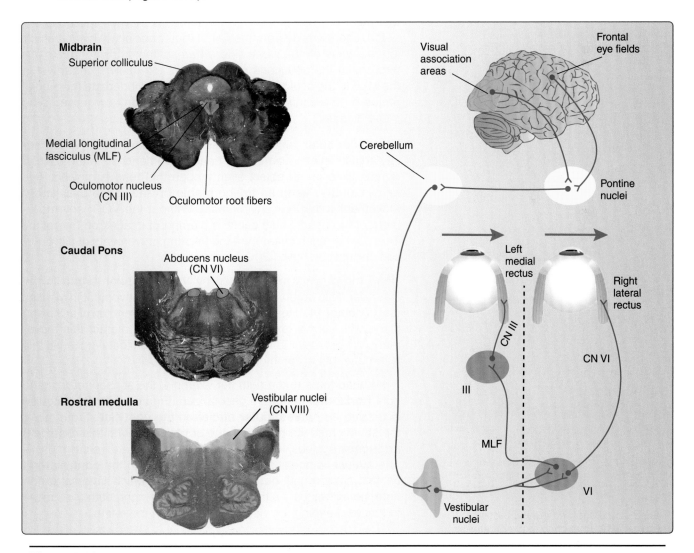

Figure 9.16
Wiring diagram for smooth pursuit. CN = cranial nerve.

Because smooth pursuit deals with keeping an object on the fovea of both eyes, **both sides of the cortex** are activated at the same time. We discuss the wiring on one side to illustrate what happens. It is important to keep in mind that pursuit movements do not orient the eyes toward a target (the latter is achieved by saccadic eye movements). Rather, they **keep** a target on the fovea. Unlike the wiring for saccadic eye movements, the wiring for smooth pursuit movements is ipsilateral, that is, the left side of the cortex will direct the eyes to the left. Because both sides are active at the same time, we can keep an object centered on the fovea. Pursuit is only possible with slowly moving objects. When we see something moving quickly within our visual field, we perform rapid saccadic eye movements to lock onto the object and reposition it on the retina.

For smooth pursuit, the primary visual cortex and the FEFs project to **ipsilateral pontine nuclei**. From there, fibers cross the midline and enter the **contralateral cerebellum**. Fibers from the cerebellum project to the **vestibular nuclei** on the same side and from there cross the midline to innervate the **abducens nucleus**. From this point on, the circuitry is the same as that for saccades. As noted, fibers arising from the CN VI motor nuclei will innervate the lateral rectus on the same side, whereas fibers arising from interneurons in the abducens nucleus cross the midline and travel in the MLF to the oculomotor nucleus, which will innervate the medial rectus on the contralateral side. Both eyes can now move in the same direction.

4. **Vestibuloocular reflex:** The **vestibuloocular reflex (VOR)** rapidly adjusts eye movements to head movements so that gaze can remain fixed on an object even though the head is moving. This reflex can thus **keep an image stable on the retina** despite our movement within our environment. Without the VOR, any movement of the head would cause a blurring of images on the retina. The VOR is much faster than the tracking movements of slow pursuit discussed above.

Head movements are detected by the **vestibular organ** in the inner ear. This organ is specialized to detect movement in all axes (see Chapter 11, "Hearing and Balance"). Acceleration in any given plane will activate extraocular muscles to counteract this movement. We discuss the VOR in reference to horizontal gaze, as this is the one that is commonly assessed clinically.

If the head turns to the right, for example, the sensory cells in the right **horizontal canal** are depolarized (Figure 9.17). This impulse is relayed to the **vestibular nuclei** on the right side. From there, the fibers cross the midline to the **contralateral** (in this case, left) **abducens** nucleus, which innervates the left lateral rectus muscle. The abducens nucleus is linked via the MLF to the **contralateral** (in this case, right) **oculomotor** nucleus, which innervates the right medial rectus. As the head turns to the right, both eyes move to the left, keeping the visual field stable on the retina.

Other combinations of actions between the semicircular canals and the extraocular muscles enable the eyes to compensate for head movements along other axes.

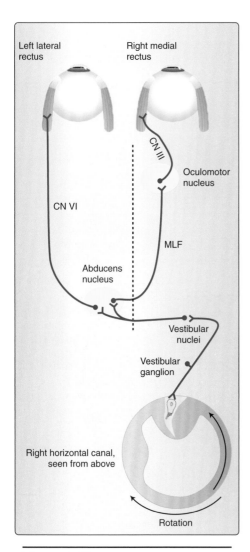

Figure 9.17
Wiring diagram for the vestibuloocular reflex (VOR). Rotation to the right results in compensatory eye movement to the left. CN = cranial nerve; MLF = medial longitudinal fasciculus.

5. **Nystagmus:** Rapid, back-and-forth, rhythmic eye movements are referred to as **nystagmus**. Nystagmus consists of a rapid component (quick "flick" of the eyes) in one direction and a slow component (slow drift) in the opposite direction. The nystagmus is named for the rapid component. Nystagmus can occur in any axis of eye movement. The most common type is **horizontal nystagmus**, which we discuss here. Nystagmus can be caused physiologically by stimulation of the vestibular system, by visual stimuli, or by various pathological processes. Here, we focus on **vestibular nystagmus** and **optokinetic nystagmus**, both of which are physiological.

Rotation of the head activates the vestibular apparatus in the inner ear, which in turn activates the VOR. When head rotation is greater than what can be compensated for by the VOR, the eyes reset with a rapid movement in the same direction as the rotation, and then, the compensatory movement of the VOR can happen again. This rapid reset followed by the slow movement mediated by the VOR is referred to as **vestibular nystagmus** (Figure 9.18). Thus, for example, head movement to the left results in slow eye movement to the right because of the VOR. The eyes then rapidly reset to the left, followed by another slow drift to the right, and so on. This is referred to as **left-beating nystagmus**.

A similar movement of the eyes can be elicited by **visual stimuli**. A rapidly moving object within the field of view or rapid movement past an object in the field of view causes the eyes to follow the object that is moving in the field of view. The focus is then rapidly reset to the next object in the field of view (e.g., passing rapidly by powerline poles as the viewer looks out the window of a car). A slow component follows one powerline pole, and then, a rapid

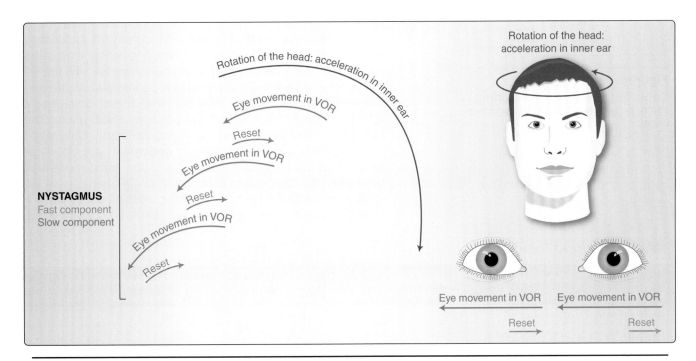

Figure 9.18
Vestibular nystagmus. VOR = vestibuloocular reflex.

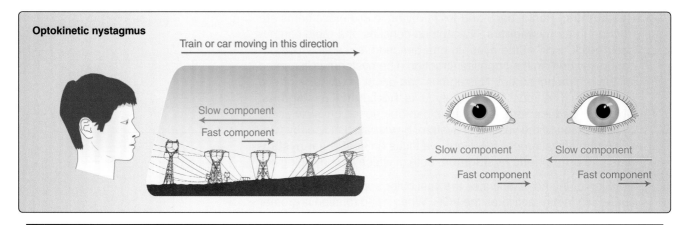

Figure 9.19
Optokinetic nystagmus.

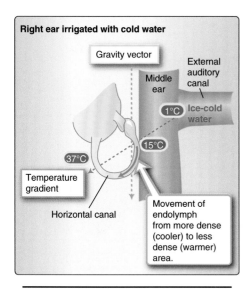

Figure 9.20
Caloric testing with cold water. Gravity vector is perpendicular to the horizontal canal.

component resets to the next one. This is referred to as **optokinetic nystagmus** (Figure 9.19).

a. **Cold water caloric testing: Vestibular nystagmus** is caused by stimulation of the vestibular organ (Table 9.1; see also Chapter 11, "Hearing and Balance"). This stimulation is due to the acceleration or movement of fluid, called **endolymph**, within the vestibular organ. The movement of endolymph can be elicited by irrigating the external auditory canal with cold water or warm water. The mnemonic **COWS** (**c**old, **o**pposite; **w**arm, **s**ame) describes the outcome of caloric testing. For example, irrigating the external auditory canal with cold water sets up a **temperature gradient** in the temporal bone (Figure 9.20). The cold water cools the part of the horizontal canal closest to the external auditory canal relative to the part farthest away from the external auditory canal. The density of endolymph depends on temperature: Cool endolymph is denser than warm endolymph. Because of gravity, cool endolymph will sink toward the warm endolymph. Caloric testing is performed with the person in the supine position, with the head tipped forward at a 30-degree angle from the neutral position, which orients the horizontal canal precisely in the horizontal plane. Cold water in the right ear causes movement of endolymph in a clockwise direction in the horizontal canal. This is the same movement seen when the head moves to the left. The VOR then causes the eyes to move to the right. The ensuing nystagmus will have a slow component to the right and a fast (reset) component to the

Table 9.1: Caloric Testing

Irrigation	Movement of Endolymph	Type of Head Movement (Simulated)	Slow Eye Movement (VOR)	Fast Eye Movement (Direction of Nystagmus)
Right ear cold	Clockwise	Left	Right	Left
Right ear warm	Counterclockwise	Right	Left	Right
Left ear cold	Counterclockwise	Right	Left	Right
Left ear warm	Clockwise	Left	Right	Left

VOR = vestibuloocular reflex.

left. Thus, the nystagmus is left beating, in the direction opposite to where the cold water was applied (**CO**WS, **c**old **o**pposite).

b. **Warm water caloric testing:** Irrigation with warm water will have the opposite effect. The temperature gradient in the horizontal canal will be reversed. Warm water in the right ear will result in the endolymph moving counterclockwise, which likewise occurs with head movement to the right. The VOR then causes the eyes to move to the left. The ensuing nystagmus will have a slow component to the left and a fast (reset) component to the right. Thus, the nystagmus is right beating, in the same direction as the side where the warm water was instilled (CO**WS**, **w**arm **s**ame).

Because caloric testing uses the same mechanisms as the VOR, it is an excellent tool to assess brainstem function in the unconscious patient.

B. Vertical gaze

Vertical gaze involves coordinated movement of the two eyes upward and downward. Vertical gaze can be in the form of a saccadic eye movement, a tracking movement, or as part of the VOR. As with horizontal gaze, the eyes must move synergistically. This synergistic movement of extraocular muscles in the vertical plane is coordinated by the vertical gaze center, a collection of nuclei located in the midbrain reticular formation and the **pretectal area** (just rostral to the superior colliculus) (Figure 9.21). There is some suggestion, based on gaze deficits that occur with pressure on the vertical gaze center (e.g., from a tumor), that the anterior portion of this region mediates downgaze, whereas the more posterior portion of this region mediates upgaze. Clinically, the first deficit observed is usually a deficit in upgaze. Here, we consider the vertical gaze center as a whole. The input to the vertical gaze center is analogous to the input to the PPRF or abducens nucleus (i.e., from the **FEFs** and from visual association areas via the **superior colliculus**). From the vertical gaze center in the midbrain, the signal travels bilaterally to the respective cranial nerve nuclei: CN III (oculomotor) to innervate the superior and inferior rectus and inferior oblique and CN IV (trochlear) to innervate the superior oblique. Together, the superior rectus and inferior oblique elevate the eye, and the inferior rectus and superior oblique depress the eye.

Gaze that involves **oblique eye movements** is generated through a combination of horizontal and vertical gaze.

C. Disconjugate gaze

When the two eyes move in opposite directions from each other, we refer to the movement as **disconjugate gaze** (Figure 9.22). This occurs physiologically when we are focusing on something in a very near or very far field of view. When focusing near, both eyes **adduct**, that is, the left and right eyes both move inward toward the nose. This is called **convergence**, which is a component of **accommodation** (see below). In contrast, when focusing on something very far in the distance, both eyes will **abduct**, that is, the left and right eyes both move outward, away from the nose. This is called **divergence**.

Information about an object in a near or far field of view is relayed to **both occipital lobes** via the two **lateral geniculate nuclei in the**

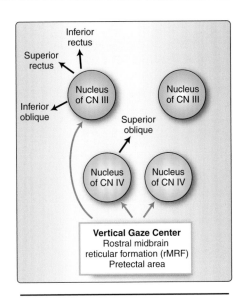

Figure 9.21
Wiring diagram for vertical gaze. CN = cranial nerve; rMRF = rostral midbrain reticular formation.

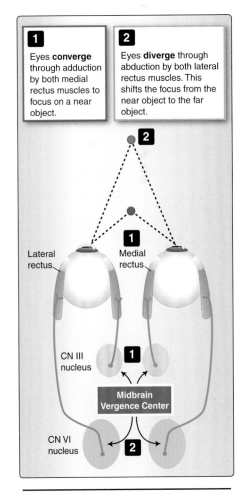

Figure 9.22
Wiring diagram for convergence and divergence. CN = cranial nerve.

thalamus. In the occipital lobes, the information from both eyes is integrated, and the decision to converge (for near objects) or diverge (for far objects) is made. This decision is then relayed to the **vergence centers** in the rostral midbrain near the CN III motor nucleus. From there, CN III is activated on both sides to innervate both medial rectus muscles for convergence, or CN VI is activated on both sides to innervate both lateral rectus muscles for divergence. A discreet switch between convergence and divergence is made when our focus goes from a near object to a far object. In order to shift the focus, the eyes must diverge so that the second (more remote) object can be brought into focus.

Accommodation happens when we **focus on a near object**. In order for accommodation to occur, allowing a near image to be projected precisely onto the **fovea**, three things must happen:

- The eyes must **converge**. Both eyes move toward the midline (**adduct**) through the activation of both medial rectus muscles.
- The **refractive power** of the lens must be increased. This occurs by **increasing the curvature** of the lens.
- The **pupils must constrict**. This increases the **depth of field**.

These three components of the accommodation response are termed the **near triad**.

Accommodation is coordinated through a structure in the rostral midbrain called the pretectal area as shown in Figure 9.23. The pretectal area sends bilateral projections to the somatic motor neurons of

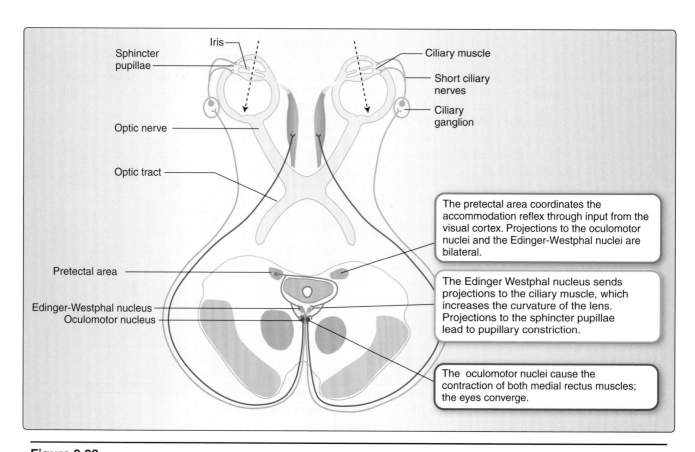

The pretectal area coordinates the accommodation reflex through input from the visual cortex. Projections to the oculomotor nuclei and the Edinger-Westphal nuclei are bilateral.

The Edinger Westphal nucleus sends projections to the ciliary muscle, which increases the curvature of the lens. Projections to the sphincter pupillae lead to pupillary constriction.

The oculomotor nuclei cause the contraction of both medial rectus muscles; the eyes converge.

Figure 9.23
Conceptual overview of accommodation. In this diagram, the orientation of the midbrain section has been flipped in relation to the orientation of the eyes.

CN III to innervate the medial rectus muscles. In addition, projections to the visceral efferent or **parasympathetic Edinger-Westphal** nuclei allow for pupillary constriction through activation of neurons projecting to the **constrictor pupillae muscles**. Excitation of the **ciliary muscles** in the eyes leads to relaxation of the zonule fibers attached to the lens, which, in turn, increases the curvature of the lens.

Input to the pretectal area comes from the occipital cortex. Here, the retinotopic maps from both eyes are analyzed and compared. Any discrepancy leads to a blur in the visual field, which is corrected through adjustments in accommodation. It is thought that the cerebellum also sends input to the pretectal area.

Clinical Application 9.1: H-test

The most common way to test the functioning of extraocular muscles and the cranial nerves that innervate these muscles is a simple eye movement test referred to as the **H-test**. In this test, each eye is tested separately, as indicated by the different colors in the figure. Most eye movements involve combinations of muscles working together. It is, however, possible to isolate the actions of single extraocular muscles by focusing on select movements. Although the H-test does not test each extraocular muscle, it does test for all cranial nerves innervating these muscles and is commonly used in clinical practice to screen for disorders in eye movement.

First, the patient is asked to look laterally, for example, to the left. The left eye abducts through the action of the lateral rectus muscle innervated by cranial nerve (CN) VI. Abduction of the eye aligns the axis of the eyeball with the axis of the orbit and the muscles attaching to the tendinous ring. When the eye is abducted, the patient is asked to look up—this tests the superior rectus, which is innervated by the superior division of CN III. The action of superior rectus can only be isolated when the patient has abducted the eye. With the eye still abducted, the patient is then asked to look down; this tests the inferior rectus muscle, which is innervated by the inferior division of CN III. The patient is then asked to look to the right, adducting the left eye. This tests the medial rectus muscle, which is innervated by the inferior division of CN III. With the eye adducted, the patient is then asked to look down. A downward movement will isolate the action of the superior oblique muscle, testing CN IV. With the eye still adducted, the patient is asked to look up. This tests the inferior oblique muscle, which is also innervated by the inferior division of CN III.

The same movements are repeated, focusing on the actions of the other eye.

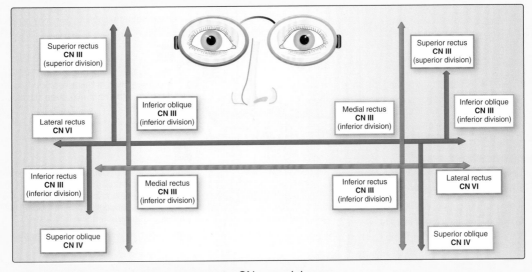

The H-test for assessing eye movements. CN = cranial nerve.

Clinical Application 9.2: Internuclear Ophthalmoplegia

A lesion of the **medial longitudinal fasciculus (MLF)** will lead to a disruption in the synergistic movement of the eyes in horizontal gaze.

As noted, interneurons within the abducens nucleus send fibers via the MLF to the contralateral oculomotor nucleus. These fibers cross the midline immediately upon exiting the abducens nucleus and travel in the contralateral MLF.

A complete unilateral lesion in one MLF (the right, for instance) will cause a problem with gaze to the contralateral side (in this case, the left). The horizontal gaze center (PPRF) provides input to the left abducens nucleus, from which gaze to the left is initiated. The left eye will **abduct** through innervation of the left lateral rectus by the left abducens nerve. However, because the communication to the right oculomotor nucleus via the right MLF is interrupted, the right eye will not move to the left or **adduct** as shown in the figure. If the lesion results in an incomplete or partial impairment of the MLF, then the right eye will adduct slowly or partially.

Because the two eyes are not properly aligned, a horizontal eye movement to the left will result in double vision (diplopia). In addition, a nystagmus is typically present with this type of lesion. The left, abducted eye will drift to the midline and essentially align itself with the right eye, which is not moving. The input from the abducens nucleus will then rapidly move the left eye back to the left. It then drifts back to the midline but again rapidly moves to the left, and so on. This is a left-beating nystagmus. One hypothesized reason for this nystagmus is that there is increased drive to the paretic adducting eye (an attempt to initiate adduction). This is accompanied by an increased innervation (CN VI) to the abducting eye, which leads to an overshoot of the abducting saccade. The abducting eye then moves slowly back to midline as a response to the overshoot. These movements present as nystagmus.

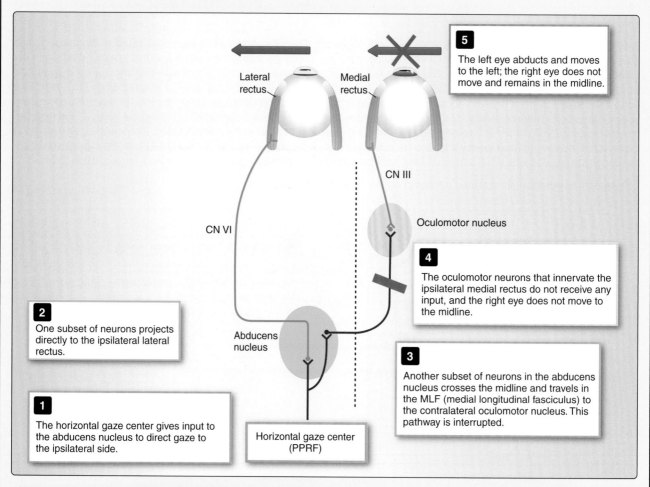

Wiring diagram showing the deficit in internuclear ophthalmoplegia. CN = cranial nerve; PPRF = paramedian pontine reticular formation.

Clinical Application 9.2: Internuclear Ophthalmoplegia continued

In order to test whether the problem is indeed with the MLF and not with the oculomotor nucleus or the peripheral nerve, the patient is asked to focus on a point in the midline. In this case, the **vergence center** sends input to both oculomotor nuclei, and both eyes move to the midline, or adduct.

If a patient can adduct the eye in vergence movements, but not in horizontal eye movements, the lesion is usually located in the MLF.

This type of internuclear ophthalmoplegia is often seen as the first sign in a demyelinating disorder such as **multiple sclerosis**. The MLF contains heavily myelinated fibers and is particularly susceptible to demyelination.

Clinical Cases:

Gaze Disorders

The following cases are all examples of gaze disorders. Correlate the symptoms to lesions in cranial nerves, their nuclei, or components of the gaze pathways.

Case Histories

Case 1

A 35-year-old female patient has a 3-day history of diplopia (double vision) when she looks to the right. An H-test is performed to assess horizontal eye movements. When the patient is asked to look to the right (saccadic eye movement to the right), the left eye moves slowly toward the midline (slow adduction) and the right eye shows a right-beating nystagmus. When the patient is asked to look to the left, the saccade is normal. The patient is asked to look at a near point and convergence is normal. Smooth pursuit extraocular movements are full. The size and reactivity of both pupils is normal (5 mm and reactive to light). The patient has a relative afferent pupillary defect (RAPD, or Marcus Gunn pupil) of the left eye with reduced visual acuity of 20/100 of the left eye. This is indicative of a lesion to the optic nerve. Visual acuity in her right eye is normal at 20/20. There is no visual field deficit. The rest of her cranial nerve examination is normal.

Case 2

A 63-year-old man comes to the clinic with weakness in the left side of his face and in his left arm. His gaze is directed to the right. His extraocular movements are normal when testing for smooth pursuit. He has trouble with volitional saccadic movement to the left. Saccades to the right are normal. The rest of his neurological exam is normal.

Case 3

A 55-year-old woman has double vision (diplopia), which she describes as a second hazy outer image. She only has this diplopia when looking to the right. She notices that when she covers her right eye, the outer hazy image (false image) disappears. The diplopia is worse with far vision. She has a mild esotropia (eye turned in; intorsion) of her right eye that increases when looking to the right and is absent when looking to the left. Her pupils are both normal in size and reactive to light.

Case 4

A 75-year-old man presents to the clinic with double vision. He describes the images as being shifted at a 45-degree angle from each other—this description indicates a separation of the images on both a horizontal and vertical axis and is referred to as oblique diplopia. He notices that when he covers either eye, his double vision disappears. His pupils are unequal in size: the right pupil is 5 mm and the left pupil is 2 mm. The eyelid on the right side is depressed (ptosis). When you perform an H-test, you notice that the right eye has limited movement toward the midline (adduction) and limited movements up and down (elevation and depression). During abduction of his right eye, the eye intorts. His left extraocular movements are normal.

Case 5

A 25-year-old man presents with oblique diplopia following head trauma. When examining his eye movements, the patient states that his double vision is worse when looking down and looking to the right. When his head is straight and he is looking forward, the left eye appears higher than the right eye (this is referred to as the left eye being hypertropic). When he looks to the right, the hypertropia increases. His head is tilted to the right. His pupils are symmetric and reactive to light. There is no ptosis and no other cranial nerve abnormalities.

Case Discussions

Case 1

The horizontal gaze deficits are indicative of a lesion to the left ascending medial longitudinal fasciculus (MLF). The right horizontal gaze center (PPRF) gives input to the right abducens nucleus to initiate the gaze to the right. The right eye will abduct through innervation of the right lateral rectus muscle by the abducens nerve. This pathway is intact. The right PPRF gives input to the left oculomotor nucleus via the left MLF. There is a lesion of the left MLF resulting in slow movement of the left eye to the right. She also has left optic neuropathy with impaired visual acuity and RAPD. An afferent pupillary defect can be detected with the swinging light test. In this test, a bright light is swung from the unaffected eye to the affected eye. The affected eye has a deficit with the perception of light or the transmission of that signal along the optic nerve. It still senses the light and the pupil constricts to some degree, but not fully. Whereas both pupils constrict when light is shone into the unaffected eye (direct and consensual), the deficit in light perception of the affected eye results in a lesser degree of pupillary constriction and both eyes appear to dilate (Figure 9.24).

MRI of her brain showed a demyelinating lesion in the medial left rostral pons. These symptoms can be seen in patients with demyelinating disorders such as multiple sclerosis (MS).

Case 2

This patient has been recovering from a right middle cerebral artery (MCA) stroke. This has affected his primary motor area on the right, especially the areas for the innervation of the face and the upper limb. The right frontal eye fields are just anterior to the primary motor area and have also been affected in this stroke. There is a gaze preference to the right, and the underlying cause for this could be (1) excitation of the left frontal eye field, (2) inhibition (loss of function) of the right frontal eye field, or (3) inhibition of the horizontal gaze center (PPRF) on the right. The right frontal eye field initiates volitional saccades to the left via the left PPRF (see Figure 9.13); it also initiates smooth pursuit to the right (see Figure 9.16).

This patient's smooth pursuit is normal; it is therefore unlikely to be a lesion that involves any of the brainstem structures involved in horizontal gaze (MLF, CNs III, IV, or VI). The presence of left face and arm weakness helps to localize the lesion for his gaze preference to the right frontal eye field. Due to the lesion of the right FEF, the patient can no longer initiate volitional saccades to the left. The contralateral (left) frontal eye fields are now predominant, which in this case causes a gaze deviation to the right. In a cortical stroke, eyes "look at the lesion"—they deviate to the side of the stroke.

Case 3

This presentation is indicative of a right abducens nerve (CN VI) palsy, which causes weakness of the lateral rectus muscle (Figure 9.25B). Right abducens nerve palsy causes double vision in horizontal eye movements, which is worse in gaze toward the affected muscle and goes away when the affected eye is covered. The worsening seen with far vision is due to the fact that this requires divergence of both eyes, which is dependent on the lateral rectus muscles. The esotropia (eye turned in; intorsion) is due to the weakness of the lateral rectus muscle, which can no longer counterbalance the effects of the superior oblique and medial rectus. It typically gets worse during horizontal gaze to the side of the lesion.

Case 4

This presentation is typical of a lesion to the right oculomotor nerve (CN III) after it emerges from the brainstem

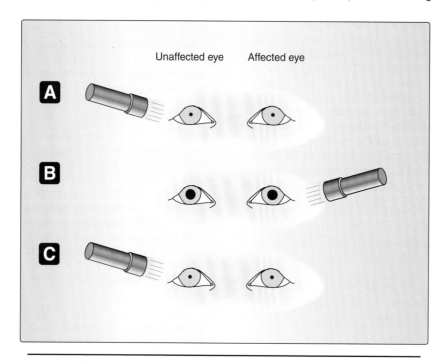

Figure 9.24
Relative afferent pupillary defect (Marcus Gunn pupil). **A.** A penlight directed at the unaffected eye causes constriction of both pupils because both direct and consensual constriction occurs. **B.** As the light is directed at the affected eye, which has reduced detection of light, the direct and consensual responses are reduced and both pupils appear to dilate. **C.** When the light is swung back toward the unaffected eye, both pupils again constrict. (From Plantz SH, Huecker M. *Step-Up to Emergency Medicine.* Wolters Kluwer Health, Philadelphia, PA: 2015.)

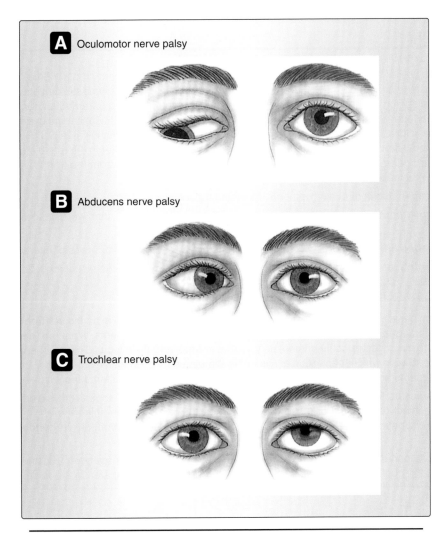

Figure 9.25
Position of the eyes after lesion to **(A)** the oculomotor nerve, CN III; **(B)** the
abducens nerve, CN VI; and **(C)** the trochlear nerve, CN IV - often the eye
is also deviated out. (Modified from Young PA, Young PH, Tolbert DL. *Basic
Clinical Neuroscience*, 3rd ed. Wolters Kluwer Health, Philadelphia, PA: 2015.)

(Figure 9.25A). A CT angiogram of the arteries that supply the
brain shows an aneurysm on the posterior communicating
artery, which compresses the right oculomotor nerve.
Compression of the oculomotor nerve causes oblique or vertical diplopia due to weakness of the medial rectus, superior
rectus, inferior rectus, and inferior oblique muscles. There is
ptosis due to weakness of the levator palpebrae muscle. The
dilated pupil, which is poorly reactive to light, is due to impairment of the parasympathetic fibers originating in the visceral
motor (Edinger-Westphal) nucleus that travel with CN III to
the orbit where they innervate the constrictor pupillae muscle.

Case 5

This presentation is typical for a lesion of the left trochlear nerve due to head trauma (Figure 9.25C). The trochlear nerve is the only nerve that exits the brainstem on the
posterior surface; it is thin and susceptible to shear injury.
The left trochlear nerve innervates the left superior oblique

muscle. Its actions depend on the position of the eye within
the orbit—in the adducted eye, it depresses the eyeball, and
in the abducted eye, it causes intorsion of the eyeball. A weak
left superior oblique muscle would cause left hypertropia with
oblique diplopia that is worse on gaze to the right. A head
tilt to the right would improve the diplopia, since this would
realign the two eyes to some extent.

***In case 4, an important part of the examination was to
check for intorsion of the right eye when that eye is
abducted. What is the clinical significance of this finding?***

When performing a clinical exam to assess eye movements,
it is important to check for possible lesions of all nerves supplying the extraocular muscles. If there is an impairment of
intorsion, it would suggest an additional lesion to the trochlear
nerve. The causes of injury to both the oculomotor and trochlear nerves are different from lesions to just the oculomotor
nerve and could be indicative of further neurological damage.

Chapter Summary

- Each eye is moved by six extraocular muscles that are innervated by three different cranial nerves. The movement of the eyes must be coordinated so that both eyes focus on the same visual field. This synergistic movement of the eyes is coordinated through gaze centers in the brainstem. The horizontal gaze center is in the paramedian pontine reticular formation and projects to the abducens nucleus (cranial nerve [CN] VI), where abduction of the ipsilateral eye is initiated. At the same time, fibers cross the midline and innervate the oculomotor nucleus (CN III), which initiates adduction in the contralateral eye. Both eyes can, thus, move in the same direction. Similarly, a vertical gaze center in the rostral midbrain coordinates the upward and downward gaze of both eyes. Cortical input to these gaze centers is from the frontal eye fields and the visual association areas via the superior colliculus.

- There are several different types of eye movements, including the following:
  - **Saccadic eye movements** are rapid coordinated movements of both eyes that we use to orient ourselves in our environment.
    - **Volitional saccades** are under cortical control and direct the eyes to objects or areas of interest.
    - **Reflexive saccades** are coordinated through the superior colliculus where the visual input is linked to a motor map of eye movements.
  - **Pursuit movements** are used to track an object, keeping it stable on the fovea.

- The vestibular system plays a key role in eye movements via a reflex pathway known as the **vestibuloocular reflex (VOR)**. This reflex allows for images to remain stable on the fovea despite head movements. When the head moves, compensatory movements of the eye muscles occur immediately so that images remain in focus. This mechanism of the VOR is used clinically in **caloric testing** in which the vestibular system is stimulated through application of cold or warm water to the ear canal, and eye movements are then analyzed. This allows for assessment of brainstem function in the unconscious patient.

- **Disconjugate gaze** occurs when the eyes move in opposite directions. This happens physiologically when both eyes converge on a near object and adduct or move medially. When the focus is then shifted to an object farther away, both eyes move laterally, or abduct. This disconjugate eye movement is coordinated through **vergence centers** in the rostral midbrain.

- **Vergence** is also a component of **accommodation**, in which the eyes focus on a near object. Both eyes move medially, the lens is rounded, and the pupils constrict. This is coordinated through the pretectal area in the midbrain.

Study Questions

Choose the ONE best answer.

9.1 During an examination, the physician asks the patient to follow his finger as he moves it to the right and left. Eye movements to the left are normal. When the patient tries to follow the physician's finger to the right, the right eye moves normally, but the left eye remains in the midline and does not follow to the right. Convergence is deficient with the left eye remaining in the midline. Where could the problem lie?

    A. A lesion to the left cranial nerve IV.
    B. A lesion to the right cranial nerve VI.
    C. A lesion to the left Edinger-Westphal nucleus.
    D. A lesion to the left cranial nerve III.
    E. A lesion to the right cranial nerve III.

The correct answer is D. Cranial nerve (CN) IV supplies the superior oblique muscle, which is not involved in lateral eye movements. The right CN VI supplies the right lateral rectus muscle that is abducting the eye normally. The Edinger-Westphal nucleus is involved in the parasympathetic supply to the pupillary constrictor and the ciliary muscle of the lens and is not involved in the problem. The left CN III to the medial rectus muscle would be needed to adduct the left eye in conjunction with the lateral rectus of the right eye, enabling the patient to follow the finger to the right. A lesion to the right CN III would prevent the right medial rectus from adducting the right eye medially.

9.2 Synergistic or coordinated movements of the two eyes and adjustments of eye position in response to movements of the head are mediated by:

A. The corticobulbar tracts.
B. The medial longitudinal fasciculus.
C. The cochlear apparatus.
D. The lateral vestibulospinal tract.
E. The corticospinal tract.

The correct answer is B. The medial longitudinal fasciculus (MLF) connects the nuclei of cranial nerves (CNs) III, IV, and VI with the vestibular nuclei to allow synergistic and coordinated movements of the eyes and adjustments of the eyes to changes in head position. Corticobulbar tracts do not influence the nuclei of CNs III, IV, and VI. These nuclei receive their input from frontal eye fields and other cortical gaze centers. The cochlear apparatus is concerned with hearing, not vision. The lateral vestibulospinal tract arises from the lateral vestibular nuclei and is important in maintaining balance and extensor muscle tone. The corticospinal tract sends motor output to the anterior horn cells of the spinal cord.

9.3 In an unconscious patient, vestibular function was assessed through caloric testing. The right auditory canal was irrigated with cold water. If the vestibular system is intact, which statement below is correct?

A. Due to movement of endolymph in a clockwise direction, the eyes would drift slowly to the right followed by a quick flick to the left.
B. Due to movement of endolymph in a clockwise direction, the eyes would drift slowly to the left followed by a quick flick to the right.
C. Due to the movement of endolymph in a counterclockwise direction, the eyes would drift slowly to the right followed by a quick flick to the left.
D. Due to the movement of endolymph in a counterclockwise direction, the eyes would drift slowly to the left followed by a quick flick to the right.
E. Due to the movement of endolymph in a counterclockwise direction, both eyes move to the side of the cold irrigation and then stay there.

The correct answer is A. Cold water in the right ear causes movement of endolymph in a clockwise direction in the horizontal canal. This is the same movement seen when the head moves to the left. The vestibuloocular reflex then causes the eyes to move slowly to the right, with a fast (reset) component to the left. Thus, the nystagmus is left beating, in the direction opposite to where the cold water was instilled (**CO**WS, **c**old **o**pposite). Cold water causes a clockwise, not a counterclockwise, movement of endolymph. Irrigation of warm water into the right ear sets up a counterclockwise movement of endolymph causing the eyes to drift slowly to the left followed by a quick flick to the right (CO**WS**, **w**arm **s**ame).

9.4 A patient presents with weakness of the left hand and upper limb, increased tone and deep tendon reflexes in the left upper limb, deviation of both eyes to the right, and an inability to initiate voluntary gaze to the left. Voluntary gaze to the right and convergence movements of the eyes as well as sensory function throughout the body are normal. A lesion resulting in these problems is most likely in the:

A. Right frontal cortex, in the area of the middle frontal gyrus, including the motor areas and the frontal eye field.
B. Left frontal cortex, in the area of the middle frontal gyrus, including the motor areas and the frontal eye field.
C. Left half of the pons, extending from the sixth nerve nucleus to the basal pons.
D. Left half of the pons, involving primarily the left paramedian pontine reticular formation and basal area.
E. Right paramedian midbrain involving the oculomotor and Edinger-Westphal nuclei.

The correct answer is A. The right frontal cortex, middle frontal gyrus, includes the primary and motor association areas innervating the hand and upper limb and also includes the right frontal eye field (FEF). The lesion to the motor areas results in upper motor neuron (UMN) signs including weakness, hypertonia, and hyperreflexia. The right FEF initiates voluntary saccades to the left. Due to reduced activity in the right FEF and normal activity in the left FEF, the eyes are deviated to the right. A lesion in the left frontal cortex would result in weakness and UMN signs on the right as well as deviation of the eyes to the left and loss of volitional gaze to the right. A lesion in the left half of the pons involving the sixth cranial nerve nucleus could result in loss of voluntary gaze to the left, as both the motor neurons for the left lateral rectus and the interneurons giving rise to the right medial longitudinal fasciculus to the right medial rectus would be affected. However, descending corticospinal fibers in the left basal pons innervate lower motor neurons (LMNs) to the right limbs. Such an extensive lesion to the entire left half of the pons would affect other structures, such as the left medial lemniscus and anterolateral system, resulting in marked sensory loss. A lesion of the left paramedian pontine reticular formation could produce the deficits in gaze described, but the eyes would not be deviated to the right.

9.5 A patient presents complaining of double vision and trouble reading. Examination of the eyes reveals that the pupils are responsive to light but fail to constrict on convergence. In addition, both upgaze and convergence movements are impaired. Individually, the eyes show intact movements in all directions. The most likely site of a lesion is the:

A. Posterior area of the caudal pons.
B. Posterior area of the rostral midbrain.
C. Right frontal lobe.
D. Right medial longitudinal fasciculus.
E. Tegmentum of the caudal pons on the right.

The correct answer is B. A lesion in the posterior area of the rostral midbrain can affect the center responsible for upward gaze. Pressure can also damage the pretectal area posterior to the cranial nerve (CN) III nucleus, resulting in deficits in convergence. The posterior area of the caudal pons is where the CN VI nuclei are located. A lesion here would result in loss of abduction and loss of conjugate gaze due to damage to the interneurons in the abducens nucleus from which the medial longitudinal fasciculus (MLF) arises. A lesion in the frontal lobe will result in loss of voluntary gaze to the contralateral side, and the eyes will be deviated to the side of the lesion due to input for gaze from the intact contralateral frontal eye field. Weakness or paralysis on the contralateral side of the body may also occur. A lesion in the right MLF will result in problems with conjugate gaze to the left: the left eye will abduct, but the right eye will remain in the midline and will not adduct. A lesion in the tegmentum of the caudal pons on the right will damage the paramedian pontine reticular formation (PPRF), resulting in loss of horizontal gaze; the eyes may be deviated away from the lesion due to input for gaze from the intact PPRF on the left, driving gaze to the left. Note the contrast among gaze palsies resulting from pontine versus midbrain versus cortical lesions: A pontine lesion results in a horizontal gaze palsy, and eyes deviate away from the lesion; a midbrain lesion results in a vertical (usually upgaze) palsy, a cortical lesion results in a horizontal gaze palsy, and eyes deviate to the side of the lesion.

Sensory and Motor Innervation of the Head and Neck

10

I. OVERVIEW

The head and major areas of the neck receive their sensory and motor innervation through **cranial nerves V, VII, IX, X, XI,** and **XII** (Figure 10.1). These are the trigeminal, facial, glossopharyngeal, vagus, accessory, and hypoglossal nerves, respectively. The specialized innervation for the control of eye movements and ocular reflexes was covered in Chapter 9, "Control of Eye Movements." The nerves associated with special senses will be discussed in subsequent chapters dedicated to those modalities.

The modalities carried by these cranial nerves are analogous to the modalities found in the spinal nerves, and just like spinal nerves, most cranial nerves are mixed nerves, carrying more than one modality. Just like the grey matter in the spinal cord is subdivided into sensory areas (posterior horn) and motor areas (anterior and lateral horns), different brainstem nuclei are associated with sensory and motor modalities. In Chapter 6, "Overview and Organization of the Brainstem," the general organization of the cranial nerve nuclei in the brainstem was reviewed, noting that sensory nuclei are lateral to and motor nuclei are medial to the sulcus limitans. Visceral afferents and efferents lie closest to the sulcus limitans (see Figure 10.1). This organization is important for understanding the innervation of the head and neck.

Cranial nerves have the same four components as the spinal nerves and an additional three. **General visceral afferent** ([**GVA**] from viscera in the core), **general somatic afferent** ([**GSA**], from receptors in the skin, muscles, joints), **general somatic efferent** ([**GSE**] to skeletal muscles), and **general visceral efferent** ([**GVE**] preganglionic autonomic fibers to core and periphery) are the components also found in spinal nerves. The components unique to cranial nerves are **special visceral afferent** ([**SVA**] taste, olfaction), **special somatic afferent** ([**SSA**] vision, balance, hearing), and **special visceral efferent** ([**SVE**] to muscles of the jaw, face, larynx, pharynx; derived from pharyngeal arches, not somites) (Table 10.1).

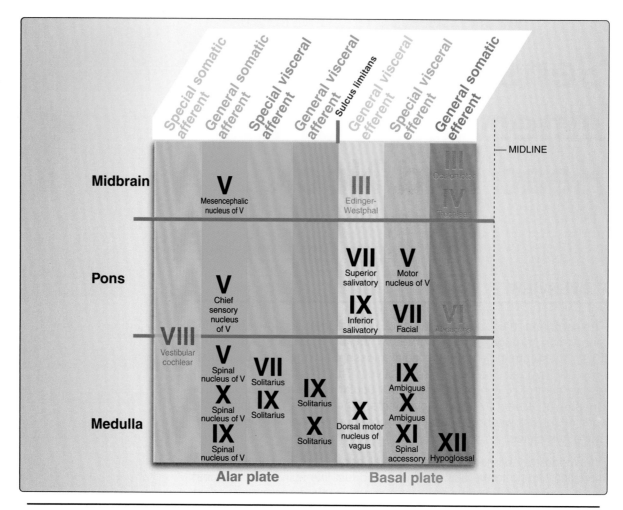

Figure 10.1
Overview of the locations and functional components of the cranial nerves. (Eye movements and special senses excluded.)

Table 10.1: Overview of the Cranial Nerves, Associated Nuclei, and Modalities

	GSA	SVA	GVA	GVE	SVE	GSE
CN V trigeminal	**Mesencephalic nucleus** **Spinal nucleus of V** **Chief sensory nucleus of V**				**Motor nucleus of V**	
CN VII facial	**Spinal nucleus of V** **Chief sensory nucleus of V**	**Nucleus solitarius**		**Superior salivatory nucleus**	**Facial nucleus**	
CN IX glosso-pharyngeal	**Spinal nucleus of V** **Chief sensory nucleus of V**	**Nucleus solitarius**	**Nucleus solitarius**	**Inferior salivatory nucleus**	**Nucleus ambiguus**	
CN X vagus	**Spinal nucleus of V** **Chief sensory nucleus of V**	**Nucleus solitarius**	**Nucleus solitarius**	**Dorsal motor nucleus of X** **Nucleus ambiguus**	**Nucleus ambiguus**	
CN XI accessory					**Spinal accessory nucleus**	
CN XII hypoglossal						**Hypoglossal nucleus**

CN = cranial nerve; GSA = general somatic afferent; SVA = special visceral afferent; GVA = general visceral afferent; GVE = general visceral efferent; SVE = special visceral efferent; GSE = general somatic efferent. This table comprises only CNs discussed in this Chapter.

Sensory:

- **GSA**, sensory information from the skin, meninges, and mucosa in the head, is carried by CNs V, VII, IX and X.

- **GVA**, sensory information from the viscera of the head and neck, is carried by CNs IX and X.

- **SVA**, carrying taste, travels in CNs VII, IX, and X. The details of taste perception and processing are covered in Chapter 21, "Smell and Taste."

- **SSA**, information for hearing and balance, is carried by CN VIII (vestibulocochlear) and is discussed in Chapter 11, "Hearing and Balance."

Motor:

- **GSE**, motor innervation to skeletal muscles derived from somites, is through CNs III, IV, and VI (oculomotor, trochlear, and abducens, respectively), which innervate the extraocular muscles (as discussed in Chapter 9, "Control of Eye Movements"), and CN XII, which innervates the tongue (as discussed below in *Hypoglossal Nerve: Cranial Nerve XII*).

- **GVE**, in this case *parasympathetic* innervation to the glands in the head and neck, is through CNs VII, IX, and X (facial, glossopharyngeal, and vagus, respectively). As discussed previously, CN III carries parasympathetic fibers to the eye.

- **SVE**, motor innervation to skeletal muscles derived from pharyngeal arches, is through CN V, which innervates the muscles of mastication, and CN VII to the muscles of facial expression as well as CNs IX and X to muscles of the pharynx and larynx and CN XI (accessory) to some muscles in the back and neck.

Cranial nerves are mixed nerves that can carry more than one modality; however, each *modality* has at least one cranial nerve nucleus associated with it. Each *cranial nerve* will, therefore, have several brainstem nuclei associated with it, and each brainstem nucleus can supply fibers or contain target neurons for more than one cranial nerve. For example, GSA information on touch and proprioception is carried by several cranial nerves (V, VII, IX, and X). All of these fibers synapse in the trigeminal nuclear complex. Similarly, SVA fibers carrying taste travel in CNs VII and IX, and their target neurons are in the nucleus solitarius.

In this chapter, we examine CNs V, VII, IX, X, XI, and XII and associated nuclei that innervate the head and neck.

II. TRIGEMINAL NERVE: CRANIAL NERVE V

The **trigeminal nerve (CN V)** emerges from the **midpons**, contains the largest **sensory ganglion** in the skull, and comprises **three major divisions** that provide motor and sensory innervation to the head (Figure 10.2). The trigeminal ganglion is located in the **middle cranial fossa**.

The **ophthalmic division of the trigeminal nerve (V$_1$)** passes through the superior orbital fissure and carries sensory information from the upper face, the orbit, and the eye.

The **maxillary division (V$_2$)** passes through the foramen rotundum and carries sensory information from the upper teeth and the midface.

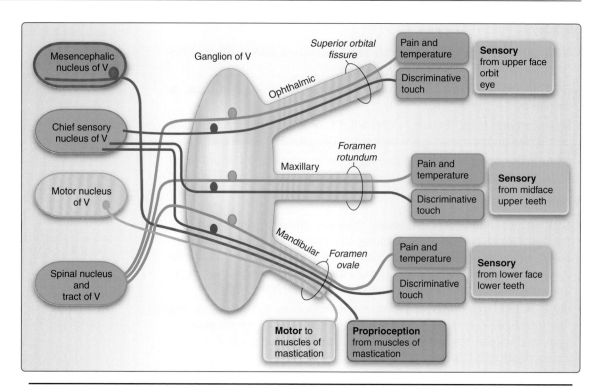

Figure 10.2
Overview of the modalities and functions of the trigeminal nerve.

The **mandibular division (V₃)** passes through the foramen ovale and carries sensory information from the lower face and lower teeth and provides motor innervation to the muscles of mastication. Proprioceptive information from these muscles also travels with this nerve.

The trigeminal nerve has **four cranial nerve nuclei** associated with it. General somatic afferent information is processed by the **chief sensory nucleus** for touch, pressure, and conscious proprioception and by the **spinal nucleus and tract of V** for pain and temperature. The **mesencephalic nucleus** in the midbrain processes nonconscious proprioceptive information. Special visceral efferent innervation to the muscles of mastication, derived from the first branchial arch, comes from the **motor nucleus of V**.

A. Sensory component of CN V

The trigeminal nerve processes sensory information from the head, especially from the face, as well as the oral and nasal cavities. Much like the spinal cord, this somatosensory information can be divided into three categories: **discriminative touch, vibration,** and **conscious proprioception**; **pain** and **temperature**; and **nonconscious proprioception**. As in the spinal cord, each one of these categories has its own pathway in the trigeminal system and synapses in a specific nucleus. The gray matter of the spinal cord and the trigeminal system have similar structure and function, and there is a gradual transition from one to the other, as summarized in Table 10.2.

GSA fibers are processed by three components of the trigeminal nuclear complex. Discriminative touch, vibration, and proprioception are processed by the **chief sensory nucleus of V**, and pain

Table 10.2: Comparison of Spinal Cord Structures to Trigeminal Nuclear Complex

	Spinal Cord	Trigeminal Nuclear Complex
Discriminative Touch, Vibration, and Proprioception		
Cell bodies	Spinal ganglion	Trigeminal ganglion, mesencephalic nucleus
Central processes	Fasciculus gracilis and cuneatus	Entering trigeminal fibers
Second-order neurons	Nucleus gracilis and cuneatus	Chief sensory nucleus of V
Ascending pathway	Medial lemniscus, crossed	Trigeminal lemniscus, crossed
Thalamic nucleus	Contralateral VPL	Contralateral VPM
Pain and Temperature		
Cell bodies	Spinal ganglion	Trigeminal ganglion
Central processes	Lissauer tract	Spinal tract of V
Second-order neurons	Posterior horn	Spinal nucleus of V
Ascending pathway	Spinothalamic tract, crossed	Trigeminothalamic tract, crossed
Thalamic nucleus	Contralateral VPL	Contralateral VPM

VPL = ventral posterolateral nucleus; VPM = ventral posteromedial nucleus.

and temperature are processed in the **spinal trigeminal nucleus.** Proprioception is processed in the **mesencephalic nucleus** of CN V in the midbrain (Table 10.3).

1. **Chief sensory nucleus of CN V:** Located in the **midpons,** the **chief sensory nucleus of V** (Figure 10.3) is the trigeminal homologue of the posterior column nuclei (nucleus gracilis and nucleus cuneatus) where primary afferent fibers synapse. The cell bodies of the primary afferent fibers are located in the **trigeminal ganglion.**

From the chief sensory nucleus, two ascending pathways project to the thalamus, from which **third-order neurons** travel in the posterior limb of the internal capsule and terminate in the face area of the primary somatosensory cortex (Figure 10.4).

Table 10.3: The Trigeminal Nerve: Cranial Nerve V

Nerve Fiber Modality	Nucleus	Associated Tract	Function
General somatic afferent (GSA)	Chief sensory nucleus of V	Second-order neurons travel in the contralateral trigeminal lemniscus and terminate in the VPM of the thalamus. Second-order neurons from afferents from inside of the mouth travel in the ipsilateral posterior trigeminothalamic tract.	Discriminative touch, vibration, conscious proprioception
General somatic afferent (GSA)	Spinal nucleus of V	Second-order neurons travel in the contralateral trigeminothalamic tract; collaterals to pain-modulating systems terminate in the VPM of the thalamus.	Pain and temperature
General somatic afferent (GSA)	Mesencephalic nucleus of V	Central processes travel to reticular formation, cerebellum, and motor nucleus of V.	Nonconscious proprioception from muscles of mastication
Special visceral efferent (SVE)	Motor nucleus of V	Afferents to motor neurons through bilateral innervation via the corticobulbar tract	Motor to muscles of mastication, tensor tympani

VPM = ventral posteromedial nucleus.

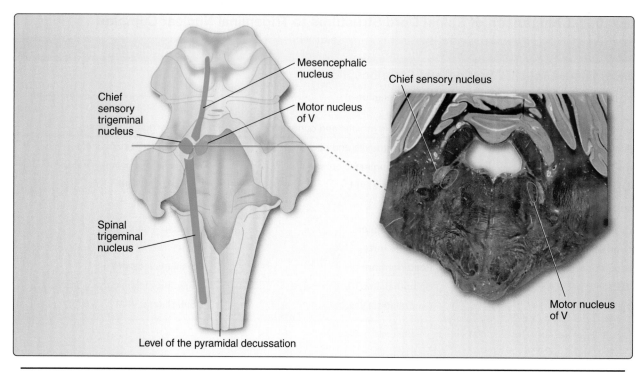

Figure 10.3
Location of the chief sensory trigeminal nucleus in the midpons.

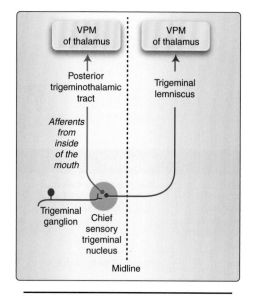

Figure10.4
Overview of tracts associated with the chief sensory trigeminal nucleus. CN = cranial nerve; VPM = ventral posteromedial nucleus.

a. **Trigeminal lemniscus:** One set of **second-order postsynaptic fibers** crosses the midline and travels lateral to the **medial lemniscus** as the **trigeminal lemniscus**. This tract terminates in the **ventral posteromedial nucleus (VPM) of the thalamus**.

b. **Posterior trigeminothalamic tract:** Another set of fibers is uncrossed and ascends with the trigeminal lemniscus as the **posterior trigeminothalamic tract** to the VPM of the thalamus. These uncrossed fibers contain somatosensory information from the inside of the mouth. They terminate in the thalamus adjacent to taste fibers, which are also uncrossed.

c. **Somatotopic arrangement:** There is a **somatotopic arrangement** of fibers within this nucleus (Figure 10.5). Fibers originating from V_1 (ophthalmic) are anterior, fibers from V_3 (mandibular) are posterior, and fibers from V_2 (maxillary) are in between. In other words, the face is represented upside down (see Figure 10.5).

2. **Spinal tract and nucleus of V:** The **spinal tract and nucleus of V** extend from the caudal end of the chief sensory nucleus to the upper cervical levels of the spinal cord (to C3). In the upper levels of the spinal cord, the spinal trigeminal nucleus and tract blend structurally and functionally with the posterior horn of the spinal cord, in particular, the **substantia gelatinosa** and the **tract of Lissauer**. Like the posterior horn of the spinal cord, the spinal nucleus and tract of V process **pain and temperature**.

a. **Pain and temperature fibers:** Fibers (Aδ and C) carrying pain and temperature from the periphery have their cell bodies in the trigeminal ganglion, and their central processes enter the brainstem through the sensory root of V in the midpons. These fibers then descend in the spinal trigeminal tract and synapse in the caudal portion of the trigeminal nucleus (Figure 10.6). This is the only instance in which sensory fibers descend upon entering the brainstem.

Much like the pain and temperature fibers traveling in the anterolateral system, the trigeminal system gives off a multitude of collaterals to the reticular formation and other brainstem structures for the descending modulation of pain (see Chapter 22, "Pain").

b. **Anterior trigeminothalamic tract:** From the spinal trigeminal nucleus, second-order neurons cross the midline and join the fibers from the **spinothalamic tract** as the **anterior trigeminothalamic tract** in the rostral medulla. Fibers terminate in the VPM of the thalamus (Figure 10.7). From there, third-order neurons ascend in the posterior limb of the internal capsule to the facial area of the primary somatosensory cortex.

c. **Somatotopic arrangement:** Fibers in the spinal nucleus and tract and the **anterior trigeminothalamic tract** have a **somatotopic arrangement** similar to that seen in the chief sensory nucleus and its associated tracts. Fibers from V₁ (ophthalmic)

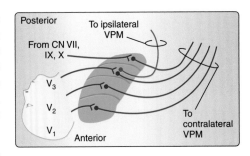

Figure 10.5
Somatotopic arrangement of fibers in the chief sensory trigeminal nucleus. CN = cranial nerve; VPM = ventral posteromedial nucleus.

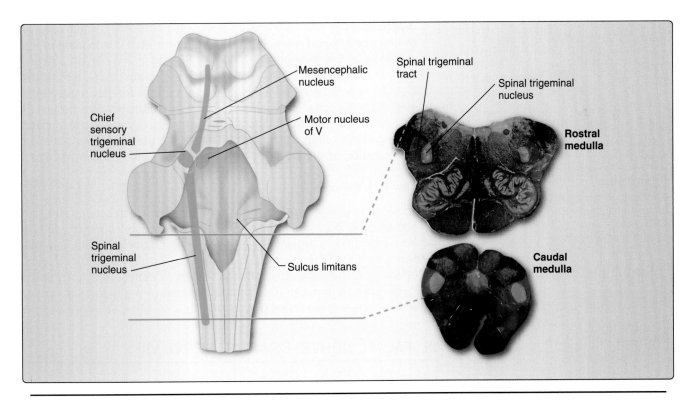

Figure 10.6
Location of the spinal trigeminal nucleus in the brainstem.

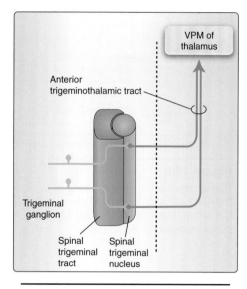

Figure 10.7
Overview of tracts associated with the spinal trigeminal nucleus. VPM = ventral posteromedial nucleus.

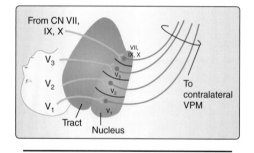

Figure 10.8
Somatotopic arrangement of fibers in the spinal trigeminal nucleus. CN = cranial nerve; VPM = ventral posteromedial nucleus.

are located most anteriorly, fibers from V$_3$ (mandibular) are located most posteriorly, and fibers from V$_2$ (maxillary) are located between the two (Figure 10.8).

3. **Mesencephalic nucleus of V:** Located in the midbrain, rostral to the chief sensory nucleus, the **mesencephalic nucleus of V** processes **nonconscious proprioception**, receiving afferents from muscle spindles, periodontal ligaments of the teeth, and the temporomandibular joint. The nucleus consists of **pseudounipolar cells** that derive from the neural crest but remain within the neural tube during development. This nucleus can be considered part of the trigeminal ganglion inside the brainstem (Figure 10.9). This is the only instance in which primary sensory neurons reside within the central nervous system instead of within a peripheral ganglion.

   In contrast to all other pseudounipolar neurons located in peripheral sensory ganglia, these neurons actually receive synaptic input from brainstem structures. Central processes from the mesencephalic nucleus of V travel to the **reticular formation** and the **cerebellum** for the processing of nonconscious proprioception. In addition, fibers are involved in sensory feedback loops to the spinal trigeminal nucleus and **motor reflex** pathways to the motor nucleus of V. Specifically, they are involved in mediating the **jaw-jerk reflex** (Figure 10.10).

B. Motor component of CN V

The motor component of the trigeminal nerve supplies the **muscles of mastication** as well as some small muscles in the head, most notably the **tensor tympani muscle**. The motor nucleus of V is located in the **midpons** at the level of attachment of the trigeminal nerve and contains the cell bodies of the motor neurons (Figure 10.11). It receives input from both cerebral hemispheres to mediate the jaw movements involved in speech and voluntary chewing, as well as input from the sensory nuclei of V. The motor fibers emerge through a **motor root**, bypass the trigeminal ganglion, and are distributed to their target muscles via the **mandibular division (V$_3$)** of the trigeminal nerve (Figure 10.12).

C. Jaw-jerk reflex

To elicit the **jaw-jerk reflex**, the chin is tapped, which causes stimulation of proprioceptors. These project to the mesencephalic nucleus from which fibers project to synapse in the motor nucleus of V. Motor fibers innervate the muscles of mastication, which leads to occlusion of the jaw. Clinically, this reflex is tested to assess the function of these brainstem nuclei and the third branch of the trigeminal nerve, which carries these fibers (Figure 10.13).

III. FACIAL NERVE: CRANIAL NERVE VII

The **facial nerve, cranial nerve (CN) VII,** emerges at the **pontocerebellar angle** together with the vestibulocochlear nerve (CN VIII). The facial nerve contains four components (Figure 10.14). It is primarily a **motor nerve**, with a **SVE** component to the muscles of facial expression and a **GVE**, or parasympathetic component, to major glands in the head. CN VII has a small sensory component (**GSA**) and, with that, a small sensory

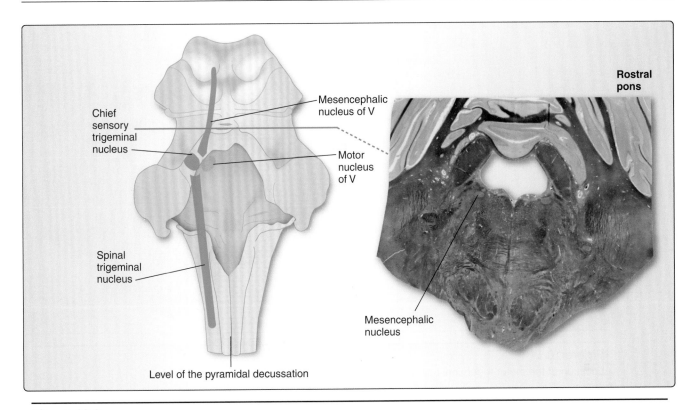

Figure 10.9
Location of the mesencephalic trigeminal nucleus in the rostral pons.

ganglion. The **SVA** component of this nerve comprises taste afferents from the anterior two-thirds of the tongue (Table 10.4). Taste is covered in more detail in Chapter 21, "Smell and Taste."

A. Motor component of CN VII

The motor component of CN VII can be divided into SVE to muscles derived from the second pharyngeal arch and parasympathetic efferents (GVE) to major glands in the head.

1. **Special visceral efferents:** These fibers innervate muscles from the second pharyngeal arch, most notably the **muscles of facial expression** as well as some other small muscles in the head. The cell bodies of the motor neurons are located in the **facial nucleus** in the tegmentum of the caudal pons. Axons project to the ipsilateral side of the face through the facial canal, where fibers to the stapedius muscle in the inner ear branch off.

    a. **Somatotopic arrangement:** The facial nucleus is arranged **somatotopically** with discrete areas dedicated to the upper face and the lower face. The corticobulbar input to the lower face is from the contralateral motor cortex, whereas input to the motor neurons supplying the upper face is from both the ipsilateral and the contralateral cortices (Figure 10.15). Interestingly, it has recently been shown that the bilateral input to the **upper face** originates from the **cingulate gyrus** and not from the primary motor area. The cingulate gyrus is part of the limbic system, and this is an indication of how closely linked our facial expressions are to our expression of emotions.

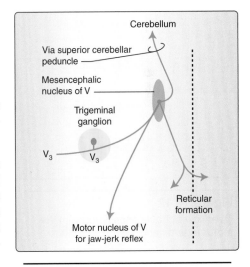

Figure 10.10
Overview of connections of the mesencephalic trigeminal nucleus with other brainstem structures and the cerebellum.

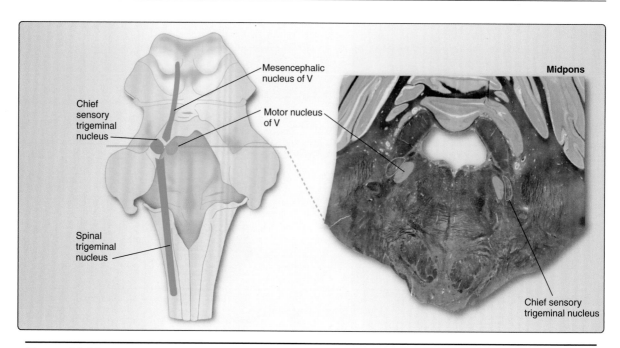

Figure 10.11
Location of the motor trigeminal nucleus in the mid-pons.

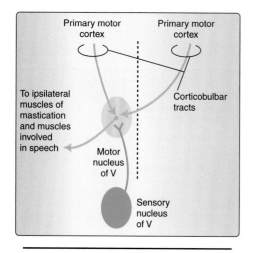

Figure 10.12
Overview of tracts associated with the motor trigeminal nucleus.

b. Lesions: A lesion to the facial nerve peripherally (to the lower motor neuron [LMN]) will compromise the innervation of both the upper and the lower face and will cause paralysis of half the face on the ipsilateral side (see Figure 10.15). This type of lesion can be seen in tumors compressing the nerve at the pontocerebellar angle (mostly gliomas) or tumors in the parotid gland, through which the nerve travels.

A central lesion to the corticobulbar fibers will compromise only the lower face on the contralateral side, due to the bilateral innervation to the upper face.

2. General visceral efferents: The other motor component in the facial nerve is the **GVE**, or **parasympathetic**, input to the lacrimal, submandibular, and sublingual glands as well as the glands of the oral, nasal, and pharyngeal mucosae. The cell bodies of these fibers are scattered throughout the **pontine tegmentum** and are collectively referred to as the **superior salivatory nucleus**. These preganglionic fibers travel to the peripheral **pterygopalatine** and **submandibular ganglia**. From there, postganglionic fibers provide visceral motor innervation to their respective glands. Input to the superior salivatory nucleus comes from the hypothalamus.

B. Sensory component of VII

The sensory component of CN VII comprises general somatic afferents and special visceral afferents.

1. General somatic afferents: The sensory component of the facial nerve is relatively small. The cell bodies of these fibers are located in the **geniculate ganglion**. GSA fibers from the concha of the external ear, the external acoustic meatus, and the external surface of the tympanic membrane project via the facial nerve to the

trigeminal nuclear complex (chief sensory trigeminal and spinal trigeminal). Second-order neurons project to the contralateral VPM of the thalamus.

2. **Special visceral afferents: SVA fibers** carry **taste** from the anterior two-thirds of the tongue. As discussed in detail in Chapter 21, "Smell and Taste," these fibers enter the brainstem, travel in the **solitary tract**, and synapse in the **nucleus solitarius**, the major brainstem nucleus for processing visceral afferent information.

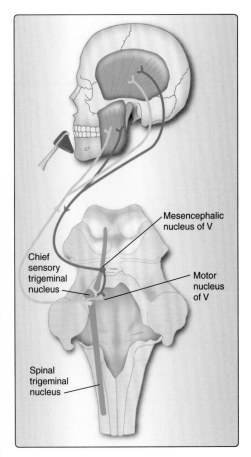

Figure 10.13
The jaw-jerk reflex.

Clinical Application 10.1: Bell Palsy

Bell palsy describes the paralysis of the **peripheral facial nerve (these are lower motor neurons)** on one side. The cause of this paralysis is unknown or **idiopathic**. Bell palsy is the most common form of facial nerve paralysis.

Patients present with paralysis of the **muscles of facial expression** on both the upper and the lower halves of one side of the face (see figure), as would be expected in a peripheral nerve lesion. In addition, the **production of tears** on that side can be impaired, due to the loss of parasympathetic efferents to the **lacrimal gland**. A decrease in salivation is usually not noticed, since the lesion is unilateral and the bulk of saliva in the mouth comes from the parotid gland, which receives its efferents through the glossopharyngeal nerve.

Due to the paralysis of the muscles of facial expression, the **blink reflex** can be impaired. Although the cornea will detect a stimulus through the trigeminal nerve, the efferent branch of the blink reflex through the facial nerve to the orbicularis oculi muscle will not function. Since the eye cannot be closed properly on that side, and there is a decrease in tear production, it is important to lubricate the eye with artificial tears and to tape the eye shut during the night in order to prevent damage to the cornea. Although the facial nerve receives taste input from the ipsilateral anterior 2/3 of the tongue, patients typically do not report any significant changes in taste. This is most likely due to the fact that the other side of the tongue has preserved innervation.

The **etiology** of Bell palsy is not known, but the most accepted theory is that it is due to a **viral infection** that causes swelling of the facial nerve, resulting in compression of the nerve as it passes through the bony facial canal in the skull. There are central brainstem symptoms as well, and about 50% of patients report symptoms attributable to other cranial nerves. The most common symptom is **facial pain**, related to the trigeminal nerve. A significant percentage of patients also suffer from **hyperacusis** or hypersensitivity to sound. This was thought to be attributable to a dysfunction of the stapedius muscle in the middle ear, which is innervated by a branch of cranial nerve VII. Recent studies were not able to corroborate this, however, because bilateral symptoms of hyperacusis often exist. It is now thought that this symptom is a central problem related to the cochlear nuclei.

The **prognosis** of Bell palsy is excellent. In the vast majority of cases, the symptoms resolve spontaneously. The **management** of this disorder is limited to observation, eye care, and corticosteroid treatment. The treatment with antiviral drugs has not been proven to be more beneficial than the treatment with corticosteroids alone. Antiviral medication is still commonly used in combination with corticosteroids if there are other symptoms suggestive of a viral infection. In some cases, surgical decompression of the facial nerve in the base of the skull may be necessary, but this treatment option remains controversial because of the damage that can be caused.

A very small cohort of patients who recover from Bell palsy develop what has been called **"crocodile tear" syndrome**. These patients start shedding

Clinical Application 10.1: Bell Palsy continued

tears when eating. This is likely due to efferents "missprouting" from the superior salivatory nucleus. Instead of innervating the salivary glands, these efferents are misdirected to the lacrimal gland (which is usually innervated from the lacrimal portion of this nucleus). The hypothalamic and olfactory inputs to the superior salivatory nucleus remain stimulatory during eating, resulting in tearing through the lacrimal gland in these cases.

In very cold climates, the facial nerve can be damaged through frostbite on the cheek, where the facial nerve travels through the parotid gland and then travels to its target muscles of facial expression. This is a transient lesion that resolves when the face warms up again.

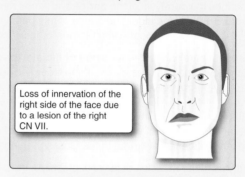

Loss of innervation of the right side of the face due to a lesion of the right CN VII.

Typical facial expression of a patient with Bell palsy. CN = cranial nerve.

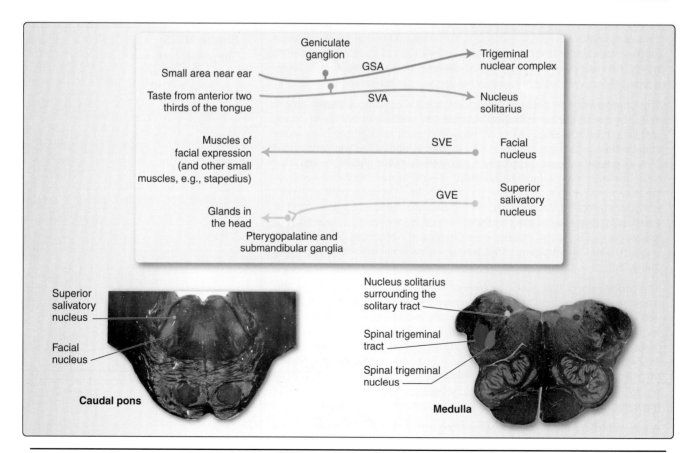

Figure 10.14

Overview of the facial nerve and its associated nuclei. GSA = general somatic afferent; SVA = special visceral afferent; SVE = special visceral efferent; GVE = general visceral efferent.

Table 10.4: Facial Nerve: Cranial Nerve VII

Nerve Fiber Modality	Nucleus	Associated Tract	Function
Special visceral efferent (SVE)	Facial nucleus	Efferents to motor neurons through contralateral innervation from the primary motor area of the cortex via the corticobulbar tract to the lower face Bilateral innervation from the cingulate gyrus via the corticobulbar tract to the upper face	Motor innervation to muscles of facial expression
General visceral efferent (GVE)	Superior salivatory nucleus	Input from hypothalamus	Parasympathetic innervation to lacrimal, submandibular, and sublingual glands
General somatic afferent (GSA)	Chief sensory trigeminal nucleus Spinal trigeminal nucleus	Second-order neurons travel in contralateral trigeminal lemniscus and terminate in the VPM of the thalamus.	Discriminative touch, pain, and temperature from the outer ear
Special visceral afferent (SVA)	Nucleus of the solitary tract	Output to ipsilateral insula	Taste from anterior two-thirds of the tongue

VPM = ventral posteromedial nucleus.

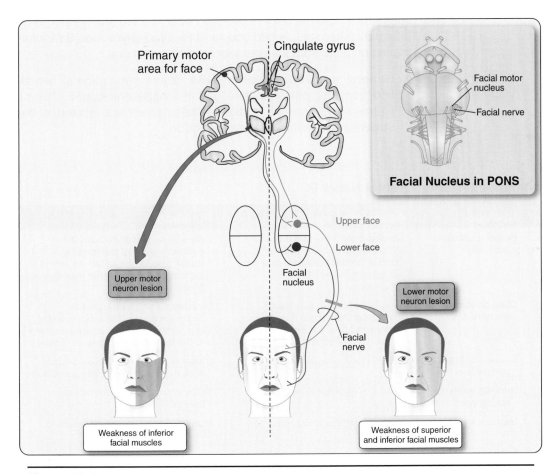

Figure 10.15

Cortical input to the facial nucleus and lesions of the central or peripheral pathways. Purple and orange bars and arrows indicate the sites of the lesions.

IV. GLOSSOPHARYNGEAL NERVE: CRANIAL NERVE IX

The **glossopharyngeal nerve** is a complex nerve carrying several different types of sensory and motor fibers to areas of the head and neck, as summarized in Table 10.5. It carries **general sensation (GSA)** and **taste (SVA)** from the posterior third of the tongue, soft palate, and pharynx, and **visceral afferents (GVA)** from the carotid body and carotid sinus. In addition, it supplies one muscle, the stylopharyngeus **(SVE)**, and sends **parasympathetic efferent (GVE)** input to the parotid gland, carotid body, and sinus.

Cranial nerve (CN) IX emerges from the brainstem as a series of rootlets between the olive and the inferior cerebellar peduncle. It leaves the posterior cranial fossa through the jugular foramen together with CNs X (vagus) and XI (accessory), as shown in Figure 10.16. CN IX has two ganglia associated with it, the **superior** and **inferior glossopharyngeal ganglia**, which contain the cell bodies of the sensory afferents.

A. Sensory component of CN IX

The somatic afferents in the glossopharyngeal nerve include **GSA** information from the back of the oral cavity and the oropharynx, **GVA** information from the carotid body and the oropharynx, and **SVA** information on taste from the posterior third of the tongue.

1. **General somatic afferents:** GSA fibers carry general sensation from the posterior third of the tongue and upper pharynx (Table 10.5). The cell bodies of these fibers are located in the **superior glossopharyngeal ganglion**.

Table 10.5: Glossopharyngeal Nerve: Cranial Nerve IX

Nerve Fiber Modality	Nucleus	Associated Tract	Function
General somatic afferent (GSA)	Spinal trigeminal nucleus	Second-order neurons travel in the contralateral anterior trigeminothalamic tract and terminate in the VPM of the thalamus.	Pain and temperature from the posterior third of the tongue, tonsil, skin of the outer ear, internal surface of the tympanic membrane, pharynx
	Chief sensory nucleus of V	Second-order neurons travel in the contralateral trigeminal lemniscus and terminate in the VPM of the thalamus.	Discriminative touch from the posterior third of the tongue, tonsil, skin of the outer ear, internal surface of the tympanic membrane, pharynx
General visceral afferent (GVA)	Nucleus solitarius, middle part	Reflex arcs to nucleus ambiguus	Chemoreceptors and baroreceptors from the carotid body, gag sensations
Special visceral afferent (SVA)	Nucleus solitarius, rostral part	Output to ipsilateral insula	Taste from posterior third of the tongue
Special visceral efferent (SVE)	Nucleus ambiguus	Bilateral corticobulbar input	Motor to stylopharyngeus muscle
General visceral efferent (GVE)	Inferior salivatory nucleus	Input from hypothalamus and olfactory system	Stimulation of the parotid gland
	Nucleus ambiguus	Input from nucleus solitarius, for reflex	Carotid body and sinus: vasodilation

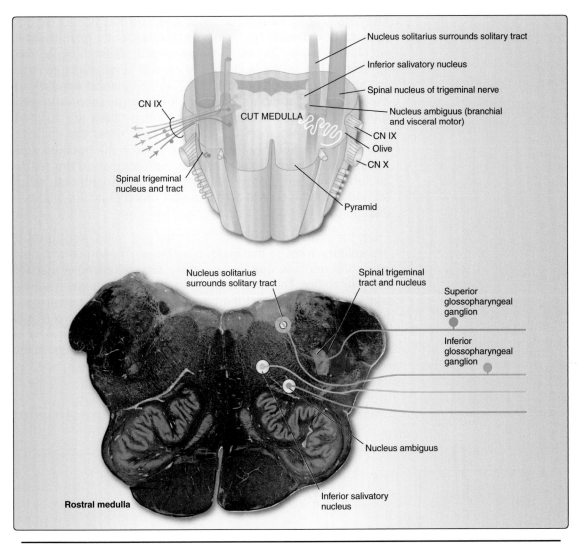

Figure 10.16
Overview of the glossopharyngeal nerve and its associated nuclei. CN = cranial nerve. (Modified from Wilson-Pauwels L, et al. *Cranial Nerves: Function and Dysfunction*, 3rd ed. PMPH: Shelton, CT, 2010.)

Axons carrying pain enter the medulla and descend in the **spinal tract of V** to synapse in the caudal part of the **spinal nucleus of V**. From here, second-order neurons cross to the contralateral side of the medulla and project mainly to the **VPM of the thalamus** (see Figure 10.7). From there, fibers travel through the posterior limb of the internal capsule to the primary somatosensory cortex.

Axons carrying discriminative touch enter the medulla and synapse in the **chief sensory nucleus** of the trigeminal nerve. As expected, second-order neurons then cross the midline to synapse in the VPM and, from there, project to the primary somatosensory cortex.

2. **General visceral afferents: GVA fibers** also travel with the glossopharyngeal nerve, carrying afferents from the **carotid body** as well as the oropharynx, where they are involved in the **gag reflex** (see *Gag reflex*, below). The cell bodies of these fibers are located in the **inferior glossopharyngeal ganglion** (see Figure 10.16).

The carotid body contains **chemoreceptors** that monitor oxygen (O_2), carbon dioxide (CO_2), and acidity/alkalinity (pH) levels in circulating blood. Similarly, **baroreceptor** (stretch receptors) nerve endings in the walls of the carotid sinus measure arterial blood pressure (Figure 10.17). These **visceral sensations** ascend in the carotid nerve to the **inferior glossopharyngeal ganglion** where the cell bodies of these fibers are located. Central processes from the ganglion cells enter the medulla, descend in **the solitary tract**, and synapse in the **nucleus solitarius**. From the nucleus solitarius, output goes to the reticular formation and the hypothalamus for the appropriate reflex responses in the control of respiration, blood pressure, and cardiac output.

From the oropharynx, afferents for the gag reflex also travel in the **solitary tract** to the nucleus solitarius. These neurons project to the **nucleus ambiguus**, where visceral efferents mediate the **gag reflex**. The gag reflex is separate from general sensation in the pharynx, and the fibers involved in these sensations have separate brainstem nuclei (chief sensory nucleus of V and nucleus solitarius, respectively). It is indeed a different quality of sensation to *feel* a bolus of food or to *gag* on a bolus of food in the oropharynx.

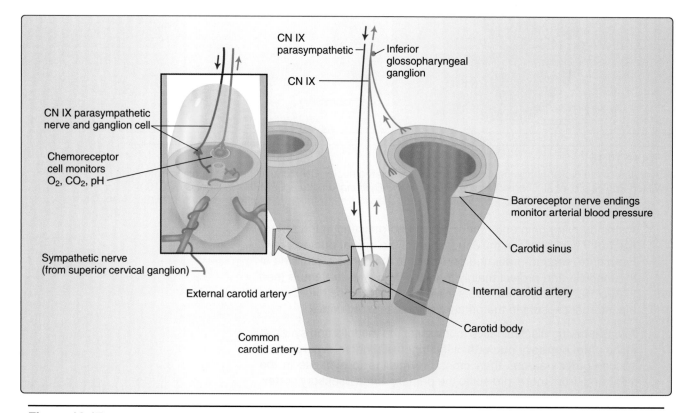

Figure 10.17

Innervation of the carotid sinus and carotid body. CN = cranial nerve. (Modified from Wilson-Pauwels L, et al. *Cranial Nerves: Function and Dysfunction*, 3rd ed. PMPH: Shelton, CT, 2010.)

3. **Special visceral afferents:** The nucleus solitarius processes **SVA** information regarding **taste** from the posterior third of the tongue. The cell bodies of these fibers are also located in the **inferior glossopharyngeal ganglion**.

   Central processes from the ganglion pass through the jugular foramen, enter the medulla, and ascend to synapse in the **nucleus solitarius**. Axons of cells in the nucleus solitarius then ascend to reach the **ipsilateral VPM**. From the thalamus, fibers take the usual route through the posterior limb of the internal capsule to reach the primary somatosensory cortex in the inferior third of the postcentral gyrus and adjacent surface of the insula, where taste is perceived.

B. Motor component of CN IX

The motor component of the glossopharyngeal nerve is relatively small: **SVE**, or **branchial efferent**, information is projected to the stylopharyngeus muscle, and **GVE**, or **parasympathetic** output, innervates the parotid gland.

1. **Special visceral efferents:** From the **nucleus ambiguus**, SVE fibers travel with the glossopharyngeal nerve to innervate the stylopharyngeus muscle, derived from the **third branchial arch**. These SVE neurons in the nucleus ambiguus receive **bilateral** innervation from the **corticobulbar tract**.

2. **General visceral efferents:** The **parasympathetic motor (GVE)** component of CN IX arises from two nuclei and has two main functions. The **nucleus ambiguus** sends fibers to the carotid body and sinus, and the **inferior salivatory nucleus** provides secretomotor supply to the parotid gland. The GVE neurons in both nuclei are influenced by output from the **hypothalamus** and **reticular formation**.

   Axons from the **nucleus ambiguus** travel with the carotid branch of CN IX to reach the **carotid body and sinus** where the preganglionic axons synapse on parasympathetic ganglia within the carotid body and in the walls of the carotid sinus. Their role is **vasodilation** of blood vessels within the carotid body (see Figure 10.17). This is the only case in which a blood vessel receives parasympathetic innervation—all other blood vessels receive only sympathetic innervation.

   Axons from the **inferior salivatory nucleus** travel with CN IX to the **otic ganglion**. From there, the postganglionic neurons supply **secretomotor input** to the **parotid gland**. Stimuli from the hypothalamus and information from the olfactory system act on the inferior salivatory nucleus, resulting in, for example, salivation in response to smelling food.

C. The gag reflex

When an object in the oropharynx is interpreted as an unpleasant sensation, this is relayed to the **nucleus solitarius** via fibers of the glossopharyngeal nerve (CN IX) as shown in Figure 10.18. From

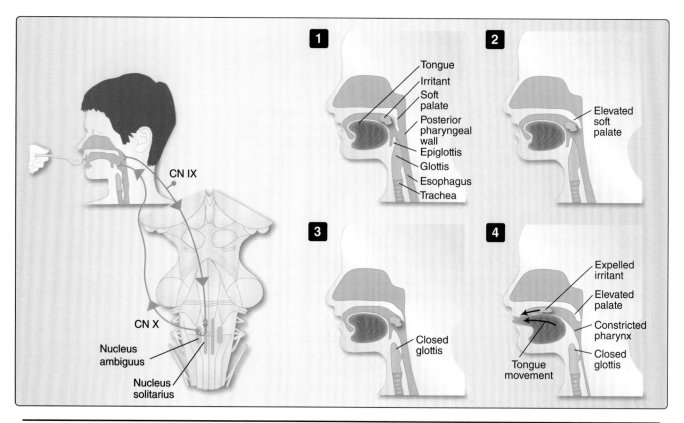

Figure 10.18
The gag reflex. CN = cranial nerve. (Modified from Wilson-Pauwels L, et al. *Cranial Nerves: Function and Dysfunction*, 3rd ed. PMPH: Shelton, CT, 2010.)

the nucleus solitarius, fibers synapse with efferents in the **nucleus ambiguus**, which then travel with the vagus nerve (CN X) to the pharynx. This causes the soft palate to elevate and close off the upper airway, the glottis to close to protect the lower airway, and the pharyngeal walls to constrict to expel the object that caused the gagging sensation.

V. VAGUS NERVE: CRANIAL NERVE X

The name of **cranial nerve (CN) X**, **vagus**, comes from the Latin word meaning "wandering," and it does just that. It emerges from the medulla of the brainstem and ends close to the left colic flexure of the large intestine, giving off many branches along the way. The vagus is *the* parasympathetic nerve of thoracic and abdominal viscera and supplies structures in the pharynx and larynx. It is also the largest visceral sensory nerve.

CN X emerges from the **medulla** as 8 to 10 rootlets just **posterior to the olive** and **anterior to the inferior cerebellar peduncle**. The rootlets are immediately caudal to the rootlets of CN IX. The rootlets converge into a single nerve that exits the skull through the **jugular foramen** along with CNs IX and XI. The vagus has two ganglia: the **superior vagal (jugular) ganglion** located on the nerve within the jugular fossa and the **inferior vagal (nodose) ganglion** located on the vagus nerve below the jugular foramen.

CN X carries both sensory and motor modalities. Its fibers are connected to four brainstem nuclei: the **spinal nucleus of V (GSA)**, the **nucleus solitarius (GVA)**, the **nucleus ambiguus (SVE** and **GVE)**, and the **dorsal motor nucleus of the vagus (GVE)**. These are summarized in Figure 10.19 and Table 10.6.

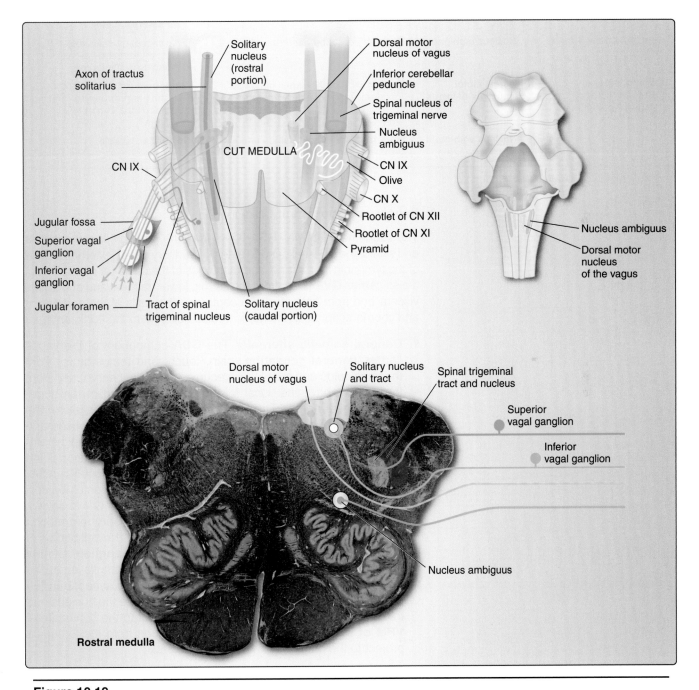

Figure 10.19

The vagus nerve and its associated nuclei. CN = cranial nerve. (Modified from Wilson-Pauwels L, et al. *Cranial Nerves: Function and Dysfunction*, 3rd ed. PMPH: Shelton, CT, 2010.)

Table 10.6: Vagus Nerve: Cranial Nerve X

Nerve Fiber Modality	Nucleus	Associated Tract	Function
General somatic afferent (GSA)	Spinal trigeminal nucleus	Second-order neurons travel in the contralateral anterior trigeminothalamic tract and terminate in the VPM of the thalamus.	Sensory from posterior meninges, concha, pharynx, and larynx
General visceral afferent (GVA)	Nucleus solitarius	To reticular formation and hypothalamus	From larynx, trachea, esophagus, and thoracic and abdominal viscera; stretch receptors in the aortic arch
Special visceral efferent (SVE)	Nucleus ambiguus	Bilateral corticobulbar innervation	Pharyngeal muscles and cricothyroid and intrinsic muscles of the larynx
General visceral efferent (GVE) or parasympathetic	Dorsal motor nucleus of vagus	Input from hypothalamus, olfactory system, and reticular formation	Smooth muscle and glands of the pharynx, larynx, and thoracic and abdominal viscera
	Nucleus ambiguus	Input from the hypothalamus and reticular formation	To cardiac muscle

VPM = ventral posteromedial nucleus.

A. Sensory component of CN X

The vagus nerve carries **GSA** fibers from the pharynx, larynx, concha and skin of the external ear and external auditory canal, external surface of the tympanic membrane, and meninges of the posterior cranial fossa. It also carries **GVA** fibers from the larynx, lower trachea, and abdominal viscera and from both stretch receptors in the walls of the aortic arch and chemoreceptors in the aortic bodies adjacent to the aortic arch.

1. **General somatic afferents:** The **GSA** component of the vagus carries **general sensation** (pain, touch, and temperature) from the vocal folds and subglottis (recurrent laryngeal nerve) and from the larynx above the vocal folds (laryngeal branch of the superior laryngeal nerve). The cell bodies for these nerves are in the **inferior vagal ganglion.** General sensation from the concha and skin of the external ear, the auditory canal, and the tympanic membrane travels in the auricular branch of the vagus. Stimulation of the auricular nerve in the external auditory meatus can result in reflex coughing, vomiting, and even fainting through activation of the dorsal motor nucleus of X (see *Motor component of CN X* below). The meningeal branch of the vagus carries sensory information from the meninges of the posterior cranial fossa. The cell bodies of these fibers are located in the **superior vagal ganglion.** Central processes from the inferior and superior vagal ganglia enter the medulla and descend in the **spinal tract of V** to synapse in the **spinal nucleus of V.** Second-order neurons leave the nucleus and travel in the **anterior trigeminothalamic** tract to the contralateral **VPM of the thalamus,** and third-order axons from the thalamus project to the primary somatosensory cortex.

2. **General visceral afferents:** Visceral sensation is not appreciated at a conscious level of awareness but rather as a vague sensation of "feeling good" or "feeling bad." Visceral sensory fibers arise from the plexuses around the viscera of the abdomen and thorax and eventually come together as the right and left vagus nerves.

Importantly, GVAs from the aortic arch carry information from the baroreceptors and chemoreceptors located there and are an integral part of the reflex pathway that maintains blood pressure.

The cell bodies are located in the **inferior vagal ganglion**, and the central processes enter the medulla to descend in the **solitary tract** and synapse in the caudal part of the **nucleus solitarius**. From the nucleus solitarius, bilateral connections are made with the **reticular formation** and the **hypothalamus**. For visceral reflexes, projections go to the dorsal motor nucleus of the vagus, the nucleus ambiguus, and the rostral medulla. These connections are important in the reflex control of cardiovascular, respiratory, and gastrointestinal function. See Chapter 4, "Overview of the Visceral Nervous System," for more information.

B. Motor component of CN X

The vagus has a **branchial motor (SVE)** component that supplies the muscles of the pharynx and larynx derived from the fourth or sixth **pharyngeal arches**. There are also **parasympathetic (GVE)** fibers in the vagus to smooth muscle and to cardiac muscle, to glands of the pharynx and larynx, and to abdominal viscera.

1. **Special visceral efferents:** Axons from premotor, motor, and other cortical areas send bilateral fibers through the posterior limb of the internal capsule to synapse on motor neurons in the **nucleus ambiguus**, which lies just posterior to the inferior olivary nucleus in the medulla. These SVE fibers innervate muscles of the pharynx and the larynx. The nucleus ambiguus also receives sensory input from brainstem nuclei, mainly the spinal trigeminal nucleus and nucleus solitarius, which initiate reflex responses (e.g., coughing, vomiting).

2. **General visceral efferents:** Together, the **dorsal motor nucleus of the vagus** and the medial portion of the **nucleus ambiguus** contain the nerve cell bodies of the parasympathetic component of the vagus nerve. The **dorsal motor nucleus of the vagus** is fairly large, extending from the floor of the fourth ventricle to the central gray matter of the closed medulla. The nucleus ambiguus is an ill-defined collection of neurons in the tegmentum of the rostral medulla.

Preganglionic neurons from the dorsal motor nucleus synapse in the thoracic visceral plexus to innervate the lungs and the prevertebral plexus in the abdomen to innervate the gut and its derivatives (liver, gallbladder, pancreas). The preganglionic GVE fibers from the nucleus ambiguus synapse in the cardiac plexus and innervate the heart. Input from the hypothalamus, olfactory system, reticular formation, and the nucleus solitarius all influence these neurons.

In the lung, postganglionic fibers cause bronchoconstriction. Throughout the gut, vagus nerve fibers synapse on ganglia in the myenteric and submucosal plexuses. Here, they promote peristalsis and fluid absorption and supply the digestive glands.

Axons supplying the **heart** arise from the medial portion of the **nucleus ambiguus**. Their role is to slow down the cardiac cycle.

The parasympathetic fibers of the vagus nerve thus have two roles: They speed up gut motility and slow down heart rate. The efferents to the gut and heart arise from two separate nuclei because they have opposite effects!

VI. ACCESSORY NERVE: CRANIAL NERVE XI

The **accessory nerve** is a pure motor nerve (SVE). It innervates two muscles derived from pharyngeal arches: the trapezius and the sternocleidomastoid muscles. It is described by some as having a cranial root and a spinal root. However, we consider the cranial root as a part of the vagus nerve, because the cranial component is separated from the vagus for only a short distance before joining with it.

Cranial nerve (CN) XI arises from a column of cells called the **spinal accessory nucleus** that extends from the first to the sixth spinal cord segments (C1–C6) in the posterolateral part of the anterior horn. The nerve emerges from the cord as a series of rootlets posterior to the denticulate ligament and ascends in the subarachnoid space through the foramen magnum of the skull. From there, it turns and runs anterolaterally to the **jugular foramen** where it joins with CNs IX and X to exit the skull (Figure 10.20).

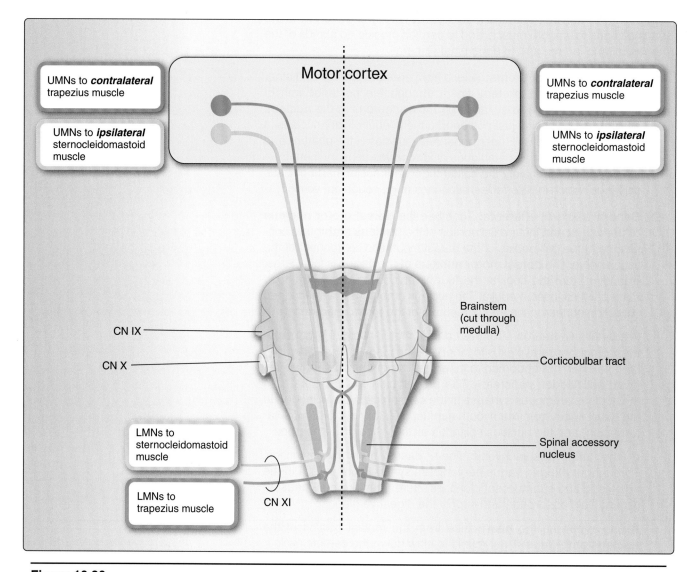

Figure 10.20

Overview of the accessory nerve. CN = cranial nerve; LMN = lower motor neuron. (Modified from Wilson-Pauwels L, et al. *Cranial Nerves: Function and Dysfunction*, 3rd ed. PMPH: Shelton, CT, 2010.)

Table 10.7: Accessory Nerve: Cranial Nerve XI

Nerve Fiber Modality	Nucleus	Associated Tract	Function
Special visceral efferent (SVE)	Spinal accessory nucleus	Corticobulbar tract ipsilateral to neurons supplying the sternoclei-domastoid; contralateral to neurons supplying the trapezius	Motor innervation of sternocleido-mastoid and tra-pezius muscles

Upper motor neuron (UMN) input to the accessory nucleus descends in the **corticobulbar** tract (even though these fibers extend into the spinal cord, the axons to the accessory nucleus are classified as corticobulbar) through the posterior limb of the internal capsule. Input for the **sternocleidomastoid muscle descends ipsilaterally** to the spinal accessory nucleus. Axons designated to supply the **trapezius muscle cross** the **midline** in the pyramidal decussation to synapse in the contralateral spinal accessory nucleus. This arrangement of ipsilateral UMN input for the sternocleidomastoid and contralateral input for the trapezius enables us to hold something in our left hand with the left trapezius muscle acting on the left shoulder, whereas the right sterno-mastoid contracts to tip the head up and to the left so we can observe the object being held. This muscle cooperation is important for **eye–hand coordination** (Table 10.7).

VII. HYPOGLOSSAL NERVE: CRANIAL NERVE XII

The **hypoglossal nerve** is the motor nerve of the tongue. Its fibers arise from the **hypoglossal nucleus**, a longitudinal cell column in the paramedian area of the **medulla** that lies deep to the hypoglossal trigone in the floor of the fourth ventricle (Figure 10.21). The nucleus extends from the caudal medulla to the pontomedullary junction. The rootlets of the hypoglossal nerve emerge from the medulla in the **anterolateral sulcus**, between the pyramid and the olive. The rootlets converge to form the hypoglossal nerve, which leaves the skull through the **hypoglossal foramen**. The hypoglossal nerve supplies all of the intrinsic muscles of the tongue and all but one of the extrinsic muscles of the tongue (Table 10.8).

Supranuclear control of the tongue is mediated by **corticobulbar fibers** that originate mainly within the lower portion of the precentral gyrus. The corticobulbar fibers controlling the **genioglossus muscle** are **crossed**, whereas the other tongue muscles receive bilateral input. A lesion above the hypoglossal nucleus (UMN lesion) of the tongue may result in weakness or paralysis of the tongue. However, because the tongue (except for the genioglossus) has bilateral control, a lesion above the decussation to the hypoglossal nucleus (UMN lesion) could result in deviation of the tongue away from the side of

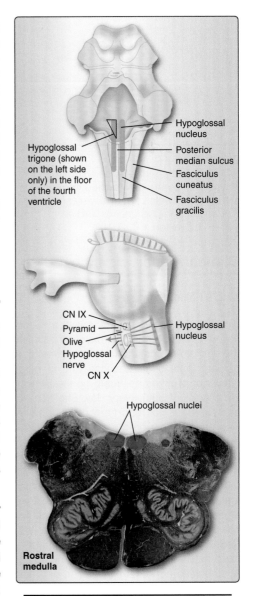

Figure 10.21
Overview of the hypoglossal nerve.
CN = cranial nerve.

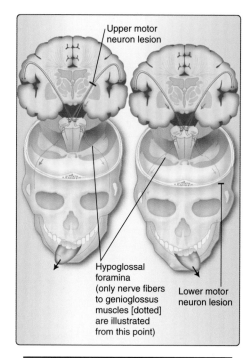

Upper motor neuron lesion

Hypoglossal foramina (only nerve fibers to genioglossus muscles [dotted] are illustrated from this point)

Lower motor neuron lesion

Figure 10.22

Central and peripheral lesions of the hypoglossal nerve. (Modified from Wilson-Pauwels L, et al. *Cranial Nerves: Function and Dysfunction*, 3rd ed. PMPH: Shelton, CT, 2010.)

Table 10.8: Hypoglossal Nerve: Cranial Nerve XII

Nerve Fiber Modality	Nucleus	Associated Tract	Function
General somatic efferent (GSE)	Hypoglossal nucleus	Bilateral innervation through corticobulbar tract to all muscles except genioglossus Innervation to genioglossus is contralateral.	Innervation of the muscles of the tongue

the lesion (toward the side of the weakness) due to the inability of the paralyzed genioglossus to oppose the action of the intact genioglossus (Figure 10.22, *right*). In a lesion anywhere between the hypoglossal nucleus and the tongue (LMN lesion), ultimately, there would be flaccid paralysis of the ipsilateral side of the tongue with fasciculation and atrophy of tongue muscles. In this case, the ipsilateral genioglossus muscle would be paralyzed and the tongue would deviate to the same side as the lesion (Figure 10.22, *left*).

Clinical Case:

Ling's Facial Pain

Ling is a 63-year-old woman with a history of intermittent, severe shooting pain (paroxysmal pain) that radiates from the ear into the cheek and jaw on the left side. Each attack of pain lasts 5–30 seconds and is triggered when Ling chews or brushes her teeth on the left side. Ling has over 20 attacks of pain every day. She has no other symptom and she does not report any facial numbness or weakness. She has no pain on the right side of her face or over her entire forehead. She has no neck or ear pain. Her speech and swallowing are normal. During her attacks of pain, there is no ptosis (droopy eyelid), miosis (small pupil), tearing of the eyes, eyelid/facial edema or reddening of the eye (conjunctival injection). She saw the dentist 2 weeks ago and her dental evaluation was normal. Ling is otherwise healthy. She has been prescribed nonsteroidal anti-inflammatory drugs (NSAIDs) and opioid medication for her pain, but these have not helped her pain. Ling's neurological examination is otherwise normal. In particular, her facial sensation and strength are normal.

Case Discussion

This is a typical presentation of trigeminal neuralgia. Trigeminal neuralgia is a facial pain condition characterized

by recurrent unilateral brief electric shock-like, shooting, or stabbing pains that are triggered by innocuous stimuli such as chewing, brushing teeth, or light touching of the face. The pain is short in duration (typically less than 2 minutes), abrupt in onset and offset, and located in one or more divisions of the trigeminal nerve (Figure 10.23).

Differential diagnoses include trigeminal autonomic cephalalgias (TACs), which are a group of headache disorders associated with cranial autonomic features, involving the autonomic innervation of the eyes, glands, or mucus membranes. Symptoms include ptosis, miosis, conjunctival injection, lacrimation, nasal congestion, and eyelid/facial edema. Ling does not have these features. Typically, V_1 (first branch of the trigeminal nerve) is involved in TACs, whereas Ling's symptoms are localized to V_3.

Migraines can also cause facial pain, but pain typically lasts more than 4 hours in adults. Migraine pain is typically unilateral and throbbing. It is moderate to severe in intensity and gets worse with activity. Ling's pain, on the other hand, is seconds in duration and very severe in intensity.

Trigeminal neuralgia typically involves the V_2 and V_3 branches of the trigeminal nerve. The root entry zone is the area where the trigeminal nerve attaches to the pons. The presumed pathophysiology of trigeminal neuralgia includes the

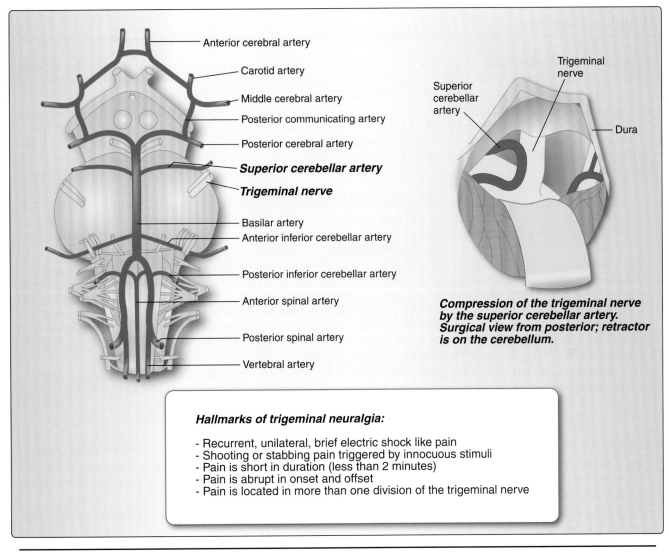

Anterior cerebral artery

Carotid artery

Middle cerebral artery

Posterior communicating artery

Posterior cerebral artery

Superior cerebellar artery

Trigeminal nerve

Basilar artery
Anterior inferior cerebellar artery

Posterior inferior cerebellar artery

Anterior spinal artery

Posterior spinal artery

Vertebral artery

Trigeminal nerve

Superior cerebellar artery

Dura

Compression of the trigeminal nerve by the superior cerebellar artery. Surgical view from posterior; retractor is on the cerebellum.

Hallmarks of trigeminal neuralgia:

- Recurrent, unilateral, brief electric shock like pain
- Shooting or stabbing pain triggered by innocuous stimuli
- Pain is short in duration (less than 2 minutes)
- Pain is abrupt in onset and offset
- Pain is located in more than one division of the trigeminal nerve

Figure 10.23
Compression of trigeminal nerve by the superior cerebellar artery as a cause of trigeminal neuralgia.

demyelination of trigeminal sensory fibers, which causes the spread of excitation between adjacent fibers. This ephaptic spread of excitation is usually prevented by the myelin sheath around neurons. Demyelination from multiple sclerosis and the compression of the entry root by an overlying vessel are the two most common causes of trigeminal neuralgia.

Treatment for trigeminal neuralgia includes pharmacological therapy and surgery. Surgical options include decompression of the entry root through resection of the overlying vessel, among other options.

What are the pharmacological treatment options for trigeminal neuralgia?

Carbamazepine is considered first-line therapy for trigeminal neuralgia. Carbamazepine is an anticonvulsant medication that can be effective for certain types of paroxysmal pain. Sodium channel blockers such as oxcarbazepine and lamotrigine will block the spread of action potentials and can be used to treat trigeminal neuralgia. Baclofen is a $GABA_B$ receptor agonist; stimulation of the $GABA_B$ receptor will lead to Cl^- influx, which moves neurons away from threshold and further blocks the generation of action potentials.

Conclusion

Ling was prescribed carbamazepine, which lead to the resolution of her pain. However, 3 years later, her pain returned and she was no longer responding to carbamazepine. An MRI scan of her brain showed that her left superior cerebellar artery was compressing her left trigeminal nerve at the root entry zone. A neurosurgeon performed microvascular decompression of her left trigeminal nerve and she has been pain-free for over 7 years.

Chapter Summary

- **Cranial nerves (CNs) V, VII, IX, X, XI,** and **XII** innervate the head and neck, carrying various **motor and sensory modalities**. Each modality has brainstem nuclei associated with it. A brainstem nucleus can be associated with more than one cranial nerve, and a single cranial nerve can be associated with more than one brainstem nucleus.

- The **trigeminal nerve (CN V)** is the major **general somatic afferent** nerve in the head. The **trigeminal nuclear complex** processes this sensory information. The **chief sensory trigeminal nucleus** is associated with discriminative touch, vibration, and conscious proprioception, the **spinal trigeminal nucleus** is associated with pain and temperature, and the **mesencephalic nucleus** is associated with nonconscious proprioception. The motor component (general somatic efferent) of CN V supplies the muscles of mastication. The cell bodies of these neurons are located in the **motor nucleus of V,** medial to the chief sensory trigeminal nucleus.

- The **facial nerve (CN VII)** provides **motor** (special visceral efferent) innervation to the muscles of facial expression. The cell bodies of these neurons are located in the **facial nucleus.** In addition, CN VII has a **parasympathetic** (general visceral efferent) component arising from the **superior salivatory nucleus** and carries special visceral afferent **taste** fibers from the anterior two-thirds of the tongue to the **nucleus solitarius.**

- The **glossopharyngeal nerve (CN IX)** carries several sensory and motor modalities to the head and neck. A **general sensory afferent** component from the tongue and pharynx projects to the **trigeminal nuclear complex.** General visceral afferents from the **carotid body** and the **gag reflex** from the oropharynx project via CN IX to the **nucleus solitarius.** Special visceral afferent **taste** fibers also project to the **nucleus solitarius.** General somatic efferent **motor innervation** to stylopharyngeus muscle is from the **nucleus ambiguus** via CN IX. The general visceral efferent (GVE) **secretomotor** innervation to the **parotid gland** comes from **the inferior salivatory nucleus.** GVE **to the carotid body** resulting in vasodilation comes from the **nucleus ambiguus** via branches of CN IX.

- The **vagus nerve (CN X)** is the major **parasympathetic nerve** in the body. Its general visceral efferent fibers arise from the **dorsal motor nucleus of vagus,** where they innervate the smooth muscles and glands of the pharynx, larynx, and thoracic and abdominal viscera (up to the left colic flexure). Innervation to **cardiac muscle** arises from the **nucleus ambiguus** and is also carried by CN X. In addition, this nerve has a small general somatic afferent component, which projects to the **trigeminal nuclear complex,** whereas general visceral afferents project to the **nucleus solitarius.** A small special visceral efferent component to muscles of the pharynx and larynx comes from the **nucleus ambiguus.**

- The **accessory nerve (CN XI)** is a pure **motor (special visceral efferent) nerve,** supplying the sternocleidomastoid and trapezius muscles of the neck. The cell bodies are located in the upper cervical levels of the spinal cord in the **spinal accessory nucleus.**

- The **hypoglossal nerve (CN XII)** is a **somatic motor (general somatic efferent) nerve** and supplies the muscles of the tongue. The cell bodies are located in the **hypoglossal nucleus.**

Study Questions

Choose the ONE best answer.

10.1 A patient comes to the clinic complaining of chronic cough and difficulty swallowing. The patient's voice is hoarse, and upon clinical examination, a loss of taste from the posterior third of the tongue and a mild weakness of the left trapezius muscle are found. Where is the lesion most likely to be?

    A. Genu of the internal capsule on the left.

    B. Left medial midbrain.

    C. Right middle cranial fossa.

    D. Left jugular foramen.

    E. Right carotid canal.

The correct answer is D. The progressively hoarse voice, coughing, and trouble swallowing are caused by damage to the vagus nerve. The loss of taste in the posterior third of the tongue on the left is due to a lesion of the left glossopharyngeal nerve. Weakness of the trapezius on the left is due to a lesion of the left accessory nerve. The only place where these nerves can be lesioned together is at their exit point from the skull, the left jugular foramen.

10.2 A patient presents with paralysis of the lower part of his face on the right. The upper face still shows symmetrical muscle movements. What is the underlying pathology?

A. Lesion of the right peripheral facial nerve.
B. Lesion of the internal capsule on the left.
C. Lesion of the primary motor cortex on the right.
D. Lesion of the facial nucleus on the right.
E. Lesion of the facial nucleus on the left.

The correct answer is B. The facial nucleus in the pons receives cortical input via the corticobulbar tract. The motor neurons that supply the lower face receive input from the contralateral primary motor cortex, in this case, from the left primary motor cortex. Fibers then descend in the internal capsule and cross over to the contralateral side in the brainstem. A lesion of the left internal capsule would, thus, result in paralysis of the right lower face. The motor neurons supplying the upper face receive their input from the cingulate gyrus from both sides of the brain. When one side is lesioned, the innervation from the contralateral side compensates for this loss. A peripheral nerve lesion (as in A, D, and E) would always result in the loss of innervation in both the upper and the lower portions of the face. Even though the central control of those neurons is intact, a lesion of the peripheral nerve will result in a loss of function on the entire side.

10.3 What statement describes the glossopharyngeal nerve?

A. It exits the skull through the stylomastoid foramen.
B. It carries secretomotor fibers to the submandibular gland.
C. It carries taste from the anterior two-thirds of the tongue.
D. It carries afferents from the inside of the mouth and lips.
E. It carries general visceral afferents from the carotid body.

The correct answer is E. General visceral afferent fibers from the carotid body are carried in cranial nerve (CN) IX. CN IX exits the skull through the jugular foramen. The submandibular gland is supplied by secretomotor fibers from CN VII. CN IX carries taste from the posterior third of the tongue, and CN VII carries the taste afferents from the anterior two-thirds of the tongue. General sensory afferent fibers from inside the mouth and lips are carried by CN V.

10.4 Which statement about the trigeminal nerve is correct?

A. The chief sensory nucleus of V is located in the caudal medulla.
B. The mesencephalic nucleus monitors conscious proprioception.
C. The trigeminal ganglion contains pseudounipolar neurons.
D. The motor nucleus of V supplies somatic motor innervation to the muscles of mastication.
E. The spinal trigeminal nucleus is analogous with the posterior horn of the spinal cord.

The correct answer is E. The spinal trigeminal nucleus extends from the pons to the upper cervical levels and processes pain and temperature. It is continuous with and analogous to the posterior horn of the spinal cord. This is where pain fibers synapse and modulation of pain can occur. The motor nucleus of V is in the midpons and contains the cell bodies of the lower motor neurons to the muscles of mastication. The mesencephalic nucleus processes nonconscious proprioception from the muscles of mastication. The chief sensory nucleus of V is in the midpons. Cranial nerve V supplies muscles derived from pharyngeal arches, and the neurons, therefore, carry special visceral efferents.

11 Hearing and Balance

I. OVERVIEW

Both hearing and balance are special senses carried by **special somatic afferent** fibers that form the **vestibulocochlear nerve (cranial nerve [CN] VIII)**.

The sensory organs and peripheral ganglia associated with CN VIII are located in the petrous part of the temporal bone in the base of the skull (Figure 11.1). The structures related to vestibular function are specialized to translate position and movement of the head into information about balance. Afferents from these structures that carry this balance information are bundled together as the vestibular division of CN VIII. The structures related to auditory function are specialized to transduce sound waves into neural signals. Afferents that carry information about sound (hearing) are bundled together as the cochlear division of CN VIII. Both divisions come together as the vestibulocochlear nerve, which travels from the receptor organs in the temporal bone through the **auditory canal** into the cranial cavity through the **internal auditory meatus**. Afferents then enter the brainstem at the pontomedullary junction (Figure 11.2), and information is then carried to higher cortical centers where it is interpreted.

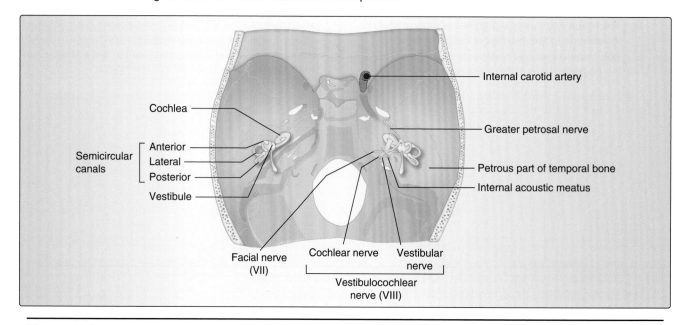

Figure 11.1
Position of the inner ear in the temporal bone of the skull.

Hearing and balance are two very different types of senses. Both the cochlear (hearing) and vestibular (balance) divisions of CN VIII receive stimuli from specialized end organs that contain mechanoreceptors called "**hair cells**" due to their appearance. Hair cells in these two divisions of CN VIII respond to mechanical stimulation that results in the opening of ion channels. The anatomical organization of these sensory organs enables them to respond to different, specific stimuli: hair cells in the cochlear division respond to mechanical stimulation caused by sound waves, whereas hair cells in the vestibular division respond to mechanical stimulation resulting from position and movement of the head in relation to gravity.

II. HEARING

For **hearing**, sound waves are interpreted in terms of their pitch, loudness, and origin of location. The human ear has the remarkable capability to distinguish a large range of sounds whose pitch can be very close together (as little as a quarter note apart) or very far apart (ranging from the low rumblings of a pipe organ to the highest notes of a piccolo flute).

Hearing is an integral component of communication. The sounds of speech are perceived, processed, and then relayed to higher centers where they are reassembled to make sense as words and phrases.

A. Structures involved in hearing

The structures involved in hearing are part of the outer, middle, and inner ear. They are specialized to bundle, amplify, and fine-tune the sounds that surround us so that we can make sense of them.

The outer ear collects and focuses sound waves onto the ear drum, or tympanic membrane. The middle ear is an air-filled space, which contains three small bones that amplify the sound energy from the tympanic membrane to the fluid-filled inner ear. The inner ear contains the cochlea, which contains the sensory organ of hearing, the organ of Corti.

1. **Outer ear:** The outer ear is the visible part of the ear on the side of the head. It is composed of the **auricle** and the **external auditory meatus**, or outer ear canal. These structures gather sound waves and focus them on the **tympanic membrane**, at the medial end of the outer ear canal. The tympanic membrane separates the outer ear from the middle ear (Figure 11.3).

   Interestingly, the external ear also reflects sound, causing it to reach the tympanic membrane in a time-delayed manner. This plays a role in sound localization, as discussed below.

   The external auditory meatus also plays a role in how sound waves are transmitted to the middle ear. Sound pressure at frequencies around 3 kHz (the frequency to which the human ear is most sensitive) is boosted in the external auditory meatus through passive resonance effects (echo).

2. **Middle ear:** The middle ear is located between the tympanic membrane and the inner ear. It is an air-filled chamber that contains three small bones, or **ossicles**, that transfer the sound energy from the tympanic membrane to the fluid-filled inner ear. The middle ear is continuous with the nasopharynx through the pharyngotympanic (eustachian) tube (see Figure 11.3). This connection is important

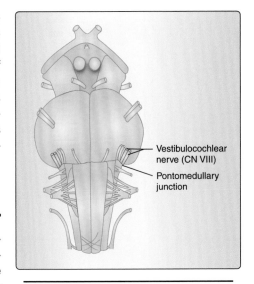

Figure 11.2
The vestibulocochlear nerve at the pontomedullary junction of the brainstem. CN = cranial nerve.

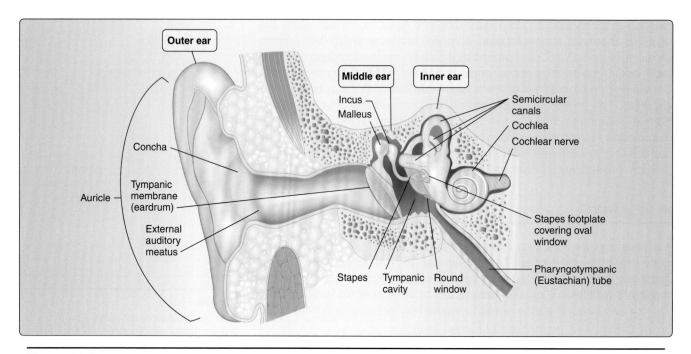

Figure 11.3
Overview of structures of the outer, middle, and inner ear.

to ensure that air pressure in the middle ear corresponds to the air pressure around us. The pharyngotympanic tube is closed most of the time but opens to let air into the middle ear to equilibrate the pressure (e.g., during a plane landing when the ears "pop"). Activities such as yawning, swallowing, and chewing open the pharyngotympanic tube and can be done purposely to allow air pressure to equilibrate when needed.

a. **Bones in the middle ear:** The ossicles in the middle ear are the malleus, the incus, and the stapes. The **malleus** is directly attached to the **tympanic membrane**. The malleus articulates with the **incus**, which is connected to the **stapes**. The stapes is connected to the **oval window** of the inner ear (see Figure 11.3). The function of these articulating ossicles is to boost the sound energy from the tympanic membrane into the inner ear. This boost is necessary so that the sound waves traveling through the air can be transferred efficiently to the fluid-filled space of the inner ear. Without a boost, the sound energy would be lost through reflection once the sound waves hit fluid. The boost is achieved through the lever action of the ossicles as well as through compression of sound waves from the large-diameter tympanic membrane to the small-diameter oval window.

b. **Muscles in the middle ear:** The middle ear also contains two muscles: the **tensor tympani**, which attaches to the malleus and is innervated by CN V, and the **stapedius muscle**, which attaches to the stapes and is innervated by CN VII. Contraction of the stapedius muscle can reduce the transmission of vibration into the inner ear, especially for low-frequency sounds, possibly to selectively filter out low-frequency background noises. These two muscles also dampen movements of the ossicles in response to loud sounds, which serves as a protective mechanism for the auditory nerve.

3. **Inner ear:** The inner ear comprises a **bony labyrinth**, which is a bony tube that sits within the petrous part of the temporal bone. It contains the **three semicircular canals** and the **vestibule**, which are specialized for vestibular function or balance, and the **cochlea**, which is dedicated to hearing. The bony labyrinth contains a fluid called perilymph. A **membranous labyrinth** is suspended within the bony labyrinth and generally follows its contours. The membranous labyrinth is also fluid-filled; the fluid here is called endolymph. The sensory receptors, the hair cells, are located in specialized areas of the membranous labyrinth.

   a. **Cochlea:** The cochlea is a bony tube that coils through two- and three-quarter turns in the shape of a snail's shell (*cochlea* is Latin for "snail"), from a relatively broad base to a narrow apex. It sits in the petrous portion of the temporal bone, with its base facing medially and posteriorly. Within bony cochlea, the **membranous labyrinth** is called the cochlear duct. It contains the **organ of Corti**, the sensory organ that mediates the transduction of sound waves into the electrical energy of a nerve impulse (Figure 11.4).

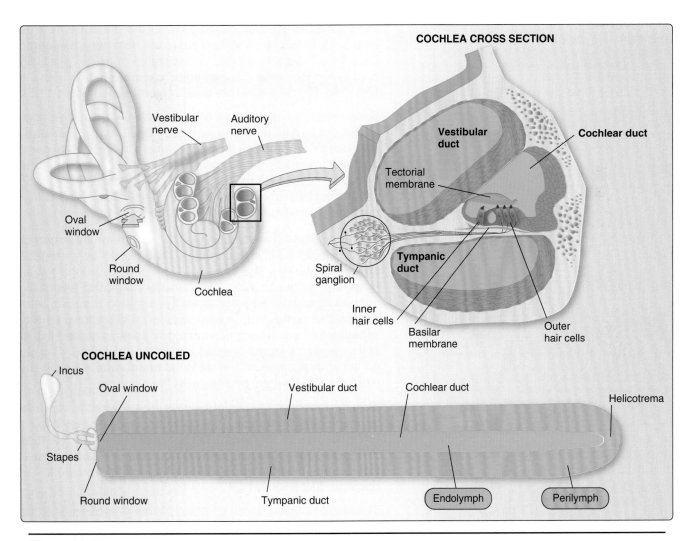

Figure 11.4
Structures of the inner ear: the cochlea.

b. **Three chambers:** Viewed in cross section, the spaces between the bony labyrinth and cochlear duct together form three chambers (or ducts) along most of their length (see Figure 11.4). The cochlear duct has a triangular shape in cross section; it forms the middle chamber. The chamber above the cochlear duct is the **vestibular duct** and is continuous with the vestibule (see below). The chamber below the cochlear duct is called the **tympanic duct** because it ends at the **round window** or secondary tympanic membrane. Both the bony labyrinth and the membranous labyrinth are filled with fluid. The fluid in the bony labyrinth (vestibular and tympanic ducts) is called **perilymph**, which is similar in composition to cerebrospinal fluid (and also to extracellular fluid). Perilymph is low in K^+ and high in Na^+. The fluid in the membranous labyrinth (cochlear duct) is called **endolymph**, which is similar in composition to intracellular fluid, and is high in K^+ and low in Na^+. Endolymph is produced by the **stria vascularis**, a layer of cells on the lateral surface of the cochlear duct (Figure 11.5). The high concentration of K^+ in the endolymph plays a critical role in signal transduction, as discussed below.

The tympanic and vestibular ducts are joined at the apex of the cochlea by a small opening called the **helicotrema** (see Figure 11.4), where perilymph can pass from one chamber to the other. The cochlear duct is separated from the vestibular duct by the **Reissner** (or **vestibular**) **membrane** and from the tympanic duct by the flexible **basilar membrane** (Figure 11.5).

Sound energy is transmitted onto the **oval window**, which displaces the fluid in the vestibular duct. Vibrations are then transmitted along the cochlea to the end, where it joins the tympanic duct and ultimately causes the **round window** at the end of the tympanic duct to bulge. The sound energy or vibrations also cause the basilar membrane, which separates the cochlear duct from the tympanic duct to vibrate (Figure 11.6).

c. **Organ of Corti:** The auditory sensory organ, or the **organ of Corti**, is located within the cochlear duct and sits on the flexible basilar membrane. The organ of Corti comprises one row of **inner hair cells** and three rows of **outer hair cells,** along with supporting cells. The hair cells are the signal-transducing cells. Their name comes from the "hairlike" microvilli, known as **stereocilia**, that are arranged symmetrically and in graded height (with the tallest toward one side of the hair cell) in a V shape on the apex of the cells. The **tectorial membrane**, a gelatinous extracellular structure, extends over the hair cells.

Both the inner and the outer hair cells are anchored to the basilar membrane. Importantly, the outer hair cells are also directly embedded in or coupled to the tectorial membrane via their stereocilia. The inner hair cells do not have direct contact with the tectorial membrane but respond to fluid movement in the cochlear duct (Figure 11.7).

The spiral ganglion, which contains the nerve cell bodies of the primary auditory afferents, sits within the turns of the cochlea,

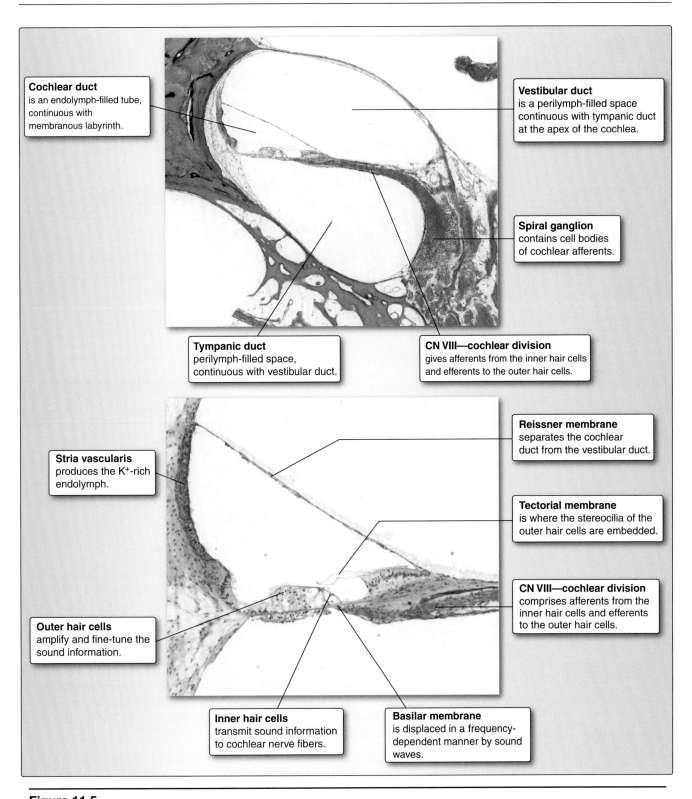

Cochlear duct
is an endolymph-filled tube, continuous with membranous labyrinth.

Vestibular duct
is a perilymph-filled space continuous with tympanic duct at the apex of the cochlea.

Spiral ganglion
contains cell bodies of cochlear afferents.

Tympanic duct
perilymph-filled space, continuous with vestibular duct.

CN VIII—cochlear division
gives afferents from the inner hair cells and efferents to the outer hair cells.

Reissner membrane
separates the cochlear duct from the vestibular duct.

Stria vascularis
produces the K⁺-rich endolymph.

Tectorial membrane
is where the stereocilia of the outer hair cells are embedded.

CN VIII—cochlear division
comprises afferents from the inner hair cells and efferents to the outer hair cells.

Outer hair cells
amplify and fine-tune the sound information.

Inner hair cells
transmit sound information to cochlear nerve fibers.

Basilar membrane
is displaced in a frequency-dependent manner by sound waves.

Figure 11.5
Histology of the cochlea. CN = cranial nerve.

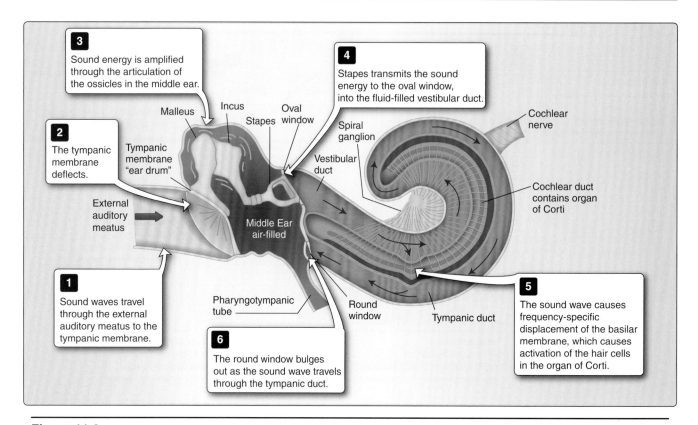

Figure 11.6
Sound transduction in the ear.

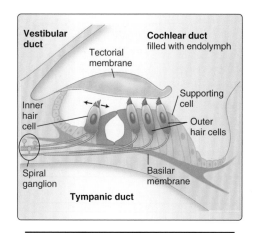

Figure 11.7
Organ of Corti.

close to the organ of Corti (see Figure 11.5). Peripheral processes are located within the cochlear duct where they receive input from the receptor cells.

B. Physiology of sound perception in the inner ear

Sound is a pressure wave that travels through the air. It is then amplified in the outer and middle ear before it reaches the fluid-filled inner ear, where the sensory organ of Corti sits. The organ of Corti transduces this pressure wave to a neuronal signal. Sound waves have different shapes and sizes. The **amplitude** of a sound wave determines its **loudness** and is measured in **decibels (dB)**. The **frequency** of a sound wave determines the **pitch** and is measured in **Hertz (Hz)** (Figure 11.8). The human ear can hear frequencies between 20 and 20,000 Hz. The lowest note on a large pipe organ is at 20 Hz, and the highest note on a piano is at 4,200 Hz (Figure 11.9). The human voice ranges between 300 and 3,000 Hz.

1. **Basilar membrane:** When a sound wave reaches the inner ear, it sets off a wave in the **basilar membrane** at the same frequency as the sound. This wave propagates from the base to the apex until it reaches a **point of maximal displacement** of the basilar membrane. This point is reached due to the geometry and flexibility of the basilar membrane. The base of the basilar membrane is narrow and stiff and is where the propagation of each sound wave begins. High-frequency sounds produce their maximal displacement at the

base. The apex of the basilar membrane, on the other hand, is wider and more flexible and is where low-frequency sounds are perceived (Figure 11.10). These mechanical properties result in the **tonotopy** of the inner ear, with distinct locations interpreting discrete frequencies. Tonotopy (analogous to the somatotopy of the sensory systems) is then carried forward throughout the auditory pathway.

Most of the sounds we hear are a combination of different frequencies. As the sound waves travel into our inner ear, they are broken up into their component parts. Each component will individually reach its point of maximal displacement on the basilar membrane.

2. **Inner and outer hair cells:** Basilar membrane vibrations create a shearing force against the stationary tectorial membrane, causing the stereocilia of the outer hair cells to be displaced in that plane (Figure 11.11). The inner hair cells are not in direct contact with the tectorial membrane and are activated through fluid movement in the cochlear duct. Stereocilia are arranged symmetrically by height. Displacement toward the tallest stereocilium causes **depolarization** of the cell, whereas displacement toward the shortest stereocilium causes **hyperpolarization** of the cell (Figure 11.12).

Depolarization of the cell occurs when cation channels open at the apex of the stereocilia. Stereocilia are connected to each other via **tip links** that transmit force to an elastic gating spring, which, in turn, opens the cation channel (see Figure 11.12). These cation channels are examples of **mechanotransduction** channels, which have the advantage of conferring immediate effects. In fact, hair cells can respond to a stimulus within 50 μs. Such a rapid response would not be possible with a slow chemical signal transduction process. Another advantage of mechanotransduction channels is that they do not require receptor potentials, thereby increasing the **sensitivity** of the response (see Chapter 1, "Introduction to the Nervous System and Basic Neurophysiology"). The sensitivity of the ion channel opening is remarkable: even small vibrations of 0.3 nm (the size of an atom) can cause channel opening.

Because the stereocilia are bathed in the K^+-rich endolymph of the cochlear duct, the opening of the cation channels will cause a rapid influx of K^+ into the cell (the driving force for K^+ uptake is about 150 mV). The hair cells then depolarize, which causes Ca^{2+} channels at the base of the cells to open. Calcium influx causes neurotransmitter-filled vesicles to fuse with the basilar membrane and release the neurotransmitter glutamate into the synaptic cleft. The afferent cochlear nerve fibers are thus stimulated and transmit this signal to the central nervous system (CNS).

The inner hair cells are responsible for hearing. About 90% of cochlear nerve fibers come from the inner hair cells. The outer hair cells amplify the signals that are then processed by the inner hair cells.

3. **Frequency selectivity:** The **frequency selectivity**, or tuning, of the basilar membrane is due to and limited by its mechanical properties. The sound wave traveling along the basilar membrane

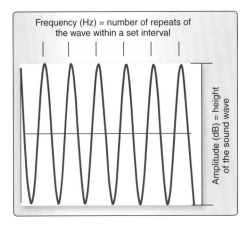

Figure 11.8
The physics of frequency and amplitude.

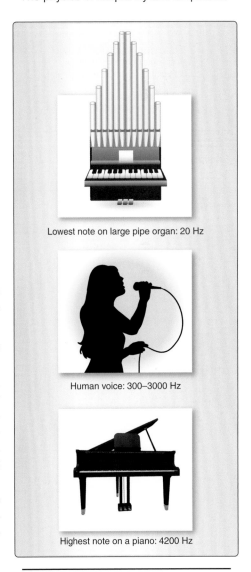

Lowest note on large pipe organ: 20 Hz

Human voice: 300–3000 Hz

Highest note on a piano: 4200 Hz

Figure 11.9
Examples of different frequencies.

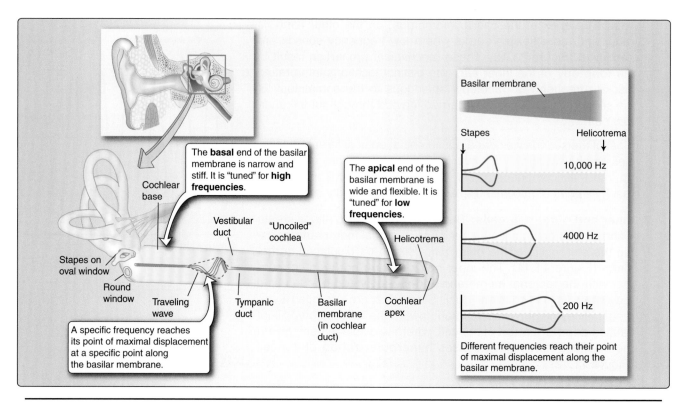

Figure 11.10
Basilar membrane tuning.

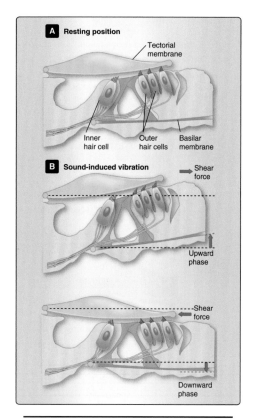

Figure 11.11
The organ of Corti during displacement by sound waves.

and its associated point of maximal displacement cannot be as selective in frequency tuning as our hearing is, which suggests involvement of an additional mechanism of sound amplification and tuning. This additional mechanism is from movement of the outer hair cells in response to specific frequencies. When the outer hair cells are depolarized, their cell bodies actively contract. When they are hyperpolarized, their cell bodies actively lengthen. High frequencies cause contraction of the outer hair cells at the base, and low frequencies cause contraction at the apex of the basilar membrane. This mechanism influences the movement of the basilar membrane in that particular segment, increasing the fluid displacement around the inner hair cells. This amplifies the magnitude of the K^+ influx into the inner hair cells, increasing the signal to the cochlear nerve. Because of this fine-tuning and amplification of the sound wave through the outer hair cells, we can both discriminate tones of neighboring frequencies with astounding accuracy and detect low-level sounds. Interestingly, the outer hair cells are also innervated by **efferents** originating from the auditory pathway (Figure 11.13). These inputs hyperpolarize, or inhibit, the outer hair cells, reducing their response to displacement of the basilar membrane through sound and allowing the central auditory pathway to influence sound amplification in the inner ear. A possible function of this mechanism is to help focus the inner ear on relevant sounds while filtering out background noises.

4. **Otoacoustic emissions:** Another interesting feature of auditory physiology is known as otoacoustic emissions. Because the motility

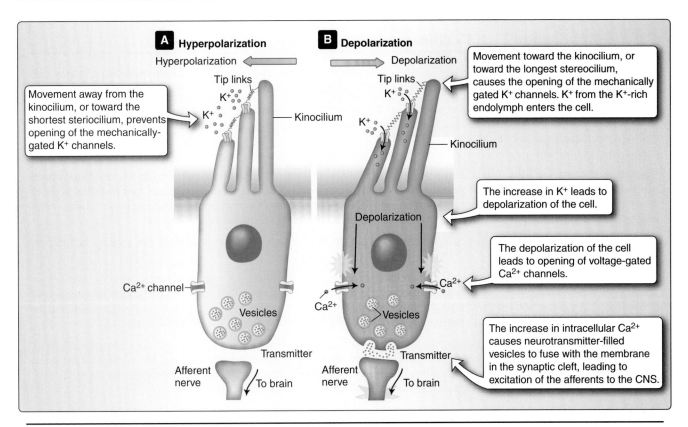

Figure 11.12
Hyperpolarization and depolarization of hair cells in the inner ear. CNS = central nervous system.

of the outer hair cells can cause the basilar membrane to move, it is conceivable that this movement could be retrograde, or backward, toward the oval window and through the middle ear via the ossicles to cause displacement of the tympanic membrane. This process would result in the ear itself producing a sound. Indeed, this is what actually happens. These sounds can be measured in the external auditory meatus as **otoacoustic emissions**. Such measurements are routinely done in infants to assess the function of the inner and middle ears.

C. Central auditory pathways

The central auditory pathways carry signals from the cochlea to the CNS. The auditory system analyzes different aspects of sound including the **frequency** (pitch), the **amplitude** (volume), and the **location** of the sound in space.

The pitch and volume of the sound travel centrally in a relatively straightforward pathway. The localization of sound, however, is more complicated, and the pathways differ depending on whether high-frequency or low-frequency sounds are being analyzed.

1. **Central pathway for pitch and volume:** The frequencies of a sound are broken down in the cochlea and then relayed to the cochlear nerve fibers innervating the hair cells at different locations along the basilar membrane. Each cochlear nerve fiber only transmits information of a specific frequency spectrum. The nerve cell

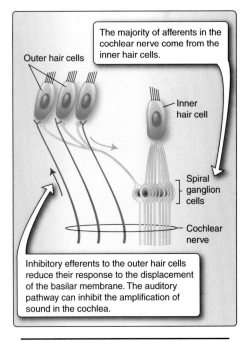

Figure 11.13
Afferents to and efferents from the hair cells in the cochlea.

Clinical Application 11.1: Cochlear Implants

Hearing loss can have several underlying causes and is divided into two main categories: **conductive** hearing loss and **sensorineural** hearing loss.

Conductive hearing loss results from obstruction of the conduction of sound energy from the outer ear to the inner ear. This can occur either in the outer ear (such as buildup of earwax or rupture of the tympanic membrane) or in the middle ear (e.g., fluid buildup or arthritis of the ossicles). A hearing aid, which amplifies sound energy, can significantly ameliorate conductive hearing loss.

Sensorineural hearing loss results from a problem in the inner ear, either with the **hair cells** or with the **cochlear nerve** itself. Hair cells are very susceptible to damage and do not regenerate in humans. Common causes for congenital hearing loss include genetic causes and prenatal infection with TORCH organisms (toxoplasmosis, other [syphilis], rubella, cytomegalovirus, and herpes), which lead to dysfunctional hair cells with an intact cranial nerve (CN) VIII.

For the latter patients, where the hair cells are damaged but CN VIII is intact, a **cochlear implant** can improve sound perception, or hearing.

A cochlear implant consists of an external and an internal component. The external component includes a microphone, a speech processor, and a transmission system. Sound information is broken down into its component parts and converted into electrical signals, which are then relayed to the internal component. The internal component includes a receiver and an electrode array. The receiver decodes the signal and delivers the electrical signals to the electrode array.

The electrode array is inserted into the cochlea through the oval window where it sits in the cochlear duct along the afferents from CN VIII. Electrical signals anywhere along the electrode array will stimulate a particular cochlear nerve afferent along the basilar membrane. The electrode array mimics the tonotopy of the basilar membrane and stimulates nerves at discrete frequencies. This information is then relayed centrally, resulting in the perception of sound.

Insertion of a cochlear implant does not automatically result in "hearing." Hearing involves the interpretation of sounds heard and is a central process that requires new neuronal connections to be made. That is, one must learn to understand speech.

For patients with damage to CN VIII, devices that will directly relay sound information from the hair cells to brainstem auditory nuclei are currently being developed.

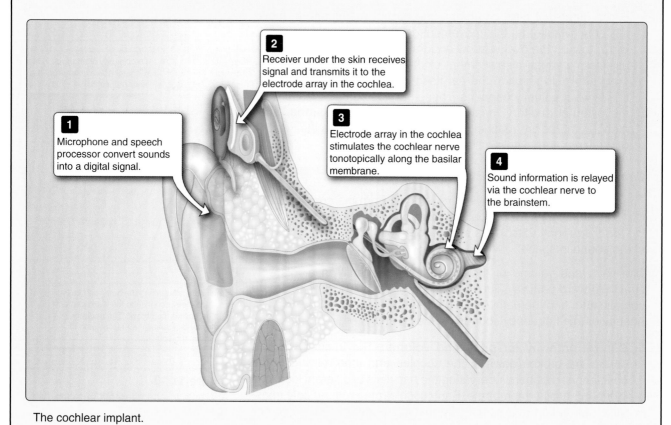

The cochlear implant.

bodies of these cochlear afferents are located in the **spiral ganglion**. The central processes of these first-order neurons synapse in the **cochlear nuclei**. The cochlear nuclei comprise columns of cells adjacent to the inferior cerebellar peduncle and can be divided into a **posterior** and **anterior nucleus**. From the cochlear nuclei, most second-order fibers then cross the midline and travel in the contralateral **lateral lemniscus** to the **inferior colliculus** in the caudal midbrain, a major integration center in the auditory pathway. From there, third-order fibers travel to the **medial geniculate nucleus (MGN)** of the thalamus via the **inferior brachium**. From the thalamus, fibers travel through the internal capsule to the **primary auditory cortex**, which is located on the superior surface of the superior temporal gyrus in the temporal lobe (Figure 11.14).

2. **Central pathways for sound localization:** We live in three-dimensional space, and sounds are perceived as coming from within this space. The auditory system can, in fact, map sounds even though space is not directly represented in the auditory system. We can map sounds in both a **vertical plane** (whether sounds come from above or below) and a **horizontal plane** (whether sounds come from left or right, in front or in back of us). Vertical

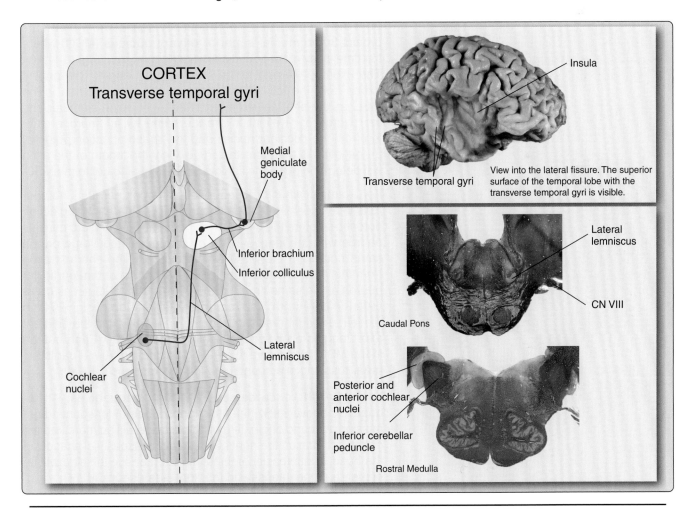

Figure 11.14
Central pathway for pitch and volume. CN = cranial nerve.

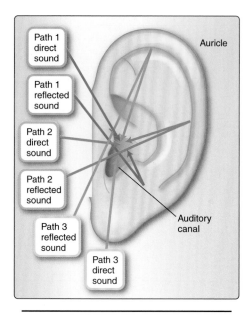

Path 1
direct
sound

Path 1
reflected
sound

Path 2
direct
sound

Path 2
reflected
sound

Path 3
reflected
sound

Path 3
direct
sound

Auricle

Auditory
canal

Figure 11.15
Vertical sound mapping through
reflection of sound in the outer ear.

analysis can be done with only one ear, whereas horizontal analysis relies on the input from both ears.

Vertical sound mapping occurs in the external ear. Sounds reach the tympanic membrane both **directly** and through **reflection** in the external ear. The brain can localize sounds in the vertical plane through analysis of the differences in the direct and reflected sound inputs. The neuronal mechanisms of this are not fully understood (Figure 11.15).

In order to achieve **horizontal spatial mapping**, the input to both ears is compared in brainstem nuclei. For low-frequency sounds, the sound waves will reach the ear farther away from the sound after they reach the ear closer to the sound, and a time difference is detected and analyzed. For high-frequency sounds, the sound waves are closer to each other, and the head forms a barrier for these waves as they travel to the ear farther away from the sound stimulus. The far ear will hear the sound at a lesser intensity than the near ear due to the "sound shadow" created by the head.

a. **Time difference detection at low frequencies:** For low-frequency sounds (below 3 kHz), the ear closer to the sound source will perceive the sound waves before the ear farther away from the source. These low-frequency sounds from both ears project to the **medial superior olivary nucleus (MSO)** where the time delay of the sound perceived is analyzed (Figure 11.16). The axons projecting to the MSO vary in length. The longest axons from the left will converge on the same neuron in the MSO as the shortest axons from the right. Axon diameter and the degree of myelination are the same for all neurons coming from the cochlear nuclei, and the speed of action potential propagation is therefore the same. Only the length of the axon will determine how long it takes for the signal to get to the MSO.

For example, when a sound reaches the left ear first, the neurons in the cochlear nucleus on the left will start sending action potentials before the neurons on the right. The cochlear nucleus neuron with the longest axon on the left will converge on the same neuron in the MSO as the one with the shortest axon from the right. The action potentials will arrive at that particular neuron at the same time. The neurons in the MSO act as **coincidence detectors**. The **temporal summation** of signals from the left and right resulting from the time delay and the different axon lengths allow for the localization of sound. Each neuron in the MSO is sensitive to a sound originating from a particular area, resulting in a sound map for low-frequency signals.

b. **Intensity difference detection at high frequencies:** At frequencies above 3 kHz, the head forms a barrier for sound transmission. A sound originating on the left side will be more intense on the left than on the right because of the acoustical shadow of the head. The intensity of the stimulation on the left side will be higher than the intensity on the right side (Figure 11.17).

The intensity of the stimulation is transmitted to the **cochlear nuclei** and from there to the **lateral superior olivary nucleus (LSO)**. At the same time, a signal encoded at the same intensity

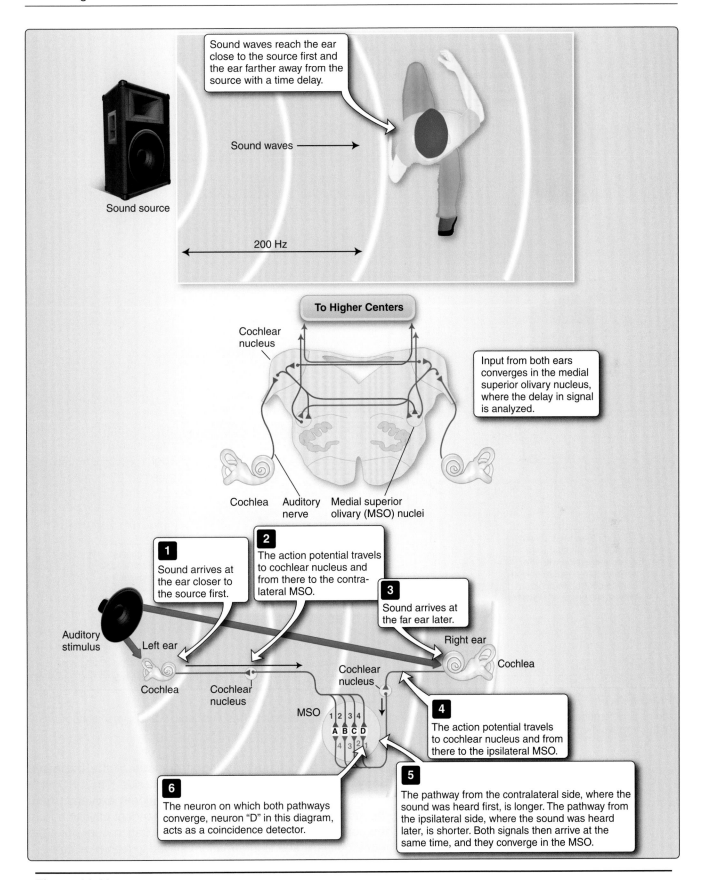

Figure 11.16
Time difference detection at low frequencies.

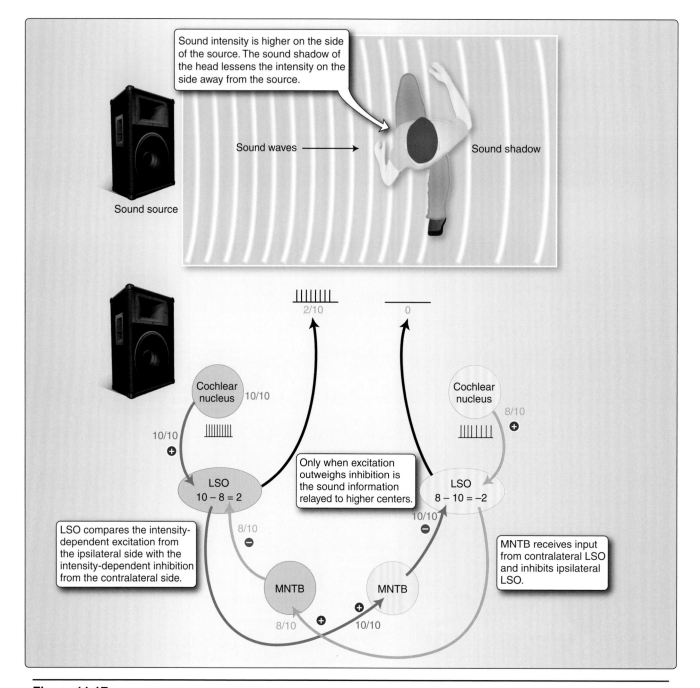

Figure 11.17
Intensity difference detection of sound at high frequencies. LSO = lateral superior olivary nucleus; MNTB = medial nucleus of the trapezoid body.

is sent to the contralateral **medial nucleus of the trapezoid body**, which will inhibit the LSO on that side. The LSO then compares the amount of intensity-dependent excitation from the ipsilateral side with the intensity-dependent inhibition from the contralateral side. Only when the excitation outweighs the inhibition is the sound information relayed to higher centers.

Each LSO can only relay information from the ipsilateral side of the soundscape. In order to get a full appreciation of the sound-filled space, both lateral superior olivary nuclei must function.

c. **Convergence of pathways:** Both the intensity level and time difference–encoded sound localization pathways converge in the **inferior colliculus**. Similar to the visual map in the superior colliculus, the inferior colliculus contains an auditory space map. Here, both the vertical and the horizontal analyses of sound are integrated, resulting in a precise sound localization. The inferior colliculus also analyzes the temporal patterns of sound. From the inferior colliculus, the information is relayed to the MGN of the thalamus. There, the frequency-analyzed component and the temporal component of sound converge in a tonotopically mapped pathway (Figure 11.18).

The signal is then relayed to the **primary auditory cortex**, which is also organized tonotopically and can interpret sounds and spatial distribution patterns. The secondary auditory or auditory association areas of the cortex are localized around the primary area and process complex sounds necessary for communication. In the human brain, the cortical area for speech comprehension (**Wernicke area**) extends from the inferior parietal lobe into the temporal lobe. It is tightly associated with and overlaps the primary and association auditory areas. See Chapter 13, "The Cerebral Cortex," for more information.

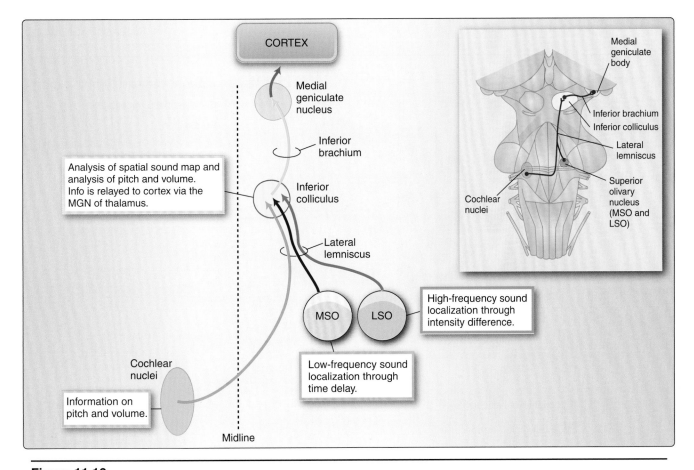

Figure 11.18
Convergence of pathways in the auditory system. MGN = medial geniculate nucleus; MSO = medial superior olivary nucleus; LSO = lateral superior olivary nucleus.

III. BALANCE

Although we have no conscious appreciation of balance, it is a key sense that interacts with many systems to ensure stable posture and coordinated movements.

Carried in the vestibular component of the vestibulocochlear nerve (CN VIII), the sense of balance allows perception of the body in motion, head position, and the orientation of the head in relation to gravity.

A. Structures involved in balance

The vestibular structures are embryologically and structurally related to the auditory structures. The part of the **bony labyrinth** related to balance is adjacent to and continuous with the cochlea in the temporal bone. It consists of **three semicircular canals**, which are roughly orthogonal (at 90 degrees) to each other and attached to the central **vestibule**. Like the cochlea, these bony structures contain a **membranous labyrinth**, which comprises the utricle and saccule in the vestibule and the semicircular ducts in their canals and is continuous with the cochlear duct of the cochlea. The membranous labyrinth contains **K+-rich endolymph**, whereas the space between the membranous and the bony labyrinth is filled with perilymph, with low K+ concentrations (Figure 11.19). Within the membranous labyrinth,

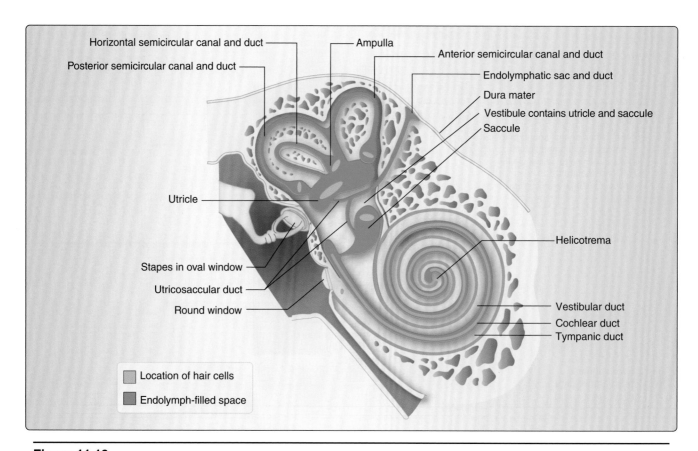

Figure 11.19
The membranous labyrinth and the location of hair cells in the inner ear.

there are receptor cells, which are hair cells analogous to those in the organ of Corti.

1. **Semicircular canals:** The semicircular canals contain the membranous **semicircular ducts**. At the base of each duct is an enlarged area called the **ampulla**, which contains the receptor hair cells. These hair cells sit on a structure call the **crista** (crista ampullaris) and, analogous to the hair cells of the cochlear duct, are embedded in a gelatinous mass, which in this location is called the **cupula** (Figure 11.20).

2. **Vestibule:** The vestibule contains two endolymph-filled sacs or enlargements of the membranous labyrinth, the **utricle** and the **saccule**. The three semicircular ducts and their ampullae are continuous with the utricle, which is connected to the saccule via the utriculosaccular duct. The utricle and saccule contain the **otolithic organ**. Each contains a **macula**, where the receptor hair cells are located. The utricular macula is on the floor of the utricle, in a horizontal plane. The saccular macula is on the medial wall of the saccule, in a vertical plane. The hair cells of the maculae are embedded in a gelatinous mass, which has an outer layer covered in calcium carbonate crystals (**otoconia**, or **otoliths**). This gives this gelatinous structure its name, the **otolithic membrane** (Figure 11.21).

B. Physiology of balance

We can move our bodies and our heads along all three axes of our three-dimensional space. We can move in a linear way, along any one of these axes (**linear acceleration**), or we can rotate around any one of these axes (**rotational acceleration**). Our movements are often a combination of linear and rotational acceleration, and the hair cells within the membranous labyrinth (semicircular canals and otolithic organ) of the inner ear detect the different components of our movements and faithfully relay them to central nuclei.

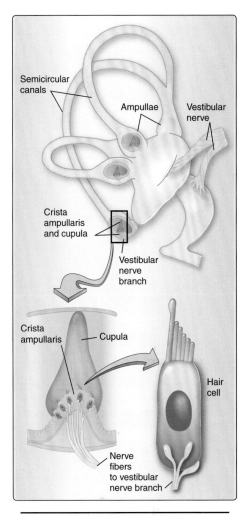

Figure 11.20
The semicircular canals with the ampullae containing crista and cupula.

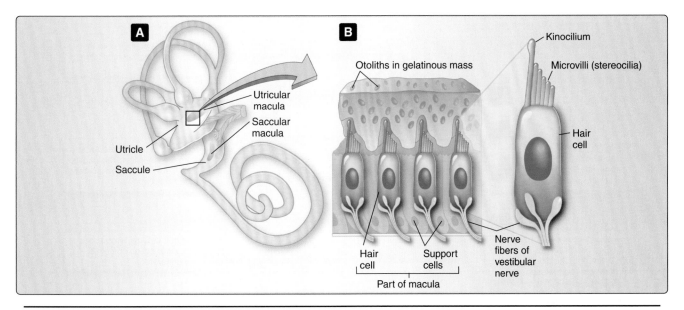

Figure 11.21
The otolithic organs: utricle and saccule.

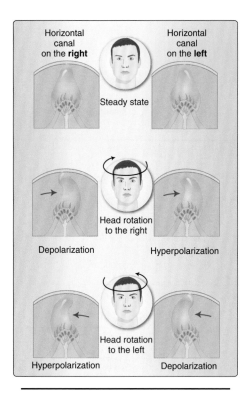

Figure 11.22
Rotational acceleration and deflection of the cupula in the horizontal semicircular canal.

The system is best equipped to detect *changes* in movement. Vestibular afferents will fire most at the beginning and end of an acceleration.

1. **Rotational acceleration:** Rotational acceleration is detected in the **three semicircular canals**, where identically oriented hair cells sit atop cristae. Each hair cell has a long **kinocilium** and several microvilli, called **stereocilia**, in graded height as in the cochlea.

   During rotational acceleration, the endolymph is set into motion. The movement of endolymph causes a deformation of the cupula (Figure 11.22). This deformation causes deflection of the stereocilia of the hair cells. Movement toward the kinocilium will cause the mechanically gated ion channels to open, resulting in depolarization of the cell and increased signal transduction in the vestibular nerve (Figure 11.23). Movement away from the kinocilium causes the cation channels to close and, thereby, hyperpolarizes the cell, with a decrease of signal transduction in the vestibular nerve. All hair cells in the ampulla on each side have the same orientation and will respond similarly to deformation of the cupula.

   The semicircular canals on one side of the head are a mirror image of the semicircular canals on the other side of the head. The two horizontal canals are in the same plane, the left posterior and right anterior canals are paired and in the same plane, and the left anterior and right posterior canals are paired and in the same plane (Figure 11.24). Rotational acceleration is detected on

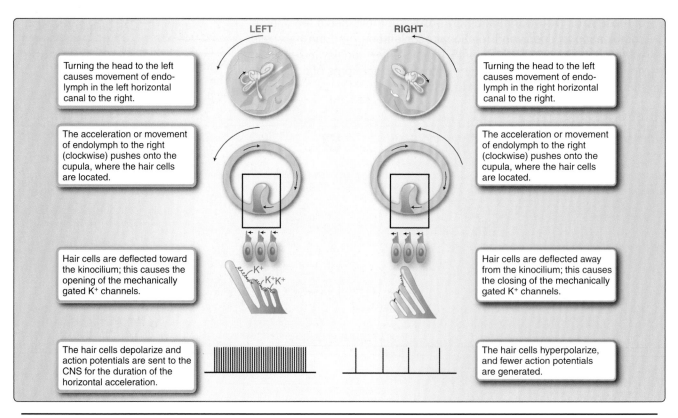

Figure 11.23
How rotational acceleration leads to signal transduction. CNS = central nervous system.

both sides and sets endolymph in motion in the same direction on both sides, but with different effects. Movement of endolymph will cause hyperpolarization of hair cells on one side and depolarization on the other side, depending on whether the stereocilia are deflected away from or toward the kinocilium, respectively (see Figure 11.23).

For example, if the head is rotated to the left (or counterclockwise), endolymph in both horizontal canals will rotate to the right (or clockwise). This will lead to displacement of the cupula on both sides. On the left side, hair cells are displaced toward their kinocilia, which causes the opening of cation channels and increased signal transduction. On the right side, hair cells are displaced away from their kinocilia, which causes the closing of cation channels and decreased signal transduction.

This system of hyperpolarization in one canal and depolarization in the paired canal works in all pairs of canals. Increased signal transduction will always occur in the canal toward which the head is rotating.

2. **Linear acceleration:** Linear acceleration is detected by the **otolithic organ**. The otoconia make the otolithic membrane heavier than the surrounding endolymph. During tilting head movements, gravity pulls on the otolithic membrane causing it to shift relative to the underlying macula and the hair cells anchored there. This shearing motion causes displacement of the hair cells and, with that, the opening or closing of the cation channels, depending on the direction of the shift. Similarly, when the head moves forward without a tilt (pure linear acceleration), the same shearing motion occurs between the otolithic membrane and the macula. Because of the weight of the otoconia, the inertia of the otolithic membrane is greater than that of the macula, and the otolithic membrane will lag behind the movement of the macula, which causes deflection of the hair cells. When pure linear acceleration ends, or there is deceleration, the otolithic membrane again lags behind. The inertia of the otolithic membrane results in its continued forward movement (Figure 11.25). Hair cells are arranged relative to the striola; some will hyperpolarize and some will depolarize. The resulting pattern is interpreted as the direction of acceleration.

The location of the macula and the orientation of the hair cells determine which type of linear acceleration can be detected. The saccular and utricular maculae on one side of the head are mirror images to those on the other side of the head. This results in opposing effects on corresponding hair cells of the two maculae, similar to that of the hair cells of two paired semicircular canals.

a. **Utricle:** In the utricle, the macula is located at the bottom of the sac. The hair cells can be divided into two groups with different orientations, separated by the **striola**, a depression in the otolithic membrane. In the utricle, kinocilia are oriented toward the striola. This enables the utricle to detect linear movement in a horizontal plane in two directions,

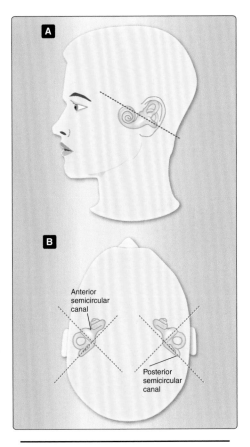

Figure 11.24
The pairing of semicircular canals and their orientation in the head relative to each other (the size of the inner ear is exaggerated for diagrammatic purposes).

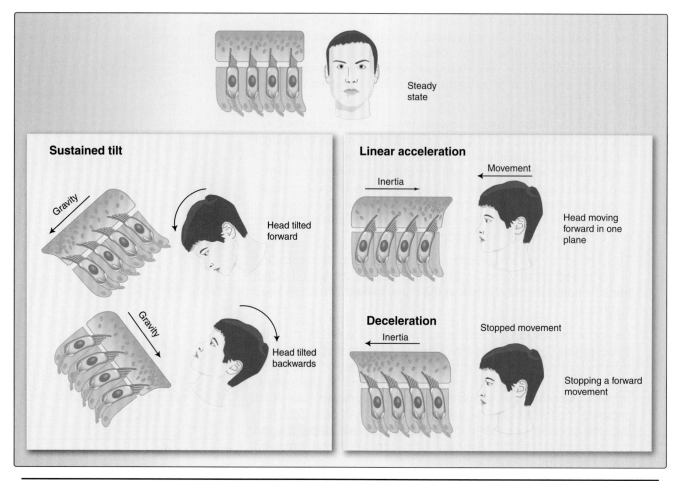

Figure 11.25
Linear acceleration and deflection of the macula in the saccule and utricle. Note that the diagram only shows one population of hair cells to demonstrate the movement of the otolithic membrane.

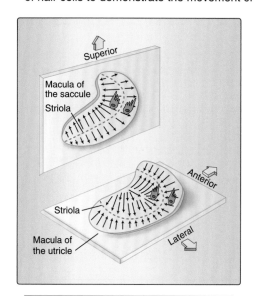

Figure 11.26
Orientation of the saccule and utricle in the inner ear and orientation of hair cells on the maculae.

such as head tilts to the right or left or rapid lateral displacements (Figure 11.26).

b. Saccule: In the saccule, the macula is located in the medial wall of the sac. Again, the striola divides the hair cells into two groups with different orientations. In the saccule, kinocilia are oriented away from the striola. The saccule detects head movement in a vertical plane, such as up and down movements or forward and backward tilts (see Figure 11.26).

C. Central vestibular pathways

The afferents from the labyrinth have their cell bodies in the **vestibular** (or **Scarpa**) **ganglion**, located close to the spiral ganglion. The central processes enter the brainstem as the vestibular portion of the **vestibulocochlear nerve** at the pontomedullary junction and project to the vestibular nuclear complex. The vestibular nuclei are located in the posterior portion of the **tegmentum**, at the junction between the pons and the medulla, adjacent to the inferior cerebellar peduncle and the cochlear nuclei. The vestibular nuclei can be subdivided into two functionally distinct groups: **lateral vestibular nuclear group** and **medial vestibular nuclear group** (Figure 11.27).

The vestibular nuclei participate in three major reflex pathways. The **vestibuloocular reflex** adjusts eye movements to head movements and stabilizes images on the retina (see Chapter 9, "Control of Eye Movements"). The **vestibulocervical reflex** is important for postural adjustments of the head, and the **vestibulospinal reflex** is important for the postural stability of the body. The vestibulospinal pathways also contribute to muscle tone.

The vestibular nuclei are integration centers that receive not only afferents from the inner ear but also feedback loops from the cerebellum as well as visual and somatosensory input. As a result, outflow from the vestibular nuclei incorporates more than just input from the inner ear.

1. **The vestibulocervical reflex:** Postural adjustments of the head are most relevant in response to **rotational movements**, which are detected in the semicircular canals.

   Afferents from the semicircular canals project to the medial vestibular nuclei. From there, fibers travel in the **descending medial longitudinal fasciculus (or medial vestibulospinal tract)** to the upper cervical levels of the spinal cord. Here, they cause postural adjustments of the head and neck muscles in response to head movements (Figure 11.28).

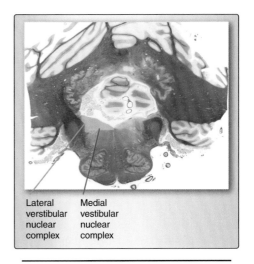

Lateral verstibular nuclear complex | Medial vestibular nuclear complex

Figure 11.27
The lateral and medial vestibular nuclei in the rostral medulla.

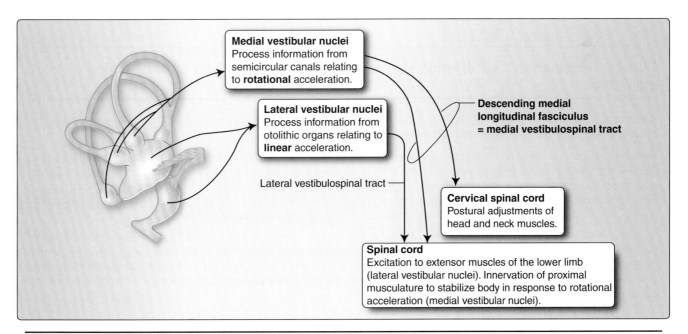

Figure 11.28
Overview of the central vestibular pathways.

2. **The vestibulospinal reflex:** Postural adjustments of the body occur in response to both linear and rotational acceleration. Linear acceleration is detected by the otolithic organ, and afferents project mainly to the lateral vestibular nuclei. Rotational acceleration is detected by the hair cells in the semicircular canals, and afferents project primarily to the medial vestibular nuclei.

Projections from both the medial and the lateral vestibular nuclei travel through the **descending medial longitudinal fasciculus (or medial vestibulospinal tract)** and **lateral vestibulospinal tract**, respectively, to the spinal cord. In the anterior horn of the spinal cord, the lateral vestibulospinal tract provides excitatory input to the extensor muscles of the legs, which are key muscles in mediating balance and postural stability in upright gait. They also influence proximal trunk musculature, particularly in response to rotational accelerations.

The vestibulospinal reflex is a direct modulator of lower motor neuron function to allow for rapid postural adjustments in response to a change in balance (see Figure 11.28).

3. **Cortical projections:** Although there is no conscious appreciation of balance, there are projections from the vestibular nuclei to the cortex via the thalamus. The cortical targets are in the **primary** and **secondary somatosensory areas**, which receive additional visual and proprioceptive inputs. These cortical areas are thought to be important for the conscious appreciation of the position of our bodies in space as well as for the perception of extrapersonal space.

Clinical Application 11.2: Benign Paroxysmal Positional Vertigo and the Epley Maneuver

Benign paroxysmal positional vertigo (BPPV) is the most common peripheral vestibular disorder. Patients report brief spells of vertigo directly related to movements of the head.

Pathophysiologically, BPPV is typically caused by altered sensitivity to gravity in the posterior semicircular canal due to the presence of floating otoliths in that canal. However, other canals (anterior and horizontal) can sometimes be involved. These otoliths are thought to have dislodged from the utricle in the vestibule and then floated into the posterior semicircular canal. Here, they "bump into" the cupula in the crista ampullaris and stimulate the hair cells in response to certain head movements. This isolated stimulation of the posterior semicircular canal on one side results in vertigo.

To assess the side from which the BPPV originates (or in which inner ear the otoliths are floating in the semicircular canal), the patient is placed in a supine position, and the head is rotated to one side and then to the other. During this procedure, the eyes are carefully observed. On the affected side, otoliths will stimulate the cupula in the semicircular canal in response to the head rotation, resulting in vertical and torsional nystagmus as well as vertigo if the posterior semicircular canal is causing BPPV.

The treatment of BPPV aims to remove the debris from the semicircular canal back into the vestibule through a sequence of head-positioning maneuvers. This sequence is referred to as the Epley maneuver and is summarized in the figure.

Clinical Application 11.2: Benign Paroxysmal Positional Vertigo and the Epley Maneuver continued

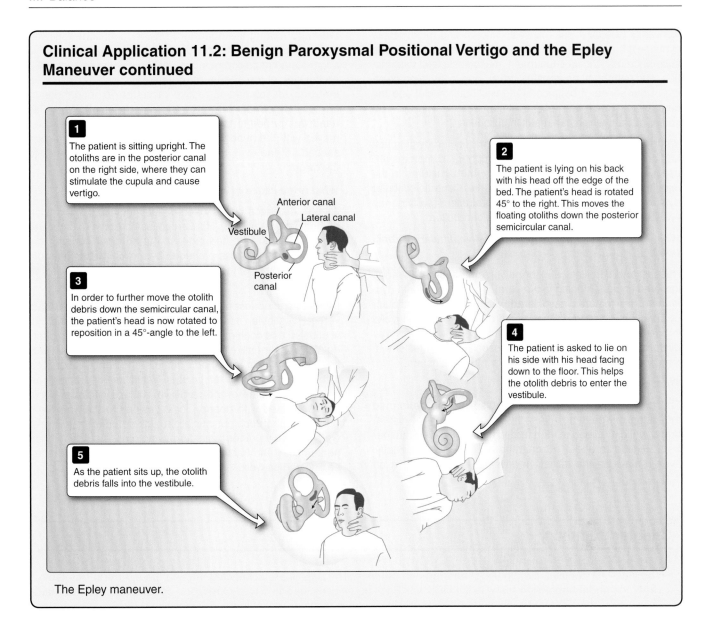

1 The patient is sitting upright. The otoliths are in the posterior canal on the right side, where they can stimulate the cupula and cause vertigo.

2 The patient is lying on his back with his head off the edge of the bed. The patient's head is rotated 45° to the right. This moves the floating otoliths down the posterior semicircular canal.

3 In order to further move the otolith debris down the semicircular canal, the patient's head is now rotated to reposition in a 45°-angle to the left.

4 The patient is asked to lie on his side with his head facing down to the floor. This helps the otolith debris to enter the vestibule.

5 As the patient sits up, the otolith debris falls into the vestibule.

Anterior canal
Lateral canal
Vestibule
Posterior canal

The Epley maneuver.

Clinical Case:

Jacob's Dizziness

Jacob is a 42-year-old man with a 2-day history of new persistent dizziness. He describes the dizziness as a spinning sensation that is worse with head movement. He has nausea but has not vomited. Two weeks ago, Jacob had a fever with a cough and runny nose (rhinorrhea). He is unaware of any other symptoms except for slight veering to the left when walking. He has a history of hypertension and takes a diuretic medication. On examination, his blood pressure is 125/80 with normal orthostatic responses. His heart sounds are normal and there are no bruits. He has full range of extraocular movements. There was right beating horizontal nystagmus that increased in amplitude with gaze to the right and decreased in amplitude on gaze to the left. On head thrusts to the left, he had corrective saccades but not on head thrust to the right. He has reduced hearing over his left ear. The rest of his cranial nerve examination is normal. His sensory, motor, and gait assessment are normal except for mild veering to the left.

Case Discussion

When obtaining a history in a person with dizziness, it is important to distinguish between nonvertiginous and vertiginous dizziness. Nonvertiginous dizziness includes lightheadedness and unsteadiness. Vertiginous dizziness or vertigo is the sensation of self-motion (head/body movement) when no self-motion is occurring or the sensation of distorted self-motion during an otherwise normal head movement.

Nonvertiginous dizziness manifests with fainting (presyncope/syncope); it can be a side effect of some medications and substance use. Visual impairment, musculoskeletal dysfunction, and peripheral neuropathy can all lead to faulty processing of our sense of balance in the vestibular nuclei and the cerebellum (which has reciprocal connections with the vestibular nuclei), which can result in dizziness.

Vertigo is classified into peripheral vestibular dysfunction and central vestibular dysfunction. Peripheral symptoms are associated with a lesion or infection of the structures in the middle ear or the vestibulocochlear nerve. Central vestibular dysfunction is due to a lesion to the brainstem involving the vestibular nuclei and their tracts or the cerebellum.

What are symptoms/signs of peripheral vestibular dysfunction?

Nausea and vomiting, auditory symptoms such as hearing loss and tinnitus (ringing in the ears), and unidirectional nystagmus are features associated with peripheral vestibular dysfunction. Additional sensory and motor symptoms involving the face and limbs are not seen in peripheral vestibular dysfunction; these are symptoms associated with lesions to the ascending and descending tracts in the brainstem.

What is a head thrust test?

The head thrust test (or head impulse test) is performed by asking the patient to keep his or her eyes on a distant target while the head is then turned quickly by the examiner from midline to about 15 degrees on one side. Normally, eyes remain on the target; however, if the eyes move off the target during the head turn (in one direction), followed by a saccade back to the target after the head turn, this suggests an impairment in the vestibuloocular reflex (VOR) on the side of the head turn (peripheral vestibular lesion on that side). Jacob had corrective saccades on head thrust to the left (abnormal) but not to the right (normal). When his head was turned to the left, his eyes moved off target (i.e., moved with the head to the left), and therefore, a corrective saccade of the eyes to the right was needed to go back to the target.

What is the cause of Jacob's dizziness?

Jacob's dizziness was caused by labyrinthitis (inflammation of the vestibular nerve). His vertigo lasted hours to days and was associated with hearing loss but no other neurological symptoms/signs. It was preceded by a viral illness; the most likely diagnosis is therefore labyrinthitis involving the left inner ear.

Differential diagnoses include benign paroxysmal positional vertigo, vestibular neuronitis, and lesions to the brainstem, vestibular, and cochlear nuclei. In this case, all of these alternatives appear unlikely. In BPPV, the triggered rotational vertigo typically lasts less than 1 minute and is not associated with hearing changes or tinnitus. Vestibular neuronitis can present with some of the same vestibular symptoms but is not associated with hearing loss. Lesions to the brainstem vestibular and cochlear nuclei through a tumor or a stroke typically involve other brainstem symptoms, as other cranial nerve nuclei and tracts in the area surrounding CN VIII may be affected as well.

Chapter Summary

- The inner ear contains the organs of **hearing** and **balance**. The organs are connected through the **membranous labyrinth**, the **endolymph**-filled space in the inner ear. Each organ uses the same type of receptor cell, the **hair cell**. The hair cells are mechanoreceptors that open ion channels in response to movement of the endolymph.

- In the cochlea, movement of endolymph occurs as a consequence of sound waves displacing the **basilar membrane** in the cochlea. The basilar membrane is organized in a **tonotopic** manner, and specific frequencies cause the basilar membrane to be displaced in discrete areas. This is further fine-tuned by the movement of **outer hair cells**, which also amplify the signal. The localization of sound requires input from both ears. For **low frequencies**, the time difference of sound waves reaching the two ears is analyzed. For **high frequencies**, the head forms a "sound shadow," and the intensity of sound between the two ears is analyzed. The brainstem nuclei analyze pitch, volume, and temporal patterns of sound, and the cortical regions assign meaning to sounds such as language, music, traffic noise, etc.

- **Balance** is analyzed according to head movements. These movements can be **rotational** or **linear**. **Rotational movements** are detected through deflection of the **cupula** in the **semicircular canals**. Each semicircular canal is coupled with a canal in the same plane on the other side of the head. The information from both sides is relayed to the vestibular nuclei. **Linear acceleration** is detected in the **otolithic organs**, which are sensitive to gravity because of the otoconia ("ear rocks") that sit on the sensory organ. Gravity will pull on the otoconia and cause a movement of the otolithic membrane, which, in turn, displaces the hair cells and either depolarizes them or hyperpolarizes them depending on their orientation on the macula. There is no conscious appreciation of balance. Rather, the vestibular nuclei interact with motor systems to ensure stable posture and adjustments of movement.

Study Questions

Choose the ONE best answer.

11.1 A patient comes to the office with symptoms of vertigo and difficulty hearing. She also reports having a dull headache that gets worse throughout the day. During the neurological examination, the clinician notices a weakness in the muscles of facial expression on the entire right side of the face. A computed tomography scan shows a tumor, which is pushing onto cranial nerves VII and VIII. Where is this tumor most likely to be localized?

A. Midpons.
B. Midbrain.
C. Caudal medulla.
D. Pontomedullary junction.
E. Rostral medulla.

Correct answer is D. Both the facial (cranial nerve [CN] VII) and the vestibulocochlear (CN VIII) nerves emerge from the brainstem at the pontomedullary junction. A lesion of the nerves themselves rather than of the nuclei associated with the nerves is the most likely location of the tumor because the brainstem nuclei associated with these nerves are located throughout the brainstem. A large lesion involving all of these nuclei would also have other symptoms (motor and/or sensory deficits). The headache is due to increased intracranial pressure and irritation of the dura mater from the tumor growth.

11.2 A child is brought to the office with otitis media, an infection involving the middle ear. Which statement about the middle ear is correct?

A. The middle ear is a fluid-filled cavity.
B. The middle ear contains three ossicles: the malleus, incus, and stapes.
C. The middle ear acts to dampen sound from the external ear.
D. The middle ear is connected to the oropharynx.
E. The middle ear lies in the frontal bone.

Correct answer is B. The three bones of the middle ear are the malleus, incus, and stapes. The middle ear cavity is filled with air. Sound energy is increased in the middle ear cavity, largely through the lever action of the ossicles. The middle ear connects to the nasopharynx by the pharyngotympanic (eustachian) tube. The middle ear cavity lies within the petrous part of the temporal bone. An infection in the middle ear usually involves accumulation of fluid. This is painful and reduces the sound energy transferred, which makes hearing more difficult with that ear.

11.3 A young man loses his hearing in one ear and now needs to learn how to localize sound effectively. Which statement about vertical sound mapping is correct?

A. Vertical sound mapping relies on input from both ears.
B. Vertical sound mapping occurs in the internal ear.
C. Vertical sound mapping measures whether sounds come from below or above.
D. Vertical sound mapping analyzes directional sound.
E. Vertical sound mapping depends on the differences between high- and low-frequency sounds.

Correct answer is C. The brain can localize sounds in the vertical plane through analysis of the differences in the direct and reflected sound inputs. Vertical sound mapping relies on input from one ear, not both ears. Vertical sound mapping occurs only in the external ear. Directional sound is measured by horizontal sound mapping. Differences between high- and low-frequency sounds are detected by horizontal sound mapping.

11.4 Benign paroxysmal vertigo is due to otoliths floating in the vestibular organ, causing stimulation of the vestibular system without head movements. Which of the following comprise(s) the otolithic organ?

A. Cupula.
B. Utricle and saccule.
C. Otoliths.
D. Ampulla.
E. Semicircular ducts.

Correct answer is B. The utricle and saccule comprise the otolithic organ. The cupula is a gelatinous mass in which hair cells are embedded. Otoliths are calcium carbonate crystals covering the gelatinous mass containing the hair cells. The ampulla is the enlargement or bulge at the base of each semicircular canal that contains the receptor cells. The semicircular ducts are part of the membranous labyrinth.

12 Brainstem Systems and Review

I. OVERVIEW

In the previous chapters, we looked at the ascending and descending pathways that travel through the brainstem as well as the blood supply to the brainstem and the cranial nerve (CN) nuclei and their connections within the brainstem.

In this chapter, we discuss the intrinsic systems of the brainstem, which are interconnected with virtually all parts of the central nervous system (CNS). The most important of these intrinsic systems is the **reticular formation** (Figure 12.1). The reticular formation consists of a diffuse network of cells that influences and modifies sensory and motor systems and plays a key role in **consciousness**. We discuss several groups or clusters of neurons within this network. One of these clusters generates patterns of movement (**central pattern generators [CPGs]**). These coordinate complex motor programs including aspects of gait, swallowing, coughing, yawning, vomiting, and breathing. Sensory input or feedback can modify the strength or frequency of the central program, but the essential motor pattern remains the same. In this chapter, we focus on the neuronal groups involved in the coordination of breathing as an important example of a complex motor program initiated by CPGs. The same neuronal groups that coordinate breathing activate the muscles of respiration during coughing, hiccuping, and vomiting. These related motor patterns are also discussed. The reticular formation also contains discrete groups of nuclei that use specific neurotransmitters and project to widespread areas of the CNS. We discuss several of these key neurotransmitter systems and their influence on consciousness, sleep and wakefulness, motivation, emotion, reward, addiction, and pain processing. At the end of this chapter, we provide a clinically oriented review of brainstem function. An understanding of the blood supply is important for the understanding of clinical symptoms due to disruption of the normal circulation. Whereas blood supply to the brainstem was discussed in detail in Chapter 6, "Overview and Organization of the Brainstem," this chapter assesses the effects of lesions of specific arteries to brainstem structures and the resulting clinical symptoms. Study questions also reflect a general review of brainstem function. This overview brings together the material covered in previous chapters (specifically, Chapters 3, 6, 7, 8, 9, 10, and 11).

240

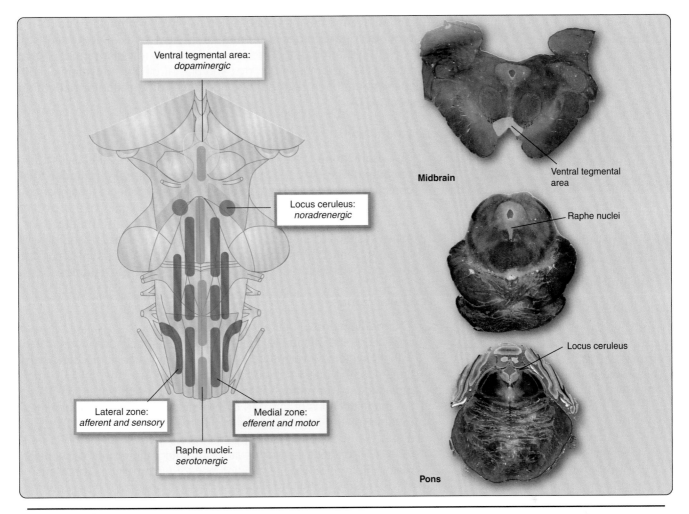

Figure 12.1
Conceptual overview of the reticular formation in the brainstem. Cross sections through the brainstem show the location of the ventral tegmental area and raphe nuclei in the midbrain and the locus ceruleus in the pons.

II. THE RETICULAR FORMATION

The **reticular formation** consists of a network of neurons deep in the **tegmentum** of the brainstem that extends throughout the brainstem (see Figure 12.1) as well as the central core of the entire spinal cord. Distinct nuclei are virtually impossible to identify, although functional units can be isolated physiologically. The vast majority of neurons in this network are **interneurons** that have multiple efferent projections, resulting in trillions of synaptic contacts. Any given neuron in the reticular formation may process information from both the ipsilateral and the contralateral side (both crossed and uncrossed information). In addition, the projections of any single neuron can be both ascending and descending. All systems in the reticular formation are influenced by projections from other brain areas and can, in turn, influence the function of these other brain areas and each other. Thus, the reticular formation is truly *the* integrator in the CNS.

The reticular formation can be subdivided into three functional components: (1) a **lateral zone** that processes afferent, sensory information;

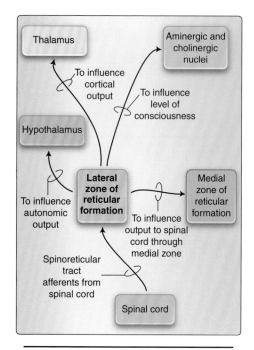

Figure 12.2
The lateral zone nuclei of the reticular formation.

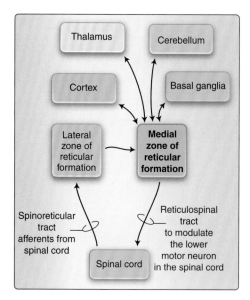

Figure 12.3
The medial zone nuclei of the reticular formation.

(2) a **medial zone** that processes efferent, motor information; and (3) the sum of **neurotransmitter systems** that project to widespread areas of the CNS. Furthermore, the projections from the reticular formation that ascend to the thalamus and cortex and play a role in modulation of consciousness are often referred to as the **ascending reticular activating system (ARAS)**.

A. Lateral zone

The **lateral zone** of the reticular formation (Figure 12.2) receives **afferent** information from the spinal cord through the **spinoreticular tract** comprising collaterals from ascending sensory systems (posterior column and spinothalamic tracts). Neurons in the lateral zone project to the medial zone to modulate motor function, to nuclei of neurotransmitter systems to influence the level of **consciousness**, and directly to the thalamus. Some ascending projections can also influence the autonomic nervous system via projections to the hypothalamus.

B. Medial zone

The **medial zone** of the reticular formation (Figure 12.3) has **efferent** projections that modulate motor output. It has reciprocal connections with all systems involved in the control of movement: the cortex and thalamus, the basal ganglia, the cerebellum, and the spinal cord. It projects to lower motor neurons via the **reticulospinal tract**. One of the main functions of this part of the reticular formation is to maintain **muscle tone** during movement and at rest, which is achieved through a balance of excitatory and inhibitory projections to lower motor neurons. This balance is the result of the integration of all descending motor information with ascending sensory information. The result of this is muscle tone that reflects our level of arousal, stress, and general state of mind.

C. Neurotransmitter systems

A series of parallel networks of **neurotransmitter systems** projecting to widespread areas of the CNS influence the level of consciousness, wakefulness and sleep and play a role in pain processing, motivation, emotion, reward, and addiction. The most important neurotransmitter systems include those involving dopamine (DA), norepinephrine (NE), and serotonin (5-HT). These three **aminergic systems** are the focus of this chapter because of their immense clinical importance. Other systems are discussed briefly, including those involving acetylcholine and histamine.

1. **Dopaminergic systems:** Dopaminergic neurons in the brainstem are located in two anatomically and functionally distinct areas: the **substantia nigra** and the **ventral tegmental area (VTA)**. The substantia nigra is located in the rostral midbrain. Dopaminergic cell bodies in the substantia nigra project to the caudate nucleus and the putamen (**nigrostriatal system**) and play an important role in the control of movement. (The nigrostriatal system is discussed in Chapters 16, "The Basal Ganglia," and 18, "The Integration of Motor Control.") The **VTA**, also located in the rostral midbrain, has widespread projections to various CNS areas and plays a pivotal

role in the circuitry involved in reward, motivation, and emotion (Figure 12.4). Both natural rewards and addictive drugs release DA in structures including the nucleus accumbens, the prefrontal cortex, and other forebrain regions. Thus, addictive drugs can mimic the effects of natural rewards, can effectively increase the magnitude of dopaminergic signals, and can shape behavior. In addition, the dopaminergic neural circuitry has been implicated in depression and anxiety disorders and in some cognitive functions including executive function (the ability to organize a sequence of actions toward a goal, requiring working memory and decision making).

a. **Reward:** The VTA is located medial to the substantia nigra and anterior to the red nucleus. VTA neurons are involved in reward and emotional processing pathways of the brain and influence a variety of behaviors. Their projections are to areas of the brain critical for emotional processing such as the nucleus accumbens and ventral striatum, with additional projections to the prefrontal cortex and areas of the limbic system including the amygdala, hippocampus, hypothalamus, and olfactory tubercle. The VTA is influenced by activity in these structures while influencing them at the same time. The majority of cells in the VTA are dopaminergic; this system is therefore known as the **mesocorticolimbic dopamine system.**

The VTA–nucleus accumbens pathway has long been known to be a key aspect of the reward circuitry. Rewarding stimuli and behavior can be divided into two components: the "wanting"

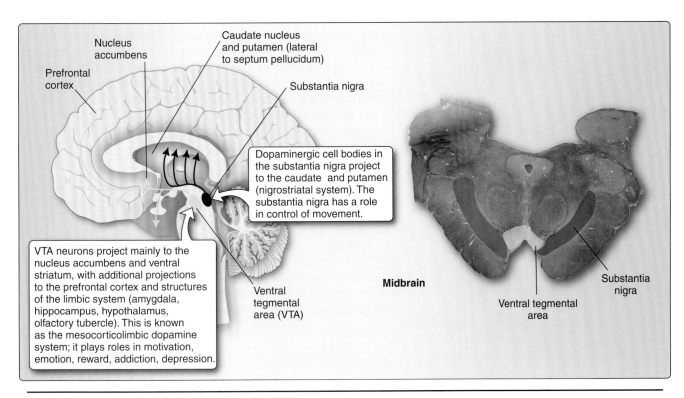

Figure 12.4
Dopaminergic projections from the ventral tegmental area and substantia nigra.

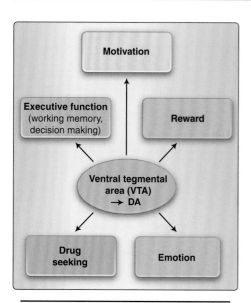

Figure 12.5
Overview of dopamine (DA) function.

of the stimulus, or appetitive motivation, and the "liking" of the stimulus, or consummatory motivation. It has been suggested that DA transmission in the nucleus accumbens mediates the assignment of "incentive salience" to rewards and reward-related cues, such that these cues can subsequently trigger a state of "wanting." The "wanting" aspect of a reward is the key to a stimulus being rewarding. DA signaling appears to be more important for appetitive motivation than for consummatory motivation. One can like something in the absence of DA but cannot use this information to motivate the behaviors necessary to obtain it. "Wanting" and "liking" are difficult to separate, however, because when we like something we tend to want it more. This positive reinforcement is also mediated by DA signaling in the VTA.

It appears that all major drugs of abuse, including nicotine, opiates, cannabinoids, and ethanol (alcohol), activate the VTA–nucleus accumbens circuit. The VTA appears to be key in **drug-seeking behavior**, a hallmark of addiction (Figure 12.5).

b. **Emotional learning and memory:** DA signaling from the VTA has also been implicated in the processing of emotional learning and memory. Any stimulus in our environment must be sorted and prioritized according to its emotional importance for us. Tagging emotional value (whether positive or negative) to stimuli, situations, and events allows us to respond in an emotionally appropriate manner when encountering similar stimuli, situations, or events. This tagging is mediated by DA signaling in the VTA–nucleus accumbens circuit. When this signaling is not functional, appropriate emotional responses are not possible. This is often seen in patients with schizophrenia in whom emotional responses can be either abnormally potentiated or severely blunted. An alteration in VTA signaling, specifically an increased sensitivity to DA, may be involved in this inappropriate emotionality. Medications that influence DA signaling, such as antipsychotic drugs (e.g., chlorpromazine) that antagonize DA by binding to the DA D_2 receptor, may be effective in treating schizophrenia, particularly in the early phases of the disease.

See further discussion on the role of dopamine in reward and addiction in Chapter 20, "Overview of the Limbic System."

2. **Noradrenergic (norepinephrine, NE) systems:** Neurons using **NE** as their principal neurotransmitter are clustered in the pons next to the fourth ventricle in the **locus ceruleus ([LC]** Latin for "blue spot," named for its blue appearance on brain sections). Additional noradrenergic neurons are scattered throughout the lateral tegmentum of the brainstem.

Noradrenergic neurons project to widespread areas of the CNS, both ascending to forebrain structures and descending to spinal cord neurons (Figure 12.6). The activity of these noradrenergic neurons can be either **tonic**, that is, at a constant and continuous level, or **phasic**, that is, the firing rate is increased periodically and temporarily.

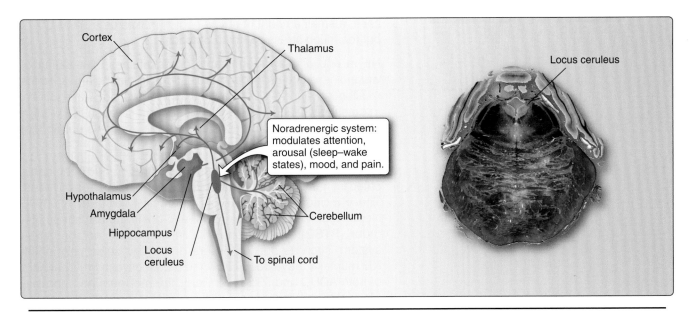

Figure 12.6
Noradrenergic projections from the locus ceruleus.

The main function of these neurons is to modulate attention, arousal (sleep–wake states), mood, and pain. They work in conjunction with other brainstem neurotransmitter systems, such as the serotonergic and dopaminergic systems.

In order to navigate a world full of stimuli effectively, it is necessary to detect and filter these stimuli so that we can direct our attention to the most relevant stimuli. The tonic firing of LC neurons determines our general level of arousal and attention through projections to the CNS that modulate synaptic activity. NE helps synapses work more effectively. A phasic increase in LC neuron firing happens when attention needs to be directed to a specific stimulus. This phasic firing helps us to focus our attention on a specific task while suppressing distracting stimuli (NE has both stimulatory and inhibitory neuromodulatory functions here). Whether or not a stimulus is rated as relevant or "interesting" depends on external factors, body homeostasis, and experience. For instance, the stimulus of food will become relevant when we are hungry, and experience will help direct our attention to trusted sources of food (Figure 12.7).

a. **Wakefulness:** The tonic firing rate of LC neurons can increase or decrease, and thus, our level of arousal can vary (e.g., we can be hypervigilant or drowsy). A moderate phasic increase of LC activity helps us to direct focus on a specific task when needed. However, too great an activation of LC neurons decreases our ability to focus on one specific task. The inputs to LC neurons that determine their activity levels arise from widespread areas, although the details are still not fully understood. The influence of stress through the **catecholamines** and **glucocorticoid hormones** (i.e., stress hormones such as epinephrine from the adrenal medulla and cortisol from the adrenal cortex, respectively), however, is well established. In this

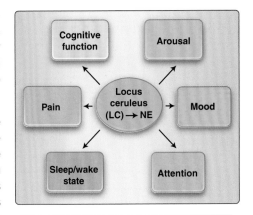

Figure 12.7
Overview of norepinephrine (NE) function.

case, a small amount of stress (which makes us alert) is good, but too much stress can disrupt focus.

NE is one of the determinants of wakefulness. Mediation of wakefulness occurs through projections to the thalamus, which is silenced during sleep, so that stimuli do not wake us up. Interestingly, bladder distension leads to activation of LC neurons, which in turn increases arousal, or wakes us up, so that the bladder can be emptied.

b. **Attention disorders:** Noradrenergic signaling has been implicated in many disorders related to attention, arousal, and mood. These include attention deficit hyperactivity disorder (ADHD), sleep disorders, panic disorders, and posttraumatic stress disorder (PTSD). Medications that influence the noradrenergic system, such as antidepressant drugs that selectively inhibit NE reuptake, have been shown to be effective in the treatment of both ADHD and PTSD. Similarly, the **monoamine hypothesis** has been the dominant one in the depression field for many years. This hypothesis states that a deficiency in noradrenergic and/or serotonergic transmission underlies the symptoms of depression (Clinical Application Box 12.1). Indeed, many

---

Clinical Application 12.1: Monoamines and Depression

The monoamines, including serotonin and norepinephrine, are known to be involved in mood, especially in depression and anxiety disorders. The majority of current antidepressant drugs target these monoaminergic systems specifically, either individually or in combination. Although serotonergic and noradrenergic pathways are key players in the pathology of affective disorders, it has become apparent that going beyond monoamines is necessary to understand the neurobiology of depression. Up to 50% of people who experience depression do not respond to current antidepressant drugs such as the selective serotonin reuptake inhibitors. Furthermore, deficits in monoamine activity are not always observed in clinically depressed patients, and facilitation of monoamine neurotransmission is only one component of antidepressant activity. Moreover, despite inducing a fairly rapid elevation in synaptic monoamine levels, monoaminergic drugs typically show a long latency for clinical effect, suggesting that depression may involve neurobiological systems other than the monoamines.

Our increasing knowledge of brain areas and neural circuits involved in depression and anxiety disorders has provided the basis for a new look at the neurobiology of depression as well as for new approaches to drug development. A focus on disturbances in hypothalamic regulation of neuroendocrine function is among the most widely accepted of these new approaches. In particular, studies have shown that alteration in activity and regulation of the hypothalamic–pituitary–adrenal (HPA) axis (see Chapter 19, "The Hypothalamus") is one of the most consistently described biological abnormalities in major depression. This is typically normalized by successful antidepressant therapy, and conversely, continued HPA disturbances are associated with increased risk for relapse. Drugs targeting the HPA axis and other neuroendocrine abnormalities, including those of growth hormone and thyroid hormone, are currently under development. These are likely to be used in conjunction with drugs targeting monoaminergic systems.

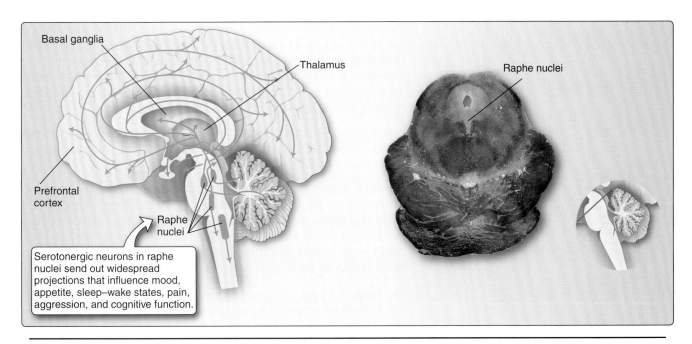

Figure 12.8
Serotonergic projections from the raphe nuclei.

antidepressant drugs act by increasing availability of either or both of these amines by decreasing their degradation or inhibiting their reuptake into presynaptic terminals.

c. **Alzheimer disease:** In **Alzheimer disease (AD)**, LC neurons are very susceptible to neurodegeneration and a loss of these neurons affects NE signaling in the entire CNS. Interestingly, NE suppresses neuroinflammation, and its signaling can activate **microglia** (see Chapter 1, "Introduction to the Nervous System and Basic Neurophysiology"). Microglial activation is key for the clearance of neurofibrillary tangles and Aβ deposits, and neuroinflammation is thought to be one of the pathological mechanisms in AD. Furthermore, the loss of noradrenergic neurons from the LC in AD exacerbates the neuroinflammation and hinders the clearance of debris by microglia.

3. **Serotonergic systems:** Serotonergic neurons in the brainstem are located in the **raphe nuclei**, a collection of neurons in the midline along the entire length of the brainstem and spinal cord. These neurons project to widespread areas of the forebrain, including limbic forebrain structures, such as the **prefrontal cortex**, as well as to the **thalamus**, **basal ganglia**, and **cranial nerve nuclei** (Figure 12.8). In addition, they project to and interact with other neurotransmitter systems of the reticular formation, most notably the noradrenergic system (discussed above). Serotonin is an important **neurotrophic factor** in development, and serotonergic signaling appears to be involved in a broad range of functions. These include the regulation of mood, appetite, and sleep as well as modulation of pain, state of wakefulness, aggression, and some cognitive functions, including memory and learning (Figure 12.9). Modulation of serotonin at synapses is a major action of several

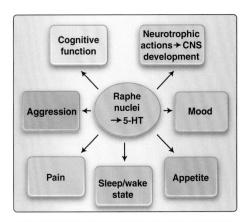

Figure 12.9
Overview of serotonin (5-HT) function. CNS = central nervous system.

pharmacological classes of antidepressants. These include the selective serotonin reuptake inhibitors (SSRIs). SSRIs increase extracellular levels of serotonin by inhibiting its reuptake into the presynaptic cell and thus increasing the levels of serotonin available to bind to the postsynaptic receptor. SSRIs are currently among the most widely prescribed antidepressants.

An increase in serotonergic signaling to widespread forebrain areas is associated with a "quiet" waking state, with decreased food intake and decreased sex drive.

Overall, serotonin signaling enhances our mood and decreases anxiety and aggression. Interestingly, serotonergic neurons are also **thermosensitive**. They participate in cooling down processes when the body is overheated. Conversely, their warmth-related signaling has been associated with the feeling of well-being in warm environments, such as in a sauna or hot bath.

a. **Pain:** As stated above, serotonergic signaling has been established in the central modulation of pain. Descending fibers from the raphe nuclei directly modulate pain transmission in the posterior horn of the spinal cord. Projections to the spinal cord also influence local circuit neurons in the anterior horn, where they regulate motor activity and play a role in the motor response to pain.

b. **Sudden infant death syndrome:** Serotonin signaling has recently been under scrutiny for its involvement in **sudden infant death syndrome (SIDS)**. Serotonergic neurons in the medulla act as **chemoreceptors** and can **stimulate respiration**. Therefore, a disruption of this and other serotonergic brainstem systems is thought to be a factor in SIDS. Neither a diagnostic paradigm nor therapeutic interventions have so far been established.

4. **Other neurotransmitter systems:** In addition to the **aminergic** systems (DA, NE, 5-HT), there are a number of systems using different types of neurotransmitters that interact with each other and project to widespread areas of the CNS, influencing the overall arousal and functioning of the CNS.

a. **Cholinergic neurons: Cholinergic** neurons, which use **acetylcholine** as their neurotransmitter, are located in the tegmentum of the pons (Figure 12.10) and have a neuromodulatory role by enhancing the functioning of synapses. Cholinergic projections to the thalamus appear to strengthen the excitatory output from the thalamus to the cortex and, thereby, play an important role in arousal and motor function.

b. **Histaminergic neurons:** Histaminergic neurons can be found in the tegmentum of the midbrain (see Figure 12.10). They are functionally related to the cluster of histaminergic neurons in the posterior hypothalamus. Projections of histaminergic neurons appear to play a role in general arousal and alertness. In fact, antihistamine drugs that can cross the blood–brain barrier and block central histamine release cause drowsiness.

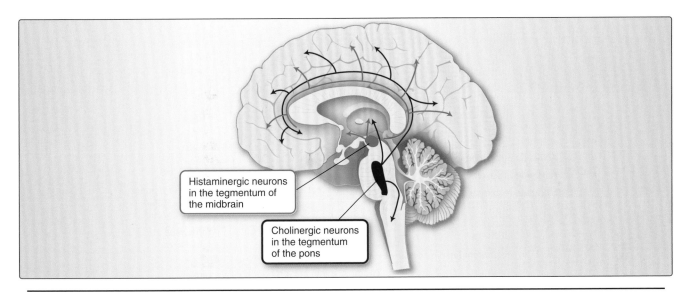

Figure 12.10
Cholinergic and histaminergic projections from the tegmentum of the brainstem.

III. BREATHING CENTER

Clusters of neurons responsible for coordination of breathing are located in the medulla and pons. These neurons are responsible for establishing an automatic rhythm of breathing and must be able to adjust the rhythm in response to metabolic, postural, and environmental changes. They influence muscles of inspiration and expiration as well as valve muscles that control airflow.

A. Central pattern generator

The cluster of neurons that generates the breathing rhythm at rest is located bilaterally in the medulla and is referred to as the **CPG**. The exact location of this CPG is unknown, but it appears that multiple sites in the brainstem are combined into a network that coordinates breathing.

The respiratory neurons can be grouped into an anterior and a posterior group. The **posterior respiratory group** is located bilaterally around the nucleus solitarius (Figure 12.11). The main function of these neurons is to modulate respiratory patterns. They receive sensory input (afferents) from peripheral chemoreceptors and stretch receptors in the lung. The motor output (efferents) from the posterior group coordinates the innervation of the muscles of inspiration (diaphragm and external intercostals). The vagus nerve innervates the upper airway.

The **anterior respiratory group** is anterior to the posterior group in the medulla (see Figure 12.11). It coordinates the innervation of both inspiratory and expiratory muscles. At rest, expiration is a passive process, whereas forced expiration requires the use of the abdominal muscles and the internal intercostals.

The respiratory neurons are under the influence of the CPG and are closely linked to other systems in the reticular formation. Changes in breathing patterns can indicate damage to the brainstem, and

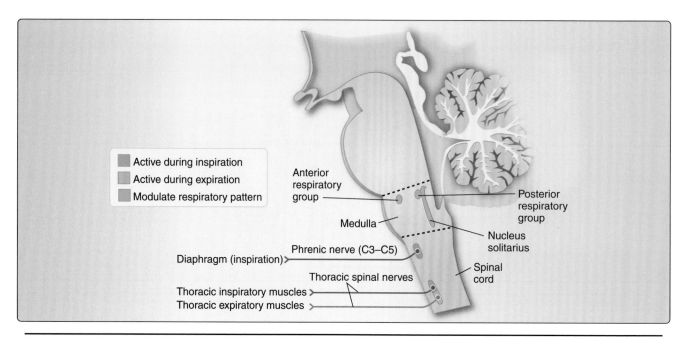

Figure 12.11
Neural control of breathing.

compression of the medulla can lead to compression of these respiratory neurons ("breathing center"), resulting in depression of breathing and death. This can occur, for example, during **tonsillar herniation**. A subtentorial mass, as it expands, can cause herniation of the cerebellar tonsils through the foramen magnum (see Chapter 17, "The Cerebellum"). Pressure on the brainstem will result in respiratory irregularities and, ultimately, respiratory arrest, as well as depression of the level of consciousness from compression of the ARAS.

A number of drugs have effects on respiratory neurons. Most notably, opioids, such as codeine, can depress the drive to breathe, a mechanism that is used in cough suppressants.[1]

B. Nonrespiratory functions of respiratory neurons

The same neuronal systems that coordinate breathing activate the muscles of respiration during coughing, hiccuping, and vomiting. These systems are relevant here, because they often manifest as the first symptoms in brainstem disorders, such as tumors or strokes.

1. **Neuronal control of emesis:** Vomiting (emesis) is an important protective mechanism that enables the expulsion of potentially harmful substances taken up through the digestive tract. There is no single vomiting center in the brainstem. However, the impulse to vomit and the different muscle groups involved in emesis need to be coordinated. This coordination appears to happen through activation of a network of neurons in the posterior medulla, adjacent to the nucleus solitarius.

[1]See *Lippincott Illustrated Reviews: Pharmacology.*

The impulse to vomit can be triggered by cognition (memories, expectations); emotion (disgust); vestibular disturbances (dizziness); vagal afferents (from the GI tract); or activation of the **area postrema** (in the medulla at the inferior edge of the fourth ventricle) through external influences, such as drugs or toxins (Figure 12.12). The blood–brain barrier that normally separates the neuronal environment from the blood is fenestrated in the area postrema, allowing toxic substances in the blood to activate these neurons, thereby initiating emesis. Vomiting requires the opening of the esophageal sphincter and a reversal of normal peristalsis as well as a coordinated effort to prevent the aspiration of particles into the lung. The same muscle groups active during respiration, coordinated by the CPG, work differently here to contract the abdominal muscles (retching) while relaxing the diaphragm and opening the esophagus to allow vomiting. At the same time, the valve muscles of the upper airway constrict to prevent aspiration. This process is thought to occur through activation of the CPG neurons in a manner different from what occurs during normal respiration (Figure 12.13).

2. **Hiccuping and coughing:** These processes involve stimulation of peripheral parts of the respiratory system (diaphragm for hiccups and upper airway for coughing), but the motor patterns that result are different from those that occur with normal respiration. In a sense, hiccuping and coughing represent abnormal breathing patterns.

Hiccuping can have a number of etiologies, ranging from gastrointestinal problems to irritation of the diaphragm or even myocardial infarction (heart attack). A hiccup results from disruption of the normal coordination of the breathing cycle, such that activity of inspiratory and expiratory muscles and the synchronous closing of the upper airway valves are no longer coordinated (see Figure 12.13). Importantly, damage in the medulla, where the CPGs for breathing are located, can result in hiccupping as a first symptom.

Coughing is also an abnormal breathing pattern caused by irritations in the airway. As with a hiccup, a cough can also be triggered by pathology in the brainstem associated with the breathing centers.

IV. CLINICALLY ORIENTED REVIEW OF THE BRAINSTEM

Because the brainstem is so complex, it cannot be covered in a single chapter. We introduced the brainstem in Chapter 2, "Overview of the Central Nervous System," discussed it in more detail in Chapter 6, "Overview and Organization of the Brainstem," and in Chapters 7 through 11 discussed the major ascending and descending tracts that travel through the brainstem as well as the location, function, and interconnections of the major CN nuclei. Here, we now provide a clinically oriented review of the brainstem. An understanding of the blood supply is

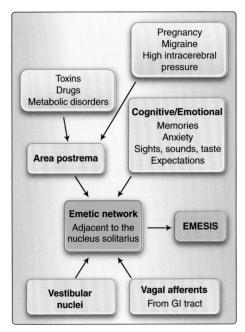

Figure 12.12
Neural control of emesis.
GI = gastrointestinal.

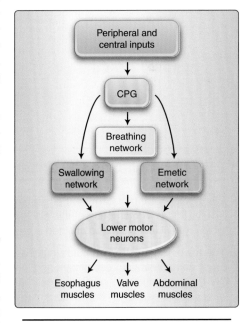

Figure 12.13
Nonrespiratory functions of respiratory neurons. CPG = central pattern generator.

important for the understanding of clinical symptoms due to disruption of the normal circulation. Blood supply to the brainstem was discussed in detail in Chapter 6. This overview of the brainstem provides an opportunity to bring together and integrate the material covered in the previous chapters.

As described in Chapter 6, the blood supply to the brain comes from both an anterior (internal carotid) system, which arises from the internal carotid arteries, and a posterior (vertebral–basilar) system, which arises from the vertebral arteries (see Figure 6.21). The circle of Willis interconnects the anterior and posterior systems. The anterior system and the circle of Willis are discussed in detail in Chapter 13, "The Cerebral Cortex." The brainstem is supplied by the posterior system. Note that for consideration of the blood supply, the brainstem in cross section can be divided into a paramedian area, a lateral area, and a posterolateral or posterior area (see Figure 6.22). The medulla receives its blood supply from the anterior spinal artery (paramedian areas) and posterior spinal artery (posterolateral areas) as well as the vertebral arteries (lateral areas). The posterior inferior cerebellar artery (PICA) provides additional supply to the posterolateral areas of the rostral medulla, including the inferior cerebellar peduncle. The anterior aspect of the rostral medulla and the basal pons are supplied by branches of the basilar artery. The tegmentum and posterior areas of the pons are supplied by the basilar artery (long circumferential branches) as well as the anterior inferior (AICA) and superior cerebellar arteries. The cerebellar peduncles are supplied by the superior cerebellar arteries (superior cerebellar peduncle) and the AICAs (middle cerebellar peduncle). The remaining aspects of the midbrain are supplied mostly by branches from the posterior cerebral arteries, with varying involvement of the basilar and superior cerebellar arteries.

Clinical Application Box 12.2 provides an example of deficits that result from occlusion of PICA.

Clinical Application Boxes 12.3 and 12.4 describe the deficits that may result from occlusion of branches of the posterior cerebral artery.

Clinical Application 12.2: Lateral Medullary Syndrome (Wallenberg Syndrome)

Blood supply to the lateral and posterolateral medulla is through the vertebral artery and the posterior inferior cerebellar artery (PICA). A disruption of this blood supply due to trauma or a stroke will lead to a typical combination of symptoms, which reflect lesions to the affected structures.

The major diagnostic symptoms include a loss of pain and temperature sensation from the face, ipsilaterally, and from the body, contralaterally. Loss of pain and temperature in the face on the ipsilateral side is due to a lesion of the spinal trigeminal tract and nucleus, which process pain and temperature from the ipsilateral face. Concomitantly, the loss of pain and temperature sensation on the contralateral side of the body is due to a lesion of the spinothalamic tract, which contains crossed fibers for pain and temperature. Disruption of PICA also causes a lesion of the ipsilateral spinocerebellar tract as it travels through the lateral medulla or as it enters the cerebellum through the inferior cerebellar peduncle, resulting in gait ataxia. A concurrent loss of the descending sympathetic fibers that travel in the lateral tegmentum of the brainstem causes ipsilateral ptosis, miosis, and anhidrosis (collectively known as Horner syndrome). **Ptosis**, or a drooping eyelid, is due to a loss of sympathetic innervation to the superior tarsal muscle in the eyelid. Small pupil, or **miosis**, results from the loss of

Clinical Application 12.2: Lateral Medullary Syndrome (Wallenberg Syndrome) continued

sympathetic innervation to the pupillodilator muscle in the eye. The loss of sweating (**anhidrosis**) is due to a loss of sympathetic innervation to sweat glands. A lesion of the vestibular nuclei, located in the posterior area of the medulla, may also occur, resulting in vertigo, nausea, and nystagmus. The patient might also present with trouble speaking (**dysarthria**) and swallowing (**dysphagia**) due to a lesion to the nucleus ambiguus (see Chapters 6 through 11 for more information).

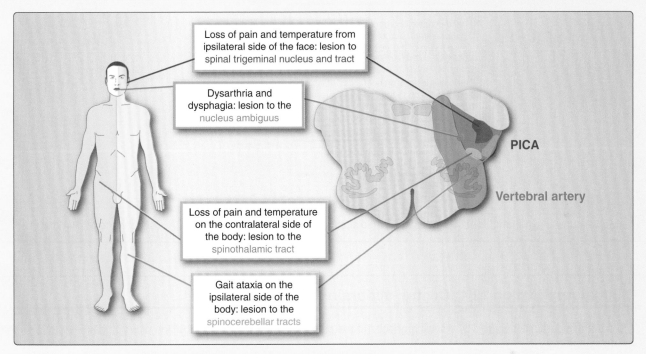

Lateral medullary syndrome. PICA = posterior inferior cerebellar artery.

Clinical Application 12.3: Medial Pontine Syndrome

The blood supply to the medial pons is through branches of the basilar artery, in particular, paramedian branches. Posterior and posterolateral areas of the rostral pons are also supplied by the basilar artery (long circumferential branches) with contributions from the superior cerebellar artery. The basilar artery also gives off smaller branches that supply the deep structures. Occlusion of the paramedian branches of the basilar artery on one side of the medial pons due to hypoperfusion or smaller emboli results in the following constellation of symptoms.

Patients present with contralateral **hemiparesis** due to involvement of the corticospinal tract, rostral to the pyramidal decussation. A lesion of the pontine nuclei and transverse fibers in the basal pons that arise from the pontine nuclei and cross to enter the contralateral cerebellum results in cerebellar symptoms, such as **ataxia** (loss of muscle coordination), on the contralateral side, although ipsilateral cerebellar signs may also be seen. Patients also show a loss of discriminative touch, vibration, and conscious proprioception on the contralateral side due to a lesion of the medial lemniscus, which carries the crossed fibers from the posterior column–medial lemniscus pathway. If the lesion extends somewhat laterally, the spinothalamic tract may also be involved, resulting in a loss of pain and temperature contralaterally. Due to involvement of the paramedian pontine reticular formation, the medial longitudinal fasciculus, and the abducens nerve (cranial nerve [CN] VI), there will be a number of gaze palsies, such as horizontal gaze palsy and esodeviation of the affected eye (deviation toward the nose). In addition, depending on the level of the lesion, fibers of the facial nerve, CN VII, may be involved, resulting in facial weakness on the ipsilateral side (see Chapters 6, 7, 8, and 9 for more information).

Clinical Application 12.3: Medial Pontine Syndrome continued

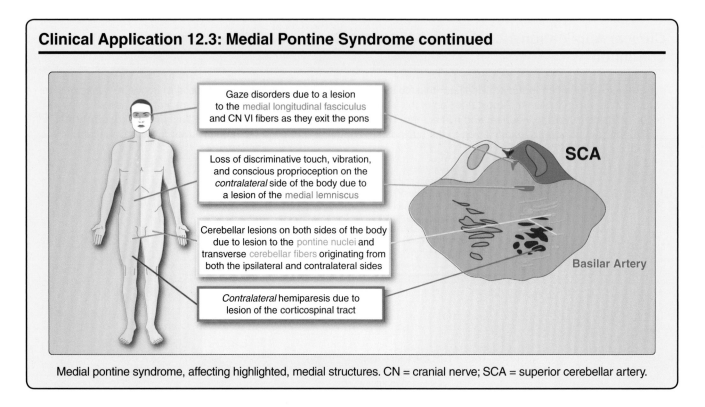

Gaze disorders due to a lesion to the medial longitudinal fasciculus and CN VI fibers as they exit the pons

Loss of discriminative touch, vibration, and conscious proprioception on the *contralateral* side of the body due to a lesion of the medial lemniscus

Cerebellar lesions on both sides of the body due to lesion to the pontine nuclei and transverse cerebellar fibers originating from both the ipsilateral and contralateral sides

Contralateral hemiparesis due to lesion of the corticospinal tract

SCA

Basilar Artery

Medial pontine syndrome, affecting highlighted, medial structures. CN = cranial nerve; SCA = superior cerebellar artery.

Clinical Application 12.4: Central Midbrain Syndrome

The central midbrain is supplied by central branches of the posterior cerebral artery, occlusion of which can lead to the following symptoms.

The patient presents with cranial nerve (CN) III (oculomotor) palsy on the ipsilateral side. The eye is abducted and rotated down, because only the lateral rectus (through CN VI [abducens]) and superior oblique (through CN IV [trochlear]) muscles are innervated (all other extraocular muscles are supplied by the oculomotor nerve). The parasympathetic component of CN III from the Edinger-Westphal nucleus is compromised as well, resulting in a loss of pupillary constriction; the pupil of the ipsilateral eye is thus dilated. Full ptosis (eyelid closed) will also occur due to loss of function of the levator palpebrae superioris muscle, which is supplied by CN III. The involvement of the red nucleus (and to a lesser degree, the substantia nigra) will manifest as a contralateral intention tremor, chorea or athetosis.

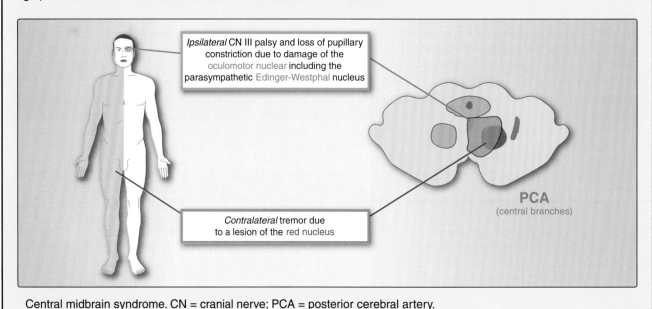

Ipsilateral CN III palsy and loss of pupillary constriction due to damage of the oculomotor nuclear including the parasympathetic Edinger-Westphal nucleus

PCA
(central branches)

Contralateral tremor due to a lesion of the red nucleus

Central midbrain syndrome. CN = cranial nerve; PCA = posterior cerebral artery.

Clinical Case:

Parkinson Disease

A 62-year-old patient has a 4-year history of right hand tremor with bradykinesia (slow movement) and rigidity in his right arm. He also has a history of poor sleep, low mood, and reduced appetite. He was diagnosed with Parkinson disease and depression. He was prescribed levodopa, a precursor to dopamine, for his Parkinson disease and a selective serotonin reuptake inhibitor (SSRI) for his depression.

Case Discussion

What are the two main dopaminergic areas of the brainstem that levodopa will affect?

Levodopa will affect dopaminergic neurons in the substantia nigra (SN) and ventral tegmental area, both located in the rostral midbrain area.

Which of these dopaminergic areas is mediating the motor symptoms observed in Parkinson disease?

The dopaminergic neurons in the substantia nigra are primarily responsible for the motor symptoms observed.

The dopaminergic neurons in the ventral tegmental area are primarily part of the mesocorticolimbic dopaminergic system. The mesocorticolimbic dopaminergic system consists of projections from the ventral tegmental area (VTA) to the nucleus accumbens, ventral striatum, prefrontal cortex, and limbic structures such as the amygdala, hippocampus, hypothalamus, and olfactory tubercle.

What is the major functional role of the mesocorticolimbic dopaminergic system?

The ventral tegmental area projects to areas involved with drug seeking, reward, motivation, emotion, and executive function (working memory and decision-making).

Why would a selective serotonin reuptake inhibitor (SSRI) help this patient?

Approximately 50% of patients with Parkinson disease experience depression. This is part of the disease, not a reaction to it. Parkinson disease affects dopaminergic cells, but also results in an imbalance of neurotransmitters, including serotonin and norepinephrine, all of which can play a role in depression. Serotonin plays a role in regulating mood (depression and anxiety), sleep, appetite, pain, aggression, and cognitive functions such as memory and learning. It is also an important neurotrophic factor for CNS development. An imbalance between dopamine and norepinephrine may also occur in Parkinson disease. Indeed, treatment of Parkinson disease may involve medications that alter both dopamine and norepinephrine levels.

Chapter Summary

- The brainstem contains intrinsic systems that are critical for the normal functioning of the central nervous system (CNS). The reticular formation integrates information from all areas of the CNS and coordinates the normal functioning of complex systems. The lateral zone of the reticular formation receives afferent input through the spinoreticular tract comprising collaterals off ascending sensory tracts, whereas the medial zone sends efferent projections through the reticulospinal tract. Together, these systems work to maintain muscle tone and postural stability.

- Neurotransmitter systems in the brainstem are an integral part of the normal functioning of our brains. These systems have widespread projections and influence virtually every aspect of central nervous system function.

- Dopaminergic projections from the ventral tegmental area give events or stimuli an emotional tag. These circuits are a critical part of reward, motivation, drug-seeking behavior, emotional learning and memory, and some cognitive functions. Noradrenergic projections from the locus ceruleus are critical for wakefulness and directing attention to a stimulus of interest.

- Serotonergic projections from the raphe nuclei enhance our mood and decrease anxiety. They are also a critical component in the central modulation of pain. Serotonergic projections to the respiratory centers of the brainstem have been implicated in **sudden infant death syndrome**. Cholinergic and histaminergic projections play a critical role in arousal and alertness.

- Breathing requires the coordination of neurons that control inspiratory and expiratory muscle groups. A **central pattern generator** coordinates the rhythm of breathing, but the exact location of this group of neurons remains unknown. These neurons are active in a nonsynchronous way during hiccuping and coughing, which can be symptoms of peripheral irritations or pathologies in the brainstem. During emesis, the neurons that usually coordinate breathing can be activated to contract the abdominal muscles in retching and close the upper airway to prevent aspiration. The trigger for vomiting can come from internal stimuli or activation of the area postrema through external stimuli.

Study Questions

Choose the ONE best answer.

12.1 Which one of the following statements about the reticular formation is true?

- A. The locus ceruleus contains dopaminergic neurons.
- B. The lateral zone of the reticular formation receives afferents through the spinoreticular tract.
- C. The raphe nuclei are located exclusively in the midbrain reticular formation.
- D. The medial zone of the reticular formation projects primarily to the cerebral cortex.

Correct answer is B. The locus ceruleus contains noradrenergic neurons, whereas the ventral tegmental area is dopaminergic. The raphe nuclei extend throughout the entire brainstem and spinal cord and contain serotonergic neurons. The lateral zone of the reticular formation receives afferents from the spinal cord through the spinoreticular tract, and the medial zone projects to spinal cord neurons through the reticulospinal tract.

12.2 The therapeutic effect of a drug that selectively increases the amount of noradrenaline and serotonin at synapses can best be described as:

- A. Decreasing drug-seeking behavior.
- B. Decreasing wakefulness.
- C. Enhancing mood.
- D. Enhancing anxiety.
- E. Enhancing sex drive.

Correct answer is C. A number of antidepressant drugs selectively alter synaptic levels of norepinephrine and serotonin. Drug-seeking behavior is associated with dopaminergic signaling. Norepinephrine increases wakefulness. Serotonin enhances mood and decreases anxiety, and it is associated with decreased food intake and sex drive.

12.3 A patient was brought to the emergency room unconscious after he had collapsed at work. After he regained consciousness, examination revealed the following: weakness of the right arm and leg, increased muscle tone and deep tendon reflexes on the right, diminished vibration and position sense on the right, dysarthria (decreased ability to articulate while speaking), and deviation of the tongue to the left when protruded. What is the most likely site of a lesion that would produce these deficits?

- A. Left lateral area of the caudal pons.
- B. Left paramedian area of the caudal medulla.
- C. Left paramedian area of the rostral medulla.
- D. Right lateral area of the caudal midbrain.
- E. Right lateral area of the rostral midbrain.

Correct answer is C. Weakness, increased muscle tone, and increased reflexes of the right arm and leg are due to a lesion of the descending corticospinal fibers (an upper motor neuron lesion) in the left pyramid. These fibers cross in the decussation of the pyramids in the caudal medulla and innervate lower motor neurons on the right side of the body. Diminished vibration and position sense on the right is due to a lesion of the medial lemniscus on the left, which is carrying information about discriminative touch, vibration, and position sense from the right side of the body. Dysarthria and deviation of the tongue to the left result from a lesion of the hypoglossal nerve (cranial nerve [CN] XII). The CN XII nucleus is located in the midline, in the posterior area of the rostral medulla, and the nerve fibers exit anteriorly, close to the midline, just lateral to the pyramid. CN XII supplies all of the tongue muscles except palatoglossus, ipsilaterally. A lesion of CN XII results in weakness of the tongue ipsilaterally so that it deviates to the side of the lesion when protruded. Weakness of the tongue can interfere with articulation during speech. The fact that the patient collapsed at work suggests a vascular problem or hemorrhage, most likely a branch of the left anterior spinal artery, or possibly the left vertebral artery, depending on the arrangement of blood vessels in this person. This lesion is known as **medial medullary syndrome** and is an example of an "alternating hemiplegia," with weakness on one side of the body and weakness of cranial nerve on the opposite side. Alternating hemiplegia can also be seen with a medial lesion of the pons (e.g., right-sided paralysis and a lesion of left CN VI) or midbrain (e.g., right-sided paralysis and a lesion of left CN III).

12.4 A 65-year-old patient presents to the emergency department with the following symptoms:

- Nausea, vomiting, and nystagmus
- Difficulty swallowing (dysphagia) and hoarseness
- Ataxia on the left

During the neurological exam, the clinician notices the following additional symptoms:

- Decreased pain and temperature sensation on the left side of the face
- Decreased pain and temperature sensation on the right side of the body
- A droopy eyelid and a small pupil on the left side

An infarct in which of the following vessels is the most likely cause for these symptoms?

A. A branch of the anterior spinal artery on the right.
B. A branch of the posterior spinal artery on the left.
C. A branch of the vertebral artery on the left.
D. A branch of the basilar artery on the right.
E. A branch of the posterior interior cerebellar artery on the right.

Correct answer is C. The vertebral artery supplies the lateral medulla. In this case, a branch of the vertebral artery on the left is affected. The nausea, vomiting, and nystagmus are due to involvement of the vestibular nuclei. The hoarseness and dysphagia are due to the involvement of the nucleus ambiguus. Ataxia on the left side is due to a disruption of connections to the cerebellum on the left. Involvement of the spinal trigeminal nucleus and tract leads to decreased pain and temperature sensation from the ipsilateral side (left) of the face. Decreased pain and temperature sensations of the body on the contralateral side are due to involvement of the spinothalamic tract, whose fibers cross in the spinal cord. The droopy eyelid and miosis (small pupil) are part of Horner syndrome, which is caused by a disruption in descending sympathetic fibers that travel in the tegmentum of the lateral medulla.

The Cerebral Cortex

13

I. OVERVIEW

In previous chapters, we provided an anatomical overview of the central nervous system (CNS). The boundaries of the cortical lobes were defined, and a cursory overview of cortical function was presented. In this chapter, we provide more detailed information on the anatomical features of the cortical lobes and explore in more depth the functional anatomy of the cerebral cortex.

Figure 13.1 gives a detailed description of the major gyri that comprise the five lobes of each cerebral hemisphere. A discussion of the anatomy of the cerebral hemispheres, including cortical histology and subcortical fiber bundles, is followed by a description of the primary functional areas of the cortex and their related association areas. We discuss the representation of functions bilaterally as well as the lateralization of clinically relevant functions, particularly language, attention, and spatial orientation. We explore the relatively new concept of the mirror neuron system and its relevance for social interaction. Finally, we provide a detailed review of the blood supply to the cortex and discuss functional deficits seen in common vascular infarcts.

II. ANATOMY OF THE CEREBRAL HEMISPHERES

The cerebral cortex covers the entire surface of the brain. The cortex and the deep nuclei comprise the gray matter of the forebrain. This is where the cell bodies of neurons are located. Histologically, the cortex is a layered structure, and the cortical cytoarchitecture, or the arrangement of cells, is a reflection of the different functional areas of the cortex.

In order to accommodate the vast number of neurons in the human cortex, the cortical surface area needs to be increased. Therefore, the cortex is highly folded, forming gyri and sulci. If the cortical mantle were to be spread out, it would form a sheet about 1 m^2 (approximately 3 sq ft).

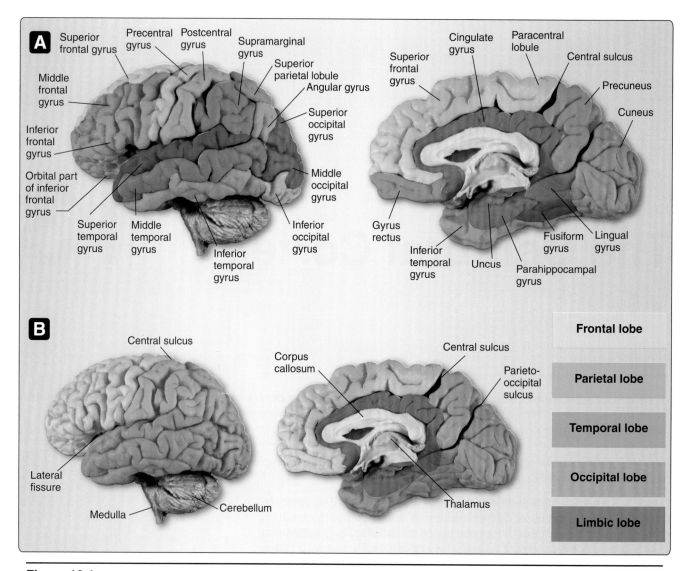

Figure 13.1
Major gyri of the forebrain.

Each cerebral hemisphere can be divided into five lobes: the frontal, parietal, temporal, occipital, and limbic lobes. These lobes and the major gyri and sulci within these lobes are illustrated in Figure 13.1. They are important surface markings for the anatomical localization of functional areas. An introduction to the lobes is given in Chapter 2, "Overview of the Central Nervous System."

Areas of the cortex are connected to one another through subcortical fiber bundles. These include association fibers that pass between areas within one hemisphere, commissural fibers that connect a functional area in one hemisphere to the comparable functional area in the other hemisphere, and ascending or descending projection fibers that travel to or from the cortex to interconnect it with more caudal areas of the CNS.

A. Histological organization of the cortex

The cells of the cortex are organized in a layered fashion, with different cell types predominating in each layer. Most of the cortex in humans is **neocortex**, which has six layers (I–VI). Older areas of the cortex have only three layers and are referred to as **paleocortex** in the olfactory bulb and **archicortex** in the hippocampus. We have all three types of cortices in the human brain, but it is the neocortex, which evolved after the other two, that underlies higher cortical function.

1. **Pyramidal and granular neurons:** There are two major neuronal cell types in the cortex: (1) **pyramidal neurons**, which are found in layers III and V, and (2) **granular neurons**, which are located in layers II and IV. Layer I is the molecular layer and contains mainly neuronal processes. Layer VI is the multiform layer and contains output neurons of varying shapes and sizes.

   Pyramidal neurons have a characteristic triangular structure, typically with one apical dendrite and abundant dendritic trees coming from the cell body. The axons of the pyramidal cells project from the cortex to other regions of the CNS, making them the main output cells of the cortex.

   Granular neurons, or stellate neurons, have shorter axons and smaller dendritic trees and remain within the cortex. They are the main **interneurons**.

   Not every area of the cortex has the same distribution of cells in all layers. The motor cortex, for example, contains a large number of pyramidal neurons that will project to the lower motor neurons via the corticospinal and corticobulbar tracts. There are not many granular neurons in this area of cortex, and it is often referred to as **agranular cortex**. The sensory cortex, on the other hand, contains few pyramidal cells but has a large number of granular neurons that process sensory information. It is referred to as **granular cortex**.

2. **Cytoarchitecture:** The cortex is organized into functional units, or cortical columns, which are specialized to process specific inputs and outputs. The cytoarchitecture of the columns will differ depending on their function, whether they are input or output columns (as indicated by the *arrows* in Figure 13.2). The typical distribution of cells throughout the six layers of the neocortex is summarized in Figure 13.2.

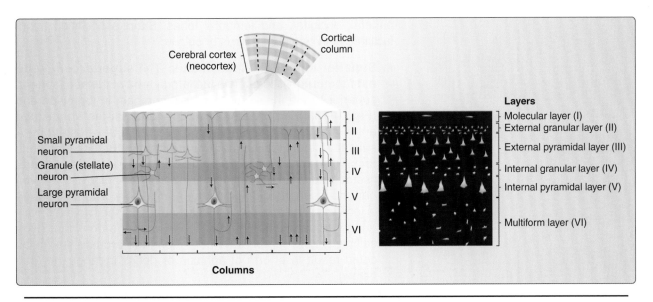

Figure 13.2
Histological organization of the neocortex.

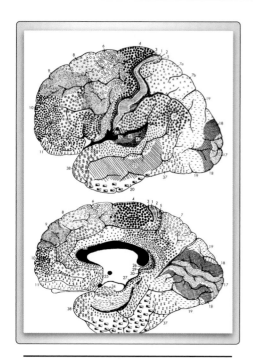

Figure 13.3
Histological mapping of the cerebral cortex by K. Brodmann in 1909.

These differences in the cytoarchitecture of the cortex led to the first mapping of the cortical mantle by Korbinian Brodmann in 1909. He assigned a numbering system to every cortical area according to its histological organization (Figure 13.3). It was only in later studies in the 20th century that Wilder Penfield found that these histologically different areas correlate with functionally different areas. This is a great example of how structure at the histological level closely correlates with function. In this book, we focus on the functional areas of the cortex, without mention of the historical Brodmann numbers.

In 2016, M. Glasser and his team described a new brain map that delineates 360 (180 per hemisphere) distinct cortical areas. This map was constructed using a combination of brain-mapping techniques that elucidated functions of different brain areas, relative density of myelin, indicative of cortical architecture, and information on neural connectivity within and between different regions. This updated map will be a tremendous resource for ongoing research to understand individual variations in brain organization and connectivity, both in health and disease.

In this chapter, we present information based on the most recent findings using the current brain maps.

B. Subcortical fiber bundles

Subcortical fiber bundles relay information to and from specific areas of the brain, depending on their classification as association, commissural, or projection fibers.

1. **Association fibers:** Association fibers interconnect areas of cortex within one hemisphere. Short association fibers connect areas in adjacent gyri, whereas long association fibers (including the superior longitudinal and inferior fronto-occipital fasciculi and the cingulum, among others) connect those areas more distant from each other. For example, the superior longitudinal fasciculus provides important sensory communication between the frontal, parietal, occipital, and temporal lobes. Short association fibers facilitate activity along a gyrus or sulcus.

   a. **Superior longitudinal fasciculus:** The **superior longitudinal fasciculus** is located in the lateral area of the hemisphere, superior to the insula, and is most compact in its midportion (Figure 13.4). It spreads out to the cortex of the frontal lobe anteriorly and to the parietal and occipital lobes posteriorly. A subset of fibers, called the **arcuate fasciculus** because of its arching shape, arches around the posterior end of the lateral fissure and enters the temporal lobe. Importantly, in the dominant hemisphere, defined as the hemisphere where language is localized, the arcuate fasciculus connects the two major language areas with each other. A lesion in the dominant (typically left) hemisphere anywhere along the arcuate fasciculus could result in a form of aphasia or deficit in language (see below).

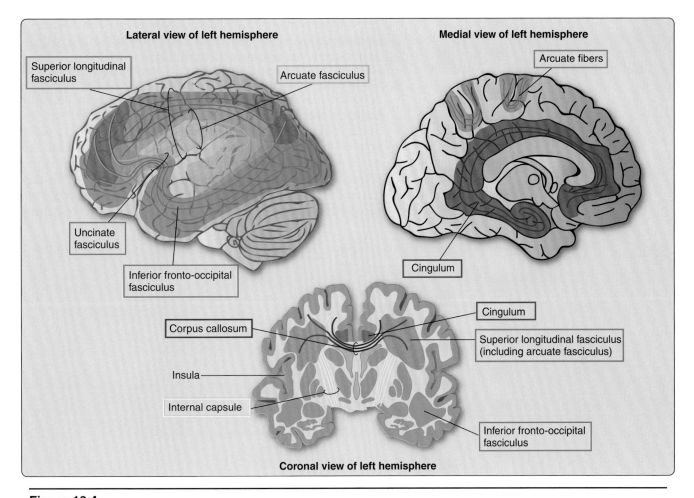

Figure 13.4
Association fibers in the forebrain.

b. **Inferior fronto-occipital fasciculus:** The **inferior fronto-occipital fasciculus** is located below the insula (see Figure 13.4). It runs from the frontal lobe through the temporal lobe to the occipital lobe and interconnects the overlying cortex in these areas. Fibers that hook around the margin of the lateral fissure to connect the frontal lobe to the temporal lobe are called the **uncinate** (*uncus* = hook) **fasciculus**. In the human brain, this is a very large bundle as it interconnects the expanded frontal and temporal lobes to more distant cortical areas.

c. **Cingulum:** This fiber bundle is located deep within the **cingulate** and **parahippocampal gyri,** which are referred to as the **limbic lobe**. It connects cortical areas overlying limbic system structures with each other (see Figure 13.4; see Chapter 20, "The Limbic System," for more information).

2. **Commissural fibers:** The **commissural fibers** connect functional areas of the cortex in one hemisphere with the same areas in the opposite hemisphere, which enables coordination of cortical activity between the hemispheres. Functional areas of the brain can be highly lateralized or bilateral: in either case, the two hemispheres

need to integrate information to function as one unit. The majority of commissural fibers in the brain cross the midline in the largest cortical commissure, the **corpus callosum** (Figure 13.5).

a. **Corpus callosum:** This structure lies deep in the interhemispheric fissure and is the main commissural bundle. Its body connects the two parietal lobes and the posterior parts of the frontal lobes with each other (see Figure 13.5). The posterior pole of the corpus callosum, called the **splenium** (Latin for *bandage roll*), interconnects the two occipital lobes and the two posterior temporal lobes. The **genu** (Latin for "knee") of the corpus callosum is located anteriorly, and its fibers connect the two frontal lobes with each other. As the fibers from the corpus

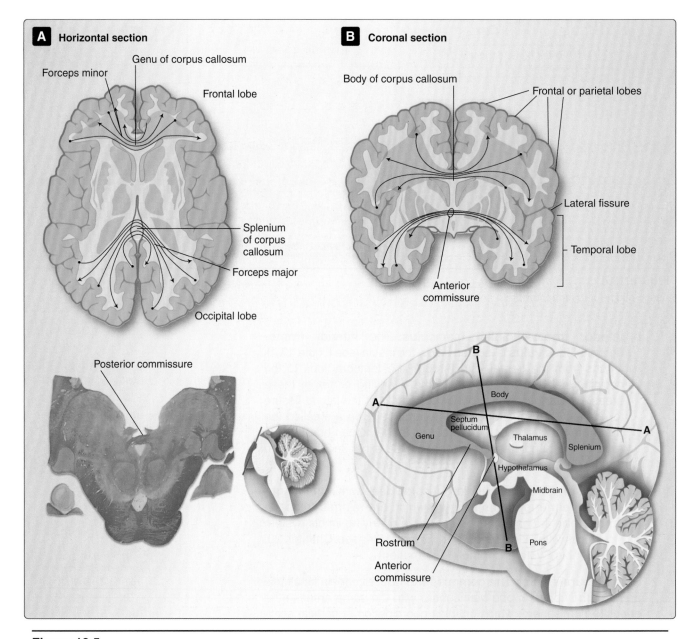

Figure 13.5
Commissural fibers in the forebrain.

callosum enter the hemispheres, they fan out to reach all parts of the cortex. At the anterior end, this radiation is called the **forceps minor** and at the posterior end, the **forceps major**. Most of the fibers in the corpus callosum are inhibitory; this ensures that the two hemispheres do not compete with each other and that cortical output is coordinated.

b. **Anterior and posterior commissures:** The **anterior commissure** is a small bundle of fibers that connects the anterior temporal lobes and the olfactory bulbs with each other. The **posterior commissure** is located in the midbrain and connects the two pretectal nuclei (see Figure 13.5).

3. **Projection fibers:** The **projection fibers** travel to or from the cortex. These fibers travel between the thalamus and cortex or descend from the cortex to the basal ganglia, the brainstem, or the spinal cord. Projection fibers come from or project to all parts of the cortex in what is called the **corona radiata.** Deep in the forebrain, between the basal ganglia and the thalamus, these fibers converge into a compact bundle called the **internal capsule** (Figure 13.6).

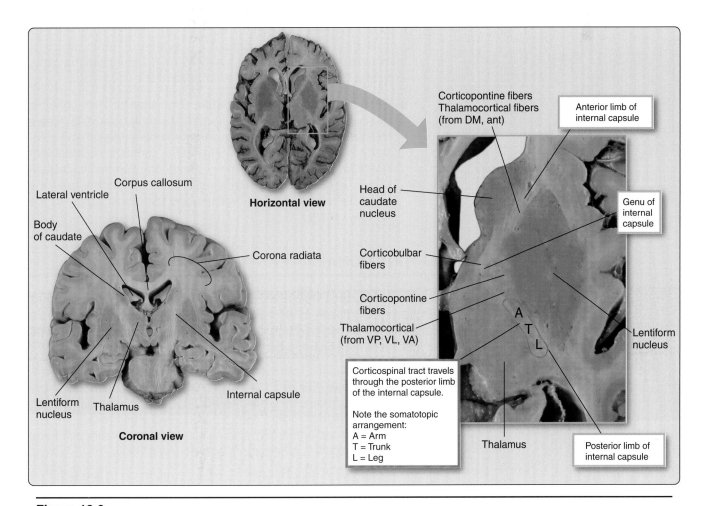

Figure 13.6
Projection fibers in the forebrain (internal capsule). DM = dorsomedial nucleus of the thalamus; ant = anterior nucleus of the thalamus; VP = ventral posterior nucleus of the thalamus; VL = ventral lateral nucleus of the thalamus; VA = ventral anterior nucleus of the thalamus.

Table 13.1: Fibers in the Internal Capsule and Their Blood Supply

	Fibers	Blood Supply
Anterior limb	**Corticopontine fibers** **Thalamocortical fibers (from dorsomedial and anterior nuclei)**	**Lenticulostriate arteries (deep branches from MCA and ACA)**
Genu	**Corticobulbar fibers**	**Lenticulostriate arteries (deep branches from MCA)** **Anterior choroidal arteries**
Posterior limb	**Corticopontine fibers** **Thalamocortical fibers (from VP, VL, and VA)** **Corticospinal fibers**	**Lenticulostriate arteries (deep branches from MCA)** **PCA, anterior choroidal arteries**

VP = ventral posterior nuclei; VL = ventral lateral nuclei; VA = ventral anterior nuclei; ACA = anterior cerebral artery; MCA = middle cerebral artery; PCA = posterior cerebral artery.

a. **Anatomy of the internal capsule:** The **internal capsule** is V shaped in horizontal section and can be divided into an **anterior limb**, which is between the caudate and the lentiform nucleus (putamen and globus pallidus); a **posterior limb**, which is between the thalamus and the lentiform nucleus; and the genu, where the two limbs meet (see Figure 13.6).

b. **Fibers of the internal capsule:** The anterior limb contains descending corticopontine fibers and thalamocortical fibers that project to the frontal cortex from the dorsomedial and anterior nuclei of the thalamus. The genu contains corticobulbar fibers, and the posterior limb contains corticopontine, corticospinal, and thalamocortical fibers. Table 13.1 summarizes the fibers within the internal capsule and the blood supply to the different limbs.

III. FUNCTIONAL AREAS OF THE CORTEX

Although the surface anatomy of the cortex provides important landmarks for locating the lobes and functional areas, it is the functional significance of the various cortical areas that is most relevant for understanding cortical function and assessing the consequences of a lesion.

Surprisingly, only a small portion of the cortical mantle is occupied by **primary cortical areas** (Figure 13.7). These primary areas consist of the primary motor, primary sensory (somatosensory), and primary areas relating to the special senses of vision, hearing, taste, and smell (olfaction). The primary cortical areas are largely symmetrical in function, although there is some lateralization in specific areas, as discussed below.

Importantly, the vast majority of the cortex is made up of **association areas**. There are specific association areas adjacent and related to each of the primary areas. These are important for higher-order processing, integration and interpretation of sensory information, and planning, integration, and initiation of motor activity. Additional large areas of the cortex have association functions of a broader nature. These relate to functions

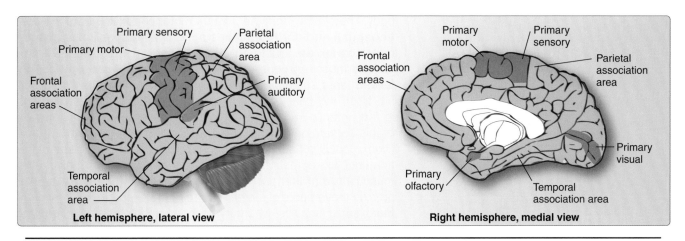

Figure 13.7
Overview of the primary (dark gray) and association (light gray) areas of the forebrain.

that define our intellect and personality, mediate complex language functions, and modulate emotions, judgment, and our sense of self in relation to society. These areas make us who we are.

The primary areas are closely interconnected with their relevant association areas. Due to differences in function, the effects of lesions in primary and association areas also differ. Whereas a lesion in a primary area leads to clearly defined deficits, such as decreased perception of stimuli or weakness and paralysis, lesions in the association areas can be much more complex because of their role in higher-order processing and in integrating and interpreting information. Lesions here can result in changes to higher cognitive abilities, emotionality, or personality.

The symmetry of the hemispheres is preserved in general terms for the association areas as well, but some of the complex tasks carried out by each region vary for the left and the right sides. Overall, the left hemisphere is associated with narrow, sharply focused attention to detail and concrete categorization of the physical world around us. The right hemisphere is associated with open vigilance and alertness and broad understanding of context and metaphor. In addition, sex differences have been described in relation to right and left hemisphere functions.

A. Primary areas and their unimodal association areas

The primary areas of the cortex are those that receive information from peripheral receptors via appropriate thalamic nuclei, with little interpretation of the meaning of that information. They include both sensory and motor areas with precise relationships to specific areas of the body. A lesion in a primary area results in a complete or partial deficit in the corresponding modality (see below). Each primary area has an association area specific to the modality of the primary area. These are unimodal association areas, in contrast to heteromodal association areas, which integrate information from more than one modality.

1. **Motor:** The **primary motor area** of the cortex is localized in the **precentral gyrus** of the frontal lobe. The precentral gyrus of the left hemisphere sends motor output to the right side of the body, and the right precentral gyrus sends output to the left side of the

body. The outflow from the primary motor cortex is part of the **corticospinal and corticobulbar tracts**. The neurons in the primary motor cortex are clustered in functional areas representing the various muscle groups they influence. This **somatotopy** carries forward as a somatotopic arrangement of fibers in the corticospinal tract and finally to the arrangement of lower motor neurons in the anterior horn of the spinal cord. A graphic representation of this somatotopy on the cortex results in a **motor homunculus** (*homunculus* is Latin for *small man*). The size of body parts of the homunculus represents the size of the neuron pool supplying the musculature of that part of the body (Figure 13.8). For example, precise movements of the hand require the separate innervation of many small muscles, whereas the innervation of the trunk requires less precise regulation. Therefore, the cortical representation of the trunk is significantly smaller than that of the hand.

a. **Primary motor area lesion:** A lesion of the primary motor cortex results in upper motor neuron signs similar to those that would be seen with a lesion anywhere along the lateral corticospinal tract, including weakness or paralysis, together with hypertonia, hyperreflexia, and other typical signs of an upper motor neuron lesion (see Chapter 8, "Descending Motor Tracts," and Chapter 18, "The Integration of Motor Control").

b. **Supplementary motor complex:** A **supplementary motor** complex is an association area that spans the area of cortex anterior to the primary motor area, from the lateral fissure extending onto the medial surface of the hemisphere (see Figure 13.8). It includes the premotor and supplementary

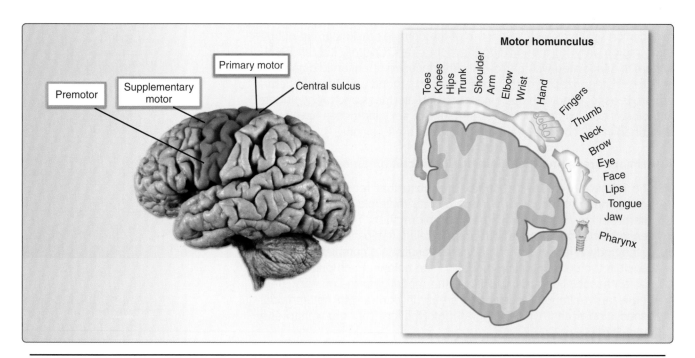

Figure 13.8
The primary motor cortex of the left cerebral hemisphere, anterior to the central sulcus and the supplementary motor complex (supplementary motor, premotor, and frontal eye fields).

motor areas, which run into each other, as well as the frontal eye fields. This complex contains motor maps for postures and coordinated limb movement.

Motor maps for postures send efferents that generally terminate bilaterally. Efferents travel via the lateral corticospinal tract to innervate the limbs and via the anterior corticospinal tract to innervate the proximal musculature for postural stabilization. Motor maps for coordinated limb movement send their efferents to contralateral lower motor neurons.

As an association area, the supplementary motor complex is also important for higher-order processing and for integrating and interpreting motor information and activity. It also appears to play a role in anticipating or "planning" voluntary movements. Support for this idea comes from studies suggesting that during voluntary movement, this area "lights up" prior to the activation of the primary motor area.

For information about the frontal eye fields, please see Chapter 9, "Control of Eye Movements."

The motor neurons in both the primary motor area and the supplementary motor complex receive input from many association areas, the primary sensory cortex and subcortical structures, and, most important, the basal ganglia and cerebellum. The sum of these afferents to the primary motor cortex determines the firing pattern of the primary motor neurons.

c. **Supplementary motor complex lesion:** A lesion in the supplementary motor complex results in a deficit in learned, skilled motor activity in the absence of paralysis. This is known as **apraxia**. Various types of apraxia include limb apraxia (e.g., brushing teeth, combing hair, hammering a nail, buttoning shirt) and buccofacial apraxia (e.g., whistling, blowing out a match, drinking from a straw).

2. **Sensory:** The **primary somatosensory cortex** is composed of the **postcentral gyrus** of the parietal lobe. Sensory afferents from contralateral peripheral receptors travel through the **posterior column–medial lemniscus system**, the **spinothalamic tract**, and the **trigeminal lemniscus/trigeminothalamic tract** to the sensory nuclei of the thalamus (ventral posterolateral [VPL] and ventral posteromedial [VPM], respectively) and then through the **posterior limb of the internal capsule** to the postcentral gyrus. Interestingly, the different types of peripheral receptors (muscle receptors, slowly and rapidly adapting skin receptors, and joint receptors) (Figure 13.9) project to distinct areas of the postcentral gyrus. These afferent projections to the postcentral gyrus preserve the **somatotopic organization** found throughout the tracts, resulting in a sensory map of the body on the cortex or a **sensory homunculus** (see Figure 13.9). The size of the cortical representation is correlated with tactile acuity, or the ability to discriminate different sensory inputs, in that part of the body. Similar to the motor area, the hand has high tactile acuity. Each receptor has a small receptive field, and the area of the cortex representing the

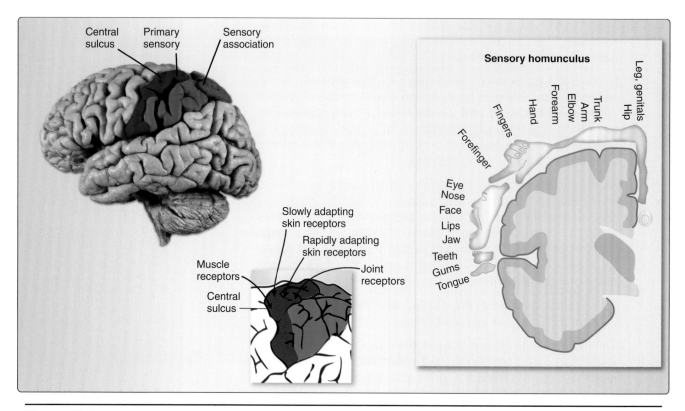

Figure 13.9
The primary and association sensory areas of the cortex of the left cerebral hemisphere.

hand is large. Conversely, the tactile acuity of the back is fairly low, the receptive fields of the skin receptors are large, and the representation on the cortex is small.

a. **Cortical plasticity:** The cortical representation of the body is highly plastic. The area of the cortex that represents any particular body area can change over time in response to the input or lack of input from a particular area of the body. When a limb is amputated, for example, the cortical representation of that limb is diminished and other adjacent body areas take over that part of the cortex. On the other hand, when the tactile acuity of a particular body part is trained and practiced, that area of the body will gain larger cortical representation. This has been shown, for example, in individuals who read Braille with their index finger or in musicians who play string instruments.

b. **Lateral inhibition:** As discussed in Chapter 7, "Ascending Sensory Tracts," the processing of sensory information occurs at every step of the sensory pathways, with the aim of increasing tactile acuity. This also occurs at the cortical level. An area that receives sensory input sends inhibitory projections to adjacent areas, thereby increasing the contrast between the area receiving input and the one not receiving input. This **lateral inhibition** increases tactile acuity at a cortical level. The primary sensory cortex sends projections to the primary motor cortex, where it influences motor output through the descending motor pathways.

c. Primary somatosensory cortex lesion: A lesion of the primary somatosensory cortex typically does not result in a complete loss of sensory perception, but rather in a deficit in the awareness of sensory input and poor localization of sensory stimuli.

d. Somatosensory association area: Immediately adjacent to the primary sensory area in the parietal lobe is the sensory association area (see "Parietal association areas," below). The somatosensory association area is critical for allowing us to interpret the significance of sensory information. A lesion in the somatosensory association area results in **tactile agnosia**, which is a deficit in the ability to combine touch, pressure, and proprioceptive input to interpret the meaning or significance of the sensory information. Another deficit that can result from such a lesion is known as **astereognosis** or the inability to recognize an object placed in the hand (i.e., using tactile information alone).

3. Visual: A comprehensive overview of the visual system is presented in Chapter 15, "The Visual System." The cortical representation of vision is discussed here.

a. Primary visual cortex: The **primary visual cortex** consists of the area of cortex on either side of the **calcarine sulcus**, on the medial side of the **occipital lobe** (Figure 13.10). Fibers from the retina project to the **lateral geniculate nucleus** of the thalamus, which, in turn, sends fibers known as the **optic radiations** to the primary visual cortex.

The right side of the cortex receives information from the left visual field and vice versa. The organization of neurons in the cortex is mapped to the distribution of neurons in the retina, a **retinotopic organization**.

b. Primary visual cortex lesion: A lesion in the primary visual cortex results in a deficit in vision in the opposite visual field. The specific nature of the deficit depends on exactly where in the primary visual cortex the lesion occurs (see Chapter 15, "The Visual System" for full details).

c. Visual association area: The area of cortex surrounding the primary visual area on the medial surface, and extending onto the lateral surface of the occipital lobe, is the **visual association area** that gives meaning and interpretation to what we see. A lesion in the visual association area results in a deficit in the ability to recognize objects in the opposite visual field despite intact vision. This is known as **visual agnosia**. In addition, such a lesion can result in a deficit in pursuit or tracking of an object ipsilaterally (see Chapter 15, "The Visual System").

4. Auditory: The **primary auditory cortex** is located deep within the lateral sulcus, on the superior surface of the **superior temporal gyrus** of the temporal lobe (see Figure 13.10). The two transverse temporal gyri, or the **Heschl gyri**, comprise the area for primary representation of auditory information from the cochlea (organ of Corti).

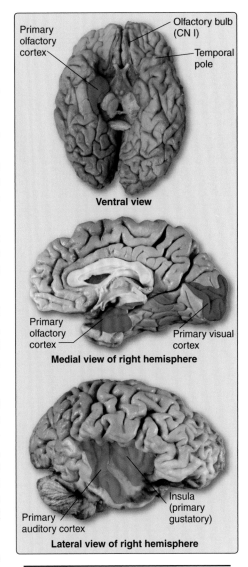

Ventral view

Medial view of right hemisphere

Lateral view of right hemisphere

Figure 13.10
Primary visual, primary auditory, primary gustatory, and primary olfactory areas of the cortex. CN = cranial nerve.

Information from the cochlea projects to the **medial geniculate nucleus** of the thalamus, which, in turn, projects to the primary auditory cortex. Ascending information from the cochlea travels both ipsilaterally and contralaterally, such that each ear is represented bilaterally on the auditory cortex. The neurons in the primary auditory cortex are organized in a **tonotopic arrangement**, similar to the tonotopy of the cochlea.

   a. **Primary auditory area lesion:** While the representation of sound in the cortex is bilateral (see further discussion in Chapter 11, "Hearing and Balance"), information from the contralateral cochlea predominates. Thus, a lesion in the primary auditory area will result in the decreased perception of sound, primarily in the contralateral ear, rather than in a loss of hearing limited to one side or the other as would occur with a lesion of the hair cells or auditory nerve on one side.

   b. **Auditory association area:** Adjacent to the primary auditory area on the lateral surface of the superior temporal gyrus is the auditory association area that enables us to interpret and give meaning to the sounds we hear. A lesion in the auditory association area can result in word deafness, or **acoustic verbal agnosia**, in which the ability to interpret what is heard is compromised, despite intact hearing. Extending more posteriorly on the superior temporal gyrus and looping around the lateral sulcus to include the supramarginal and angular gyri is the **Wernicke area**, which is critical for the understanding of language (see further discussion below).

5. **Other primary sensory areas:** The sensation of **taste** has cortical representation in the **insula**, an area of the cortex that lies deep within the lateral sulcus. The information reaches the insula from the taste receptors through the VPM of the thalamus.

The **olfactory system** also has cortical representation on the inferior and medial surfaces of the brain, in the **entorhinal cortex** as well as the inferior portions of the **temporal lobe** (see Figure 13.10).

The olfactory and gustatory systems are discussed in more detail in Chapter 21, "Smell and Taste."

B. Heteromodal association areas of the cortex

Heteromodal association areas are involved in higher-order processing and integrating and interpreting information across more than one modality. There are areas important for attention and awareness (**parietal association areas**), for planning and adapting our behavior to social contexts (**frontal association areas**), and for recognizing things and situations around us (**temporal association areas**). The characteristics that we describe as being profoundly human, such as those that determine our personality, guide our decision making, and are associated with our memories and knowledge, are mediated by the association areas.

Our knowledge about the function of these association areas comes from observations in people who have had lesions in these areas and recently from functional magnetic resonance imaging studies in which activity in brain regions is measured during specific cognitive tasks. Our understanding of the association areas for cognition and behavior is still developing. The following sections give an overview of what we know so far.

1. **Frontal association areas:** These occupy the frontal lobes anterior to the supplementary motor complex and are referred to as the **prefrontal cortex**. This area of the cortex is extensively interconnected with other brain structures and can be divided into two main regions with different connectivity and function. The **superior and lateral parts** of the prefrontal cortex are connected with the sensory and motor cortex as well as the basal ganglia and cerebellum. These regulate **attention** and **motor responses to stimuli**. The **inferior and medial portions** of the prefrontal cortex are interconnected with the amygdala, the hypothalamus, and the nucleus accumbens as well as brainstem nuclei involved in arousal (noradrenergic and cholinergic) and are responsible for the regulation of emotions. The prefrontal cortex sends input to the supplementary motor complex and the basal ganglia; these projections encode for the expression of emotions and behavior.

   a. **Size:** The frontal lobe is particularly large in primates, especially in apes and humans. The evolutionary pressure to develop such large frontal lobes is thought to come from the need for animals to transition from solving problems individually to solving them socially. Social groups can defend against predators and forage for food together, which is more effective and gives social animals an advantage over those relying on individual strategies. A social group is defined by the relationships among group members, and the frontal lobe appears to modulate these relationships and, thus, life within a social group.

   b. **Function:** The functions of the frontal lobe underlie our "personality" and let us adjust our behavior to moral and social norms. The frontal lobe appears to play a role in planning and problem solving, behavioral inhibition and working memory, as well as directing and maintaining attention on a particular situation or task (Figure 13.11). These latter functions contribute to cognitive flexibility and are sometimes grouped under the term *executive function*.

   It is through the prefrontal cortex that our attention can be directed to a specific stimulus or task in our environment, even if that particular stimulus or task is not the predominant or most salient one. The prefrontal cortex mediates our ability to decide on the relevance of the stimulus and directs our attention to it while suppressing distractions. For example, one might choose to study neuroscience although one would rather be on social media. The prefrontal cortex also plays a critical role in **working memory**. It is in this area of the cortex that information can be retained long enough to plan and execute the behavioral response to a stimulus. The prefrontal cortex allows us to make deliberate decisions about our behavior and adapt it to specific situations. It allows us to imagine our future and plan our

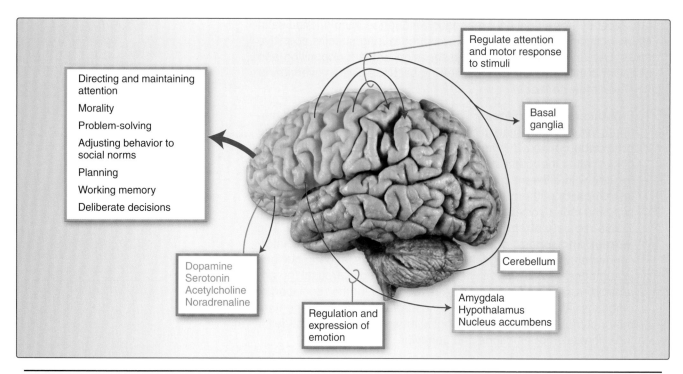

Figure 13.11
Overview of the frontal association areas in the left cerebral hemisphere.

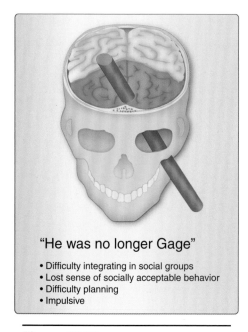

Figure 13.12
Phineas Gage: damage to the frontal association areas.

behavior for a long-term goal or outcome. Recent research suggests that the prefrontal cortex also plays a role as the seat of our **morality**, that is, the innate feeling of right and wrong, independent from social norms and cultural backgrounds.

c. **Innervation:** The frontal lobe receives extensive innervation from the **monoaminergic systems** of the brainstem, and high densities of **dopamine**, **norepinephrine**, and **serotonin receptors** can be found here. The activity of these projections is critical for the normal functioning of the prefrontal cortex. Psychiatric disorders relating to these systems, such as schizophrenia and depression, have symptoms that can be associated with alterations in the functions of the prefrontal cortex circuitry and neurotransmitters (see Chapter 12, "Brainstem Systems and Review," for more information).

d. **Lesions:** Damage to the frontal lobe leads to changes in personality and a loss of the ability to demonstrate appropriate behavior, without, necessarily, the loss of intellectual capacity. The most famous case of frontal lobe damage is the 19th century case of **Phineas Gage**. An explosion that occurred while Gage was working on a railroad blew a tamping iron through his left orbit and the frontal parts of his skull and brain (Figure 13.12). Gage recovered from the injury, and the local physician, J. M. Harlow, documented the case.

Although, before the accident, Gage appeared to have been well adapted and, by all accounts, able to make decisions and plan for the future, his personality changed after the accident:

"He was no longer Gage." Many of the attributes of frontal lobe functions disappeared or were severely diminished. He appeared to have had trouble integrating into social groups, had lost his sense of socially accepted behaviors, had trouble planning for the future, and had trouble reconciling his impulses with other people's needs.

New insight into the damage that occurred to Mr. Gage's brain comes from a study using a modern "connectomic" approach, which found that 4% of cortical gray matter and 11% of total white matter was damaged. Injured tracts included the uncinate fasciculus, cingulum, superior longitudinal fasciculus, and the inferior fronto-occipito fasciculus (see Section IIB, Subcortical fiber bundles). These findings are the first to suggest that both gray and white matter damage likely played a role in Mr. Gage's behavioral and functional problems.

2. **Parietal association areas:** Association areas of the **parietal lobe** are located posterior to the primary sensory areas in the postcentral gyri. The parietal association areas are referred to as the **posterior parietal cortex**. This is an area where somatic and visual sensations are integrated, resulting in high-level interpretation of stimuli.

   a. **Function:** The posterior parietal cortex is critical in attention and in the awareness of self and extrapersonal space. Whereas the visual cortices of the occipital lobe mediate analysis, recognition, and interpretation of visual information, the parietal association areas process information on the position and movement of objects, people, and self in space. The right posterior parietal cortex orients our attention in space, whereas the left posterior parietal cortex orients our attention in time. These areas of cortex are interconnected with the prefrontal cortex, which decides which stimulus to focus on and filters out distractions (Figure 13.13).

   b. **Lesions:** A lesion to the posterior parietal cortex in the **nondominant hemisphere** (typically the right) can lead to **contralateral neglect syndrome**. Stimuli in the environment on the side opposite to the lesion, typically the left, are ignored or "neglected." Although the sensory receptors and pathways for perceiving those stimuli are intact, there is no awareness or interpretation of these stimuli, and no attention is directed to them.

   The neglect includes both lack of awareness of extrapersonal space and lack of awareness of self on the side opposite the lesion (typically the left). Thus, patients will ignore people, voices, objects, and noises on the left and may ignore the left sides of their own bodies, failing to shave, wash, or dress the left side. Similarly, when asked to copy or draw an object from memory, the left side of that object is omitted (Figure 13.14). The fact that the left sides of objects are omitted from a drawing illustrates that the neglect is not necessarily limited to the left visual field but can expand to general *left-sidedness*.

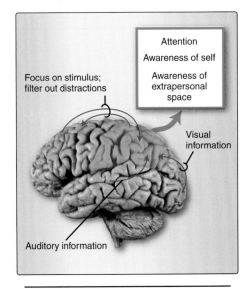

Figure 13.13
Overview of the parietal association areas of the left cerebral hemisphere.

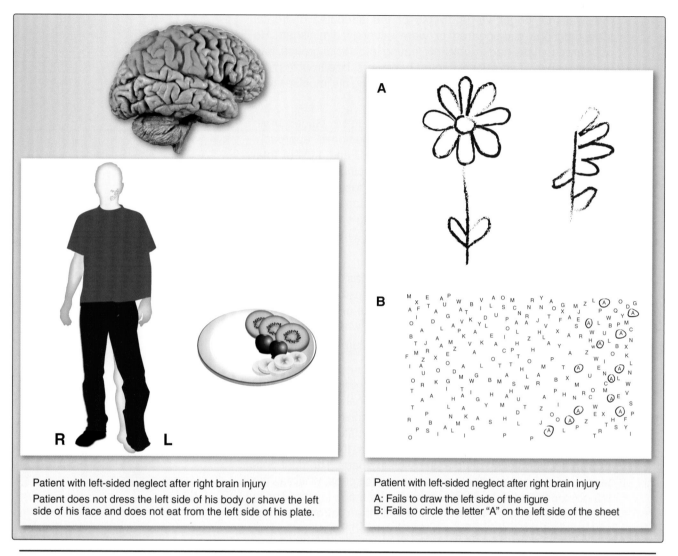

Figure 13.14
Contralateral neglect.

Interestingly, while attention to the left side of extrapersonal space and of self is controlled predominantly by the nondominant (usually right) posterior parietal lobe, attention to the right side is mediated by both the right and the left posterior parietal lobes. Thus, a lesion on one side would be compensated for by input from the other side. This may explain why a lesion in the dominant hemisphere typically results primarily in tactile agnosia, whereas a lesion on the nondominant side may have the more severe outcome of contralateral sensory neglect.

3. **Temporal association areas:** These areas of the **temporal lobe** are concerned with the task of recognizing stimuli or patterns. The medial surface of the temporal lobe, specifically the **fusiform gyrus**, is where the visual stimulus of a face or an object is linked to the recognition of its meaning or identity. The lateral surface of the temporal lobe appears to be involved in the recognition of patterns related to language (Figure 13.15).

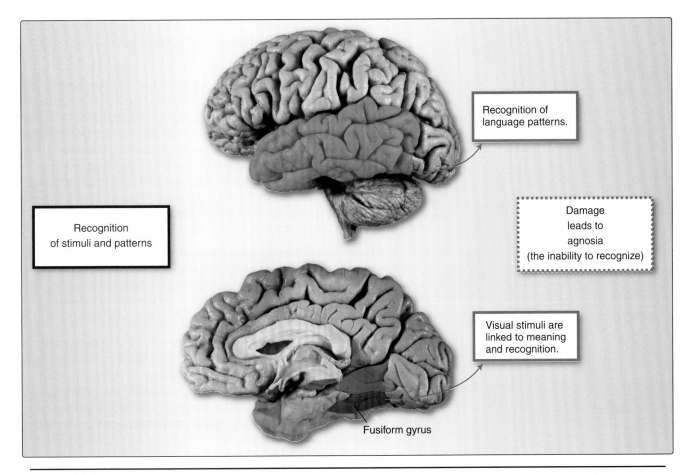

Figure 13.15
The temporal association area of the left lateral and right medial hemispheres.

Damage to the temporal lobe association areas can lead to the inability to recognize or identify objects or people. These deficits fall within the category of **agnosia**. Agnosia is distinct from neglect, because a person with agnosia can acknowledge and describe a stimulus on both the right and the left sides but is unable to recognize and name it.

The temporal association areas comprise another one of the cortical areas where lateralization occurs. The ability to recognize faces appears to be located predominantly on the right side of the inferior temporal cortex. A lesion in this area of the cortex will lead to **prosopagnosia** or the inability to recognize faces.

C. Language

The comprehension and production of language are complex tasks involving many areas of the brain in both hemispheres. The first evidence of the existence of discrete language centers came in the middle of the 19th century through the work of neurobiologists **Broca** and **Wernicke**. Their classical model of the neuroanatomy of language has remained the predominant one for the past 150 years, and it has been a useful framework for clinical diagnosis. We discuss this classical model first and then look into some of the more modern concepts and more recent evidence on how our brain tackles the complex task of language.

1. **Classical concepts of the neurobiology of language:** The areas of the brain dedicated to language can be divided into different categories, and their distribution is highly lateralized. In almost all right-handed people (approximately 98%) and most left-handed people (approximately 70%), the main centers for language are in the left hemisphere. The location of the language areas defines the dominant hemisphere: handedness may or may not be correlated with language. The right (or nondominant) hemisphere is thought to contribute more to the melody (**prosody**), rhythm, emotional expression, and accent in language.

   In the dominant hemisphere, there are two major language centers: one for the expression of language, Broca area, and one for the comprehension of language, Wernicke area. These two centers are connected to each other via a subcortical bundle of white matter, the **arcuate fasciculus**.

   a. **Broca area:** The **Broca area** is located in the **inferior frontal gyrus** of the frontal lobe, just anterior to the inferior part of the precentral gyrus (Figure 13.16). This area of cortex is the hub that links various cortical areas together for the production of language (see Figure 13.18). Language includes spoken, written, and sign language, as well as symbols (signs or words) for concepts, objects, ideas, etc.

   b. **Wernicke area:** The **Wernicke area** is located in the superior temporal gyrus and extends around the posterior end of the lateral sulcus into the parietal region (see Figure 13.16). This area of the cortex is dedicated to the comprehension of both signed and spoken language and allows us to interpret and assign meaning to symbols.

   c. **Arcuate fasciculus:** The **arcuate fasciculus,** which is the inferior part of the superior longitudinal fasciculus, connects these two areas of the cortex (see Figure 13.16). Having such a connection is logical, because we want to produce language (Broca) that make sense (Wernicke) and understand language and respond appropriately. The arcuate fasciculus is thought to be a connection that monitors speech and facilitates the repetition of words.

   d. **Lesions:** A lesion to these primary language areas leads to the inability to communicate effectively, known as **aphasia**. The type of aphasia will depend on which area of the cortex is damaged. It is important to remember that aphasias refer to language in all of its forms, whether it is spoken, written, or sign language. Aphasias are distinct from **dysarthrias** in which content (words that make sense) is intact but the production of language is impaired due to a lesion to the muscles of the pharynx, larynx, tongue, or the nerves that supply those structures.

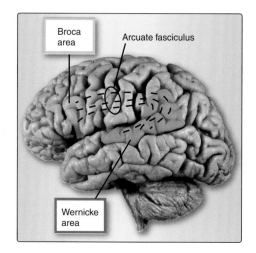

Figure 13.16
Classical model of language processing.

- *Broca aphasia:* Damage to Broca area and its connectivity leads to **expressive** or **productive aphasia** (Broca aphasia). Patients with **Broca aphasia** exhibit sparse, halting language, difficulty with syntax and grammar, word/phrase repetition, and mangled word structure. The comprehension of speech is intact, and often, these patients are very frustrated with their inability to express themselves (Figure 13.17).

- *Wernicke aphasia:* Damage to Wernicke area leads to **receptive** or **sensory aphasia** (Wernicke aphasia). Patients with **Wernicke aphasia** have no difficulty with syntax, grammar, or the structure of words, and their speech appears to be fluent and to retain melody and rhythm (see Figure 13.17). But because these patients cannot understand the language they are hearing (or seeing), the content of their speech is markedly flawed. They might use the wrong word to describe something, create new words (neologisms), or appear to be speaking "gibberish," almost as though they are speaking an unknown language. Repetition may be impaired. These patients are often not aware of their disability.

- *Conduction aphasia:* Damage to the arcuate fasciculus leads to conduction aphasia, in which the connection between Wernicke (language comprehension) and Broca (language production) areas is lost. The hallmark of **conduction aphasia** is the inability to repeat words (see Figure 13.17). Comprehension and the expression of language are intact, but patients cannot transfer the understood word to Broca area to be expressed. When asked to repeat the word "milk," for example, they might not be able to respond, or eventually say, "the white stuff." Modern studies, however, have cast some doubt on the importance of the arcuate fasciculus in conduction aphasia. A selective lesion to the arcuate fasciculus is seen only very rarely, and areas of cortex are usually involved when the clinical diagnosis of conduction aphasia is made. The prognosis for *pure* conduction aphasia, however, is good. Symptoms usually resolve over time, possibly as other association fibers take on these functions.

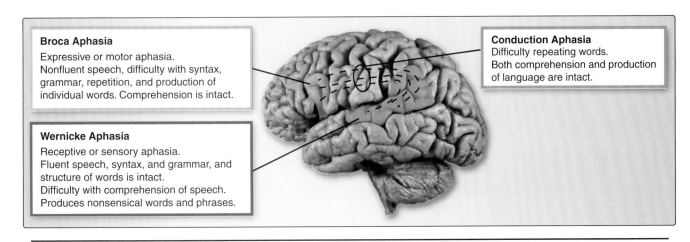

Broca Aphasia
Expressive or motor aphasia.
Nonfluent speech, difficulty with syntax, grammar, repetition, and production of individual words. Comprehension is intact.

Wernicke Aphasia
Receptive or sensory aphasia.
Fluent speech, syntax, and grammar, and structure of words is intact.
Difficulty with comprehension of speech.
Produces nonsensical words and phrases.

Conduction Aphasia
Difficulty repeating words.
Both comprehension and production of language are intact.

Figure 13.17
Broca, Wernicke, and conduction aphasias.

2. Modern concepts about the neurobiology of language:

Although the classical concepts mentioned above have had validity over the past 150 years and have proven to be good tools in clinical diagnosis, it is clear that not all pathologies can be explained through this system. In particular, the lateralization of language is not as strict as the classical model suggests. We review the more modern view of language, which is important when considering the brain's capacity to compensate for the loss of a language area and to understand the wide variety of language disorders seen clinically that do not fit into the "Broca" or "Wernicke" classification.

When we hear a sound, it is registered bilaterally in the primary auditory area (Heschl gyri), on the superior surface of the superior temporal gyrus, deep in the lateral fissure. When we recognize that the sound is actually speech or language, the superior temporal gyrus is activated. The meaning of words (**semantic processing**) is then analyzed in the cortex of the middle and inferior temporal gyri, with a distribution of meanings along the inferior temporal cortex, from *persons* anterior to *tools* posterior. All of these areas of cortex receive inputs from other parts of the brain. Parietooccipital regions and areas in the frontal lobe anterior to Broca area analyze whether the meaning assigned to a word is appropriate for the context in which it is presented as well as the grammatical structure of language and the meaning of this structure. The comprehension of language, therefore, extends well beyond the small area of cortex initially described by Wernicke (Figure 13.18).

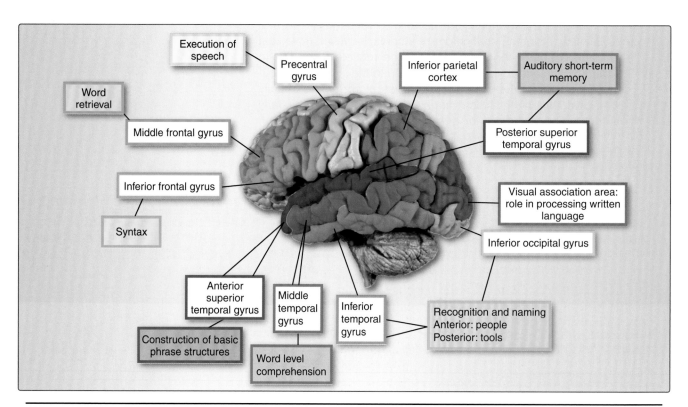

Figure 13.18

Modern understanding of language processing in the left (usually the dominant) cerebral hemisphere.

The production of language requires the use of many cortical areas to ensure that the meaning of the words is appropriate and that words are pronounced properly and put into the structure of a sentence correctly. In fact, the areas of the brain active during speech production show considerable overlap with the areas of speech comprehension. Broca area is now thought to be the hub that manipulates and forwards information across the cortical networks responsible for speech production. It also plays the coordinating role to organize sentence structure, and intact syntax. The left middle frontal gyrus has been shown to be active during word retrieval, and the left anterior insula appears to be involved in planning the articulation of speech. Initiation and execution of speech are then related to the basal ganglia (left putamen and bilateral caudate) and the anterior cingulate gyrus as well as premotor and motor areas and the cerebellum. Specific damage to the Broca area alone with no damage to its connectivity does cause Broca aphasia but results in transient mutism with good prognosis.

The role of the arcuate fasciculus has diminished as our understanding of the brain areas involved in speech perception and production has expanded. The arcuate fasciculus is still thought to play a role in repetition, which can, however, be compensated for by other association fibers.

Although many of the language-associated areas are bilateral, a dominant hemisphere (usually left) plays a major role, possibly because other higher-order processes are lateralized to the left and preferentially influence the systems related to language.

In summary, the modern understanding of language assigns comprehension of language to widespread areas of the temporal, occipital, and parietal lobes and production of language to a network of structures with Broca as the hub.

D. The mirror neuron system

The transition to social groups had a significant influence on the evolution of the brain as discussed above. Another system that evolved to allow us to live harmoniously with other people, understand their actions and emotions, and learn from them is the **mirror neuron system**.

1. **Function:** The role of the mirror neuron system is to provide a simple, direct, and quick way to understand other people's actions and emotions. The mirror neuron system is a network of neurons throughout the frontal, parietal, temporal, cingulate, and insular cortices. This network is active when we perform a particular action and, interestingly, when we observe another person performing that same action. The activity of these neurons allows us to understand action performed by another person and to predict the consequence of that action based on our own experience. We might cringe or shudder when we see another person do something that would scare or hurt us (e.g., seeing a trapeze artist perform flips high up in the air). Different areas are active depending on the motor actions we observe, whether they are independent (intransitive), relate to someone or something else (transitive), or involve the use of tools.

a. Motor learning: The mirror neuron system is also an important factor in motor learning. We can imitate the actions of others through this system. Activity in the mirror neuron system influences motor maps and modifies and fine-tunes our actions. Dancers who observe other dancers, for instance, modify their moves to imitate what they have seen.

b. Empathy: The mirror neuron system extends to the emotional understanding of others. Activity of mirror neurons in the anterior cingulate and insular cortices helps us understand and sense other people's emotions. This can be seen as a direct neuroanatomical correlate for empathy.

2. Clinical significance: Evidence has emerged suggesting that activity, circuitry, and control of the mirror neuron system may be involved in some of the behavioral changes observed in people with autism spectrum disorders. However, further research is needed to corroborate these findings and their clinical implications in relation to understanding other people's actions and emotions.

E. Sex differences in the cerebral cortex

Many neuroanatomical sex differences have been identified in both animals and humans. These differences may form at least part of the neural basis for sex-specific behaviors, both reproductive and nonreproductive, and possibly gender identity. One of the earliest studies identified a sexually dimorphic nucleus of the preoptic area of the hypothalamus. It has an elongated shape in females and a more spherical shape in males. It is more than twice as large in males than in females and contains about twice as many cells.

1. White matter tracts: White matter tracts may also show sexual dimorphism. The corpus callosum, the major bundle of fibers connecting the two cerebral hemispheres, shows a sex difference in size, particularly the splenium, which is larger in females than in males, despite larger overall brain size in males.

2. Cortex: Sex differences in cortical asymmetry are well described in both animal and human studies. However, brain structural differences are not necessarily reflected in differences at the behavioral and skill levels and there are no known differences in intellectual capacities across the sexes. Interesting examples of functional asymmetries have been reported but demonstrate that although male and female brains may process information differently, outcomes are ultimately the same. One interesting example relates to sex differences in cortical laterality of language and visuospatial processing. During phonological tasks, which involve detecting and discriminating differences in speech sounds, brain activation in males was lateralized to the left inferior frontal gyrus, whereas in females, the pattern of activation involved both left and right inferior frontal gyri. A subsequent study found that males were more left lateralized during phonological tasks and showed greater bilateral activity during the visuospatial task, whereas females showed greater bilateral activity during the phonological task and were more right lateralized during the visuospatial task.

The mechanism for these sex differences is not fully known but likely involves both organizational and activational effects of the sex hormones, estrogen and testosterone. Organizational effects of hormones occur during fetal development and act both to influence the development of the sex organs and to masculinize or feminize the brain in a relatively permanent way, including differences in axon growth, targeting, and connectivity, and in females, the substrate for hormonal cyclicity. Activational effects of hormones occur during puberty and later in life and depend on estrogen and testosterone acting on the masculinized or feminized brain. Secretion of androgens and estrogens are important in the development of secondary sexual characteristics as well as sexually dimorphic patterns of behavior, and the emergence of hormonal cyclicity in females, regulated by the hypothalamus.

IV. BLOOD SUPPLY TO THE CORTEX

The blood supply to the cortex is provided by the cerebral arterial circle (**of Willis**), which was introduced in Chapter 2, "Overview of the Central Nervous System." The **internal carotid arteries** and the **vertebral basilar system** form a circular **anastomosis** at the base of the brain. It is this circle from which the **anterior cerebral arteries**, **the middle cerebral arteries**, and the **posterior cerebral arteries** branch (Figure 13.19). These three major arteries and their smaller branches supply the forebrain.

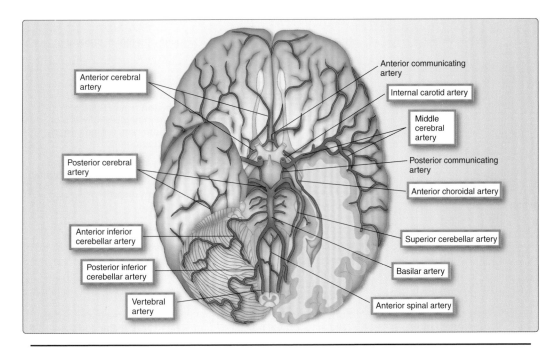

Figure 13.19
Circle of Willis on the ventral surface of the brain (the temporal lobe has been dissected to demonstrate the course of the anterior choroidal artery).

All of the vessels supplying the forebrain are **end arteries** that do not have connections with other arteries and show little or no overlap in their territories. The area where two perfusion areas come together is called a **watershed area**. This area is most vulnerable when a person's blood pressure drops or there is a bleed in a central vessel, as there is then not enough pressure to perfuse the entire territory of that vessel.

The following section gives an overview of the perfusion territories of these vessels, and Table 13.2 summarizes the major symptoms seen during an occlusion of these vessels.

Table 13.2: Functional Deficits Seen in Common Vascular Infarcts

Artery	Branches	Structures Supplied	Function	Clinical Deficits/Syndromes
Anterior cerebral artery	Superficial	Motor cortex (lower limb)	Controls movement of contralateral lower limb	Upper motor neuron weakness, contralateral lower limb
		Sensory cortex (lower limb)	Receives sensory input from contralateral lower limb	Sensory deficits, contralateral lower limb
		Supplemental motor complex (dominant hemisphere)	Planning of motor activity	Mild motor deficits, contralateral
		Prefrontal cortex	Volition, motivation, planning, and organizing of complex behavior	Frontal lobe behavioral abnormalities
	Lenticulostriate and Recurrent artery of Heubner	Internal capsule (anterior limb)	Corticopontine and thalamocortical (from anterior and DM nuclei) fibers	Possible changes in emotional behavior (thalamocortical); mild motor changes related to cerebellar function (corticopontine)
Middle cerebral artery	Left superficial	Broca area	Expressive speech area; integrates with other language areas	Nonfluent (Broca) aphasia
		Wernicke area	Receptive speech area; integrates with other language areas	Fluent (Wernicke) aphasia
	Left and right superficial	Motor cortex, supplementary motor complex	Movement of contralateral head, neck, arm, and trunk	Upper motor neuron weakness, contralateral face, neck, arm, and trunk. Possible left hemineglect if lesion on nondominant (usually right) side (variable)
		Sensory cortex	Sensation from the left head, neck, arm, and trunk	Sensory deficits from the head, neck, arm, trunk contralaterally
	Deep (lenticulostriate)	Striatum (caudate and putamen)	Receives cortical inputs relayed to basal ganglia; initiation and control of movement	Movement disorders
		Globus pallidus	Site of origin of output from basal ganglia to substantia nigra and thalamus; initiation and control of movement	Movement disorders
		Internal capsule (anterior limb)	Corticopontine and thalamocortical fibers	Possible changes in emotional behavior (thalamocortical); mild motor changes related to cerebellar function (corticopontine)
		Internal capsule (genu)	Descending fibers of the corticobulbar tract	Upper motor neuron cranial nerve signs

Table 13.2: Functional Deficits Seen in Common Vascular Infarcts (*continued*)

Artery	Branches	Structures Supplied	Function	Clinical Deficits/Syndromes
Internal carotid artery	Anterior choroidal	Internal capsule (posterior limb, lower part), globus pallidus, optic tract, temporal lobe, hippocampus and amygdala	Descending corticospinal fibers; thalamocortical fibers (sensory information); visual pathway; limbic structures related to memory, fear, and emotion	Contralateral upper motor neuron hemiparesis; sensory abnormalities; contralateral homonymous hemianopia; changes in memory, emotions
Posterior cerebral artery	Superficial	Occipital lobe	Primary and secondary visual areas; reception and interpretation of visual input	Contralateral homonymous hemianopsia, alexia without agraphia
		Splenium of corpus callosum	Carries commissural fibers connecting right and left visual association cortices	Possible deficits in interpreting visual images, alexia without agraphia
		Inferior and medial parts of the temporal lobe	Recognition and interpretation of faces	Deficit in recognition and interpretation of faces (prosopagnosia)
		Hippocampus, amygdala		Limbic problems related to memory, fear, and emotion
	Deep	Thalamus	Relay center for descending and ascending information; integration of cerebral cortex and rest of the CNS	Thalamic lesions can mimic cortical lesions
		Internal capsule (posterior limb)	Descending fibers of the lateral and anterior corticospinal tracts	Contralateral upper motor neuron hemiparesis

CNS = central nervous system.

A. Anterior cerebral artery

The **anterior cerebral artery (ACA)** branches off the circle of Willis and travels into the **interhemispheric fissure**, along the superior surface of the corpus callosum (Figures 13.19 and 13.20). The ACA supplies the medial surface of the frontal and parietal lobes as well as approximately 1–2 cm of the lateral surface of those two lobes, as the end branches reach over onto the lateral side of the brain. The deep branches of the ACA contribute to the lenticulostriate arteries and supply the head of the caudate nucleus and the anterior limb of the internal capsule.

B. Middle cerebral artery

The **middle cerebral artery (MCA)** is the direct branch of the internal carotid artery and the largest artery supplying the cerebral hemispheres. The MCA travels through the lateral fissure onto the lateral hemisphere where it divides into its end branches that supply almost the entire lateral surface of the cortex (Figures 13.21 and 13.22). The deep branches of the MCA in conjunction with some deep branches of the ACA are called the **lenticulostriate arteries**, which supply the basal ganglia as well as the anterior and posterior limbs of the internal capsule (Table 13.1 and Figure 13.23). These are small vessels coming off the large MCA and ACA. They are very vulnerable to rises in blood pressure, because their weaker arterial walls may not be able

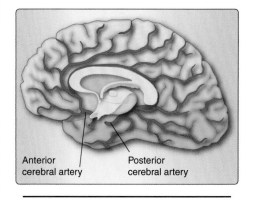

Anterior cerebral artery
Posterior cerebral artery

Figure 13.20
The anterior and posterior cerebral arteries on the medial surface of the right cerebral hemisphere.

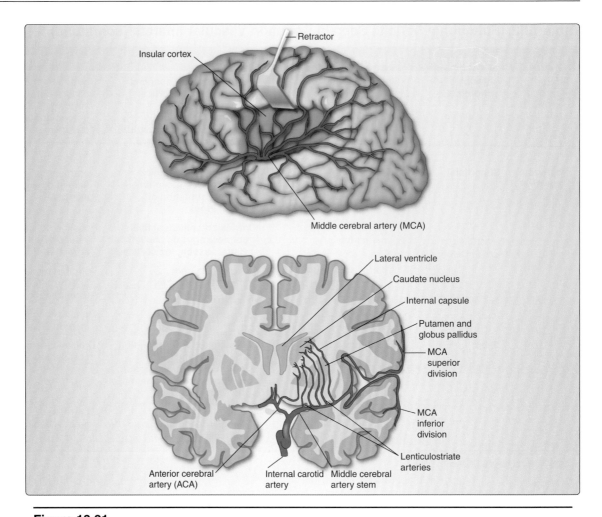

Figure 13.21
The middle cerebral artery branches into superior and inferior divisions in the lateral fissure to supply the lateral aspect of the frontal/parietal lobes and temporal lobe, respectively.

to withstand high pressure. This can lead to a hemorrhagic stroke in the internal capsule and basal ganglia.

C. Anterior choroidal artery

The **anterior choroidal artery** arises from the internal carotid artery (see Figure 13.19). It travels to the choroid plexus in the lateral ventricle and provides blood supply to deep structures of the temporal and occipital lobes, including the **hippocampus**, the inferior part of the posterior limb of the **internal capsule**, the **globus pallidus**, the **optic tract**, and the **tail of the caudate nucleus** (see Figure 13.23). It is a long thin artery coming off the large, high-pressure internal carotid artery. Similar to the lenticulostriate arteries, it is very vulnerable to rises in blood pressure, because its smaller arterial wall may not be able to withstand high pressures, leading to a hemorrhagic stroke. This is important to understand, because this vessel supplies critically important fiber tracts in the inferior portion of the posterior limb of the internal capsule.

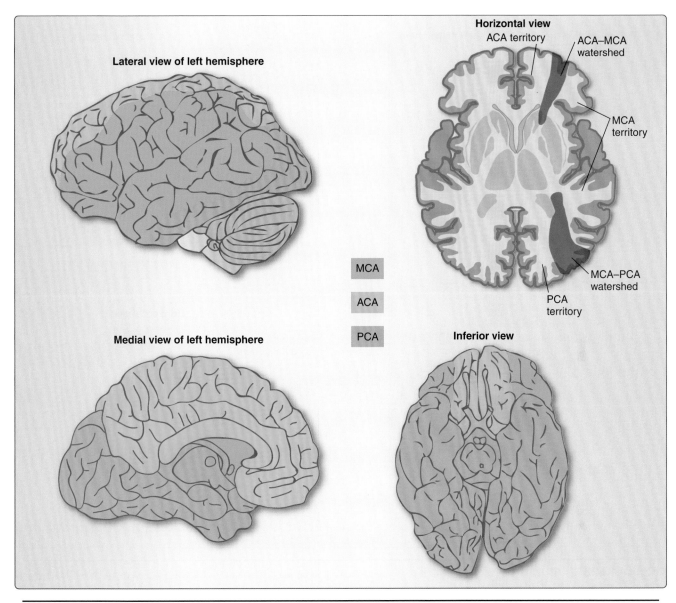

Figure 13.22
Perfusion areas of the middle cerebral artery (MCA), anterior cerebral artery (ACA), and posterior cerebral artery (PCA) on the surface of the cerebral hemispheres.

D. Posterior cerebral artery

The basilar artery, part of the vertebral basilar system, gives off two branches at its rostral end: the two **posterior cerebral arteries** (PCAs) (see Figure 13.19). These are connected to the internal carotid system through the **posterior communicating arteries**. The PCA swings back around the midbrain and supplies the medial surface of the occipital lobe and the inferior and medial parts of the temporal lobe, as well as the splenium of the corpus callosum (see Figures 13.20 and 13.22). The deep branches of the PCA also supply the thalamus, parts of the midbrain, and part of the posterior limb of the internal capsule (see Figure 13.23).

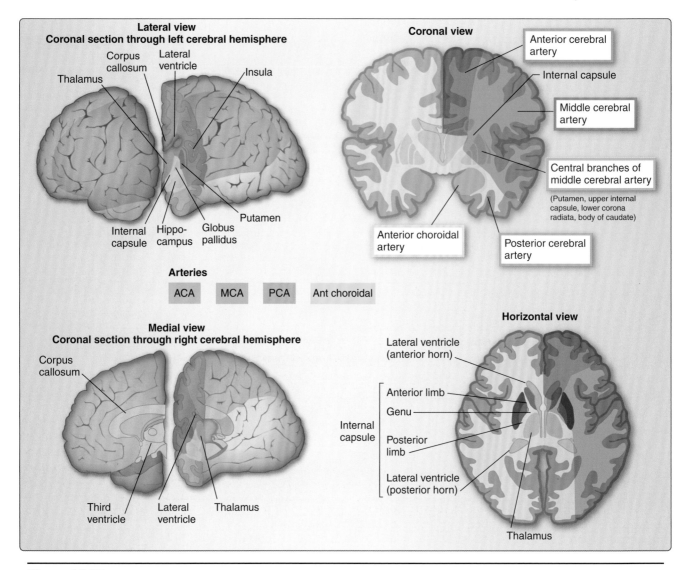

Figure 13.23
Perfusion areas of the middle, anterior, and posterior cerebral arteries as well as the anterior choroidal artery deep in the forebrain, in coronal and horizontal cross section.

Clinical cases: In the stroke clinic

Anterior Cerebral Artery Infarct

A 52-year-old woman presents to the clinic with a sudden onset of impaired dexterity of her left hand and gait difficulties. She has difficulty combing her hair and brushing her teeth with her left hand, and she also has trouble putting her pants on. She has a history of hypertension and atrial fibrillation. On examination, her cranial nerve function is normal. She has slow fine finger movement of her left hand but normal strength in both her hands and right leg. Strength in her left leg is slightly reduced (4/5). Her tone is increased in the left leg but normal in the right leg and both arms. Her knee and ankle reflexes are brisker on the left than on the right. Reflexes in both arms are normal. Toes are downgoing in both feet. Facial sensation and strength are normal. General sensation in the limbs is normal. Speech and language are normal. She has difficulty initiating her gait but otherwise her gait is normal. An MRI of her brain shows a stroke to her right anterior cerebral artery (ACA) territory.

What are the clinical features associated with the infarction of the brain tissue supplied by the superficial branches (distal to the lenticulostriate arteries) of the anterior cerebral artery (ACA)?

An ACA stroke causes more weakness of the leg than the face or arm because the motor homunculus for the leg is medial and supplied by the ACA. The motor cortex for the head, neck, face, and arm is supplied by the MCA. The ACA also supplies the inferior and medial frontal association areas that connect with the amygdala, hypothalamus, and nucleus accumbens. A lesion to this region causes apathy and disinhibition.

What is the name of the condition that causes impairment in learned skilled motor tasks such as dressing and brushing teeth?

This type of condition is referred to as **apraxia**. Apraxia is the inability to perform learned skilled motor tasks in the absence of paralysis. Examples include limb apraxia (e.g., combing hair, brushing teeth, saluting) and buccofacial apraxia (e.g., whistling, blowing out a match).

Lesions to association areas of cortex result in a number of clinical deficits

Anomia is the inability to name. Anomia can occur with or without aphasias. Wernicke, Broca, and global aphasias all have anomia.

Alexia is the inability to read. Alexia can occur with or without agraphia. Alexia without agraphia can be due to lesion of the medial occipital/parietal area and splenium. Alexia without agraphia is a disconnection syndrome involving impairment of association fibers.

Agraphia is the inability to write. Agraphia without alexia can occur with lesions to the angular gyrus. Lesion of the angular gyrus may also cause finger agnosia, right/left confusion, and acalculia.

Aphasia is the impairment of language or inability to communicate effectively.

APHASIA CASES

There are different types of aphasias: global aphasia, Wernicke aphasia, Broca aphasia, conduction aphasia, transcortical sensory aphasia, and transcortical motor aphasia. Each of the aphasias is distinguished based on the presence or absence of naming, repetition, fluency, and comprehension. The transcortical aphasias have normal repetition (Figure 13.24).

Match the presentations below to each of the following types of aphasia: Broca aphasia, Wernicke aphasia, conduction aphasia, and global aphasia.

Presentations

A. The patient has aphasia with an MRI of the brain showing a tumor to the inferior frontal gyrus.

B. The patient has fluent speech and creates new words. He can name watch and eyeglasses but not knuckle, dial, or lens. He can repeat tip-top and huckleberry but cannot repeat no, ifs, ands, or buts. He can close his eyes to command and stick out his tongue but cannot point to the light in the room on command.

C. A right-handed patient has a large left middle cerebral artery (MCA) stroke affecting both the superior and inferior divisions.

D. A patient presents with impaired repetition and naming. Her speech is fluent and her comprehension is intact.

Answers

A—Broca aphasia: The patient has aphasia with an MRI of the brain showing a tumor in the inferior frontal gyrus. The Broca area is located in the inferior frontal gyrus, and lesions of the Broca area cause expressive aphasia (Broca aphasia) with sparse language, difficulty with grammar and syntax, and impaired repetition but normal comprehension.

B—Wernicke aphasia: The patient has fluent speech and creates new words (*neologism*). He can name watch and eyeglasses but not knuckle, dial, or lens (*partial anomia*). He can repeat tip-top and huckleberry (*simple words*) but cannot repeat no, ifs, ands, or buts (*impaired repetition of complex or longer phrases*). He can close his eyes when asked and stick out his tongue but cannot point to the light in the room on command, which indicates *impaired comprehension of complex tasks*. MRI indicates a lesion to the left superior temporal lobe.

C—Global aphasia: This right-handed patient has a large left middle cerebral artery (MCA) stroke affecting both the superior and inferior divisions. The dominant hemisphere, where language is localized, is typically the left hemisphere. The MCA supplies both Broca and Wernicke areas. A large MCA stroke can cause a global aphasia with impairment of both expressive (Broca) and receptive (Wernicke) language functions.

D—Conduction aphasia: This patient has a lesion affecting the arcuate fasciculus. The arcuate fasciculus is a branch of the superior longitudinal fasciculus that connects Broca and Wernicke areas. Damage to the arcuate fasciculus causes impaired repetition and naming with normal fluency and comprehension. Associated cortical areas are likely involved.

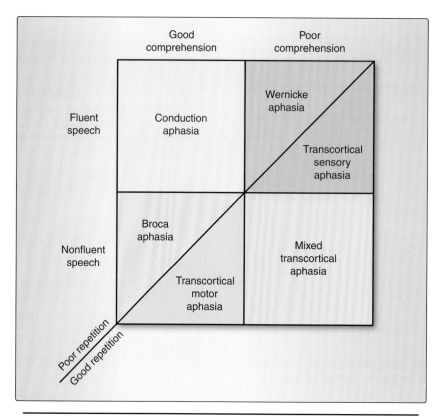

Figure 13.24
The aphasia square is used to differentiate six common aphasias. Broca,
conduction, and Wernicke aphasias are all characterized by poor repetition.

Chapter Summary

- The two cerebral hemispheres comprise the cerebral cortex, subcortical fiber bundles, and the deep nuclei. In this chapter, we focused on the functional anatomy of the cerebral cortex.

- Histologically, the cortex is a layered structure with two main types of neurons: **pyramidal cells** and **granule cells**. The vast majority of the cerebral cortex is the six-layered **neocortex**.

- Subcortical fiber bundles connect areas of the cortex with each other and with other brain areas. **Association fibers** connect cortical areas within one hemisphere. **Commissural fibers** cross the midline and connect cortical areas in the two hemispheres. **Projection fibers** connect the cortex with subcortical as well as brainstem and spinal cord structures.

- The cortex comprises primary and association areas, which are localized in both the right and the left hemispheres. The hemispheres are mostly symmetrical, especially the primary areas. However, there are large differences between the two hemispheres in areas that mediate language and in the function of several association areas. In the majority of right-handed and in most left-handed people, the left hemisphere is dominant; dominance is determined by the location of the language areas.

- The **primary motor area** contains the neurons that will give rise to the corticospinal tract, which projects to the contralateral lower motor neurons. The **primary sensory cortex** receives input through the posterior column–medial lemniscus system and the anterolateral system from the contralateral side of the body. The primary areas for vision, hearing, olfaction, and taste are the first targets for the processing of these special senses.

- The **association areas** make up the major part of the cortex. These are involved in many higher-order processes and interpret, integrate, and modulate information the brain receives and motor actions it produces.

Chapter Summary continued

- ○ The **frontal association areas** are important in the expression of our personality, for living in social groups, and adapting our behavior to specific situations. A lesion of the frontal association areas leads to a change in personality, and these patients have difficulty integrating into social situations and acting appropriately.
- ○ The **parietal association areas** are critical in the awareness of self as well as our surroundings. A lesion to the non-dominant parietal association areas leads to **neglect** of the contralateral side.
- ○ The **temporal association areas** are important for the recognition of objects, people, and language. A lesion of the temporal association area will lead to **agnosia** or the inability to recognize objects, people, and words.
- Language is a complex process that requires many cortical areas. **Broca area** in the frontal lobe is the center critical for the production of language. A lesion in this area leads to difficulty producing language, called **Broca aphasia**, or expressive, motor aphasia.
- **Wernicke area** at the junction between the parietal and the temporal lobes is the main center for the understanding of language. A lesion here will lead to difficulty understanding language, called **Wernicke aphasia**, or receptive, sensory aphasia.
- These two areas are connected by the **arcuate fasciculus**. A lesion to this subcortical fiber bundle will lead to **conduction aphasia**, the hallmark of which is the inability to repeat words.
- Modern understanding of language has made it clear that much larger areas of the frontal, parietal, and temporal lobes are involved in both the reception and expression of language. The classical concept of Broca and Wernicke areas remains relevant for clinical diagnosis.
- The **mirror neuron system** is a network of neurons throughout primary and association areas that is active when we observe motor activity and emotions in others. This system allows us to understand and relate to the actions of others without engaging complex, higher-order processing.
- The blood supply to the forebrain comes from the **internal carotid** and **vertebral basilar systems**. These two systems join to form the **circle of Willis** on the inferior surface of the brain. The **anterior cerebral artery** supplies the medial side of the frontal and parietal lobes, the **middle cerebral artery** supplies most of the lateral surface of the brain, and the **posterior cerebral artery** supplies the occipital and inferior temporal lobes. The **anterior choroidal artery** branches directly off the internal carotid artery and supplies deep structures, such as the internal capsule and the basal ganglia, together with the **lenticulostriate arteries**, which are the deep branches of the middle and anterior cerebral arteries.

Study Questions

Choose the ONE best answer.

13.1 A 65-year-old man suffers a stroke that affects most of his primary motor cortex in the left hemisphere. Which one of the following statements best describes the consequence of this lesion?

   A. The entire corticospinal tract is affected.
   B. The stroke affects the postcentral gyrus, where the primary motor cortex is located.
   C. The posterior horn of the spinal cord has lost its innervation.
   D. The afferents from the primary sensory cortex to the primary motor cortex have lost their target.
   E. The motor deficits will be on the left side of the body.

The correct answer is D. The primary sensory cortex sends association fibers to the primary motor cortex. The majority of fibers in the corticospinal tract are from the primary motor cortex, but both the supplementary cortex and premotor cortex also contribute fibers. The primary motor cortex is located in the precentral gyrus of the frontal lobe. The corticospinal tract projects to the anterior horn of the spinal cord. The deficits of a lesion to the left primary motor cortex will be on the right side of the body, because the corticospinal tract crosses the midline in the caudal medulla.

13.2 A patient is brought to the emergency department following an ischemic stroke in the perfusion territory of the right middle cerebral artery. Which one of the following syndromes best describes a possible outcome of this lesion?

    A. Broca aphasia.

    B. Paralysis of the left leg.

    C. Left-sided neglect.

    D. Personality changes.

    E. Inability to recognize faces.

> The correct answer is C. The middle cerebral artery supplies the parietal association cortex, which is important for the awareness of self and our surroundings. A lesion of the right (nondominant hemisphere) parietal association cortex leads to left-sided neglect. Broca aphasia results from a lesion in the dominant hemisphere, which, in most people, is the left hemisphere. The anterior cerebral artery supplies the medial side of the brain and a 2-cm-wide strip on the lateral surface. The medial surface of the hemisphere is where the motor representation of the leg is localized. However, the fibers cross to the contralateral side in the corticospinal tract, resulting in paralysis of the right leg. Personality is mostly defined by areas of the frontal lobes, which are supplied by the anterior and middle cerebral arteries. An infarct in the posterior cerebral artery will impact the occipital and temporal lobes, impacting vision and visual processing.

13.3 A patient presents with a sudden inability to comprehend speech and uses nonsensical words and sentences. The clinician immediately suspects an infarct in a branch of what artery?

    A. Anterior communicating artery.

    B. Basilar artery.

    C. Posterior cerebral artery.

    D. Middle cerebral artery.

    E. Anterior choroidal artery.

> The correct answer is D. The middle cerebral artery supplies the lateral surface of the brain where the Wernicke area is located. A sudden infarct in a branch of the middle cerebral artery to this area would result in these symptoms. The anterior communicating artery connects the two anterior cerebral arteries, the basilar artery lies on the anterior surface of the pons, the posterior cerebral artery supplies primary and association visual cortices, and the anterior choroidal artery supplies the internal capsule.

13.4 On a magnetic resonance image, a radiologist notices a discreet bleed in the genu of the internal capsule. Which of the following sets of fibers are impacted by this lesion?

    A. Corticobulbar.

    B. Thalamocortical.

    C. Corticospinal.

    D. Corticopontine.

    E. Reticulospinal.

> The correct answer is A. Corticobulbar fibers are concentrated in the genu of the internal capsule. Thalamocortical fibers run in the anterior and posterior limbs of the internal capsule, corticospinal fibers and corticopontine fibers are in the posterior limb of the internal capsule, and the reticulospinal fibers begin below the internal capsule.

13.5 A patient is diagnosed with a dementia mainly involving the frontal lobe. The magnetic resonance image shows degenerative changes in the frontal lobes on both sides. Which one of the following options describes the deficits this patient experiences?

    A. Loss of hearing.

    B. Changes in language comprehension.

    C. Inability to plan for the future.

    D. Loss of visual memories.

    E. Loss of short-term memories.

> The correct answer is C. The frontal lobe defines our personality and our ability to live in a social context. The frontal lobe appears to play a role in planning and problem solving as well as directing and maintaining attention on a particular situation or task ("executive function"). The primary auditory area is located in the superior temporal lobe, and language comprehension spreads to many areas of the brain but is concentrated in the temporal and parietal lobes. Visual memories are mediated by the occipital and temporal association areas. Short-term memories are primarily dependent on the hippocampus (part of the limbic system).

The Thalamus

14

I. OVERVIEW

It is impossible to talk about the functioning of the brain and the pathways that allow for sensory perception or motor output without mention of the thalamus (Figure 14.1). Indeed, the thalamus is mentioned in almost every chapter of this book. It is the target of all sensory information (except olfaction) on its way to the cortex, and subcortical structures project to the cortex via the thalamus to influence upper motor neurons for motor output. In addition, the thalamus connects cortical areas with each other, integrating, modulating, and gating the flow of information

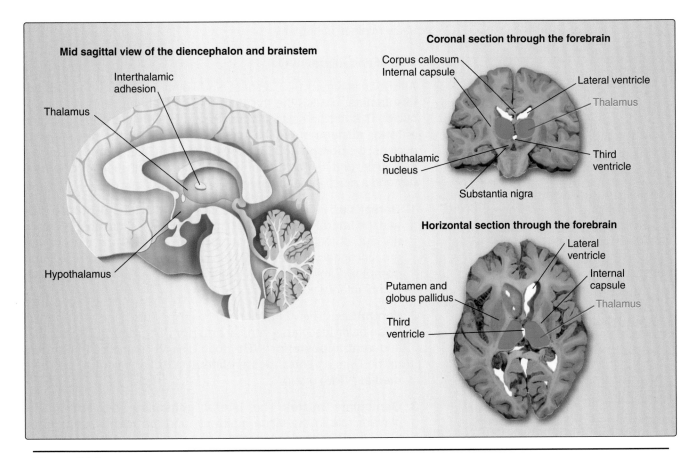

Figure 14.1
The thalamus in sagittal, coronal, and horizontal sections.

from one part of the cortex to another. The thalamus can be described as both a "gatekeeper" and "integrator," because it regulates the flow of information that will ultimately reach the cortex. Bidirectional communication between the cortex and nuclei of the thalamus provides an additional layer of information processing.

In this chapter, we look at the functional anatomy of the thalamus, describing the various nuclei and their projections. The physiology of the thalamic neurons is also covered, shedding light on some of the mechanisms by which the flow of information to the cortex is regulated. We also explain the critical role that the thalamus plays in focusing attention, in arousal, and in consciousness.

Throughout the chapter, we look at the clinical implications of thalamic lesions. A lesion in the thalamus can result in symptoms or deficits that resemble those that would occur with a lesion to the cortex. Even small lesions may have widespread functional sequelae. Numerous clinical reports demonstrate the adverse impact of specific thalamic lesions.

II. ANATOMY

The **thalamus** is a paired structure located on both sides of the third ventricle. In most people, the thalamus on the left side is connected with the thalamus on the right side through the **interthalamic adhesion** (see Figure 14.1), although the functional importance of this connection is probably minimal in humans.

A. Medial and lateral nuclei

As shown in Figure 14.2, a sheet of white matter, the **internal medullary lamina**, divides the thalamus into medial and lateral groups of nuclei. This lamina splits into two leaves anteriorly and encloses the **anterior nucleus** of the thalamus. The **medial group** has only one nucleus: the **dorsomedial nucleus (DM)**. The **lateral group** has several nuclei and can be further divided into a small superior or **dorsal tier** and a much larger inferior or **ventral tier** of nuclei.

1. **Dorsal tier:** The dorsal tier comprises the **lateral dorsal (LD)** and the **lateral posterior (LP)** nuclei as well as the most posterior nucleus of the thalamus, the **pulvinar**. The LP and the pulvinar are functionally related and often summarized as the **LP–pulvinar complex**. Similarly, we can consider the LD as functionally related to the anterior nucleus.

2. **Ventral tier:** The ventral tier makes up the bulk of the lateral group. It can be divided into a **ventral anterior (VA)**, **ventral lateral (VL)**, and **ventral posterior (VP)** group. The VP group can be further divided into a **ventral posterolateral (VPL)** and **ventral posteromedial (VPM)** group.

3. **Geniculate nuclei:** The **lateral geniculate nucleus (LGN)** plays a role in the visual pathway, and the **medial geniculate nucleus (MGN)** plays a role in the auditory pathway. Both are located at the posterior pole of the thalamus, inferior to the pulvinar nucleus.

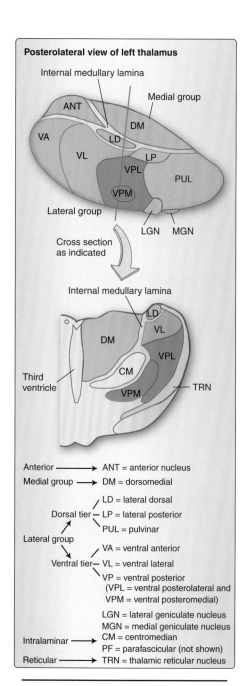

Posterolateral view of left thalamus

Internal medullary lamina

Medial group

ANT
VA
DM
LD
VL
LP
VPL
PUL
VPM
Lateral group
LGN   MGN

Cross section as indicated

Internal medullary lamina

LD
VL
DM
VPL
CM
Third ventricle
VPM
TRN

Anterior ⟶ ANT = anterior nucleus
Medial group ⟶ DM = dorsomedial
Dorsal tier ⟨ LD = lateral dorsal
LP = lateral posterior
PUL = pulvinar
Lateral group
Ventral tier ⟨ VA = ventral anterior
VL = ventral lateral
VP = ventral posterior
(VPL = ventral posterolateral and VPM = ventral posteromedial)
LGN = lateral geniculate nucleus
MGN = medial geniculate nucleus
Intralaminar ⟶ CM = centromedian
PF = parafascicular (not shown)
Reticular ⟶ TRN = thalamic reticular nucleus

Figure 14.2
Overview of the thalamic nuclei.

B. Intralaminar nuclei

The **intralaminar nuclei** are collections of neurons within the internal medullary lamina. They include the **centromedian (CM)** and **parafascicular (PF)** nuclei, which are involved in arousal and goal-oriented behavior.

C. Surrounding nuclei

The thalamus is surrounded by a sheet of neurons, which forms the **thalamic reticular nucleus** (TRN). The name is derived from its net-like appearance. It is important in modulating projections between the thalamus and the cortex.

III. FUNCTIONS OF THE THALAMIC NUCLEI

Thalamic nuclei have discrete functions in regulating the access of information to the cortex. They are categorized as relay nuclei, association nuclei, or "other" nuclei. **Relay nuclei** (motor, sensory, and limbic) receive input from their respective systems and then process and relay that information to the cortex. **Association nuclei** connect areas of the cortex with each other. The "other" category includes the **intralaminar nuclei**, which are interconnected with the functions of the basal ganglia and the limbic system, and the **reticular nucleus (TRN)**, which plays a critical role in enabling the conscious appreciation of stimuli and events by synchronizing activity in the thalamus with cortical activity. All of the thalamic nuclei have bidirectional communication with the cortex.

Most of the neurons in the thalamus are **projection neurons** (greater than 75%), and the rest are inhibitory **interneurons**.

A. Inputs to thalamic nuclei

The inputs to the thalamus can be divided into two categories: **specific inputs (drivers)** and **regulatory inputs (modulators)** as shown in Figure 14.3. Specific inputs are those that contain information that must be forwarded to the cortex. These are primarily excitatory glutamatergic inputs. Regulatory inputs are those that modulate the information and regulate whether or not it will be forwarded to the cortex. These regulatory inputs arise from cortical areas and the TRN as well as from projection systems of the brainstem, including cholinergic, noradrenergic, and serotonergic inputs. An overview of the major inputs and outputs of the thalamic nuclei can be found in Table 14.1.

Regulatory inputs are much more abundant than specific inputs, pointing to the important role of the thalamus in gating, prioritizing, and modulating information rather than being just another step in forwarding information.

B. Firing patterns

Thalamic neurons have two firing patterns. **Tonic firing** is suited for a linear transfer of information to the cortex, and **burst firing** is suited for the detection of new or changing stimuli (Figure 14.4). Whether a neuron will fire in tonic or burst mode depends on its membrane potential (i.e., sustained levels of depolarization or hyperpolarization). These

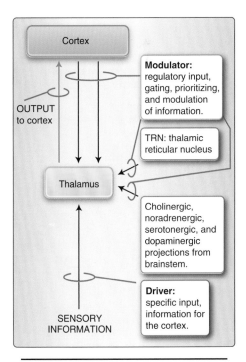

Figure 14.3
Conceptual overview of inputs to the thalamus.

Table 14.1: Major Inputs and Outputs of the Thalamic Nuclei

	Nucleus		Input	Output
Relay	Sensory	VPL	Body: dorsal column–medial lemniscus, anterolateral system	Somatosensory cortex
		VPM	Face: trigeminal lemniscus, anterior trigeminothalamic tract	Somatosensory cortex
		MGN	Auditory: inferior colliculus	Auditory cortex
		LGN	Visual: optic tract	Visual cortex (via optic radiations)
	Motor	VA	Basal ganglia, cerebellum	Primary motor cortex, premotor cortex
		VL		
	Limbic	ANT	Ipsilateral mammillary bodies (via mammillothalamic tract)	Cingulate cortex, prefrontal cortex, parietal cortex
		LD	Entorhinal cortex	Cingulate cortex, parietal cortex
Association	Pulvinar-LP		Retina; visual association cortices; auditory association cortices (superior temporal gyrus); parietal, temporal, frontal, and occipital association areas	Visual association cortices; auditory association cortices (superior temporal gyrus); parietal, temporal, frontal, and occipital association areas
	DM		Prefrontal cortex, entorhinal cortex, basal ganglia, limbic system	Prefrontal cortex, entorhinal cortex, basal ganglia, limbic system
Intralaminar	CM/PF		Cholinergic and dopaminergic input from brainstem	Striatum, cortex (diffuse and nonspecific)
Reticular Nucleus	TRN		Collaterals from corticothalamic and thalamocortical projections	Entire thalamus

VPL = ventral posterolateral nucleus; VPM = ventral posteromedial nucleus; MGN = medial geniculate nucleus; LGN = lateral geniculate nucleus; VA = ventral anterior nucleus; VL = ventral lateral nucleus; ANT = anterior nucleus; DM = dorsomedial nucleus; LP = lateral posterior nucleus; CM = centromedian nucleus; PF = parafascicular nucleus; TRN = thalamic reticular nucleus; LD = lateral dorsal.

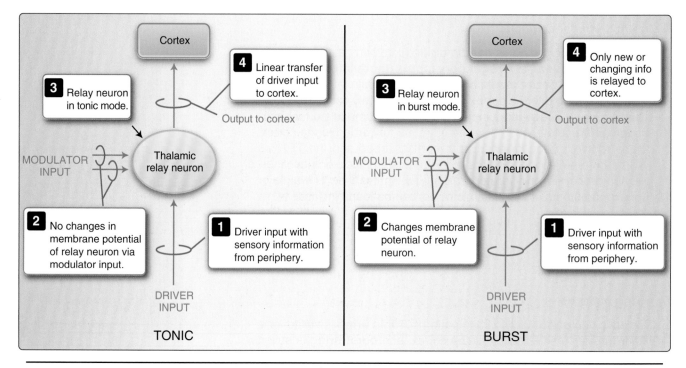

Figure 14.4
Conceptual overview of the firing patterns found in thalamic relay neurons.

membrane potentials are determined by the activity of modulator inputs acting on metabotropic neurotransmitter receptors. The activation of metabotropic receptors results in sustained, longer membrane depolarizations compared to the shorter and transient depolarizations that occur with activation of ionotropic receptors (see Chapter 1, "Introduction to the Nervous System and Basic Neurophysiology"). Thus, one way that modulator inputs determine the transfer of information to the cortex is by allowing either the steady stream of detailed information (tonic firing) or by focusing only on new and changing stimuli (burst firing).

Burst firing is seen during sleep. This activity pattern effectively blocks the linear transfer of information through tonic firing and, thereby, reduces the amount of information relayed to the cortex. It is not exclusive to sleep cycles, however, and can be seen during wakefulness as well, when it serves to reduce the amount of detailed information flow to the cortex in order to allow new and changing stimuli to be detected.

C. Relay nuclei

Relay nuclei receive **specific (driver) input** particular to the cortical area to which they connect, as well as **nonspecific (modulator) input** from other parts of the brain. Together, these inputs help the thalamus decide which information to relay to the cerebral cortex.

The relay nuclei can be divided into three functional groups: (1) **sensory relay nuclei** that receive input from peripheral sensory receptors through their respective sensory pathways and project to sensory areas of the cortex, (2) **motor relay nuclei** that interconnect with motor structures and project to motor areas of the cortex, and (3) **limbic nuclei** that interconnect with the different structures of the limbic system.

1. **Sensory relay nuclei:** These nuclei receive their input from the major sensory pathways.

    a. **Ventral posterior nucleus:** Sensory information from the **body** travels through the posterior column–medial lemniscus and the anterolateral systems, which project to the **VPL** nucleus of the thalamus. From there, information is relayed to the somatosensory and sensory association cortices. Sensory information from the face travels with the trigeminal lemniscus and the anterior trigeminothalamic tract to the **VPM** nucleus of the thalamus and from there to the somatosensory and sensory association cortices (Figure 14.5) (see Chapters 7, "Ascending Sensory Tracts," and 10, "Sensory and Motor Innervation of the Head and Neck," for more information). Information from the sensory relay nuclei of the thalamus reaches the cortex through the posterior limb of the internal capsule.

    b. **Medial geniculate nucleus:** The auditory pathway has a specific nucleus in the thalamus, the **MGN**. Auditory information ascends to synapse in the inferior colliculus and from there projects to the MGN via the inferior brachium. In turn, the MGN relays that information to the primary auditory and auditory association cortices (Figure 14.6) (see Chapter 11, "Hearing and Balance," for more information).

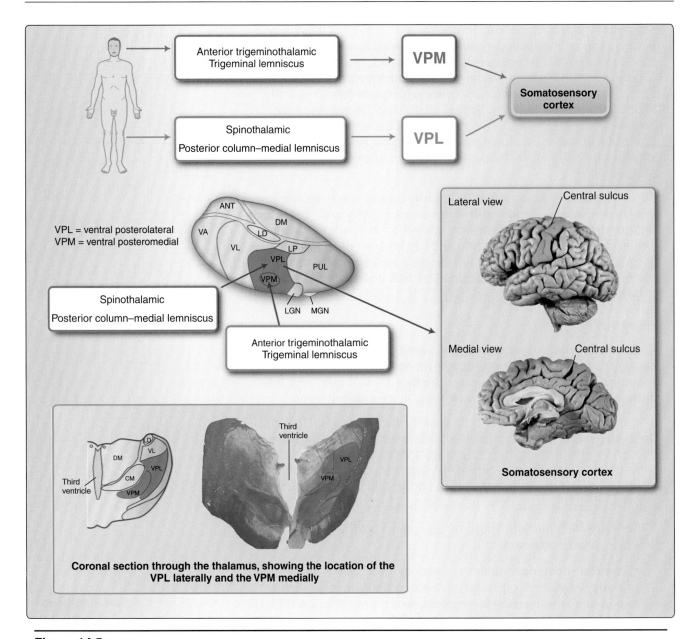

Figure 14.5
The ventral posteromedial (VPM) and ventral posterolateral (VPL) nuclei of the thalamus and their connections.

c. Lateral geniculate nucleus: The visual pathway also has a thalamic nucleus associated with it, the **LGN**. Information carried in the optic tract synapses in the LGN and from there is relayed to the primary visual and visual association cortices via the optic radiations (see Figure 14.6) (see Chapter 15, "The Visual System," for more information).

d. Modulatory system: All sensory relay nuclei of the thalamus also receive extensive input from the **modulatory system**, which filters information that will reach the cortex. Modulatory inputs arise from the areas of cortex that are the primary projection areas of the sensory relay nuclei as well as associated cortical areas. This reciprocal connection plays a critical role in awareness of a stimulus as well as consciousness, as discussed below.

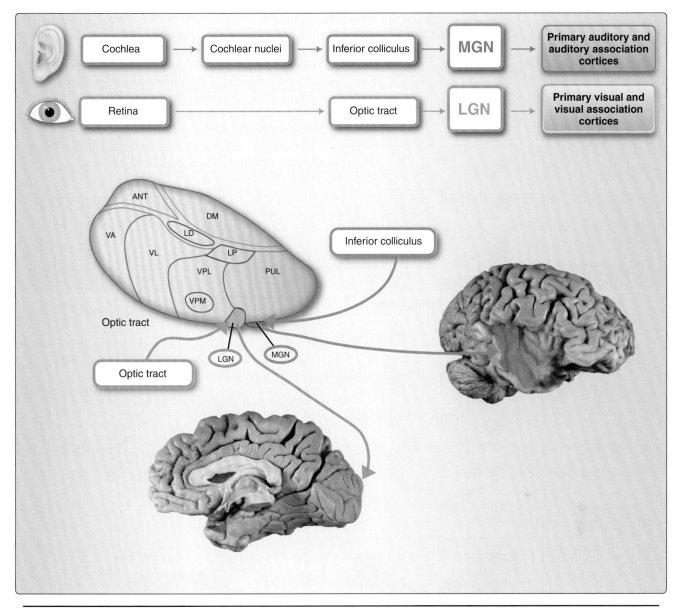

Figure 14.6
The medial geniculate nuclei (MGN) and lateral geniculate nuclei (LGN) of the thalamus and their connections.

2. **Motor relay nuclei:** These are the **VA** and ventral lateral (**VL**) nuclei. They receive their driver input from the **basal ganglia** and the **cerebellum** and project to the **primary motor** and **motor association (supplementary motor complex) cortices**. The VA and VL nuclei are normally under **tonic inhibition**.

The thalamus not only relays information to the cortex but also integrates the information it receives from multiple sources, allowing for the modulation and adjustment of ongoing motor patterns and behaviors. Information from VL and VA reaches the cortex through the posterior limb of the internal capsule.

a. **Output from the basal ganglia:** Activity in the basal ganglia will release or enhance this tonic inhibition and influence the activity of the primary motor cortex. This is the motor loop of the basal ganglia circuitry, discussed in Chapter 16, "The Basal Ganglia."

Thalamic motor nuclei relay the output from the basal ganglia to frontal cortical areas and provide direct feedback to the **striatum** (caudate and putamen) (Figure 14.7). The striatum receives cortical input to facilitate a movement through the basal ganglia circuitry. Output from the basal ganglia (specifically, the globus pallidus) to the thalamus ultimately results in inhibition or disinhibition of the thalamus. Bidirectional communication between the thalamus and the cortex fine-tunes the information, and feedback is ultimately sent back to the striatum from the thalamus ("Mission accomplished" or not).

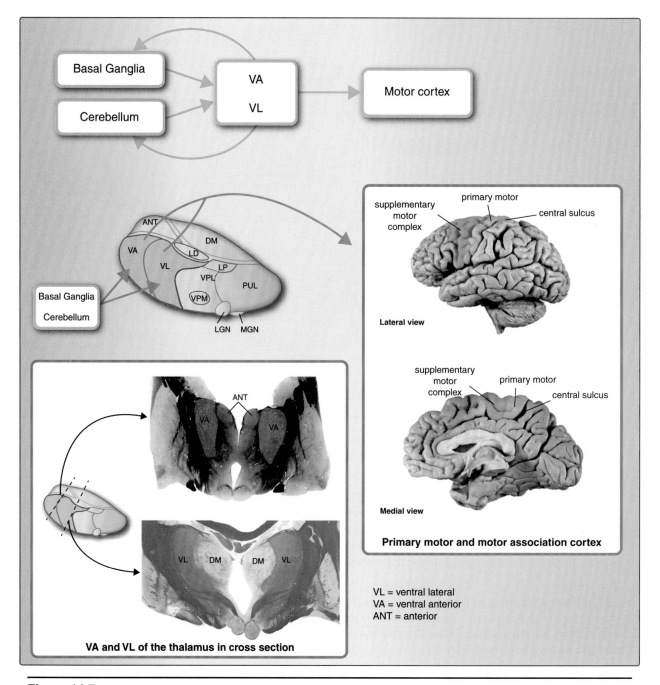

Figure 14.7
The ventral anterior and ventral lateral motor nuclei of the thalamus and their connections.

Clinical Application 14.1: Thalamic Ataxia Syndrome

Strokes occurring in the ventral lateral (VL) and ventral posterior (VP) nuclei can produce a clinical picture of contralateral "cerebellar" dysfunction and sensory loss. Weakness is typically absent or if it occurs is almost always transient. The cerebellar signs (lesions in VL) include **ataxia** (problems with coordination of gait), deficits in hand–eye coordination (**dysmetria**), and the inability to coordinate agonist–antagonist movements of the extremities (**dysdiadochokinesia**) (see Chapter 17, "The Cerebellum," for more information).

Contralateral sensory loss (lesion in VP) is caused by interruption of the ascending posterior column–medial lemniscus system and the spinothalamic tract, which synapse in the VPL, as well as the trigeminothalamic fibers, which synapse in the VPM nuclei. Lesions in the VP can also result in thalamic pain syndrome (Dejerine-Roussy syndrome). In this syndrome, which is very rare, pain and temperature sensations are diminished initially but over time return and become highly abnormal. Stimuli that we would consider innocuous or even pleasant, such as a soft touch, can be horribly painful. A pinprick might produce an agonizing or burning sensation. Even the pressure of normal clothing may be perceived as painful. The cause of this syndrome is not known but may be partly the result of altered balance of information flow between the thalamus and the cortex. Treatment with analgesics, anticonvulsants, antidepressants, or phenothiazines, or sodium channel blockers (e.g., the anticonvulsant lamotrigine) generally in combination, may provide some relief. Deep brain stimulation, motor cortex stimulation, and stereotaxic surgery are newer procedures that have been shown to provide some relief in patients with this syndrome.

b. **Output from the cerebellum:** A similar pattern holds for the cerebellum. Output from the cerebellum to the primary motor cortex is relayed via the thalamic motor relay nuclei. Bidirectional communication between the cortex and thalamus then occurs, and feedback is sent back to the cerebellum from the thalamus to modulate cerebellar output.

3. **Limbic relay nuclei:** The limbic system has connections with three major nuclei of the thalamus: the **anterior nucleus**, the **LD** nucleus, and the dorsomedial (**DM) nucleus**. The anterior and LD nuclei are functionally related and sometimes referred to as the anterior–LD complex. The DM nucleus also has significant association functions (see below). All three nuclei are closely tied to limbic system function and regulation.

a. **Anterior and LD nuclei:** The **anterior nucleus** of the thalamus receives its driver input from the ipsilateral **mammillary body** through the **mammillothalamic tract**. The mammillary bodies are part of the limbic system circuitry and are directly linked to the hippocampus through the fornix (see Chapter 20, "The Limbic System"). Information from the anterior nucleus of the thalamus is then relayed to the **cingulate cortex**, the main cortical area dedicated to limbic system function, with some projections to the **prefrontal** and **parietal cortices**. The **LD** nucleus of the thalamus receives its driver input primarily from the **entorhinal cortex**, and similar to the anterior nucleus, projects to the **cingulate**, **prefrontal**, and **parietal cortices** (Figure 14.8).

Together, the anterior and LD nuclei play a role in consolidation of memories, motivation, and direction of attention to a specific stimulus.

b. **Dorsomedial nucleus:** The **DM** has connections with various structures of the limbic system and influences motivation

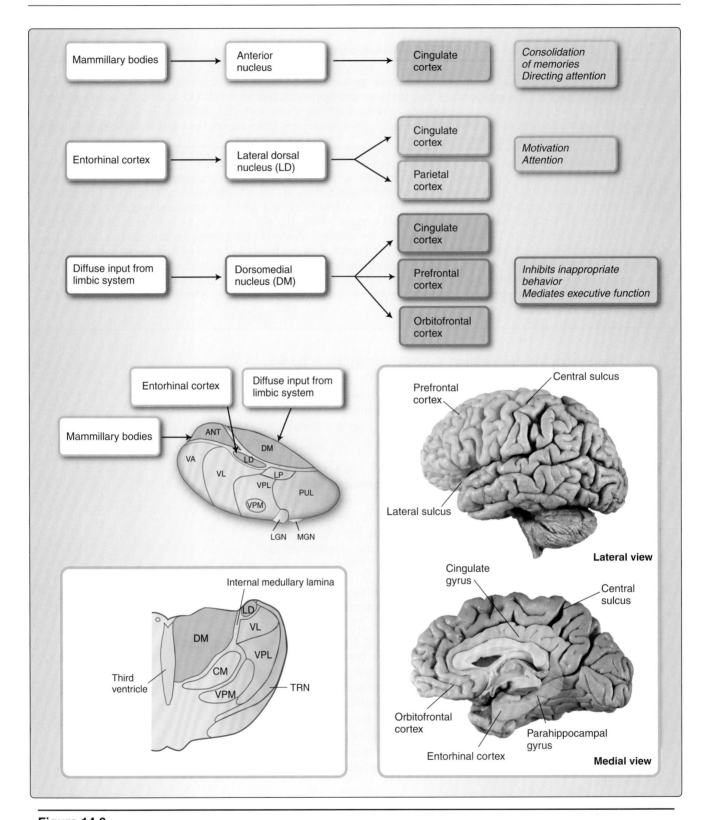

Figure 14.8
The limbic nuclei (anterior, lateral dorsal, and dorsomedial) of the thalamus and their connections.

through connection with the **cingulate gyrus**. It inhibits inappropriate behavior and mediates executive function through connections with the **prefrontal** and **orbitofrontal cortices** (see Figure 14.8).

D. Association nuclei

The **association nuclei** of the thalamus are critical modulators in interactions of one part of the cortex with another. The thalamus "gates" information transferred among cortical areas. In addition, the association nuclei receive input from subcortical structures, such as the basal ganglia, the brainstem reticular formation, brainstem nuclei, and parts of the limbic system. The thalamus uses the information from these subcortical structures to modulate the interaction of cortical areas with each other. These connections from the cortex to the thalamus and back to the cortex (**transthalamic**) exist in addition to direct connections between cortex and cortex (**corticocortical connections**), as shown in Figure 14.9. It appears, however, that the conduction velocity of the transthalamic connections is faster than that of the corticocortical connections. The role of the transthalamic connections could be to enhance the **temporal summation** of the signal to a given cortical area. The thalamus will relay a signal to its target cortical area that consists of integrated information from subcortical structures and other cortical areas. This gives the thalamus a critical role in higher cortical function that goes well beyond that of a simple relay nucleus.

1. **Pulvinar:** The **pulvinar** is the major association nucleus involved in transthalamic (cortico-thalamo-cortical) circuits. It has extensive reciprocal connections with parietal, temporal, frontal, and occipital association areas. It thus has a major role in higher cortical functioning, which is often underappreciated. The **medial pulvinar** appears to be anatomically and functionally connected to the **DM**

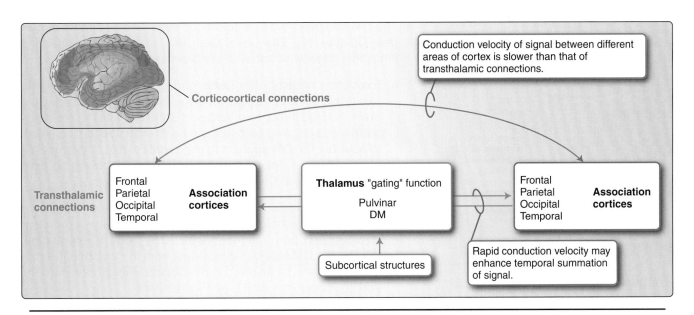

Figure 14.9
The corticocortical and transthalamic association connections. DM = dorsomedial.

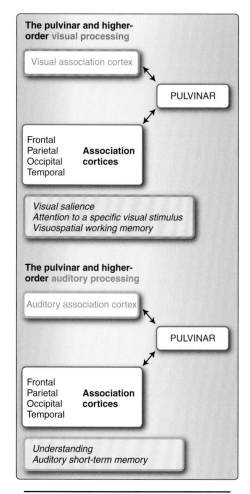

Figure 14.10

The role of the pulvinar in visual and auditory information processing.

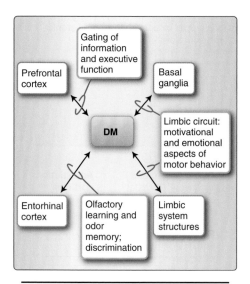

Figure 14.11

The dorsomedial (DM) nucleus of the thalamus and its functional connections.

and has the same connections with the limbic system, as well as the prefrontal and other association cortices (see below).

a. **Visual processing:** One of the main functions of the pulvinar is related to visual processing. The pulvinar receives some inputs directly from the **retina** and more extensive inputs from the **visual association cortex**. It appears to be important in integrating this information with other cortical association areas to allow for evaluation of which particular visual stimulus is the most important (**visual salience**) and to direct attention to a specific visual stimulus. The pulvinar also appears to play a critical role in **visuospatial working memory** (Figure 14.10).

b. **Auditory processing:** Another important function of the pulvinar is related to auditory processing and language. There are reciprocal connections with **auditory association cortex**, especially the **superior temporal gyrus**, which is important in understanding auditory information and auditory short-term memory (see Figure 14.10). Therefore, the pulvinar plays a role in coordinating and engaging the different cortical areas important in language tasks (see Chapter 13, "The Cerebral Cortex," for more information).

2. **Dorsomedial nucleus:** The major role of the **DM** is as an association nucleus. It has extensive reciprocal connections with the **prefrontal cortex** and is critical in the gating of information to and from this area. It also receives afferents from subcortical structures such as the **basal ganglia** and the **amygdala** (part of the limbic system) as well as other cortical areas. Most of the efferents from the DM project to the **prefrontal cortex**, but there are also connections to the **cingulate gyrus** and **entorhinal cortex**, with some connections to motor and sensory areas (Figure 14.11).

The DM also plays a critical role in the basal ganglia **limbic circuit** (see Chapter 16, "The Basal Ganglia"), where motivational and emotional aspects of behavior are processed.

a. **Executive control:** The DM is part of the circuitry underlying **executive control**. Aspects of complex goal-directed behaviors are monitored through connections of the DM with the prefrontal cortex, and ongoing behavior can be coordinated and adjusted. In addition, the DM appears to play a critical role in the process of encoding new information into memory, particularly declarative memory (see Chapter 20, "The Limbic System").

b. **Olfactory processing:** Recent studies have also implicated the DM as the thalamic nucleus where olfactory information is processed and integrated with other cortical information. Indeed, the DM receives direct input from the entorhinal cortex and appears to be involved in olfactory learning, odor memory, and odor discrimination (see Figure 14.11).

Clinical Application 14.2: Lesion to the Dorsomedial Nucleus of the Thalamus

A lesion in the DM nucleus can result in symptoms that resemble damage to the prefrontal cortex. If the lesion is due to a stroke in branches of the posterior cerebral artery that supply DM, patients may undergo a sudden and often dramatic change in personality. They may become lethargic, apathetic, and forgetful and yet be unconcerned about their symptoms. Neuropsychological testing may reveal an impairment of complex executive function including attention and working memory. One patient was reported to be "fatuous, giggling, and treating questions as a joke." They may be histrionic and tearful one moment and joking or smiling the next. They may become depressed or manic and show poor self-care. Symptoms such as apathy, lethargy, and sleepiness may be due to extension of the lesion to the intralaminar nuclei, which play a role in cortical arousal.

E. Intralaminar nuclei

The intralaminar nuclei are collections of neurons within the internal medullary lamina. The most important of these nuclei are the **CM** and the **PF**, which form a functionally related complex (CM/PF).

1. **Arousal function:** The CM/PF complex receives abundant **cholinergic innervation** from the brainstem as well as **dopaminergic input** from the ventral tegmental area. The output to the cortex is diffuse and nonspecific and can influence its overall functioning. The intralaminar nuclei are thought to play a critical role in arousal and in facilitating awareness and vigilance by virtue of their influence on the cortex (Figure 14.12).

2. **Goal-oriented behavior:** The CM/PF also sends projections to the **striatum**. These inputs help prioritize and select information, which is an important component in **facilitating goal-oriented behavior** when attention has been drawn to that goal (see Figure 14.12). Recently, the CM/PF connection to the basal ganglia, which is rich in dopaminergic neurons, has received considerable attention due to its possible role in Parkinson disease (see Chapter 16, "The Basal Ganglia").

F. Thalamic reticular nucleus

The **TRN** is a sheet of neurons surrounding the thalamus (see Figure 14.2). Collaterals from the reciprocal connections between the cortex and the thalamus are sent to the TRN, which, in turn, projects back to the thalamus to the specific area from which its afferent input arose (Figure 14.13). All neurons of the TRN are **GABAergic** and send their inhibitory projections to the thalamus, thereby negatively modulating the excitatory projections between the thalamus and the cortex. Thus, the TRN can be considered the **"gatekeeper of the gatekeeper"** (thalamus).

Through its modulatory connections, the TRN plays a critical role in **selective attention**. It influences the flow of information between the thalamus and the cortex. It also interconnects the thalamic nuclei, which enables them to modulate each other's activity.

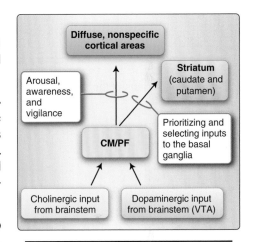

Figure 14.12
The intralaminar nuclei of the thalamus (centromedian [CM] and parafascicular [PF]) and their functional connections. VTA = ventral tegmental area.

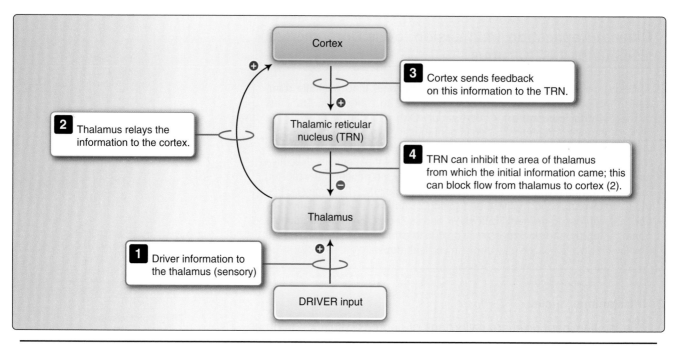

Figure 14.13
The reticular nucleus of the thalamus and its role in controlling access to the thalamus.

G. Consciousness

It has been suggested that activity of thalamus is an important determinant of consciousness, because it ultimately controls access to the cortex. For conscious awareness, the connections between the cortex and the thalamus must be synchronized. Synchronized activity is the mechanism by which the brain interprets events as being related to each other. Synchronization of the activity of the cortex with the activity in the thalamus links incoming signals to cortical activation. Consciousness means that the external and the internal experiences of an event are one or are synchronized in time and space. Through this synchronization, the thalamus creates consciousness or awareness of incoming sensory stimuli and cognitive processes and thus links the external and internal milieu. Some studies suggest that the TRN can be considered a central hub for attention and consciousness due to its modulatory influence on the thalamus and its connections. Intralaminar nuclei are also implicated in conscious awareness due to the influence on cortical arousal.

IV. BLOOD SUPPLY

The blood supply to the thalamus is primarily through branches from the **posterior cerebral, posterior communicating, and basilar arteries**, which supply discrete parts of the thalamus (Figure 14.14). A contribution from the anterior choroidal artery has been described clinically but is highly variable. Even a small vascular lesion in the thalamus can lead to select deficits in thalamic processing and may mimic cortical lesions.

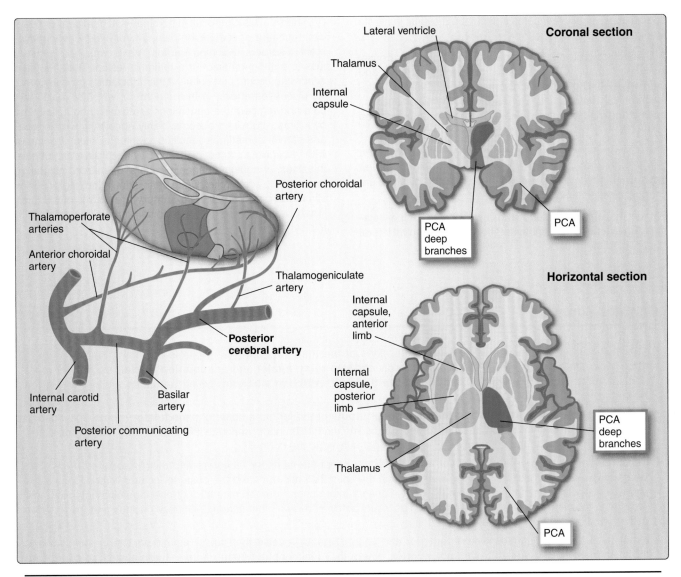

Figure 14.14
Overview of the blood supply to the thalamus through the posterior cerebral artery. PCA = posterior cerebral artery.

Clinical Cases:

Thalamic Stroke

Two patients with a history of thalamic strokes are seen in the stroke clinic. One patient is a 61-year-old man with a lacunar infarction affecting the blood supply to the dorsomedial nucleus of the thalamus. The other patient is a 73-year-old woman with a lacunar infarction affecting the blood supply to the right VPL and VPM nuclei of the thalamus.

Case Discussion

Initially, after his stroke, the 61-year-old man with the infarction to his DM nucleus is sedated, but his level of consciousness improves over the first 48 hours. His cognitive assessment after 1 month shows impairment in orientation as well as delayed recall. His wife states that he minimizes the degree of his memory deficits, he is agitated at times, and he occasionally fabricates stories about events in his life (confabulation).

A bilateral lesion to the DM can result in decreased arousal including coma. Confabulation, impaired learning and memory, including temporal disorientation and loss of autobiographical memory, as well as altered behavior such as apathy and inhibition are typical symptoms seen in a lesion to the DM.

The 73-year-old woman with the infarction to her right VPL and VPM nuclei presented to the clinic with pain in her left face, arm, and leg. Her pain was partly reduced by a sodium channel blocker, lamotrigine.

Strokes that affect the inferior and lateral aspects of the thalamus can extend to the ventral anterior (VA) and ventral lateral (VL) nuclei. This can result in hemiparesis, hemisensory deficits, hemiataxia, poste-lsion pain, and/or hyperkinetic movement disorders including tremor. Ataxia with hemisensory loss is pathognomonic for a thalamic lesion. Thalamic pain syndrome occurs particularly if the right thalamus is involved. As discussed above, in this syndrome, which is quite rare, pain and temperature sensations are diminished initially but over time return and become highly abnormal, with innocuous stimuli sometimes resulting in intense pain. Treatment with analgesics, anticonvulsants, antidepressants, phenothiazines, or sodium channel blockers, often in combination, may provide some relief. Deep brain stimulation, motor cortex stimulation, and stereotaxic surgery are newer procedures that have been shown to provide some relief in patients with this syndrome.

Chapter Summary

- The thalamus is a collection of nuclei on both sides of the third ventricle that controls the flow of information to the cortex. The nuclei of the thalamus can be divided into **relay nuclei** and **association nuclei**, as well as more diffusely projecting **intralaminar nuclei** and the **thalamic reticular nucleus**, which regulates activity within the thalamus.

- The fibers projecting to the thalamus are either **driver inputs** or the far more numerous **modulator inputs**. The specific driver inputs contain information destined for the cortex. The modulator inputs integrate information from many central nervous system areas and then determine which inputs get relayed to the cortex.

- Sensory relay nuclei receive their driver inputs from the peripheral sensory receptors and their associated tracts. The ventral posterolateral nucleus receives input through the spinothalamic tract and the posterior column–medial lemniscus pathway with sensory information from the body. Sensory information from the face through the trigeminothalamic tracts projects to the ventral posteromedial nucleus. The visual pathway relays in the lateral geniculate nucleus and the auditory pathway in the medial geniculate nucleus.

- Motor nuclei of the thalamus (ventral anterior and ventral lateral) receive their driver input from the cerebellum and the basal ganglia and project to the motor cortex, where they influence the upper motor neuron system to coordinate movement.

- The limbic nuclei of the thalamus include the anterior and lateral dorsal nuclei, which are primarily relay nuclei, and the dorsomedial nucleus, which has both relay and association functions. Through these nuclei, limbic functions are integrated and coordinated. The limbic system also interfaces with other circuits, which allows motivational and emotive processing to influence motor output and cognition.

- The association nuclei connect areas of cortex with each other and integrate the information from many cortical areas.

- The intralaminar nuclei receive cholinergic and dopaminergic input from the brainstem and project diffusely to the entire cortex, giving them a critical role in arousal and awareness. They also interface with the basal ganglia circuits where they facilitate goal-oriented behavior.

- The thalamic reticular nucleus (TRN) is a sheet of neurons that surrounds the thalamus. It sends inhibitory projections to the specific nuclei of the thalamus related to activity in their respective areas of cortex. The TRN plays a critical role in consciousness as it synchronizes the activity in the thalamus coming from the driver input with the activity in the cortex. Only when these patterns are synchronized do we become conscious of a stimulus.

- The posterior cerebral and posterior communicating arteries supply blood to the thalamus. There is also a variable supply from the anterior choroidal artery.

Study Questions

Choose the ONE best answer.

14.1 Which one of these statements about the thalamus is true?

A. The ventral posteromedial nucleus receives input from the spinothalamic tract.

B. The ventral posterolateral nucleus receives input from the medial lemniscus.

C. The ventral anterior nuclei receive input from the limbic system.

D. The dorsomedial nucleus receives input from the cerebellum.

E. The medial geniculate nucleus receives information from the retina.

Correct answer is B. The ventral posterolateral nucleus receives input from the posterior column–medial lemniscus system and the spinothalamic tracts concerning discriminative touch, proprioception, and vibration sense as well as pain and temperature. The ventral posteromedial nucleus receives input from the trigeminothalamic tracts. The ventral anterior nucleus receives input from the cerebellum and basal ganglia. The dorsomedial nucleus is an association nucleus closely related to the limbic system. The medial geniculate nucleus receives information from the cochlea for hearing.

14.2 Which one of these statements about the thalamus is correct?

A. Tonic firing allows for linear transfer of information.

B. Driver inputs to the thalamic nuclei outnumber modulator inputs.

C. Burst firing is only observed during sleep.

D. The thalamic reticular nucleus receives monoaminergic afferents from the brainstem.

E. Thalamic relay nuclei receive specific driver input but do not receive nonspecific modulator input.

Correct answer is A. Tonic firing allows for direct and linear transfer of information, whereas burst firing will only relay new and changing information to the cortex. Modulator inputs outnumber driver inputs and are critical in gating or prioritizing information sent to the cortex. Although it is true that burst firing is seen during sleep, it is also found during wakefulness and appears to be important in detecting new stimuli. The thalamic reticular nucleus receives afferents from the cortex and projects to the thalamus. The monoaminergic projections from the brainstem go to association nuclei and, more importantly, to the intralaminar nuclei. Relay nuclei receive both specific driver input and nonspecific modulator input. Together, these inputs enable the thalamus to "decide" which information will be relayed to the cortex.

14.3 A patient presents to the emergency room. She complained that she was at home and noticed that the left side of her face started to feel numb, and within a short time, the whole left side of her body felt numb. When she tried to walk to the phone to call for help, she felt very unsteady and almost fell. When the emergency room physician examined the patient, she found that pinprick and vibration sense were absent on the left side of her face and body. In addition, the patient showed marked problems with hand–eye coordination. The most likely site of a lesion that would produce these symptoms is:

A. Infarct involving the lenticulostriate arteries.

B. Infarct involving the anterior choroidal artery on the left.

C. Infarct involving branches of the posterior cerebral artery that supply the right midbrain tegmentum and cerebral peduncle.

D. Infarct involving branches of the posterior cerebral and posterior communicating arteries on the right.

E. Superficial branches of the middle cerebral artery on the right supplying the somatosensory and motor cortices.

Correct answer is D. Infarct of branches of the posterior cerebral and posterior communicating arteries on the right would damage the ventral tier thalamic nuclei, ventral anterior (VA), ventral lateral (VL), and ventral posterior (VP). This would result in motor signs similar to what would be seen with cerebellar damage, including ataxia and dysmetria. Damage to VP (ventral posterolateral and ventral posteromedial) would result in decreased sensations of touch, vibration, pain, and temperature on the face and body contralaterally. The lenticulostriate arteries supply the lenticular nucleus, resulting in deficits in movement contralateral to the lesion. However, the motor signs observed here do not reflect basal ganglia problems, and there would be no sensory involvement with an infarct of the lenticulostriate artery. An infarct of the anterior choroidal artery on the left would affect the inferior two-thirds of the internal capsule on the left, likely resulting in upper motor neuron (UMN) signs and decreased perception of sensations on the right side of the body. There might or might not be some minor effects on the left thalamus, resulting in sensory or motor changes, but symptoms would be seen on right side of the body. An infarct of the posterior cerebral artery in the right midbrain could result in sensory loss on the left. However, a lesion of the cerebral peduncle would cause a contralateral UMN sign. An infarct of the middle cerebral artery supplying the motor and sensory cortices on the right would cause sensory loss and signs of a UMN lesion on the left. It would not cause dysmetria and ataxia.

14.4 What clinical features could you expect to see in a 73-year-old woman with a lesion to her VPL and VPM?

   A. Change in personality.

   B. Visual field deficit.

   C. Thalamic pain syndrome (poststroke pain).

   D. Loss of visual salience.

   E. Impaired executive function.

> The correct answer is C. Lesions to the VPL and VPM areas can impair sensory processing and could ultimately result in modulation of pain input. A change in personality is a common feature of lesions involving the medial aspects of the thalamus (e.g., DM). A lesion of the LGN would cause a visual field deficit. Visual salience is modulated by the pulvinar. The DM nucleus is also part of the circuitry underlying executive function.

14.5 A lesion of the pulvinar nucleus is likely to result in

   A. Inability to direct attention to a specific visual stimulus.

   B. Problems with goal-directed behavior and declarative memory.

   C. Lethargy, sleepiness, and lack of awareness and vigilance.

   D. Diminished hearing in the contralateral ear.

   E. Problems with selective attention.

> Correct answer is A. The pulvinar receives input from the visual association areas of the cortex and plays a role in the ability to attend to specific or important visual stimuli. The pulvinar also receives input from the auditory association cortex, and a lesion in the pulvinar could disrupt the ability to interpret what is heard or in language. However, it would not result in deafness. Problems with goal-directed behavior and declarative memory result from lesions to the dorsomedial nucleus. Lethargy, sleepiness, and suppressed levels of arousal result from a lesion in the intralaminar nuclei. Problems with selective attention would most likely result from a lesion in the reticular nucleus, which also coordinates activity among the various thalamic nuclei.

The Visual System

15

I. OVERVIEW

The visual system is one of our most important sensory systems. In this chapter, we explore the functional anatomy of this intricate system that allows for the visual perception of our surroundings, including color, form, and motion in both bright and dim lighting conditions. We look at the optical system of the eye (Figure 15.1), where the incoming light is focused and projected onto the retina, which is the neural cell layer that contains the photoreceptor cells at the back of the eye. We explore phototransduction in the retina where the first steps of information processing occur and then follow the fibers that arise from the retina and project to the lateral geniculate nucleus (LGN) of the thalamus and, from there, to the primary visual or calcarine cortex. Finally, we discuss how visual processing occurs in several different parallel streams, which consolidate into

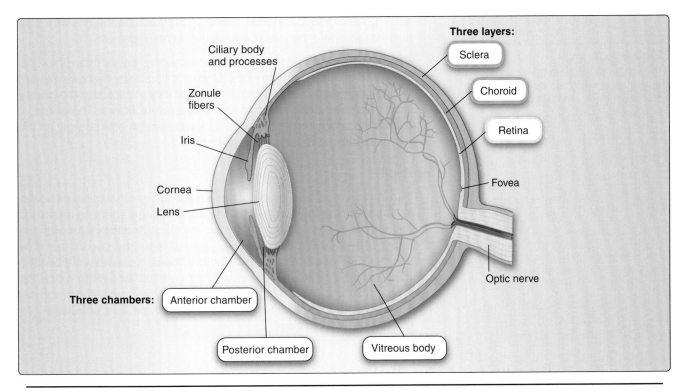

Figure 15.1
Overview of the eye.

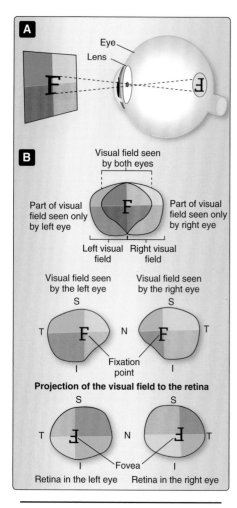

Figure 15.2
Refraction of light and the projection of the visual field onto the retina. S = superior; T = temporal; I = inferior; N = nasal.

dorsal (superior) and ventral (inferior) streams, allowing for simultaneous extraction of information and identification of stimuli in our visual field, to perceive motion, and to orient ourselves in our surroundings.

II. THE EYE

The eye is a three-layered and three-chambered structure. The innermost neuronal layer is the **retina**, which contains the light-sensitive photoreceptor cells. The middle layer contains three structures: the **uveal tract**, which contains the **choroid**, a vascular layer containing the blood vessels that supply the retina; the **ciliary body**, which can adjust the shape and, thus, the refraction of the lens; and the **iris**, which can adjust the aperture of the pupil. The outermost layer is the **sclera**, a tough connective tissue layer that at the front of the eye is specialized to form the transparent **cornea** (see Figure 15.1).

A. Cornea

Light is **refracted**, or bent, as it travels through the cornea. Light then crosses the three chambers. In the **anterior chamber**, light beams are bundled as they go through the aperture of the iris. Light then goes through the **posterior chamber** and the lens where it is refracted further and finally passes through the **vitreous body** (in the third chamber) to reach the retina. Because of the refraction of light through the cornea and lens, the projection of the visual fields (area of environment that is perceived by each eye) onto the retina is reversed and inverted: the upper visual fields project onto the lower part of the retina; the lower visual fields project onto the upper part of the retina; the lateral visual fields project onto the medial side of the retina; and the medial visual fields project onto the lateral side of the retina (Figure 15.2).

Interestingly, the cornea has greater refractive power than the lens. However, the lens has the advantage that its shape can be changed through contraction or relaxation of the **ciliary muscles** in the **ciliary body**, which is attached to the lens by **zonule fibers**. The ability of the lens to change shape in order to focus near objects sharply on the retina is referred to as **accommodation**.

B. Pupil

The iris is specialized to adjust the amount of light that falls on the retina by changing the diameter of the pupil. A large aperture increases the amount of stray light that falls on the retina and interferes with the depth of field and focus, whereas a small aperture eliminates stray light and increases focus and depth of field. However, in the dark, the pupil must dilate to allow a sufficient amount of light to reach the retina. Although our visual system works relatively well in dim lighting conditions, it is best suited for daylight or bright light conditions, as discussed further below.

III. RETINA

The **retina** is a multilayered structure at the back of the eye that contains the photoreceptor cells. The retina is the first site of processing of visual information. The retina is capable of adapting to different lighting conditions and can sharpen the image perceived in time and space through the use of different types of ganglion cells that respond to different types of visual information. The intricate structure of the retina and the presence of many types and subtypes of neurons, however, suggest that processing in the retina goes beyond the perception of an image that is then conveyed to the cortex for more detailed analysis. It appears that some of the processing that was always thought of as being a cortical function actually occurs within the retina. The retina participates in the process of sorting and reducing the amount of information the receptor cells receive to focus on the most important features in our visual environment.

A. Layers of the retina

We can think of the retina as being organized around a vertical chain of neuronal cells consisting of photoreceptor cells (rods and cones) and two types of neurons, bipolar cells and ganglion cells. In addition, a horizontal system of **interneurons** (horizontal and amacrine cells) integrates information at the synapses between the photoreceptors and the bipolar cells and between the bipolar cells and the ganglion cells. There are also supporting modified glial cells (Müller cells), and finally, there are pigment epithelial cells. Light falls onto the photoreceptor cells, either **rods** or **cones**, which then transmit information to **bipolar cells**, which in turn contact the **retinal ganglion cells**. The axons of these ganglion cells comprise the **optic nerve**. The **horizontal cells** and the **amacrine cells**, which provide horizontal integration of information, can influence the signal at a particular synapse based on activity in other synapses and, thereby, allow for better resolution of the signal.

The membranes and cells of the retina and their processes are arranged in 10 layers (Figure 15.3). Light must pass through all of these layers of cells to reach the photoreceptors, which are located in the innermost layer of the retina. In this innermost layer, light is transformed into an electrical signal, which is then transmitted back through a two-neuron chain (bipolar cells and ganglion cells). The signal thus travels back through all layers of the retina.

1. **Pigment epithelial layer:** The outermost layer is the **pigment epithelium**, which separates the choroid layer from the neuronal layer of the retina. The pigment epithelium sits on **Bruch membrane**, a very dynamic basement membrane that allows access of nutrients to the neurons in the retina. The pigment epithelial cells contain melanin, which absorbs and helps prevent the scatter of light as it passes through the retina (see Figure 15.3).

2. **Photoreceptor layer:** The **layer of rods and cones** contains the outer segments of the photoreceptor cells and is the layer where **phototransduction** takes place. Light activates a signaling cascade within the individual receptor cells. The distribution of cells in this layer will vary depending on the location within the retina. In the fovea, for example, only cones are found, which allows for very high visual acuity (see Figure 15.3).

3. **Outer limiting membrane:** The external ends of the Müller cells form the outer limiting membrane, through which photosensitive processes of rods and cones pass to connect with their cell bodies (see Figure 15.3).

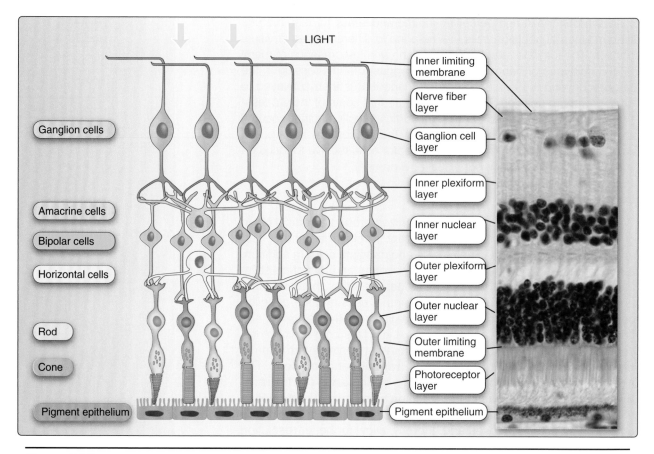

Figure 15.3
The 10 layers of the retina. Note: Müller cells not shown.

4. **Outer nuclear layer:** This layer contains the cell bodies of the rods and cones (see Figure 15.3).

5. **Outer plexiform layer:** This is a synaptic layer, where the rods and cones make contact with the **bipolar cells**. Bipolar cells will integrate the information from several photoreceptor cells. Horizontal integration of information is achieved in this layer through the processes of the **horizontal cells**, which span multiple synaptic terminals between multiple cells (see Figure 15.3).

6. **Inner nuclear layer:** The inner nuclear layer contains the cell bodies of the bipolar cells. Interspersed among the bipolar cells are cell bodies of the horizontal cells and the other major interneuron type, the **amacrine cells**. Cell bodies of the **Müller cells**, the major supporting cells in the retina, are also located primarily in the inner nuclear layer (see Figure 15.3; note: Müller cells not shown).

7. **Inner plexiform layer:** This too is a synaptic layer, similar to the outer plexiform layer, where synaptic contacts between the bipolar cells and the ganglion cells occur. These contacts are also integrated horizontally, similar to what occurs in the outer plexiform layer, but here, this occurs through the processes of the amacrine cells (see Figure 15.3).

8. **Ganglion cell layer:** This layer contains the cell bodies of the ganglion cells (see Figure 15.3).

9. **Nerve fiber layer:** In the nerve fiber layer, axons from the ganglion cells gather and travel to the **optic disc** or **optic papilla**, where they form the optic nerve. The optic disc lacks photoreceptors and is the **blind spot** in our field of view. The axons of the ganglion cells are unmyelinated until they emerge from the eye. Because myelin is highly refractile, it is an advantage to have unmyelinated axons within the retina to reduce the amount of scattered light (see Figure 15.3).

10. **Inner limiting membrane:** The apical ends of the Müller cells form the inner limiting membrane. This membrane separates the retina from the vitreous body (see Figure 15.3).

B. Photoreceptor cells

There are two types of photoreceptor cells: rods and cones. The functions of these two types of receptors enable us to see under both bright and dim lighting conditions. The heterogeneity of the cones allows us to distinguish color.

Each photoreceptor cell has an **outer segment** where the detection of light occurs. This outer segment is embedded in the pigment epithelium. The outer segments of rods and cones differ in their morphology, but each contains discs that contain a vitamin A–linked photopigment (rhodopsin in rods, iodopsin in cones). Activation of this photopigment by the absorption of light initiates the signal transduction cascade. The inner segment is where the mitochondria, which provide the energy for the cell, and the nucleus are located. The rods and cones synapse with the bipolar and horizontal cells.

1. **Rods:** Rods are extremely sensitive to light and allow us to see in dim lighting conditions, such as twilight or moonlight, which is referred to as **scotopic vision**. Their extreme sensitivity to light enables them to detect movement, that is, changing light patterns, in the periphery of the visual field. Within the retina, many rods converge onto a single bipolar cell, and many bipolar cells contact a given amacrine cell. This arrangement allows for sensitivity but results in a loss of spatial resolution.

   The photopigment in rods is **rhodopsin**, which responds to light in the same way at all frequencies and does not allow us to distinguish color.

   The outer segment of rods is longer than that of cones and is subdivided into about 1,000 free-floating discs that contain the light-sensitive rhodopsin (Figure 15.4).

2. **Cones:** Cones are specialized for visual acuity and color vision. They are relatively insensitive to light, and more than 100 times the amount of light needed to activate a rod is needed to elicit a response in a single cone cell. Within the retina, there is much less convergence of cones onto bipolar cells. Indeed, in the area of highest visual acuity, the fovea, the transduction from cone to bipolar cell is 1:1. These receptors allow us to see with high visual acuity in daylight or bright light conditions, which is **photopic vision**. The outer segment of cones containing the photopigment is shorter than that of rods and consists of a series of discs that are connected to the cone cell membrane (Figure 15.5A).

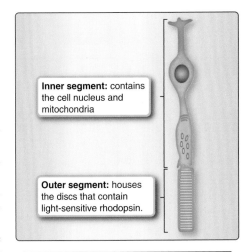

Inner segment: contains the cell nucleus and mitochondria

Outer segment: houses the discs that contain light-sensitive rhodopsin.

Figure 15.4
Structure of a rod photoreceptor.

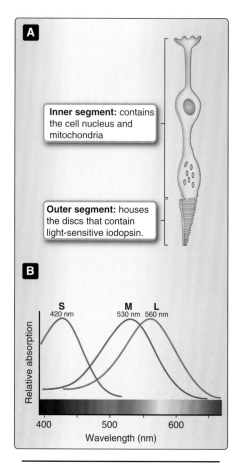

Figure 15.5
Structure of a cone photoreceptor and the spectra absorbed by cones.

Cones can be subdivided into three different populations that contain different versions of the photopigment **iodopsin**. The various iodopsins allow for the detection of light at different frequencies and, thus, for color vision (see Figure 15.5B). Together, these cones give us three channels for color vision (red, green, and blue), which allow us to detect the millions of colors visible to the human eye (see below). Some heterogeneity of the cone cell population has been described resulting in an altered ability to see colors. For example, most people who are color impaired have only two functioning types of cone cells. By contrast, the possibility of a fourth type of cone cell, which would result in greatly enhanced color vision, is being investigated.

a. **L-type cones:** L-type cones are the most abundant type of cone in the retina and detect low-frequency light waves, with maximal absorption in the red spectrum (λ_{max} = 555–565 nm).

b. **M-type cones:** M-type cones detect middle-frequency light waves and have a maximal absorption in the green spectrum (λ_{max} = 530–537 nm).

c. **S-type cones:** S-type cones detect suprafrequency light waves and have their maximal absorption in the blue spectrum (λ_{max} = 415–430 nm). S-type cones constitute only about 5% of all cones in the retina and are located mostly in the periphery of the retina.

3. **Distribution of rods and cones in the retina:** Rods and cones are not distributed uniformly throughout the retina. The fovea contains only cones, which allows for high visual acuity in daylight conditions. The fovea is not very effective in dim light, but the peripheral areas rich in rods are more sensitive to low light conditions. Figure 15.6 shows the distribution of rods and cones in the retina.

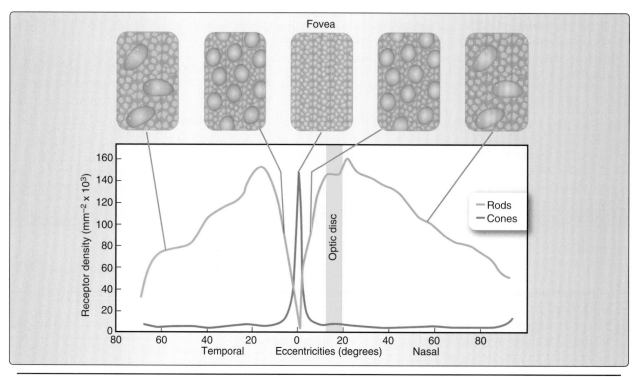

Figure 15.6
Distribution of rods and cones in the retina.

C. Retinal specialization

Specialized areas can be identified within the retina. Nerve fibers, or axons of the ganglion cells, converge to form the optic nerve and leave the eye in the **optic disc**, or **papilla**. This is the so-called **blind spot** in our visual field, because there are no photoreceptors in this area. The **macula lutea**, usually referred to simply as the **macula**, is a circular area near the lateral edge of the optic disc, in which many of the cells contain a yellow pigment (*macula lutea* is Latin for "yellow spot"). In the center of the macula is a depression about 1.5 mm in diameter, which is the **fovea** (Figure 15.7). This area is rich in cones and is, therefore, the area of highest visual acuity, to which most of the light from the center of our visual field projects. In the fovea, all cell layers of the retina are pushed aside, so that light can project directly onto the photoreceptor cones.

D. Phototransduction

The principles of the **phototransduction cascade** (Figure 15.8) are similar in rods and cones: In both receptor cells, the cascade begins with light that is absorbed by the **vitamin A–linked photopigment (opsin)**. The vitamin A derivative 11-cis-retinal undergoes a conformational change to all-trans-retinal, which induces a conformational change of the opsin molecule into its activated state. Each activated opsin then activates several molecules of the **G protein transducin**. The activated transducin molecule then activates a phosphodiesterase, which, in turn, hydrolyzes cyclic guanosine monophosphate (cGMP) to GMP. cGMP opens Na^+ channels, whereas GMP closes Na^+ channels. As the phototransduction cascade induces the ratio of cGMP:GMP to shift in favor of GMP, more and more Na^+ channels close, and the cells hyperpolarize. This enzyme cascade allows for significant amplification of the signal.

This sensory transduction system is unique in that a stimulus (i.e., light) causes the **hyperpolarization** of cells and less neurotransmitter (glutamate) release. In darkness, or during the absence of the stimulus, the cells are depolarized and constantly release neurotransmitter in what is known as the "**dark current**." Light suppresses this dark current.

The activated opsins must then be deactivated rapidly so that the cell can be ready for the next stimulus. This occurs through a series of quenching reactions that bring the photopigments back to their resting, dark-adapted state.

IV. VISUAL PATHWAY

From the photoreceptor cell, the visual pathway consists of a three-neuron chain that processes visual information and conveys it to the cortex. The first two neurons in the chain are in the retina: the bipolar cells and the retinal ganglion cells. From the retina, the visual pathway projects to the third neuron, which is located in the **LGN of the thalamus**. Axons from the thalamus project via the optic radiations to the

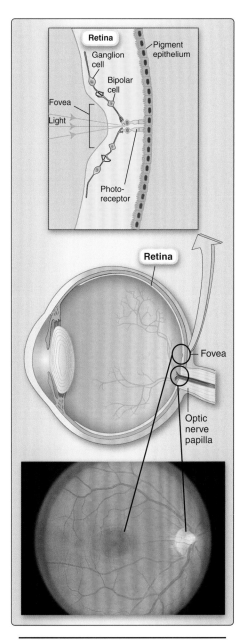

Figure 15.7
Overview of the retina showing the fovea and the papilla.

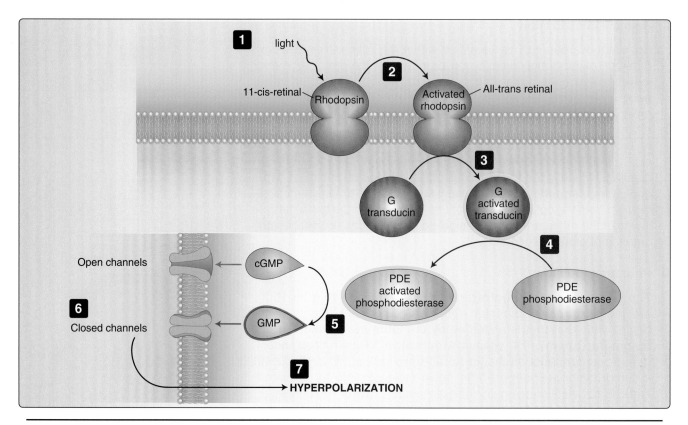

Figure 15.8
Phototransduction. cGMP = cyclic guanosine monophosphate; GMP = guanosine monophosphate.

primary visual (striate) cortex. The normal visual pathway is described below. Specific lesions of the visual pathway are shown in Clinical Application Box 15.1.

A. Bipolar cells

Bipolar cells have synaptic contact with rods and cones in the outer plexiform layer of the retina. Some bipolar cells are contacted only by rods, others only by cones, and some by both rods and cones. As discussed above, photoreceptor cells are constantly releasing the neurotransmitter glutamate in the dark (dark current), and light results in the hyperpolarization of the rods and cones, leading to less glutamate release. The synapse between photoreceptors and bipolar cells must, therefore, be adapted to this constant release of glutamate and is called a **ribbon synapse**. Ribbon synapses can transmit information tonically and in a graded fashion. In conventional synapses, transmission is achieved through modulating the frequency of **action potentials (APs)**, but in the visual pathway, the dynamic range of signals is too broad (ranging from low light conditions that stimulate rods to bright light conditions that stimulate cones) and cannot be encoded in different AP frequencies. The stimulus intensity at the ribbon synapse is encoded by the changes in transmitter release: more light results in less transmitter release.

The bipolar cells can respond to these changes in two ways: (1) a decrease in glutamate at the synapse (which occurs in the light) can

Clinical Application 15.1: Lesions to the Visual Pathway

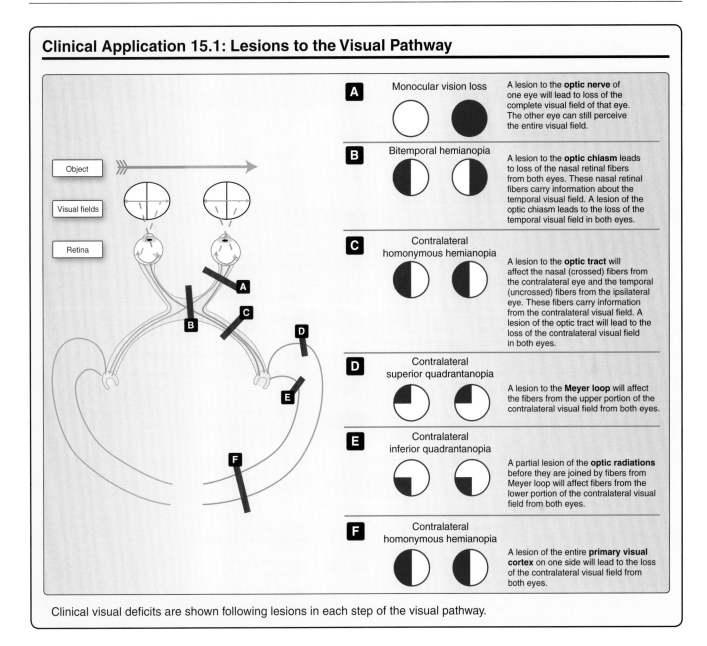

A Monocular vision loss

A lesion to the **optic nerve** of one eye will lead to loss of the complete visual field of that eye. The other eye can still perceive the entire visual field.

B Bitemporal hemianopia

A lesion to the **optic chiasm** leads to loss of the nasal retinal fibers from both eyes. These nasal retinal fibers carry information about the temporal visual field. A lesion of the optic chiasm leads to the loss of the temporal visual field in both eyes.

C Contralateral homonymous hemianopia

A lesion to the **optic tract** will affect the nasal (crossed) fibers from the contralateral eye and the temporal (uncrossed) fibers from the ipsilateral eye. These fibers carry information from the contralateral visual field. A lesion of the optic tract will lead to the loss of the contralateral visual field in both eyes.

D Contralateral superior quadrantanopia

A lesion to the **Meyer loop** will affect the fibers from the upper portion of the contralateral visual field from both eyes.

E Contralateral inferior quadrantanopia

A partial lesion of the **optic radiations** before they are joined by fibers from Meyer loop will affect fibers from the lower portion of the contralateral visual field from both eyes.

F Contralateral homonymous hemianopia

A lesion of the entire **primary visual cortex** on one side will lead to the loss of the contralateral visual field from both eyes.

Clinical visual deficits are shown following lesions in each step of the visual pathway.

result in a depolarization of the bipolar cell (known as an **ON bipolar cell**), or (2) it can result in a hyperpolarization of the bipolar cell (known as an **OFF bipolar cell**) (Figure 15.9). In the dark, when there is more neurotransmitter release (dark current), the ON bipolar cell hyperpolarizes and the OFF bipolar cell depolarizes.

These responses are achieved through the use of different types of postsynaptic glutamate receptors.

1. **OFF bipolar cells:** This type of cell expresses **ionotropic glutamate receptors**. In the light, less glutamate in the synaptic cleft leads to less cation conductance through the ion channels, and the OFF bipolar cell hyperpolarizes (see Figure 15.9). In the dark, more glutamate in the synaptic cleft leads to more cation conductance, and the OFF bipolar cell depolarizes. The OFF bipolar cell signals in the dark.

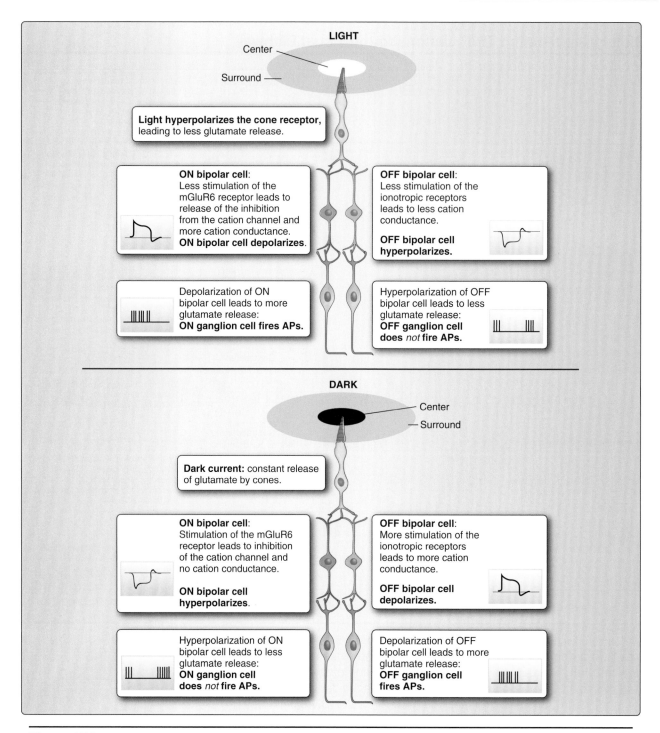

Figure 15.9
The physiology of ON and OFF bipolar cells. AP = action potential.

2. **ON bipolar cells:** These cells express the **metabotropic glutamate receptor** mGluR6 that is negatively coupled to a cation channel. When glutamate binds to this receptor, the ion channel is blocked, and the cell is hyperpolarized. In the dark, more glutamate leads to inhibition of the cation channel and hyperpolarization of the ON bipolar cell. In the light, less glutamate leads to less stimulation of these metabotropic receptors, which releases the

inhibition of the cation channels and results in more cation conductance and depolarization of the ON bipolar cell (see Figure 15.9). The ON bipolar cell signals in the light.

3. **Parallel processing:** These two types of bipolar cells are the first step of a parallel processing system that can respond rapidly and specifically to increases in light (ON bipolar cell, which depolarizes in light) and decreases in light (OFF bipolar cell, which depolarizes in darkness).

4. **Spatial distribution:** In addition, these cells are organized in a spatial distribution that feeds the signal into the ganglion cell layer, where ON and OFF ganglion cells can be found. The ON ganglion cells receive input from a receptive field of ON bipolar cells. The ganglion cells will fire more rapidly when the light source comes from the center of their receptive fields as opposed to the periphery ("surround"). The OFF ganglion cells similarly receive input from a receptive field of OFF bipolar cells and will fire more rapidly when input is received from the center rather than the surround of their (dark) receptive fields. Together, these two pathways increase the contrast of what we see, and this ON/OFF organization is present throughout the entire visual system.

B. Ganglion cells

The **ganglion cell** is the second neuron in the chain. The heterogenous population of these cells includes up to 17 different types of ganglion cells, each of which encodes different aspects of the visual information.

Here, we will focus on the three ganglion cell types that appear to play a prominent role throughout the visual pathway and are the sources of three processing pathways that operate in parallel to each other. In each of these pathways, ON and OFF ganglion cells can be found. This enhances contrast and enables us to react to increasing and decreasing levels of light. There are likely more parallel circuits processing visual information in the visual pathway, but for the scope of this chapter, we focus on the three major pathways that help us understand the visual analysis of motion, form, and color (Figure 15.10).

1. **Midget ganglion cells:** These comprise the majority (greater than 70%) of cells that project to the LGN, and they are the origin of the **parvocellular pathway**. These cells encode for a high spatial but low temporal resolution because of their small receptive fields and slow axonal conduction velocity. These cells analyze inputs from the red and green channels and are an important component in color vision (see Figure 15.10). Due to their poor temporal resolution, midget ganglion cells are well suited for analysis of static images but not for analysis of motion.

2. **Parasol ganglion cells:** These cells make up only about 10% of all cells projecting to the LGN and are the origin of the **magnocellular pathway**. These cells encode for a low spatial but high temporal resolution because of their large receptive fields and fast axonal conduction velocity. These cells encode for an achromatic (colorless) signal and are best suited for detecting contrast, motion, and shape (see Figure 15.10).

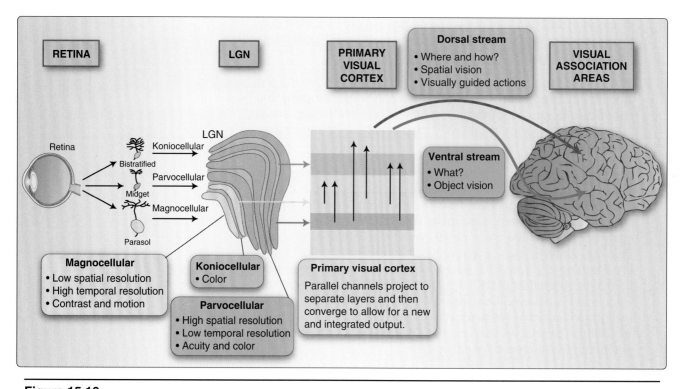

Figure 15.10
The parallel pathways and streams of the visual system from the retina to the cortex. LGN = lateral geniculate nucleus of the thalamus.

3. Bistratified ganglion cells: These comprise about 8% of all cells projecting to the LGN and are the origin of the **koniocellular pathway**. They project to their own koniocellular layer in the LGN and are scattered throughout the other layers of the LGN dedicated to the parvocellular and magnocellular pathways. These cells analyze blue and yellow color signals (see Figure 15.10 and below "Color Vision").

C. Optic chiasm and optic tract

As discussed in previous chapters, many aspects of brain function are lateralized, and many pathways cross the midline. The same principle applies to the visual pathway. In this case, the **visual field** (area of the environment) perceived by each eye is lateralized in the cortex. That is, the right visual field projects to the left primary visual cortex, and the left visual field projects to the right primary visual cortex.

Although each eye receives input from both the left and right visual fields, the projection of the visual fields onto each retina is reversed and inverted. Here, let us consider just the lateral and medial visual fields (upper and lower visual fields discussed below). The lateral (temporal) visual field projects onto the medial (nasal side) of each retina, and the medial visual field projects onto the lateral side of each retina. Thus, the medial portions of the retina receive input from the *ipsilateral* visual field, whereas the lateral portions of the retina receive input from the *contralateral* visual field (Figure 15.11). Fibers from the medial or nasal halves of the retina cross the midline in the optic chiasm on the inferior surface of the brain, whereas fibers from the temporal or lateral portions of the retina do not cross over. This partial decussation of fibers means that axons traveling in the **optic tract** to the LGN contain

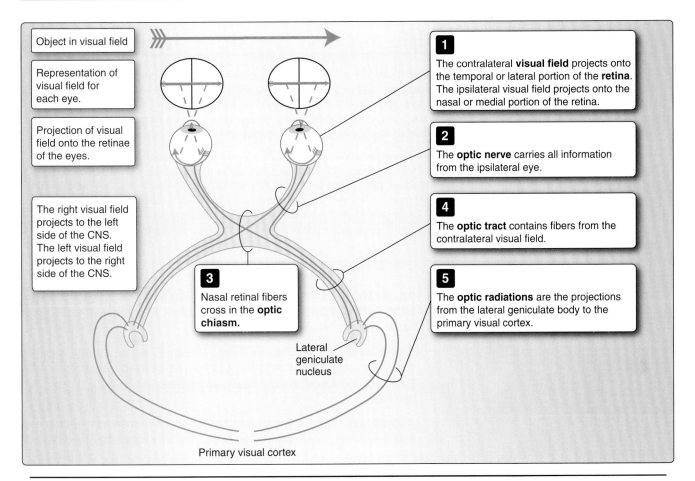

Object in visual field

Representation of visual field for each eye.

Projection of visual field onto the retinae of the eyes.

The right visual field projects to the left side of the CNS. The left visual field projects to the right side of the CNS.

1
The contralateral **visual field** projects onto the temporal or lateral portion of the **retina**. The ipsilateral visual field projects onto the nasal or medial portion of the retina.

2
The **optic nerve** carries all information from the ipsilateral eye.

4
The **optic tract** contains fibers from the contralateral visual field.

5
The **optic radiations** are the projections from the lateral geniculate body to the primary visual cortex.

3
Nasal retinal fibers cross in the **optic chiasm.**

Lateral geniculate nucleus

Primary visual cortex

Figure 15.11
The organization of fibers in the visual system.

information from the contralateral visual field. The *right* optic tract, for example, carries information that originated in the *left half* of the visual fields of both the left and the right eyes. Thus, each side of the brain ultimately deals with the contralateral visual field.

D. Lateral geniculate nuclei

Each LGN receives information from the contralateral visual field from all types of ganglion cells. The different types of information, or the different channels described above, project to the LGN in parallel and are segregated into the layers of the LGN. The magnocellular pathway will project to the **magnocellular layers**, the parvocellular pathway to the **parvocellular layers**, and the koniocellular pathway to the interspersed **koniocellular layers** (as well as to both the magnocellular and parvocellular layers [see Figure 15.10]).

In addition, all fibers from the retina maintain their **retinotopic organization**. By the time the fibers reach the LGN, the visual fields are lateralized. The left visual field projects to the right side of the brain and vice versa (see above). The organization of fibers in the LGN reflects the visual field, and the projection into different layers accounts for motion, contrast, color, and form.

Several sets of fibers bypass the LGN. As discussed in Chapter 9, "Control of Eye Movements," some axons from the visual pathway

Clinical Application 15.2: Blindsight

Blindsight is a phenomenon that can be observed in patients with damage to their primary visual cortex. These patients consider themselves blind and there is a complete lack of visual awareness, but a residual visual function persists and allows for nonconscious visual perception. Shape, movement, and even color of objects in the visual space can be perceived and result in behavioral consequences. These patients may avoid obstacles in their pathway, direct their gaze through saccadic eye movements to objects in the visual space, or have an emotional response to stimuli of which they have no visual awareness.

The underlying neurobiology can be traced back to connections from the LGN and the pulvinar to areas of the cortex other than the primary visual cortex, such as areas of the parietal and temporal association areas, which are part of the dorsal and ventral streams of visual processing. In addition, projections from the LGN to the superior colliculus can account for processing of visual space without visual awareness. In a classical, hierarchical understanding of the visual system, the primary visual cortex is critical for all visual processing; a more modern understanding sees the primary visual cortex as a hub with reciprocal connections to the thalamus and various association areas, which together create our conscious visual awareness.

bypass the LGN and project directly to the superior colliculus via the superior brachium. In the superior colliculus, they are involved in saccadic eye movements.

As discussed in Chapter 14, "The Thalamus," some axons from the visual pathway that bypass the LGN synapse in the pulvinar. There, they help modulate cortical association areas, mostly visual association areas.

Another set of fibers bypasses the LGN to synapse in the pretectal nucleus, where they are involved in the pupillary light reflex (see below).

E. Optic radiations

From the LGN, axons project to the primary visual cortex via the **optic radiations**. The fibers in the optic radiations are organized in a retinotopic fashion and in the various channels discussed above. The optic radiations on each side contain information from the contralateral visual field, which can be further divided into an upper and lower visual field. The fibers from the upper and lower visual fields take different routes toward the primary visual cortex (Figure 15.12).

1. **Lower visual field:** The lower half of the visual field projects to the upper half of the retina and then to the LGN. From the LGN, these fibers project in the superior part of the optic radiation, which travels in the parietal lobe, to the area of the calcarine cortex superior to the calcarine sulcus.

2. **Upper visual field:** In contrast, the upper half of the visual field projects to the lower half of the retina. From the LGN, these fibers project to the area of the calcarine cortex inferior to the calcarine sulcus. In order to get there, the fibers must swing around the inferior horn of the lateral ventricle in the temporal lobe. These lower fibers form the **Meyer loop**.

3. **The center of the visual field:** The central portion of the visual field projects to the fovea, and these fibers take a direct route from the LGN through the parietal lobe to the most posterior area of the calcarine cortex, both above and below the calcarine sulcus. The fovea has the largest representation in the primary visual cortex.

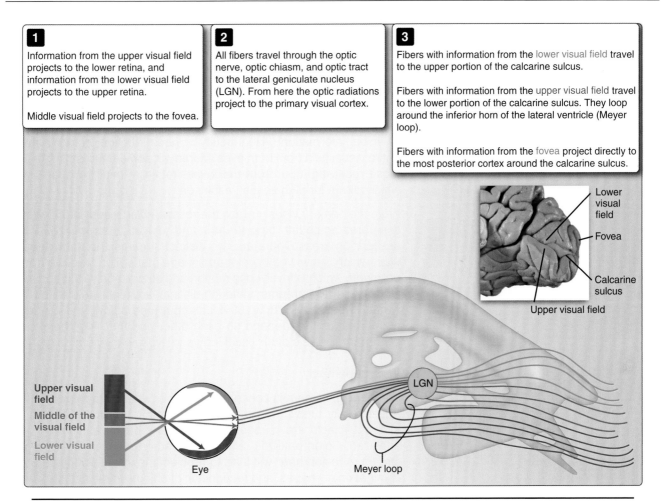

1

Information from the upper visual field projects to the lower retina, and information from the lower visual field projects to the upper retina.

Middle visual field projects to the fovea.

2

All fibers travel through the optic nerve, optic chiasm, and optic tract to the lateral geniculate nucleus (LGN). From here the optic radiations project to the primary visual cortex.

3

Fibers with information from the lower visual field travel to the upper portion of the calcarine sulcus.

Fibers with information from the upper visual field travel to the lower portion of the calcarine sulcus. They loop around the inferior horn of the lateral ventricle (Meyer loop).

Fibers with information from the fovea project directly to the most posterior cortex around the calcarine sulcus.

Lower visual field

Fovea

Calcarine sulcus

Upper visual field

Upper visual field

Middle of the visual field

Lower visual field

Eye

LGN

Meyer loop

Figure 15.12
The optic radiations.

V. CORTICAL PROCESSING OF VISION

The sensory cues that have been extracted from the light that stimulates photoreceptors in the retina must be assembled into a coherent and unified perception of the visual world. As discussed, the first steps of this process occur in the retina through the different types of ganglion cells that send information in parallel streams containing various aspects of visual information. In the primary visual cortex, this information is reassembled and then sent for further analysis and higher-order processing in association areas of the cortex.

A. Primary visual cortex

The **primary visual cortex** (**striate cortex**) is located superior and inferior to the **calcarine sulcus** in the **occipital lobe**. This is where over 90% of axons from the LGN terminate. The organization found throughout the visual system is maintained in the primary cortex through **retinotopic columns** of neurons. The parallel pathways carrying information on contrast, motion, acuity, and color project into distinct layers of the visual cortex, onto separate pools of neurons (see Figure 15.10). From there, these signals are integrated and converge onto the output neurons in the primary visual cortex. In addition, input from each eye, carrying information about the same area of the visual field, converges onto individual neurons in the striate cortex. It is thought that this is the basis for **stereopsis**, the perception of depth when looking at near objects. This organization allows the output from the primary visual cortex to the visual association areas to contain all visual information (visual acuity, contrast, motion, color) from the entire visual field from both eyes. Moreover, some neurons, especially those receiving input from the fovea, retain a small receptive field, which allows for high spatial resolution.

From the primary visual cortex, the reassembled visual information is separated again into new streams and relayed to specialized visual association areas. There are two distinct streams: the **ventral pathway**, which is involved in color and shape perception, and the **dorsal pathway**, which is involved in motion and spatial analysis (see Figure 15.10). The two streams, although distinct, are highly integrated. The "what," "where," and "how" of visual input (see below) are all pieces of information that depend on each other and that, together, give a congruent analysis of all visual inputs to evoke behavioral responses.

B. Ventral pathway

The **ventral stream** is concerned with the "what" of the visual inputs. Anatomically, the ventral stream projects to the **temporal lobe**. In this stream, objects are recognized, and their spatial relationships are analyzed. Color, patterns, and shapes are processed in the ventral stream. A lesion here will lead to problems in visual orientation, discrimination of shapes, and recognition of objects and faces, as well as deficits in attention to visual cues.

C. Dorsal pathway

The **dorsal stream** is concerned with the "where" and "how" of visual inputs. Anatomically, the dorsal stream projects mainly to the **parietal lobe**. In this stream, our visual space is analyzed, and navigation through that space is facilitated. In addition, this stream allows for the interaction with objects, including manipulation of objects and facilitation of directed eye movements and eye–hand coordination.

Both of these streams receive the same types of input from the visual system, but with different behavioral goals.

VI. COLOR VISION

The populations of cone photoreceptors that detect light at different frequencies form the basis for color vision in humans. As mentioned above, there are **L-type cones**, which have a maximal absorption in the red spectrum; **M-type cones**, which have a maximal absorption in the green spectrum; and **S-type cones**, which have a maximal absorption in the blue spectrum (see Figure 15.6). This results in **trichromacy**, or color vision based on three channels. A single cone does not distinguish color, but it is more likely to respond to light of a specific frequency.

A. Color information processing

The comparison of responses in different cones allows for the extraction of color information.

The first step in color processing occurs in two types of **opponency neurons** in the retina—single opponency and double opponency (Figure 15.13). **Red–green cells** compare L activation to M activation (or red to green) and are **single opponency** neurons. They are part of the parvocellular pathway.

Blue–yellow cells compare S (blue) activation to a combination of L and M activation (which makes yellow). They are **double opponency** neurons because they analyze L to M and the outcome of this (L/M = yellow) to S (blue). They are part of the koniocellular pathway.

The activation of both single and double opponency neurons has a similar ON/OFF pattern as occurs for the perception of contrast. In this case, the ON and OFF cells are separated into color groups: "red ON," "green OFF," and "green ON," "red OFF;" and "blue ON," "yellow OFF" and "blue OFF," "yellow ON." These ganglion cells project to the LGN. However, the processing that occurs in the thalamus remains unclear.

From the LGN, the information is relayed to the primary visual cortex. Here, the information arrives in separate channels (see above) and projects to separate layers. In the primary visual cortex, color signals are then compared across the visual field, and more color comparison is extracted through convergence onto another set of double opponency neurons (see Figure 15.13). These cells contribute to color analysis and, to a lesser degree, to the analysis of form, with low resolution because of the relatively large receptive fields. They

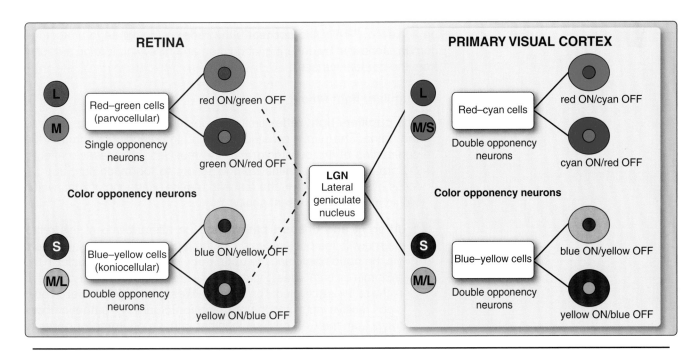

Figure 15.13
Single and double opponency neurons for color vision. L = L-type cones; M = M-type cones; S = S-type cones.

do not contribute to the distinction of contrast or motion. Downstream areas in the association cortices then take in more information and calculate detailed aspects of color such as hue.

Red–cyan cells compare input from the retinal L (red) channel to a combination of M and S (which results in cyan) channels.

Blue–yellow cells compare the retinal S (blue) channels to a combination of L and M (which results in yellow) channels.

B. Influence of color on behavior

Color analysis influences behavior by interfacing with other cortical areas. Different colors are associated with emotions and preferences. A recent cross-cultural study, for example, has shown that women prefer colors in the red spectrum, whereas men do not show that preference. Colors allow us to react quickly and efficiently in our environment (red traffic lights signal "stop," and green lights signal "go"). The figures in this book are, in fact, easier to navigate because of their color, insofar as more complex information can be assimilated and integrated with color.

VII. OPTIC REFLEXES

The visual system must be able to both adapt rapidly to changing light conditions and maintain focus on an area or object of interest. In order to achieve these things efficiently and rapidly, a number of reflex pathways exist. The **pupillary light reflex** adjusts the aperture of the pupil to control the amount of light let through to the retina. The **pupillodilator reflex** is an emotional reflex, in which sympathetic stimulation causes the pupils to dilate. The **accommodation reflex** adjusts the rounding of the lens and initiates convergence of the eyes so that focus on a near object can be achieved. Finally, the **corneal blink reflex** is designed to protect the cornea of the eye by ensuring lubrication and removing foreign particles from the corneal surface.

A. Pupillary light reflex

The **pupillary light reflex** limits the amount of light that can fall onto the retina. The pupils constrict in bright light conditions through the contraction of the constrictor pupillae muscles. This has two effects: (1) protection of the retina from exposure to too much light and (2) focus of light falling onto the retina to reduce scattering of light, which would interfere with visual acuity.

1. **Afferent and efferent pathways:** The afferent limb of this reflex begins with the photoreceptors of the retina. Information is then sent via the **optic nerve** and the **optic tract** to the **pretectal nuclei** in the midbrain on both sides (Figure 15.14). From here, the efferent branch of the reflex is activated. The **Edinger-Westphal (E-W) nuclei** on both sides receive input, and the **preganglionic parasympathetic fibers** travel from the E-W nucleus together with the **oculomotor nerve** to the orbits, where they synapse in the **ciliary ganglia.** From the ciliary ganglia, postsynaptic neurons form the short ciliary nerves that innervate the **constrictor pupillae muscles**. The constrictor pupillae are

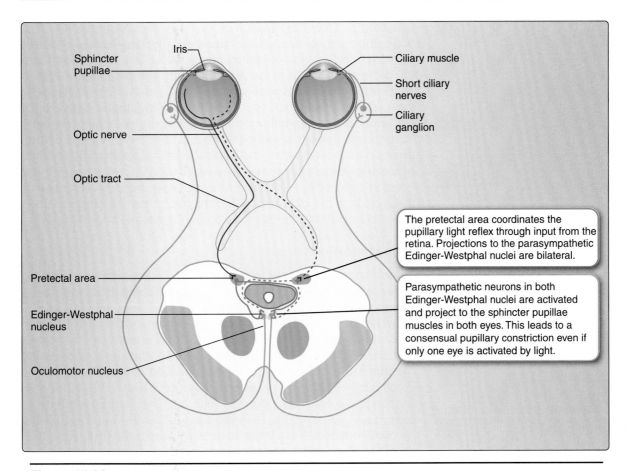

Figure 15.14
The pupillary light reflex.

circular muscles in the iris; contraction of these muscles causes the pupils to constrict. The parasympathetic fibers from the E-W nucleus are the most superficial layer of axons in CN III, located on the outer surface of the nerve. Any compression of CN III will thus affect the parasympathetic fibers first, and the pupils will dilate partially or fully due to loss of parasympathetic input and the ability to constrict. Deficits in the pupillary light reflex can therefore give insight into possible compression of CN III through rises in intracranial pressure.

2. **Direct versus consensual response:** The fibers from each pretectal nucleus project to the E-W nuclei on both sides. Therefore, shining a light on one eye will cause the pupils in both eyes to constrict. The response of the eye onto which light is directed is the **direct response**. The response becomes bilateral through transmission of information through the posterior commissure, which connects the two pretectal nuclei. Constriction of the other pupil, called the **consensual response**, thus occurs at the same time.

B. Pupillodilator reflex

The **pupillodilator reflex** occurs naturally in an emotional response that activates the sympathetic nervous system ("check your lover's eyes," the pupils should be dilated). It can be tested clinically through pinching the skin under the eye—this pain will result in sympathetic activation and therefore dilation of the pupil. Physiologically, the afferent component is through the **posterior hypothalamus**, which is stimulated in response to strong emotional states. Fibers then travel through the brainstem to the **preganglionic sympathetic** neurons in the lateral horn of the spinal cord at spinal level T1. Fibers synapse in the lateral horn and then travel within the **sympathetic chain** to the **superior cervical ganglion,** which is located at the base of the skull. Postsynaptic fibers then leave the ganglion and travel with the **carotid artery** into the skull, where they typically join the **ophthalmic division of the trigeminal nerve** (cranial nerve [CN] V, V$_1$), which takes these fibers into the orbit. Fibers enter the eye with the long ciliary nerves that innervate the **dilator pupillae muscle** of the iris (Figure 15.15). Contraction of the dilator pupillae muscle causes the pupil to dilate.

C. Accommodation

Three things must occur when we want to focus on something in our near field of view: (1) The eyes must **converge**. Both eyes move toward the midline (**adduct**) through the activation of both medial rectus muscles. The medial rectus muscles are innervated by the general somatic efferent (GSE) fibers of the oculomotor nerve. (2) The **refractive power** of the lens must be increased. This occurs by **increasing the curvature** of the lens through contraction of the ciliary muscles in the ciliary body. (3) The **pupils must constrict** through the contraction of the constrictor pupillae muscles. This increases the depth of field and visual acuity. Both the ciliary muscles and the constrictor pupillae muscles are innervated by parasympathetic fibers (GVE) from the Edinger-Westphal nucleus that travel in the oculomotor nerve to the orbit. These three components of the accommodation response are termed the **near triad.**

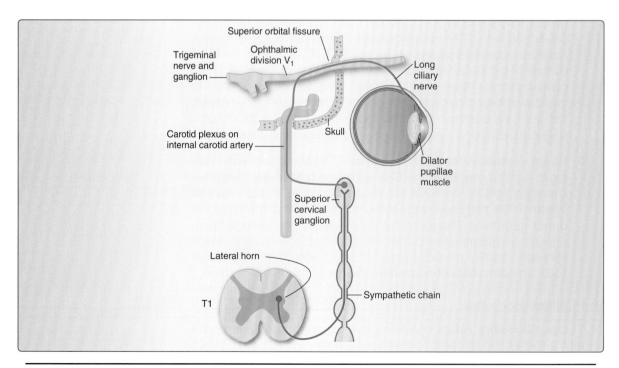

Figure 15.15
The pupillodilator reflex. T1 = thoracic vertebral level 1.

1. **Afferent and efferent pathways:** The afferent component of this reflex is the entire visual pathway. An object is seen, and the information goes all the way to the **primary visual cortex**. From there, the information is relayed to the **association cortex**, which then identifies the visual input as an object or area of interest. The **superior colliculus** and **pretectal nucleus** in the midbrain are then activated and send projections to the **oculomotor nuclear complex**, including both the somatic and the parasympathetic (**Edinger-Westphal**) components. These, in turn, comprise the efferent component of the accommodation reflex (see Figure 15.14 for the parasympathetic projections to the eye).

   The constriction of the pupils and the rounding of the lens are achieved through **parasympathetic innervation** of the **constrictor pupillae muscles** and the **ciliary muscles**. Contraction of the ciliary muscles causes the zonule fibers attached to the lens to relax, and the lens becomes thicker and rounder (see Figure 15.1).

2. **Vergence center:** The somatic component of the oculomotor nuclear complex is activated through the **vergence center** in the midbrain. It is in the pretectal area and sends bilateral projections to the somatic motor neurons of CN III to innervate both **medial rectus muscles**, and the eyes converge. (See Chapter 9, "Control of Eye Movement.")

D. Corneal blink reflex

The **corneal blink reflex** is a protective reflex that ensures the lubrication of the cornea and helps to remove foreign particles from the surface of the eye. If either cornea is touched by a foreign object, both eyes blink. The afferent limb of this reflex is through the **ophthalmic division of the trigeminal nerve** (V_1), which provides sensory innervation to the cornea (Figure 15.16). Afferent sensory information then synapses primarily in the nucleus of the

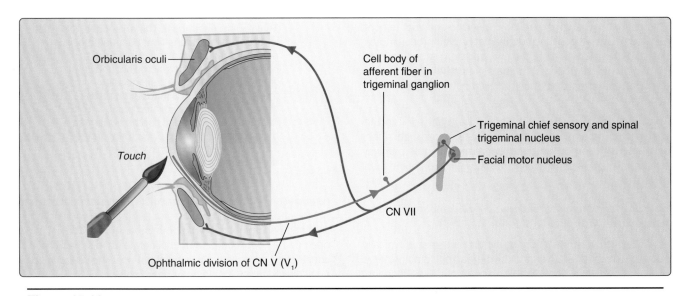

Figure 15.16
The corneal blink reflex. CN = cranial nerve.

spinal trigeminal tract and to a small extent in the **chief sensory nucleus of V** (if something touches the cornea, there is a predominant nociceptive component). Spinal trigeminal interneurons then project bilaterally (via relays in the reticular formation) to motor neurons in the **facial motor nucleus** (CN VII). Axons from the facial motor nucleus innervate the **orbicularis oculi** muscles, and both eyes close or "blink."

Clinical Case:

Anisocoria

A 31-year-old woman presents with a 3-week history of asymmetric pupil size (anisocoria) (right pupil larger than left). She has mild light sensitivity with slightly blurred vision of the right eye. She is otherwise healthy and takes no medication. She has no headache or pain in her eyes. Examination shows a 7-mm pupil of the right eye and a 3-mm pupil of the left eye. Her left pupil does not constrict when light is shone in either her right or left eye, but her right pupil constricts to light shone in either eye. Both pupils constrict during accommodation. She has no ptosis (droopy eyelid) or conjunctival injection (red eyes). Her visual acuity and visual fields are normal. Her extraocular movements are normal and the rest of her neurological examination is normal.

Case Discussion

Anisocoria is unequal pupil size. Physiological anisocoria can occur normally. However, in pathological anisocoria, the degree of anisocoria changes with varying illumination. If the anisocoria is worse in the light, it means that the large pupil is the abnormal pupil, and it fails to constrict in response to light. If the anisocoria is worse in the dark, it means that the small pupil is the abnormal pupil, and it fails to dilate in the dark (Figure 15.17).

This patient had a large pupil that failed to constrict to light. Causes of dilated pupil include cranial nerve III (oculomotor) palsy, Adie pupil, trauma to the eye, and drug exposure. Adie pupil is due to disruption of the parasympathetic supply from the ciliary ganglion or its postganglionic fibers (short ciliary nerves).

The most likely reason for the abnormal right pupil in this case is an impairment of the parasympathetic fibers that cause constriction of the pupil. Impairment in this pathway results in failure to constrict and therefore dilation of the pupil (mydriasis).

How do you distinguish between an oculomotor (III) nerve palsy and an Adie pupil?

An Adie pupil is a rare neurological disorder that is typically seen in women 20–40 years old, whereas oculomotor nerve palsy can occur in anyone. In an Adie pupil, one sees near-light

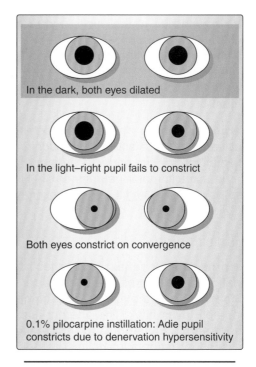

In the dark, both eyes dilated

In the light—right pupil fails to constrict

Both eyes constrict on convergence

0.1% pilocarpine instillation: Adie pupil constricts due to denervation hypersensitivity

Figure 15.17 (represents eyes of person facing the examiner)

Tonic pupil (Adie pupil). The anisocoria is more obvious in the light because the right pupil fails to constrict. Both pupils constrict with convergence. This unilateral light-near dissociation distinguishes the tonic pupil from a third nerve palsy or pharmacologically dilated pupil. The diagnosis of tonic pupil can be confirmed by instillation of dilute (0.1%) pilocarpine. The tonic right pupil constricts due to denervation hypersensitivity.

dissociation (the pupil slowly constricts to accommodation and not to light), whereas in oculomotor nerve palsy, constriction to light and accommodation are both impaired. In addition, a low dose (0.1%) of pilocarpine (a cholinergic agonist) will result in constriction of the tonic Adie pupil but not a normal pupil or a pupil dilated due to oculomotor nerve palsy.

Why does near-light dissociation occur?

There are 30 times more accommodation fibers than pupillary constriction fibers arising from the ciliary ganglion. Because of this 30:1 ratio, a lesion to the ciliary ganglion or short ciliary nerves results in a greater chance of damaging the fibers for constriction to light than for accommodation.

Conclusion

In both Adie pupil and oculomotor nerve palsy, pupillary constriction is lost and the pupil is therefore dilated. However, these two conditions can be distinguished from each other as discussed above. Anisocoria is a condition in which the pupils of the two eyes are unequal in size. It can be an entirely harmless condition or a symptom of a lesion.

Chapter Summary

- The visual system allows us to appreciate the visual world around us in color, form, and motion and with visual acuity. This is achieved through an optical system in the eye that refracts light onto the retina. The **retina** is a 10-layered structure containing several layers of neurons and is where the first step of visual processing occurs. Fibers from the retina travel in the optic nerve to the **optic chiasm**. Here, the nasal retinal fibers cross the midline, whereas the temporal fibers remain ipsilateral. This results in a lateralization of the visual field: The left visual field projects to the right side of the brain, and the right visual field to the left side. Fibers then travel in the optic tract to the **lateral geniculate nucleus (LGN)** of the thalamus. Fibers in the LGN are organized in a retinotopic fashion, reflecting the visual fields, and in parallel pathways, which process motion, form, and color. From the LGN, fibers travel through the **optic radiations** to the **primary visual cortex**. In the optic radiations, fibers from the upper visual field form **Meyer loop** around the inferior horn of the lateral ventricle in the temporal lobe and project to the primary visual cortex inferior to the **calcarine sulcus**; fibers from the lower visual field travel through the parietal lobe and project to the primary visual cortex superior to the calcarine sulcus. From the primary visual cortex, visual information is split into two main streams: a **dorsal stream** that analyzes spatial information and visually guided actions ("Where?" "How?") and a **ventral stream** that is dedicated to object recognition ("What?"). Both streams are highly integrated and interdependent.

- Color vision begins in the retina where different types of cones are sensitive to light of different frequencies. Color is analyzed through the comparison of cell activations in the retina and in the primary visual cortex.

- Several **visual reflexes** can regulate the amount of light that projects onto the retina: the **pupillary light reflex** results in the constriction of the pupils via parasympathetic input in response to light shone on the eye, and the **pupillodilator reflex** causes the pupils to dilate due to an emotional response that activates the sympathetic nervous system. **Accommodation** is a more complex process that requires vergence movements as well as pupillary constriction and the adjustment of the lens through action of the ciliary muscle. The **corneal blink reflex** protects the eye from foreign particles. The afferent limb involves the ophthalmic division of the trigeminal nerve [V_1] and the efferent limb involves the facial nerve (cranial nerve VII).

Study Questions

Choose the ONE best answer.

15.1 When light enters the eye, it is refracted before it reaches the retina. A loss of refractive power can lead to a loss in visual acuity. A surgical procedure to restore the main refractive power of light entering the eye involves the:

A. Choroid.
B. Uveal tract.
C. Anterior chamber of the eye.
D. Vitreous body.
E. Cornea.

Correct answer is E. The choroid is a vascular layer within the uveal tract. The uveal tract serves to provide a means by which blood vessels can travel within the wall of the eye. The anterior chamber of the eye is a fluid-filled chamber separated from the posterior chamber of the eye by the iris. The vitreous body is the gelatinous mass of the eye posterior to the lens. The cornea is the site of the main refractive power of light entering the eye. Degenerative changes in the cornea can lead to loss of visual acuity in that eye. A corneal transplant involves the grafting of a donor cornea onto the recipient's eye.

15.2 Which one of the following is the area of the eye with the highest visual acuity?

A. Retina.
B. Fovea.
C. Lens.
D. Papilla.
E. Ciliary body.

Correct answer is B. Together with its surrounding macula, the fovea is the area of highest visual acuity to which information from the visual field projects. The retina is the multilayered structure at the back of the eye that contains the photoreceptor cells. The lens assists in refraction of light entering the eye. The papilla is the site where the optic nerve fibers bundle together to leave the eye and therefore is the blind spot, as there are no photoreceptors present. The ciliary body contains the zonule fibers that maintain tension on the lens. Relaxation of these fibers allows the lens to round up for near accommodation.

15.3 During a clinical neurological exam all cranial nerves are tested. The examiner touches the cornea lightly with a cotton swab, and the patient blinks. Which of the following nerves is the afferent limb of the corneal blink reflex?

A. Facial nerve.
B. Glossopharyngeal nerve.
C. Trigeminal nerve.
D. Optic nerve.
E. Oculomotor nerve.

Correct answer is C. The trigeminal nerve, specifically the ophthalmic branch of cranial nerve V, V_1, senses touch, and blinking is a protective mechanism. CN V also senses dryness of the eye, ensuring that one will blink to wash tears across the eye and lubricate the cornea. The glossopharyngeal nerve is mainly sensory to the pharynx, the facial nerve is the primary motor nerve of the face, and the optic nerve carries visual information to the brain. The oculomotor nerve is the motor nerve to some of the muscles that move the eye.

15.4 A patient is brought to the emergency department after a motor vehicle accident. He is unconscious. The attending physician shines a light into the patient's eyes to see whether the pupils constrict. Which one of the following cranial nerves contains the fibers that mediate the constriction of the pupils?

A. Oculomotor nerve (cranial nerve III).
B. Optic nerve (cranial nerve II).
C. Trigeminal nerve (cranial nerve V).
D. Abducens nerve (cranial nerve VI).
E. Facial nerve (cranial nerve VII).

Correct answer is A. The efferent limb of the pupillary light reflex travels in the oculomotor nerve. The Edinger-Westphal nuclei of both sides receive input from both pretectal nuclei. From these nuclei, their fibers travel in the oculomotor nerve to the orbit where they synapse in the ciliary ganglion. Postganglionic fibers innervate the constrictor pupillae muscles that constrict the pupil, thereby limiting the amount of light that can fall on the retina. The optic nerve carries the visual sense from the retina to the brain. The trigeminal nerve is the principal sensory nerve of the face. The abducens nerve innervates the lateral rectus muscle, which turns the eyeball laterally. The facial nerve is motor to the muscles of facial expression and visceral motor to some glands in the head.

15.5 A patient complains about a loss of his peripheral vision. Upon clinical examination, you notice that there is a loss of his temporal visual field on both sides. Which diagnosis would explain this deficit?

A. An astrocytoma in the right lateral geniculate nucleus of the thalamus.
B. A retinal detachment involving the fovea of the right eye.
C. An ischemic infarct in the left temporal lobe compromising the Meyer loop.
D. A pituitary tumor pushing onto the optic chiasm.
E. An ependymoma in the cerebral aqueduct pushing onto the pretectal nucleus.

Correct answer is D. As each side of the brain will ultimately deal with the contralateral visual field, fibers from the ipsilateral field must cross the midline, and they do so in the optic chiasm. In the optic chiasm, the fibers from the nasal retina carrying information from the temporal visual field cross the midline. A pituitary tumor would put pressure onto the optic chiasm and result in a loss of function for these nasal retinal fibers and cause a loss of the temporal field of view. The lateral geniculate nucleus (LGN) receives information from the contralateral visual field from all types of ganglion cells. A lesion to the LGN would only affect one visual field. The fovea is the area of highest visual acuity in each eye. A lesion of the fovea in one eye would lead to functional blindness in that eye, because only the peripheral retina with less visual acuity would remain functional. The other nonaffected eye would, however, still perceive both visual fields. The Meyer loop is formed by fibers sweeping around the horn of the lateral ventricle to reach the calcarine sulcus. These fibers come from the LGN and are lateralized for the opposite visual field. A lesion in the Meyer loop would affect the upper portion of the contralateral visual field. The pretectal nucleus is involved in the pupillary light reflex, not higher-order visual processing.

The Basal Ganglia

16

I. OVERVIEW

Movement is controlled by the upper motor neuron (UMN) system in the cortex. The signaling pattern of the UMNs is tightly controlled and regulated by two distinct systems: the basal ganglia and the cerebellum. These systems influence UMNs so that a precisely planned and executed motor command can be conveyed to the LMNs in the brainstem and spinal cord and from there to the target muscles. Chapter 17, "The Cerebellum," will cover the important role of the cerebellum in the control of movement. In this chapter, our focus is on the complex role of the basal ganglia in the control and integration of cortical output. The basal ganglia mainly encode for:

- The decision to move
- The direction of movement
- The amplitude of movement
- The motor expression of emotions (Figure 16.1)

In addition, we explore the broader role of the basal ganglia in encoding cognitive processes and their behavioral output. Information from widespread cortical and subcortical areas is funneled through the basal ganglia through a series of circuits (motor, oculomotor, cognitive, and limbic). Together, these circuits bring multiple influences to bear on behavior. Behavior is always a *motor* output.

II. ANATOMY

The basal ganglia nuclei are large masses of gray matter deep within the cerebral hemispheres. They include the caudate, putamen, and globus pallidus that lie lateral to the thalamus; the substantia nigra (SN) in the rostral midbrain; and the subthalamic nucleus (STN)

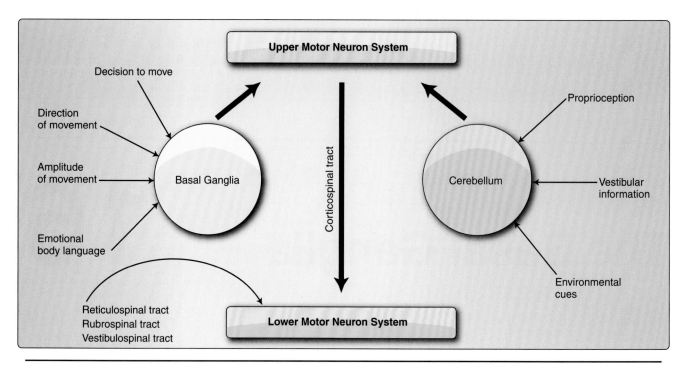

Figure 16.1
Conceptual overview of motor control.

located inferior to the thalamus (Figure 16.2). Complex connections between these nuclei influence the activity of the thalamus and thereby control the activity in the cortex. The basal ganglia are components of a series of parallel circuits that integrate cortical activity into one behavioral output.

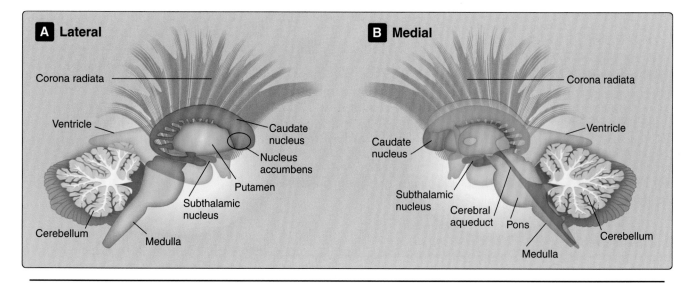

Figure 16.2
Midsagittal section through the diencephalon and brainstem showing the basal ganglia and the cerebellum from lateral and medial views.

A. Caudate nucleus

The caudate nucleus is a tadpole-shaped nucleus. The head of the caudate is located in the floor of the lateral ventricle and the body arches over the thalamus in a C shape tapering off into a tail located in the roof of the inferior horn of the lateral ventricle (Figure 16.3).

B. Putamen

The **putamen** is the most lateral of the basal ganglia and embryologically is connected to the caudate nucleus. Together, the putamen and the caudate are referred to as the **striatum**. The anterior limb of the internal capsule separates the caudate and putamen leaving only a few bridging fibers between these two nuclei (see Figure 16.2). These connecting fibers give the striatum its "striped" appearance.

The putamen and the caudate nuclei are the input nuclei of the basal ganglia. They receive mainly excitatory input from cortical and subcortical structures.

C. Globus pallidus

The **globus pallidus** is medial to the putamen and lateral to the thalamus. It can be subdivided into an external part (GPe) and an internal part (GPi). These two parts of the globus pallidus are functionally different and have distinct connections within the basal ganglia.

The globus pallidus is the output nucleus of the basal ganglia, sending inhibitory projections to the thalamus.

Together, the globus pallidus and the putamen look like a lens in coronal section, which is the reason they are given the name "**lentiform**" or "lenticular" nucleus. The internal capsule separates the lentiform nucleus from the caudate nucleus and the thalamus (see Figure 16.3).

D. Nucleus accumbens

The **nucleus accumbens** is the anterior and inferior part of the striatum where the head of the caudate and the putamen are continuous with each other (see Figure 16.2). It receives extensive dopaminergic input and is an integral part of the limbic system and reward circuitry.

E. Subthalamic nucleus

The **subthalamic nucleus** is a biconvex nucleus that lies inferior to the thalamus and superior to the tegmentum of the midbrain (Figure 16.4; also see Figure 16.2). Its primary afferents originate in the SN and the external part of the globus pallidus. The STN is under tonic inhibition, and its output is excitatory through glutamatergic projections to the internal segment of the globus pallidus as well as a reciprocal connection with the SN. It plays a central role in basal ganglia connectivity and can be described as the "pacemaker" of the basal

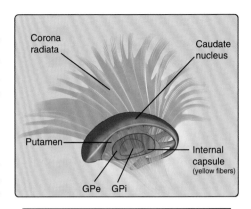

Figure 16.3
Caudate nucleus (red), putamen (green), and globus pallidus (outlined in blue) shown in their relationship to the internal capsule (medial view). GPe = globus pallidus, external segment; GPi = globus pallidus, internal segment.

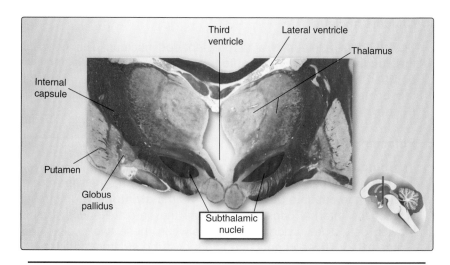

Figure 16.4
Coronal section through the forebrain showing the subthalamic nuclei inferior
to the thalamus.

ganglia, defining the output rhythm. Recent studies show that this
rhythm can be influenced through direct cortical inputs.

F. Substantia nigra

The **substantia nigra** is located in the rostral midbrain within the
cerebral peduncle, just posterior to the descending motor fibers. It
contains dopaminergic neurons that project to the putamen and caudate nucleus as well as the STN (Figure 16.5).

G. Cross-sectional anatomy

In a cross-section, important landmarks are the **ventricles** and the
insula (Figures 16.6 and 16.7).

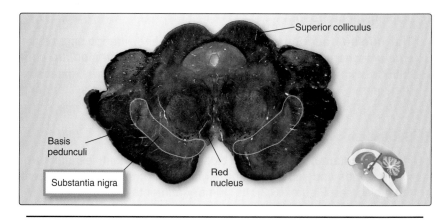

Figure 16.5
Coronal section through the rostral midbrain showing the substantia nigra
located in the cerebral peduncle at the level of the superior colliculus.

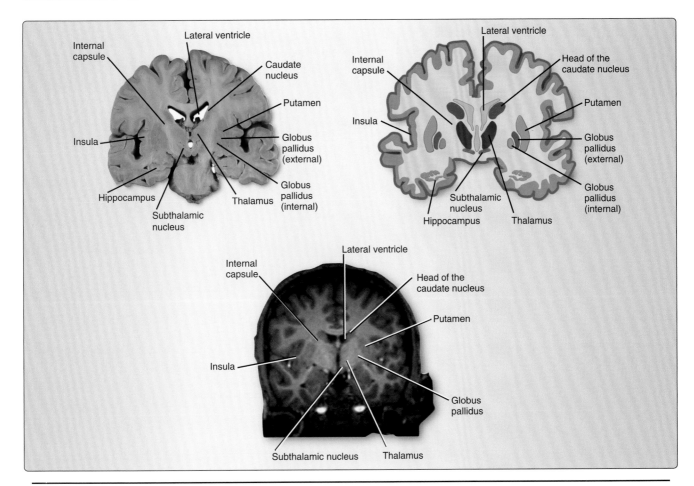

Figure 16.6
Location of the basal ganglia in coronal section.

From the insula, moving medially, the first basal ganglia structure is the putamen, the most lateral of the basal ganglia (see Figures 16.6 and 16.7). Medial to the putamen is the globus pallidus with its external (GPe) and internal (GPi) parts. Lying medial to the globus pallidus is the anterior limb of the internal capsule consisting of projection fibers to and from the cortex. The head and body of the caudate are located in the floor of the lateral ventricle and are separated from the putamen by the internal capsule. The tail of the caudate is located in the roof of the inferior horn of the lateral ventricle.

III. TRACTS

The basal ganglia receive information from widespread cortical areas. This information is funneled through the basal ganglia circuitry and results in the regulation of thalamic activity, which in turn regulates cortical activity. Here, we will look at the inputs to and outputs from the basal ganglia and, importantly, the internal wiring that allows for measured regulation of cortical output.

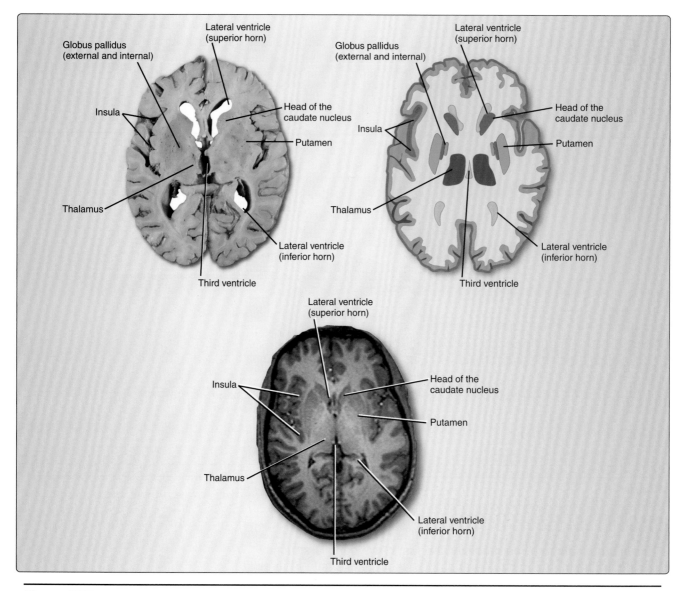

Figure 16.7
Location of the basal ganglia in horizontal section.

A. Input to the basal ganglia

Input to the basal ganglia goes to the striatum (caudate nucleus and putamen). The caudate nucleus and putamen each receive input from distinct cortical and subcortical regions. This input is topographically organized with each region projecting to a specific area of the striatum.

Input to the striatum enables the basal ganglia to integrate information from different cortical and subcortical areas. A single cell in the striatum receives input from multiple sources. This makes every cell in the striatum an **integrator**. This integration of input allows the basal ganglia to encode for the decision to move, the direction of movement, the amplitude of movement, and the motor expression of emotions.

B. Output from the basal ganglia

The output from the basal ganglia is inhibitory via GABAergic neurons. These projections arise from the internal part of the globus pallidus (GPi) and project to various nuclei of the thalamus. Activity in the basal ganglia circuits will determine which projections to the thalamus are inhibited and which are released from inhibition. This results in measured and precise input to the thalamus and thereby to the cortex.

C. Internal circuits of the basal ganglia

Through complex wiring of the basal ganglia, the input is well integrated and the output highly regulated. A balance of inhibitory and excitatory pathways regulates the activity of the thalamus, which then influences the cortex. The thalamus is under **tonic inhibition**. This means that unless the inhibition is removed, there is decreased signaling to the cortex. The output from the basal ganglia can either increase or release the tonic inhibition of the thalamus via two internal pathways (Figure 16.8). This is referred to as the **release inhibition model** of basal ganglia activity.

The **direct pathway** facilitates target-oriented and efficient behavior. The **indirect pathway** suppresses superfluous behaviors that are not

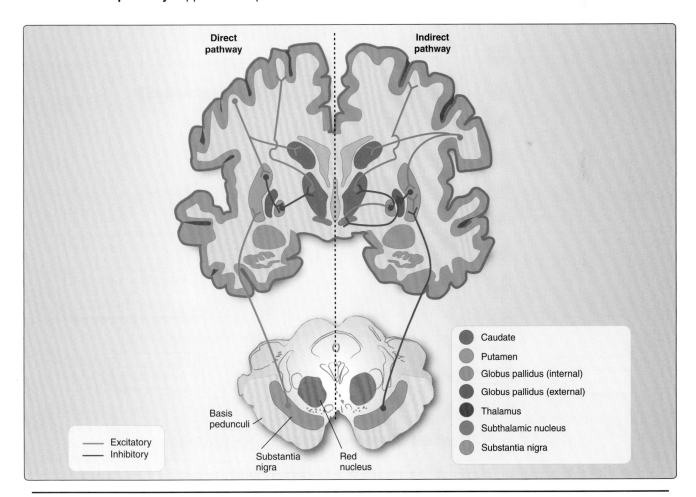

Figure 16.8
Wiring diagram for the internal circuits of the basal ganglia: the direct and indirect pathways.

related to the targeted behavior. Together, these pathways ultimately streamline our behavioral output. Dopaminergic signaling influences both the direct and indirect pathways. The net effect of dopamine signaling in the basal ganglia is to increase cortical activity (dopamine = movement).

The projection from the cortex to the striatum is referred to as the **corticostriatal pathway**. The connection between the SN and the striatum is the **nigrostriatal pathway**. The connection between the STN and the globus pallidus is the **subthalamic fasciculus**. The **thalamic fasciculus** is the projection from the globus pallidus to the thalamus (Figure 16.9).

1. **Direct pathway:** This wiring circuit releases the thalamus from tonic inhibition. Removal of this tonic inhibition leads to more excitation of the cortex and in turn more cortical output (Figure 16.10). Activity in the direct pathway facilitates target-oriented and efficient cortical output.

   Excitatory projections (corticostriate) from the cortex to the striatum (caudate and putamen) synapse with inhibitory neurons. Inhibitory neurons project directly to the GPi, where they inhibit the inhibitory projection to the thalamus. *Inhibition of the inhibition* releases the tonic inhibition of the thalamus, resulting in increased output. (Two negatives do make a positive in this case!) The thalamus then

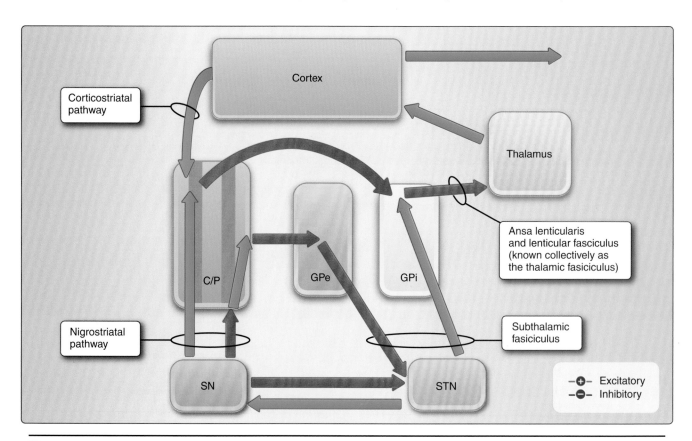

Figure 16.9
Direct and indirect pathways and the anatomical names of their connecting fibers. GPe = globus pallidus, external segment; GPi = globus pallidus, internal segment; C/P = caudate/putamen; STN = subthalamic nucleus; SN = substantia nigra.

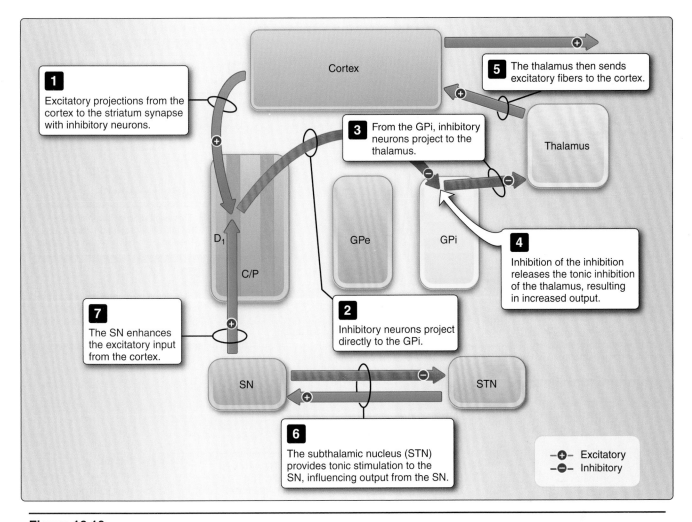

Figure 16.10
Wiring diagram for the direct pathway. GPe = globus pallidus, external segment; GPi = globus pallidus, internal segment; C/P = caudate/putamen; D_1 = dopamine receptor type 1; SN = substantia nigra; STN = subthalamic nucleus.

sends excitatory fibers to the cortex. Increased excitation of the cortex results in an increase in cortical output. At the same time, the striatum is also influenced by input from the SN. Dopaminergic neurons project to the striatum where they excite (via D_1 receptors) inhibitory neurons, which project to the GPi. The SN enhances the excitatory input from the cortex, thereby strengthening the direct pathway. The STN provides tonic stimulation to the SN, thus influencing output from the SN.

2. **Indirect pathway:** This second wiring circuit in the basal ganglia, the indirect pathway, suppresses superfluous cortical activity and motor output that competes with the projections facilitated through the direct pathway (Figure 16.11).

The indirect pathway strengthens the inhibition of the thalamus through an excitatory projection from the STN. The STN is under chronic inhibition through inhibitory projections from the GPe. Thus, a critical step in the indirect pathway is to release the inhibition of the STN.

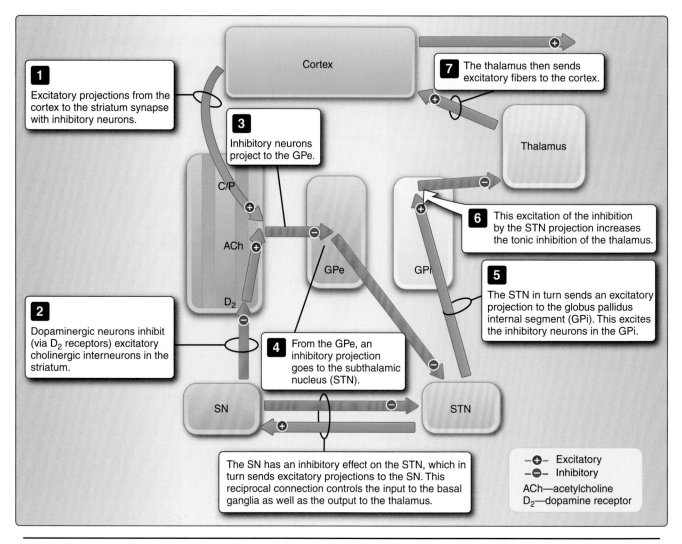

Figure 16.11
Wiring diagram for the indirect pathway. GPe = globus pallidus, external segment; C/P = caudate/putamen; SN = substantia nigra; STN = subthalamic nucleus.

Excitatory projections from the cortex to the striatum synapse with inhibitory neurons, which project to the GPe. The result of this projection is the release of the inhibition from the GPe to the STN. The STN is now disinhibited and can send an excitatory projection to the GPi. This projection stimulates the inhibitory neurons in the GPi, which project to the thalamus and increase its inhibition. This *excitation of the inhibition* by the STN projection increases the tonic inhibition of the thalamus. The inhibited thalamus suppresses cortical output.

At the same time, the striatum is influenced by the SN. Dopaminergic neurons inhibit (via D_2 receptors) excitatory cholinergic interneurons in the striatum. These interneurons in turn project to inhibitory neurons going from the striatum to the GPe. The SN counteracts the excitatory input from the cortex to the striatum, effectively weakening the indirect pathway. This results in a less effective inhibition of superfluous, competing cortical output and likely contributes to personality and fluidity of our movements. Reciprocal connections between the SN and the STN control the input to the basal ganglia as well as the output to the thalamus.

Clinical Application 16.1: Parkinson Disease

Parkinson disease (PD) is a motor disorder characterized by **hypokinesia** (a decrease in movement), bradykinesia (slowed movement), or akinesia (a loss of movement). This effect on movement is due to a degeneration of dopaminergic neurons in the substantia nigra (SN). The net effect of dopamine in both the direct and indirect pathways is to facilitate movement or cortical output; an imbalance of this system underlies Parkinson disease (PD).

As discussed, the SN influences the direct and indirect pathway in the basal ganglia circuits. Dopaminergic neurons from the SN have an excitatory input (via D_1 receptors) to the direct pathway, enhancing the input from the cortex. Loss of excitatory input from the SN to the striatum decreases the amplification of the cortical input to the striatum, resulting in less excitatory input going into the direct pathway. This causes the indirect pathway to have more influence.

Dopaminergic neurons from the SN send inhibitory input (via D_2 receptors) to a cholinergic interneuron in the striatum. A loss of this dopaminergic signaling causes a loss of inhibition of this excitatory cholinergic interneuron, which increases stimulation of an inhibitory neuron projecting to the external globus pallidus (GPe). The GPe is inhibited, and there is less inhibitory input to the subthalamic nucleus (STN). In addition, there are inhibitory projections from the SN to the STN, and loss of these fibers results in further loss of inhibition of the STN. This, in turn, causes more excitatory (glutamate) output from the STN to the internal globus pallidus (GPi). A stimulated GPi sends more inhibitory input to the thalamus, thereby increasing the tonic inhibition of the thalamus, which results in less cortical stimulation and less cortical output.

Based on the known disease etiology, the two approaches to treating the motor symptoms of PD involve drug therapy or procedures that influence the basal ganglia circuitry through direct stimulation of the STN. Drug therapy can restore the dopamine levels in the basal ganglia through the administration of levodopa (L-Dopa) or dopamine agonists. Alternatively, the cholinergic neurons in the striatum can be inhibited through anticholinergic drugs.

L-Dopa is a precursor of dopamine that will cross the blood–brain barrier and is converted to dopamine centrally. It has been shown to alleviate the hypokinetic and akinetic movement disorders as well as the rigidity in patients with PD. Anticholinergic drugs are more effective in controlling tremor than other symptoms of PD. Amantadine is a drug that increases the release of dopamine and blocks cholinergic receptors, along with an inhibitory effect on glutamatergic synapses through blockade of the NMDA receptor. Amantadine is effective for dyskinesia, which can develop in later stages of PD when patients have been on dopaminergic agents for some time.[1]

Deep brain stimulation of the STN is another treatment option for PD in selected patients. When the inhibitory influence of the SN on the STN is lost, the STN has an aberrant firing pattern resulting in excessive output, ultimately inhibiting the output from the cortex. Deep brain stimulation of the STN restores its tonic firing pattern, alleviating the hypokinetic and akinetic symptoms in patients with PD.

 [1]See Chapter 8, Section IV in *Lippincott's Illustrated Reviews: Pharmacology*.

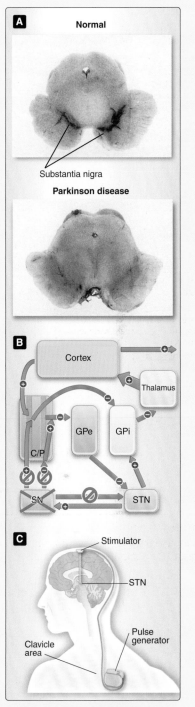

Parkinson disease (PD): **A.** Comparison of a normal substantia nigra (*top*) with substantia nigra in a patient with PD (*bottom*; note the marked loss of black pigmentation, indicative of a loss of dopaminergic neurons). **B.** Schematic illustration of changes to the pathways in PD. **C.** Electrodes implanted in the subthalamic nucleus for deep brain stimulation in a patient with PD. GPe = globus pallidus, external segment; GPi = internal segment; C/P = caudate/putamen; SN = substantia nigra; STN = subthalamic nucleus.

Clinical Application 16.2: Ballism

Ballism is characterized by sudden uncontrolled flinging movements of the proximal extremities. It usually occurs following a stroke affecting the sub-thalamic nucleus (STN), and the symptoms occur on the contralateral side (hemiballism).

The underlying pathophysiology is a loss of the contralateral STN and with that the loss of the excitatory fibers projecting to the internal globus pallidus (GPi). This results in less stimulation of the inhibitory output from the GPi to the thalamus. With loss of the indirect pathway, there is no suppression of superfluous competing movements.

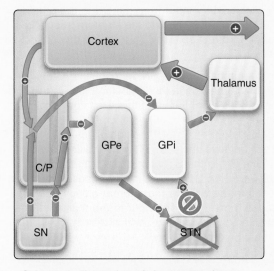

Schematic illustration of changes to the pathways in ballism. GPe = globus pallidus, external segment; GPi = internal segment; C/P = caudate/putamen; SN = substantia nigra; STN = subthalamic nucleus.

Clinical Application 16.3: Huntington Disease

Huntington disease (HD) is an autosomal dominant hereditary disorder that usually manifests in the fifth decade of life. It is characterized by deficits in cognition, behavior, and a characteristic hyperkinetic movement disorder.

Patients with Huntington disease typically have chorea. Chorea is a hyper-kinetic movement described as a nonpurposeful rapid irregular movement that randomly flows from one body part to another. Chorea can affect speech, gait, and coordinated movement.

Pathophysiologically, there is a marked degeneration of the striatum (caudate and putamen) as well as degeneration of temporal and frontal association cortices. The underlying cause is a mutation causing CAG repeats on chromosome 4, resulting in abnormal levels of the protein huntingtin.

With degeneration of the striatum, the processing of cortical inputs is impaired.

Clinical Application 16.3: Huntington Disease continued

Loss of striatal neurons leads to a loss of both the direct and indirect pathways. Individuals lose target-oriented movements facilitated by the direct pathway and are left with superfluous competing movements that are no longer inhibited by the indirect pathway.

Loss of inhibitory projections to the external globus pallidus (GPe) results in more inhibition of the subthalamic nucleus (STN), which decreases the excitatory output from the STN to the internal globus pallidus (GPi). This, in turn, leads to less inhibition from the GPi to the thalamus. The output from the thalamus is disinhibited, resulting in more excitation from the thalamus to the cortex, leading to the excessive movement in HD.

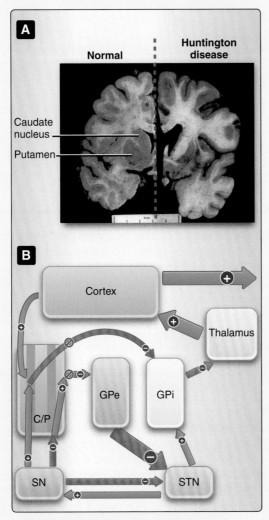

Huntington disease (HD): Coronal section through the forebrain of a patient with HD (note the degeneration of both the caudate and putamen and the enlargement of the lateral ventricles) and schematic illustration of changes to the pathways in HD. GPe = globus pallidus, external segment; GPi = internal segment; C/P = caudate/putamen; SN = substantia nigra; STN = subthalamic nucleus.

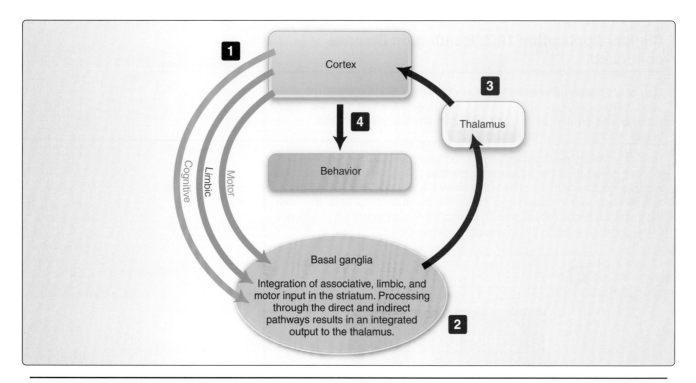

Figure 16.12
Conceptual overview of the integrative function of the basal ganglia resulting in a behavior.

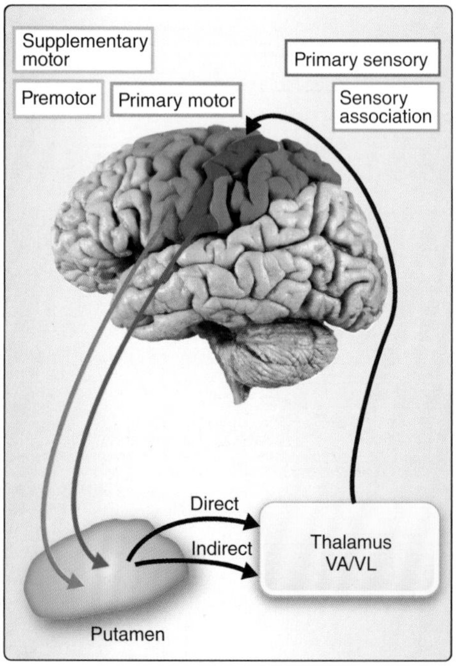

IV. FUNCTIONAL RELATIONSHIPS

The input to the basal ganglia can be described as three parallel streams of information from the cortex: motor, cognitive, and limbic. The striatum integrates these inputs, and from the striatum, the activity of the thalamus is determined via the direct and indirect pathways. The thalamus then sends projections back to the cortex. The basal ganglia integrate the sum of cortical activity into one behavioral output (Figure 16.12).

A. Motor circuit

The **motor circuit** is the best known of the three circuits and plays the key role in motor performance and in the regulation of eye movements.

Inputs related to motor performance come from widespread areas of the cortex, including both primary and association areas for motor and sensory processing, all of which are integrated in the putamen. The motor circuit is mediated through both the direct and indirect pathways within the basal ganglia, as described above (Figure 16.13). During acquisition of habits and skills, dopaminergic input reinforces behaviors that result in a desired goal-oriented outcome. The motor circuit becomes active once habits and skills are well established. These are no longer dopamine-dependent and resilient against unlearning.

A separate **oculomotor** aspect of the motor circuit is critical in regulating gaze and orientation of the eyes (Figure 16.14). Inputs from prefrontal and posterior parietal areas of the cortex are integrated

Figure 16.13
Schematic representation of the motor circuit. VA = ventral anterior nucleus; VL = ventral lateral nucleus.

in the basal ganglia, and outputs go to the thalamus (VA and dorso-medial [DM]). The VA projects to the frontal eye fields where gaze is initiated. The DM is intimately related to the limbic system and helps to direct our gaze to a salient or rewarding stimulus. The SN also connects to the superior colliculi, where it can play a role in coordinating and directing eye movements (see Chapter 9, "Control of Eye Movements").

B. Cognitive circuit

The **cognitive circuit** plays an important role in higher cortical function and motor learning. Input from the frontal, parietal, and temporal association areas travels to the caudate nucleus and the nucleus accumbens. Information is integrated in the basal ganglia and outputs go to the thalamus (VA and centromedian [CM]). The VA projects back to motor and prefrontal association areas of the cortex and the CM projects diffusely to the cortex where it influences overall cortical arousal and function (Figure 16.15).

The cognitive circuit participates in the planning of complex motor activity. When new habits and skills have been practiced and well learned, and dopaminergic input has strengthened these behaviors, activity in the cognitive circuit decreases, and the motor circuit becomes active instead.

Similarly, the cognitive circuit helps to prioritize and streamline higher cortical activity. Initially, the salience of stimuli is neutral. Experience and learning help to increase or decrease the salience of stimuli and, through the basal ganglia, influence behavioral output.

C. Limbic circuit

The **limbic circuit** is an important pathway through the basal ganglia that is involved in the regulation of emotional, motivational, and affective aspects of behavior. Inputs originating from frontal association areas, limbic lobe, hippocampus, and amygdala (see Chapter 20, The Limbic System") project to the nucleus accumbens and to inferior (ventral) aspects of the caudate and putamen (collectively called the ventral striatum). The output projects to the thalamus (VA and DM). A thalamocortical pathway then projects to the anterior cingulate and orbitofrontal areas of the cortex (Figure 16.16).

In addition, these cortical areas project directly to the STN to influence thalamic output.

The limbic circuit is important in the motor expression of emotions. Postures, gestures, and facial expressions related to emotion are mediated by this basal ganglia circuit. This circuit is rich in dopaminergic projections, and a loss of dopaminergic neurons in Parkinson disease (PD) results in a "mask face," a decrease in spontaneous gestures, and affective disturbances that appear as PD progresses.

This circuit is also particularly important in facial expressions. In fact, the UMN innervation to the upper part of the face comes from the cingulate gyrus rather than from the primary motor cortex and is directly regulated by the limbic circuit (see Chapter 10, "Sensory and Motor Innervation of the Head and Neck" for details on facial innervation).

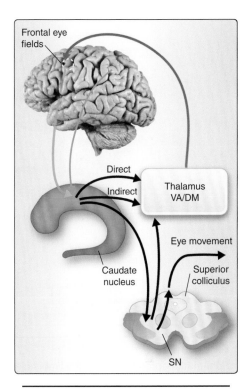

Figure 16.14
Schematic representation of the oculomotor circuit. VA = ventral anterior nucleus; DM = dorsomedial nucleus; SN = substantia nigra.

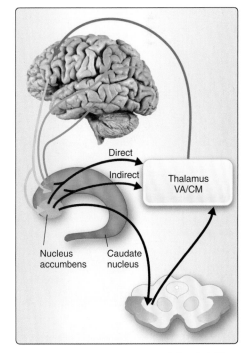

Figure 16.15
Schematic representation of the cognitive circuit. VA = ventral anterior nucleus; CM = centromedian nucleus.

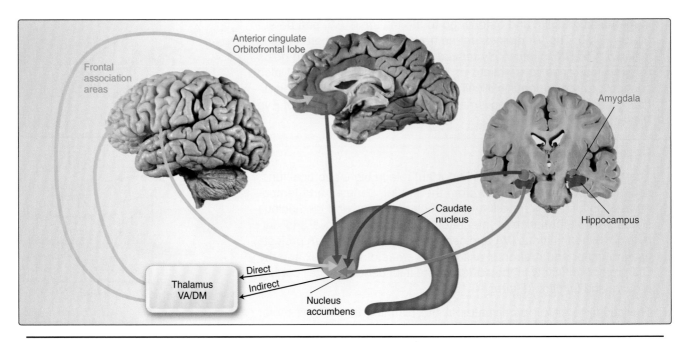

Figure 16.16
Schematic representation of the limbic circuit. VA = ventral anterior nucleus; DM = dorsomedial nucleus.

V. BLOOD SUPPLY

Blood supply to the basal ganglia arises from the penetrating branches off the cerebral arterial circle.

The cerebral arterial circle, also known as the **circle of Willis**, is located on the inferior aspect of the brain. Its supply comes from the vertebral basilar system and the internal carotid arteries (see Chapter 2, "Overview of the Central Nervous System," and Chapter 13, "The Cerebral Cortex," for details). Figure 16.17 shows the four main arteries of these systems (Table 16.1).

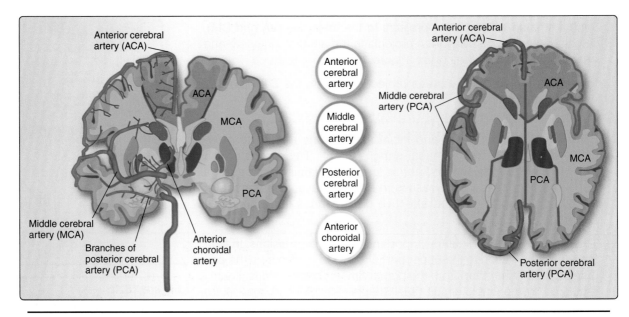

Figure 16.17
Blood supply to the basal ganglia shown in a coronal and horizontal section.

Table 16.1: Summary of Blood Supply to the Basal Ganglia

	Internal Carotid	ACA	MCA	PCA
Caudate head		Deep penetrating branches		
Body			Lenticulostriate arteries	Deep penetrating branches (minor supply)
Tail	Anterior choroidal artery			Deep penetrating branches (minor supply)
Putamen anterior		Deep penetrating branches	Lenticulostriate arteries	
Putamen posterior	Anterior choroidal artery		Lenticulostriate arteries	
Globus pallidus anterior		Deep penetrating branches	Lenticulostriate arteries	
Globus pallidus posterior	Anterior choroidal artery		Lenticulostriate arteries	
Substantia nigra				Deep penetrating branches
Subthalamic nucleus				Deep penetrating branches

A. Anterior choroidal artery

Before branching into anterior and middle cerebral arteries, each internal carotid artery gives off an **anterior choroidal artery** that supplies the tail of the caudate nucleus and the posterior portions of the globus pallidus and putamen.

B. Anterior cerebral artery

The **anterior cerebral artery (ACA)** gives off deep penetrating branches (medial lenticulostriate arteries) that go to the head of the caudate nucleus and the anterior part of the putamen and globus pallidus.

C. Middle cerebral artery

The **middle cerebral artery (MCA)** also gives off deep penetrating branches, which divide into medial and lateral lenticulostriate arteries. These supply the putamen and the globus pallidus as well as the body of the caudate nucleus.

D. Posterior cerebral artery

Deep penetrating branches of the **posterior cerebral artery (PCA)** and the posterior communicating artery supply the SN and the STN. They may also supply the most posterior parts of the caudate nucleus as well as the entire thalamus.

Clinical Case:

Mrs. Smith

Mrs. Smith is a 71-year-old woman who comes to your office with a 2-year history of progressive right hand tremor. It is present at rest and improves when she uses the hand. She has also noticed that her gait is not a brisk as previously. She occasionally gets light-headed when getting up out of bed in the morning. She has chronic constipation and requires medication for it. She denies any changes in her mood or cognition, but her husband later tells you that he is concerned about her mood. She is less interactive and socially withdrawn, but there is no suicidal ideation. Mrs. Smith has noticed that she has been having trouble with her short-term memory over the past few months; her friends have noticed this change. She has also noticed that her handwriting has become very small.

Mrs. Smith tells you that her food tastes bland and she has not been able to appreciate the flavors in her food for about 10 years. Her husband comments that her sleep is unsettled and she kicks around in her sleep and appears distressed. She also has to get out of bed several times during the night to empty her bladder.

All of her symptoms have worsened in the past 6 months.

On examination, she has reduced animation in her facial expression (mask face) with reduced eye blinking. Her cranial nerve and sensory examinations are normal. She has rigidity in her right arm with cogwheeling at the wrist and normal tone elsewhere. She has reduced fine finger movement and foot tapping on the right. Her muscle bulk, strength, and deep tendon reflexes (DTRs) are otherwise normal and her toes down going (plantar responses). She has a pin-rolling resting tremor in her right hand. Coordination is normal. She has reduced right arm swing when walking. She has a slow, hesitant gait with a slightly stooped posture.

Her blood pressure is 130/80 lying down and 105/70 standing up.

She is given a prescription of levodopa (L-dopa), and her follow-up examination while on L-dopa showed improvement in her tremor, bradykinesia/hypokinesia, and rigidity.

What is Mrs. Smith's Medical Condition?

This is likely a case of idiopathic Parkinson disease. There are four classic features of Parkinson disease: (1) tremor,

(2) rigidity, (3) akinesia, and (4) postural instability. A likely diagnosis of PD can be made when two of the first three symptoms (tremor, rigidity, and akinesia) are found with no other cause identified. In addition, PD typically presents with an asymmetric onset—in fact Mrs. Smith's symptoms are more pronounced on the right side. Her response to L-dopa is another clue that her symptoms are caused by a decrease in dopaminergic signaling due to the degeneration of dopaminergic neurons.

What is the Reason for Her Nocturia, Constipation, and Reduced Sense of Smell?

The pathological manifestation is a deposit of Lewy bodies (Figure 16.18) in neurons in the forebrain and brainstem as well as the enteric nervous system (ENS). Lewy body accumulation in the olfactory system can result in hyposmia and commonly precedes the motor symptoms of Parkinson disease.

Degeneration of neurons of the visceral nervous system and the enteric nervous system results in the autonomic symptoms seen. A degeneration of sympathetic ganglia results in orthostatic hypotension. The combined degeneration of the sympathetic system and the parasympathetic neurons especially in the dorsal motor nucleus of vagus results in symptoms such as constipation and nocturia (waking up at night to urinate.)

Why is Mrs. Smith's Mood Changing?

In Parkinson disease, there is a progressive loss of a heterogeneous population of neurons. Symptoms are often focused on the lack of dopaminergic projections from the substantia nigra, which result in the typical bradykinesia. In addition, degeneration of aminergic nuclei in the brainstem reticular formation can cause symptoms. These changes affect the

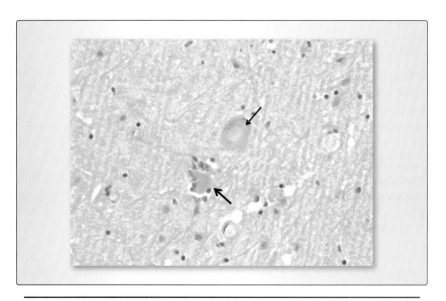

Figure 16.18

Degeneration in the substantia nigra neurons: pale bodies are pale eosinophilic structures (*thin arrow*) that are not as well defined as the classical Lewy bodies. Free classical Lewy bodies are also seen (*thick arrow*) in place of a degenerated neuron and are surrounded by the residual pigment both free and in the macrophages (H&E; original magnification: 400×). (Reproduced from Gokden M. *Neuropathologic Evaluation: From Pathologic Features to Diagnosis.* Wolters Kluwer Health and Pharma, 2013.)

noradrenergic neurons in locus ceruleus, the serotonergic raphe nuclei, and the dopaminergic ventral tegmental area.

The deposits of Lewy bodies (see Figure 16.18) throughout her forebrain lead to dementia, which she has noticed as changes to her short-term memory. The imbalance of major aminergic neurotransmitter systems (dopaminergic, noradrenergic, and serotonergic) has profound consequences for the activation of forebrain and limbic circuits that affect mood. (See Chapter 12, "Brainstem Systems and Review," for a detailed description of the effects of these neurotransmitter systems.)

Chapter Summary

- The basal ganglia nuclei are large masses of gray matter deep within the cerebral hemispheres. They include the caudate, putamen, globus pallidus, substantia nigra, and subthalamic nucleus. These nuclei are interconnected with each other and with other nuclei within the diencephalon and the midbrain. The basal ganglia function primarily as components in a series of parallel circuits and have a complex role in the control of movement and in encoding cognitive processes and their behavioral output.

- Input to the basal ganglia is to the striatum (caudate nucleus and putamen). The caudate nucleus and putamen each receive input from distinct cortical and subcortical regions. A single cell in the striatum receives input from multiple sources, which makes every cell in the striatum an **integrator**. This integration of input allows the basal ganglia to encode for the decision to move, the direction of the movement, the amplitude of the movement, and the motor expression of emotions. The output from the basal ganglia arises from the internal part of the globus pallidus (GPi) and the substantia nigra and projects to the ventral anterior and ventral lateral nuclei of the thalamus.

- The thalamus is under tonic inhibition. The *output* from the basal ganglia can either decrease or increase the tonic inhibition of the thalamus via two internal pathways. The direct pathway releases the thalamus from tonic inhibition. Removing this tonic inhibition leads to more excitation of the cortex and in turn more cortical output. The indirect pathway inhibits the output from the thalamus, leading to less excitation of the cortex and less cortical output. Thus, the indirect pathway counterbalances the effects of or puts "brakes" on the direct pathway. An imbalance of this system underlies motor disorders, such as Parkinson and Huntington diseases.

- The *input* to the basal ganglia can be described as three parallel streams of information from the cortex: motor, cognitive, and limbic. The motor circuit is the best known and plays the key role in motor performance and in the regulation of eye movements. The cognitive circuit plays an important role in cognitive processes and learning. The limbic circuit is involved in the regulation of emotional, motivational, and affective aspects of behavior. The striatum integrates these inputs and determines the activity of the thalamus via the direct and indirect pathways. The basal ganglia therefore integrate sensory, motor, emotional, and motivational inputs that result in a final common pathway, which determines the complex behavior we display.

- Blood supply to the basal ganglia arises from the penetrating branches off the circle of Willis.

Study Questions

Choose the ONE best answer.

16.1 Which statement about lesions to the basal ganglia is correct?

A. Degeneration of dopaminergic neurons in the substantia nigra leads to inhibition of activity in the indirect pathway, which increases the inhibition of the thalamus.

B. A loss of the subthalamic nucleus leads to decreased inhibition of the thalamus, which results in hyperkinetic movements on the ipsilateral side.

C. Degeneration of the striatum (caudate and putamen) leads to increased inhibition of the thalamus.

D. Degeneration of dopaminergic neurons in the substantia nigra leads to decreased inhibition of the subthalamic nucleus and, therefore, increased inhibition of the thalamus.

E. Degeneration of the striatum leads to decreased inhibition of the subthalamic nucleus and, therefore, decreased inhibition of the thalamus.

Correct answer is D. Degeneration of dopaminergic neurons in the SN is seen in Parkinson disease. This leads to a decreased amplification of the cortical input to the striatum and less activity in the direct pathway. The facilitation of target-oriented movements is impaired. In addition, excitatory cholinergic interneurons in the striatum lose their inhibition, which results in stimulation of inhibitor neurons to the external part of the globus pallidus (GPe). This, in turn, causes more excitatory output from the subthalamic nucleus (STN), which increases the inhibition of the thalamus through the internal part of the globus pallidus (GPi). A loss of the STN leads to decreased inhibition of the thalamus, which results in hyperkinetic movements on the contralateral side because the output from the cortex through the corticospinal tract crosses the midline. Degeneration of the striatum is seen in Huntington disease. The degeneration of the striatum leads to a loss of the inhibitory projections to the GPe, which *increases* the inhibition of the STN. Decreased excitation from the STN to the GPi decreases the inhibition of the thalamus, which results in more motor output.

16.2 Which one of the following statements about the motor and oculomotor circuits is correct?

    A. The motor circuit inputs to the basal ganglia come exclusively from motor areas of the cortex.

    B. The motor circuit only involves the facilitation of movement through the direct pathway.

    C. Oculomotor circuit outputs from the dorsomedial nucleus of the thalamus help to direct our gaze to a salient stimulus.

    D. Oculomotor inputs from prefrontal and posterior parietal cortices terminate in the putamen.

    E. The motor circuit is active during the planning of motor activity.

> The correct answer is C. The motor circuit receives input from both motor and sensory areas of the cortex. Through a balance of the direct and indirect pathways, target-oriented movements are facilitated, and excessive movements are inhibited, which results in balanced motor output. The motor circuit is not active during the planning of motor activity. Planning is facilitated through the cognitive circuit. The oculomotor circuit receives its input from the prefrontal and posterior parietal areas of the cortex. It is involved in initiating gaze as well as directing gaze toward a salient stimulus (via the dorsomedial nucleus of the thalamus).

16.3 An infarct to the deep branches of the cerebral arteries can have devastating effects on deep forebrain structures such as the basal ganglia. Which of the following is true in relation to the blood supply to the basal ganglia?

    A. The medial and lateral lenticulostriate branches of the middle cerebral artery supply the putamen, globus pallidus, and body of the caudate nucleus.

    B. The anterior choroidal artery supplies the head of the caudate nucleus.

    C. The substantia nigra receives its blood supply from deep branches of the middle cerebral artery.

    D. The posterior communicating artery supplies the posterior parts of the putamen and globus pallidus.

    E. The middle cerebral artery supplies the subthalamic nucleus.

> The correct answer is A. The anterior choroidal artery supplies the posterior portions of the putamen and globus pallidus. The substantia nigra receives its blood supply from the posterior cerebral and posterior communicating arteries. The posterior communicating artery and the deep branches of the posterior cerebral artery supply the substantia nigra and the subthalamic nucleus.

16.4 Parkinson disease is a motor disorder characterized by hypokinesia or akinesia. This effect on movement is due to:

    A. Increased excitatory input from the substantia nigra to the striatum, which increases the amplification of the cortical input to the striatum.

    B. Reduced inhibition of the globus pallidus (GPe), which results in greater inhibitory input to the subthalamic nucleus.

    C. Inhibition of the globus pallidus (GPi) results in less inhibitory input to the thalamus, thereby decreasing the tonic inhibition of the thalamus.

    D. A loss of dopaminergic signaling from the substantia nigra to the striatum, which results in increased stimulation of an inhibitory neuron projecting to the globus pallidus (GPe).

    E. A loss of dopaminergic input to the cholinergic interneuron in the striatum (caudate and putamen), which results in decreased stimulation of the globus pallidus (GPi).

> The correct answer is D. There is loss of excitatory input from the substantia nigra to the striatum that decreases the amplification of the cortical input to the striatum, resulting in less excitatory input going into the direct pathway. The globus pallidus (GPe) is inhibited, and there is less inhibitory input to the subthalamic nucleus. A stimulated globus pallidus (GPi) sends more inhibitory input to the thalamus, thereby increasing the tonic inhibition of the thalamus.

The Cerebellum

17

I. OVERVIEW

The cerebellum plays a critical role in the coordination and prediction of movement and mediates skilled manipulation of muscles. It receives information from the peripheral nervous system concerning the position of the body (proprioceptive information), the position of the head (vestibular information), muscle tone, and exteroceptive (visual, auditory, tactile) information from the environment.

The cerebellum compares and integrates this information with the plans for movement received from the cortex. The cerebellum can predict the consequence of movements through feed-forward mechanisms and can modulate ongoing movement patterns. This prediction of the consequences of movement applies not only to oneself but, importantly, to others as well. This is why we do not bump into each other on a busy street (Figure 17.1). The cerebellum calculates trajectories and likely behavioral outcomes and then modulates ongoing movement in an appropriate way.

Interestingly, the cerebellum also plays a significant role in cognition, particularly language, and helps to coordinate and predict mental activity.

In contrast to the basal ganglia, which integrate cortical activity into behavioral output, the cerebellum acts to **coordinate and predict** cortical output, thereby improving the **efficacy and precision** of our movement and cognition. These cerebellar functions are achieved through a series of circuits that link the cerebellum with the spinal cord, brainstem, and forebrain.

An important concept to remember when thinking about the cerebellum is that spinal inputs and outputs to the cerebellum related to truncal stability are bilateral and those related to limb movement are ipsilateral. Lesions to the cerebellum result in truncal instability or ipsilateral limb deficits.

II. ANATOMY

The cerebellum is located in the posterior cranial fossa, separated from the cerebrum by the tentorium cerebelli. It lies over the fourth ventricle and is connected to the brainstem by three cerebellar peduncles. All tracts to and from the cerebellum travel through the cerebellar peduncles (Figure 17.2).

Figure 17.1
Shibuya intersection in Tokyo at night.

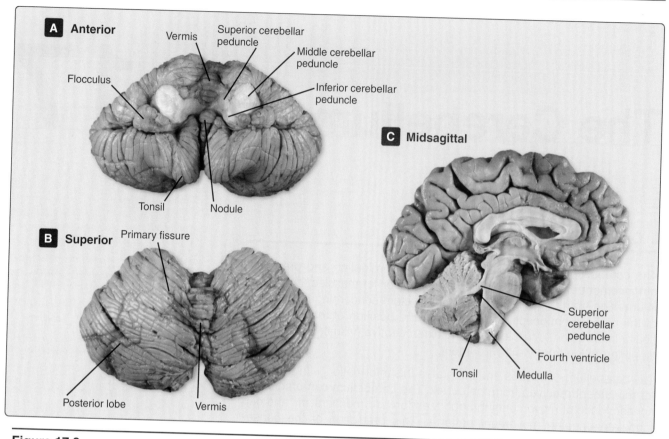

Figure 17.2

Overview of the cerebellum showing the cerebellar peduncles, vermis, and cerebellar hemispheres and the cerebellum attached to the brainstem overlying the fourth ventricle.

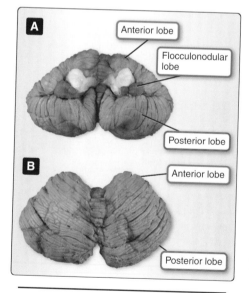

Figure 17.3

Lobes of the cerebellum in anterior (**A**) and superior (**B**) views, respectively.

The cerebellum is roughly triangular in shape with three surfaces: a superior surface touching the tentorium, an inferior surface touching the inferior cranial fossa, and an anterior surface touching the brainstem. There are right and left hemispheres and a midline structure called the **vermis**. The cerebellum is extensively folded. These folds are referred to as **folia**.

A. Lobes

The cerebellum is divided into three lobes. On the superior surface, the **anterior lobe** is separated from the **posterior lobe** by the primary fissure. The **flocculonodular lobe** is visible on the anterior surface. The nodule is the most anterior part of the vermis and is connected to the flocculus. Together, they form the flocculonodular lobe (Figure 17.3).

The most medial area of the inferior surface of the cerebellum comprises the **cerebellar tonsils.** These are located just above the foramen magnum.

B. Cerebellar deep nuclei

There are four paired deep cerebellar nuclei that serve as relay and processing stations for information coming from the cerebellar cortex to targets outside the cerebellum (Figure 17.4). Most of these nuclei also receive input from afferents to the cerebellum, which likely serves

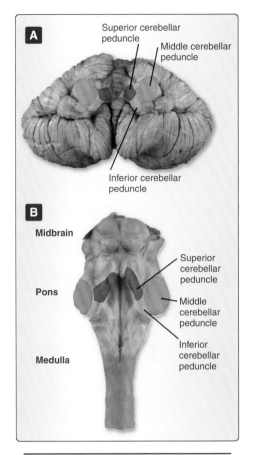

Clinical Application 17.1: Tonsillar Herniation

The cerebellar tonsils can be pushed into the foramen magnum when there is an increase in intracranial pressure affecting the infratentorial (posterior fossa) compartment. Increased pressure can result from an expanding intracranial mass in the posterior fossa or increased intracranial pressure above the tentorium that pushes supratentorial contents against the tentorium. The resulting pressure on the medulla compromises the respiratory center and puts traction on the blood vessels that can lead to hemorrhage in the midbrain and pons, called Duret hemorrhage. Tonsillar herniation is a neurosurgical emergency that is often fatal if not treated quickly.

the purpose of better coordinating and regulating the output from the cerebellum. The largest deep nucleus is the **dentate nucleus**. From lateral to medial, the nuclei are the **d**entate, **e**mboliform, **g**lobose, and **f**astigial (mnemonic: **d**on't **e**at **g**reasy **f**ood).

C. Cerebellar peduncles

The cerebellum is connected to the brainstem via three cerebellar peduncles (superior, middle, and inferior) that carry all tracts to and from the cerebellum (Figure 17.5). Table 17.1 summarizes the fibers traveling within each of these peduncles.

D. Cerebellar cortex

Similar to the cerebral cortex, representations of areas of the body can be mapped on the cerebellar cortex. This map, called a **homunculus** (Latin for "little human"), has its trunk in the midline on the vermis and its extremities on the cerebellar hemispheres. The anterior lobe has

Figure 17.5
Cut surfaces of the cerebellar peduncles in anterior (**A**) and posterior (**B**) views, respectively.

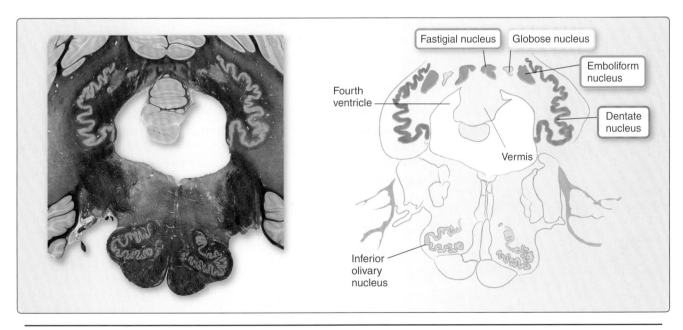

Figure 17.4
Overview of the deep cerebellar nuclei. Section through the rostral medulla and cerebellum.

Table 17.1: Overview of Afferent Tracts to and Efferent Tracts from the Cerebellum

	Afferents	Efferents
Superior cerebellar peduncle	Anterior spinocerebellar tract Acoustic and optic information	Dentatorubrothalamic tract Dentatothalamic tract
Middle cerebellar peduncle	Pontocerebellar tract	—
Inferior cerebellar peduncle	Vestibulocerebellar tract Olivocerebellar tract Posterior spinocerebellar tract Cuneocerebellar tract	Cerebellovestibular tract Cerebelloolivary tract

representation of the extremities, and the posterior lobe has mirror image representations of both the head and the extremities. The trunk is always in the midline. In order to better understand the functional areas of the cerebellar cortex, the cortex can be "peeled off" and spread out and the homunculus mapped on this flattened cortex (Figure 17.6).

III. TRACTS

The cerebellum receives and interprets proprioceptive information. It coordinates balance through its close linkage to the vestibular nuclei. Connections with the forebrain allow for the coordination of limb movement as well as fine movement and eye–hand coordination. We first describe the afferents to and the efferents from the cerebellum (see Table 17.1), and in the next section, we link them together into functional loops of information.

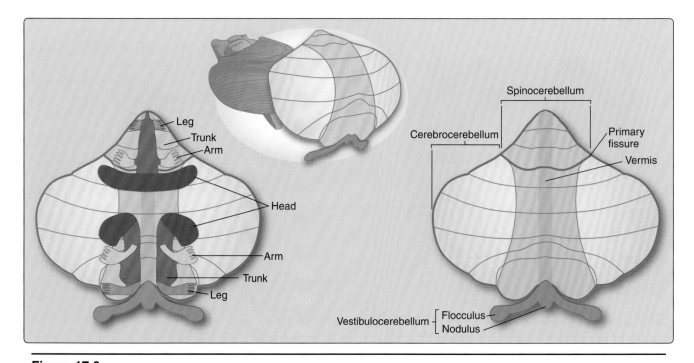

Figure 17.6
The cerebellar cortex mapped on a flat surface. *Left panel* shows the somatotopic representation of the body (homunculus). *Right panel* shows the three functional areas of the cerebellar cortex.

A. Afferents

The cerebellum receives input about balance from the vestibular system, proprioceptive information from the spinal cord, and cortical information via the pontine nuclei. All of these fibers comprise the **mossy fibers** in the cerebellar cortex. Input from the inferior olivary nuclear complex is mainly contralateral (with some bilateral projections) and helps the cerebellar cortex to modulate and coordinate its output. The olivocerebellar fibers are the **climbing fibers** of the cerebellar cortex. All information to the cerebellum travels through the cerebellar peduncles (Figure 17.7).

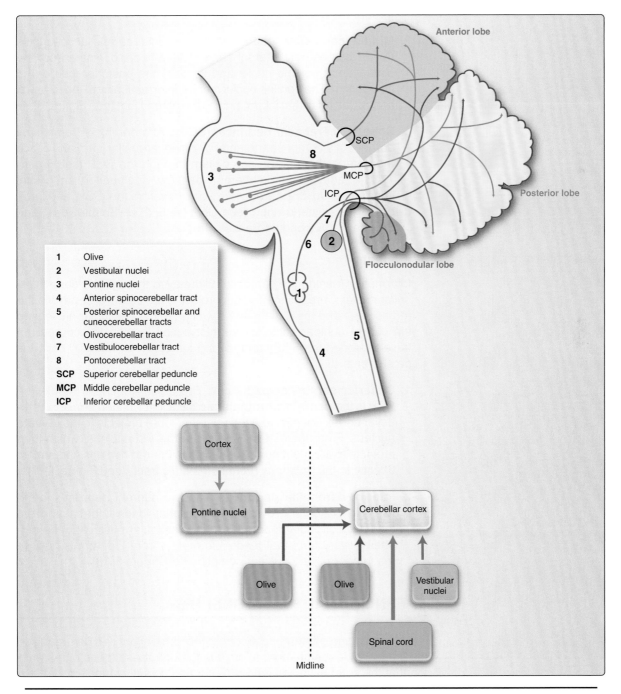

Figure 17.7
Cerebellar afferents.

1. **Inferior cerebellar peduncle:** The inferior cerebellar peduncle (ICP) carries a variety of afferent fibers to the cerebellum. Vestibular information from both ipsilateral and contralateral vestibular nuclei travels through this peduncle to the flocculonodular lobe. The ipsilateral input is the more prominent input.

   Proprioceptive input from the Clarke column (lower limbs and lower trunk) and the accessory cuneate nucleus (upper limbs and upper trunk) comprising the posterior spinocerebellar tract and cuneocerebellar tract, respectively, enters the cerebellum through the ICP and terminates primarily in the anterior lobe. Olivocerebellar fibers also travel through the ICP and terminate as climbing fibers throughout the cerebellar cortex (see Figure 17.7).

2. **Middle cerebellar peduncle:** The middle cerebellar peduncle (MCP) carries only afferent fibers. These fibers originate from the contralateral pontine nuclei and travel to the posterior lobe of the cerebellum. The pontine nuclei receive information from the cerebral cortex and relay that information to the cerebellum (see Figure 17.7).

3. **Superior cerebellar peduncle (SCP):** Afferent fibers from spinal border cells comprise the anterior spinocerebellar tract and reach the anterior lobe of the cerebellum through the SCP (see Figure 17.7). In addition, some afferent fibers carrying auditory and visual information to the cerebellum also travel in the SCP and project to the posterior lobe (see Chapter 5, "The Spinal Cord," and Chapter 7, "Ascending Sensory Tracts").

B. Efferents

Information leaving the cerebellum is relayed through the deep cerebellar nuclei. Cerebellar efferents project to the cerebral cortex via the red nucleus and the thalamus, as well as to the vestibular nuclei and the olivary nuclear complex. All information leaving the cerebellum travels through the SCPs and ICPs (Figure 17.8). The MCP contains no efferents.

1. **Superior cerebellar peduncle:** The most prominent tract in the SCP is the **dentatorubrothalamic tract**. Information from the cerebellar cortex (mostly from the posterior lobe) projects to the dentate nucleus. From there, it can project to the red nucleus, which then projects to the thalamus or can bypass the red nucleus and project directly to the thalamus (dentatothalamic tract) (see Figure 17.8).

2. **Inferior cerebellar peduncle:** The two efferent pathways in the ICP are the cerebellovestibular tract that feeds information from the flocculonodular lobe back to the vestibular nuclei and the cerebelloolivary tract that feeds information from the vermis back to the inferior olivary nuclear complex (see Figure 17.8).

IV. FUNCTIONAL RELATIONSHIPS

The afferents and efferents to and from the cerebellum can be organized into functional loops of information. It is through these loops of information that the cerebellum influences motor and cognitive function.

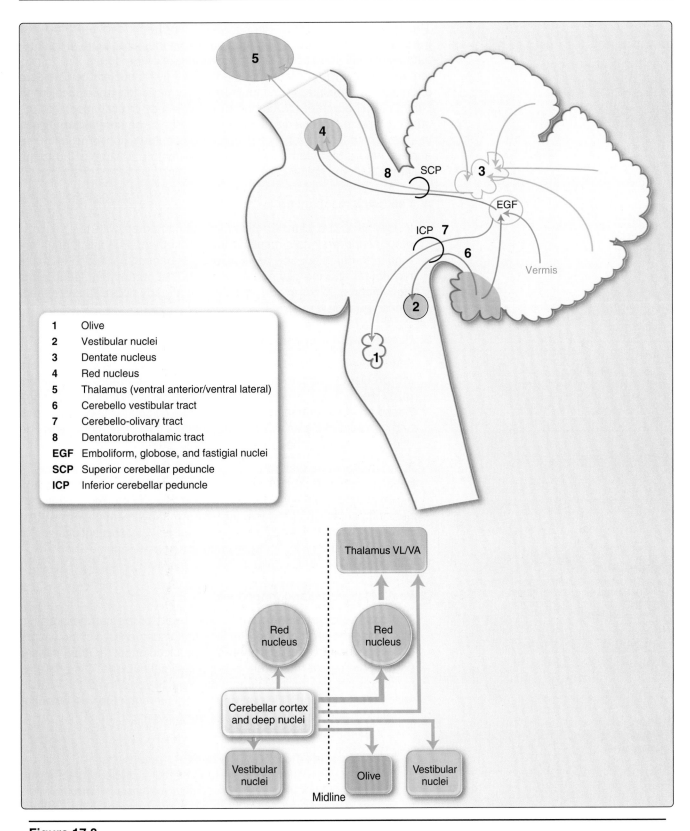

Figure 17.8
Cerebellar efferents.

The cerebellum operates through both **feedback** and **feed-forward mechanisms**. The inputs to and outputs from the cerebellum are segregated, and each functional loop is independent of the others. The important roles of the red nucleus and the inferior olivary nuclear complex are often underappreciated. The inputs from the inferior olivary nuclear complex go to all areas of the cerebellum where they directly modulate the output from the Purkinje cells, the output cells of the cerebellum. These inputs play a role in all functional loops. Many outputs converge in the red nucleus where the information is integrated before it is projected to the thalamus or to the spinal cord.

A. Feedback mechanism

In feedback, the plan for movement is compared with the sensory, proprioceptive signals resulting from movement. Feedback does not require learning but only an accurate comparison between the actual movement and the planned movement. This means that feedback is slow. It can only correct errors that have already occurred and cannot predict and prevent errors.

B. Feed-forward mechanism

Feed-forward on the other hand allows quick reaction because errors are predicted and prevented before they can happen. This system relies on learning from previous experience (associative learning). Based on our experiences, the cerebellum calculates likely outcomes and predicts the sensory consequences, therefore correcting the ongoing movement before an error can occur.

This feed-forward system functions not only for our own movements but also for movements of objects and people around us. When walking on a busy street, for example, we calculate the movement trajectories of those around us and adjust our own movement pattern so that we do not bump into each other (see Figure 17.1). This anticipation of movement also applies to static images where we can see and anticipate movement, which gives these images life.

C. Vestibulocerebellar connections

The **vestibulocerebellum** is the oldest part of the cerebellum. It is also called the **archicerebellum**. The main components of the vestibulocerebellum are the vestibular nuclei, flocculonodular lobe, inferior parts of the paravermal area, and the fastigial nucleus. The afferents to the cerebellum provide information about the position of the head in space and help in orienting eye movements through the vestibuloocular reflex (see Chapter 9, "Control of Eye Movement," and Chapter 11, "Hearing and Balance").

1. **Afferents and efferents:** The wiring of this system is relatively straightforward. Afferents from the vestibular nuclei project to the flocculonodular lobe and inferior paravermal area, where the information is processed. From there, efferents are sent to the fastigial nucleus. From the fastigial nucleus, bilateral projections are sent to the vestibular nuclei and the reticular formation. From these structures, this information projects through the vestibulospinal and reticulospinal tracts to spinal cord motor neurons to adjust truncal (axial) stability and balance (Figure 17.9).

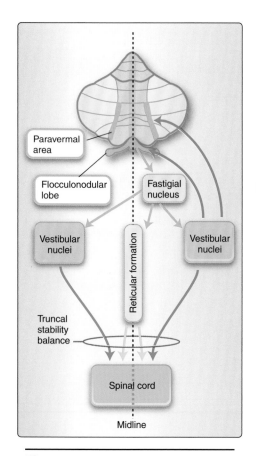

Figure 17.9
Vestibulocerebellar connections.

Clinical Application 17.2: Flocculonodular Lobe Syndrome

This syndrome is the result of a lesion of the flocculonodular lobe or its afferents or efferents. It is most commonly seen in children with a medulloblastoma. The syndrome is characterized by truncal ataxia due to the inability to stabilize or balance the core musculature. This results in a wide-based stance as well as swaying. In addition to truncal ataxia, a nystagmus is often diagnosed due to damage to the vestibuloocular pathways.

2. **Additional input:** Additional input (not shown in the figure) to the vestibulocerebellar loop comes from the contralateral inferior olivary nuclear complex providing motor information and from the contralateral basal pontine nuclei, which relay visual information from the cortex.

This system contains both feed-forward and feedback loops through the cerebellum from both the locomotor system and the vestibular system. This provides a continuous correction to and anticipation of changes in **stability and balance**.

D. Spinocerebellar connections

The **spinocerebellum** is the second oldest part of the cerebellum. It is also referred to as the **paleocerebellum**. It comprises the anterior lobe, vermis (without the nodule), and superior paravermal area. The anterior lobe is composed almost entirely of vermis and paravermis.

1. **Inputs:** Ia and II fibers from muscle spindles and Ib fibers from Golgi tendon organs carry proprioceptive information to the Clarke column and the accessory cuneate nucleus. From there, the fibers travel as posterior spinocerebellar and cuneocerebellar tracts through the ICP to the spinocerebellum.

Spinal border cells receive input from the upper motor neuron system. The axons of the spinal border cells form the anterior spinocerebellar tract. This tract projects to the spinocerebellum through the SCP. This is how the cerebellum receives a "copy" of the motor information sent to the lower motor neuron system of the spinal cord.

2. **Functions:** Body movement can be broadly divided into truncal movements and limb movements (primarily related to gait) (Figure 17.10). The cerebellum coordinates both of these through the spinocerebellar loop.

   a. **Truncal movement:** Truncal movements are processed mainly in the vermis (see homunculus Figure 17.6), and the output to truncal musculature is through the fastigial nucleus. From the fastigial nucleus, **bilateral** projections are sent to the vestibular nuclei, the red nuclei, and the reticular formation. From these structures, this information projects **bilaterally** through the vestibulospinal, rubrospinal, and reticulospinal tracts to spinal cord motor neurons to adjust and fine-tune truncal movements.

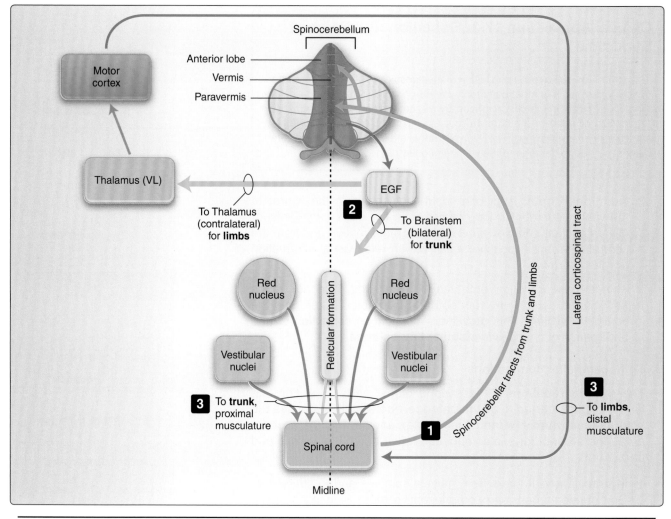

Figure 17.10
Spinocerebellar connections. VL = ventral lateral nucleus; EGF = emboliform, globose, and fastigial nuclei.

b. Limb movement: Limb movement as it relates to gait is coordinated mainly in the anterior lobe. Output goes through the emboliform and globose nuclei to the ventral lateral thalamus (VL). Projections from the VL to the motor cortex allow for the adjustment of ongoing limb movement through the lateral corticospinal tract. The coordination of limbs is lateralized: the cerebellum controls the ipsilateral side of the body through the projection to the contralateral motor cortex. For example, the left cerebellum projects to the right motor cortex, which, through the lateral corticospinal tract, projects back to the left anterior horn of spinal cord (the loop crosses the midline twice).

The deep cerebellar nuclei also receive information from the contralateral olivary nuclear complex, which provides further information on movement. By integrating ipsilateral proprioceptive information and motor plans with contralateral movement information from the olivary nuclear complex, output can regulate synergistic limb movements. This precise regulation ensures the fluidity of limb movements while maintaining the stability of the trunk.

Clinical Application 17.3: Anterior Lobe Syndrome

This syndrome is the result of a lesion of the anterior lobe or its afferents or efferents. It is characterized by gait ataxia due to the inability to process proprioceptive information from the limbs.

An example of a lesion to the anterior lobe is ethanol-induced gait ataxia. Ethanol is toxic to Purkinje cells, particularly in the anterior lobe. Acute intoxication will result in typical gait ataxia with the inability to walk stably on a line (typically tested in roadside alcohol checks). In addition, truncal ataxia and nystagmus can also be present, probably due to involvement of the vermis. These symptoms resolve when ethanol is cleared from the system. In chronic alcoholics, however, the damage to the cells in the anterior lobe can be permanent, resulting in anterior lobe degeneration with atrophy of all layers of the cerebellar cortex and a loss of Purkinje cells. The clinical presentation is ethanol-induced gait ataxia that does not resolve when ethanol is cleared from the system.

E. Cerebrocerebellar connections

The **cerebrocerebellum** is the newest addition to the cerebellum and is also called the **neocerebellum**. It comprises the lateral aspects of the posterior lobes. Input to the neocerebellum comes from the pontine nuclei. Broad areas of cortex project to the ipsilateral pontine nuclei. From the pontine nuclei, fibers cross the midline (transverse fibers of the pons) and enter the contralateral neocerebellum through the MCP. Additional afferent information comes from the contralateral olivary nuclear complex (climbing fibers of the cerebellar cortex).

1. **Reciprocal connections:** Output from the neocerebellar cortex is mainly to the dentate nucleus, which in turn projects to the red nucleus and from there to the VL of the thalamus; this is the **dentatorubrothalamic tract**. There are also direct projections from the dentate nucleus to the thalamus, the **dentatothalamic tract**. From the thalamus, information projects back to motor and sensory areas of the cortex (Figure 17.11).

   These reciprocal connections with the cerebral cortex put the cerebellum in a position to coordinate and streamline motor output from the cortex.

2. **Functions:** The neocerebellum is necessary for eye–hand coordination. It uses visual input and calculates the trajectory of movement needed to reach or manipulate a target. This involves both feedback and feed-forward mechanisms that allow learning and experience to influence movement.

   a. **Sensory consequence:** The neocerebellum also predicts the sensory consequence of a movement through comparison with past experience. This is the reason you cannot tickle yourself: the neocerebellum has already predicted the sensory consequence of this self-generated motor command and has attenuated the response in the sensory cortex.

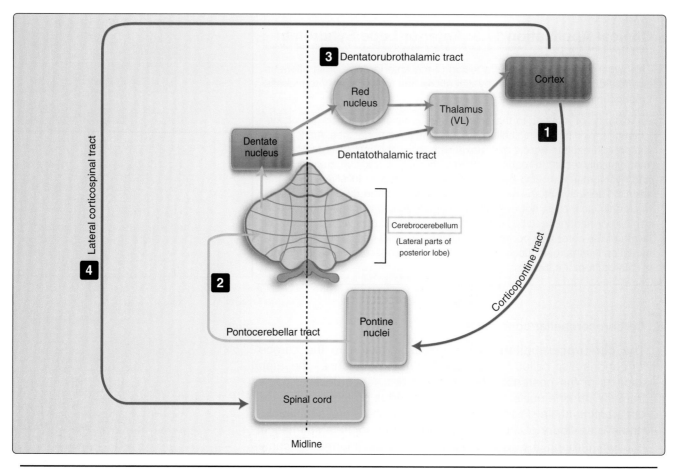

Figure 17.11
Cerebrocerebellar connections.

b. **Voluntary movement:** The neocerebellum is involved in the planning of voluntary movements, making them more automatic. It is responsible for the fine-tuning of motor patterns, such that with practice a new skill comes to be performed automatically. Examples of this are handwriting and playing the piano. Both are learned skills that have become automatic with practice. We do not think how individual letters should be written, but we think about concepts. We do not think about the order and spacing of the piano keys and the movement necessary to strike them, but we think about the musical phrasing. This automatization by the cerebellum effectively frees up the cerebrum for higher-order cognitive activity.

c. **Coordination of motor activity and cognition:** Importantly, the input to the neocerebellum is not only from motor areas but also from cortical areas related to cognitive and sensory function, and it can therefore automatize not only motor but also sensory and cognitive skills. The neocerebellum modulates but does not generate language and cognition. Through its connections, it is an interface between cognition and motor output. Language is an example of a function requiring both mental and motor activity that is coordinated by the cerebellum. It is responsible for linguistic coordination, fluidity of language, automatization of syntax and grammar, as well as the prediction of sentence structure and flow. Patients with a cerebellar dysfunction can present with cerebellar mutism.

V. CYTOARCHITECTURE OF THE CEREBELLAR CORTEX

Three layers of the cerebellar cortex can be differentiated: closest to the surface is the molecular layer, then the Purkinje cell layer, followed by the granular layer (Figure 17.12). Afferents to the cerebellum are classified as mossy fibers and climbing fibers. Efferent information is carried by the Purkinje cells. Afferents and efferents travel through the white matter. The cerebellar nuclei are located deep in the white matter.

A. Cells and fibers of the cerebellar cortex

The cerebellar cortex contains five types of cells: granule, Golgi, Purkinje, basket, and stellate cells, and three types of fibers: mossy, climbing, and parallel fibers (Figure 17.13).

1. **Cells of the cerebellar cortex:** Granule cells are very abundant and are located in the **granule cell layer** (see Figure 17.13). Their axons travel to the molecular layer where they branch into a T formation to form the **parallel fibers**.

   Golgi cells are also located in the granule cell layer. Their processes radiate into all other layers (see Figure 17.13).

   Purkinje cells are the largest cells in the cerebellar cortex and are located in the Purkinje cell layer (see Figure 17.13). Their dendrites fan out in one plane into the molecular layer.

   Basket cells and stellate cells are located in the molecular layer (see Figure 17.13). Their branching is perpendicular to the Purkinje cell dendritic tree. One basket cell synapses with about 70 Purkinje cells.

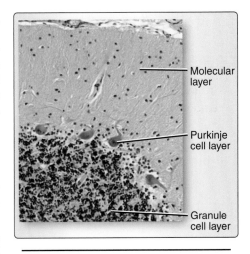

Figure 17.12
Histology of the three layers of the cerebellar cortex.

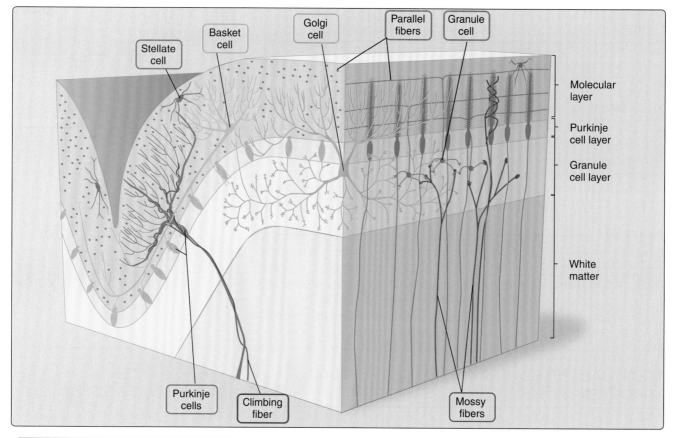

Figure 17.13
Cytoarchitecture of the cerebellar cortex shown in a 3D view.

2. **Fibers of the cerebellar cortex:** Mossy fibers comprise **cerebellar afferents** originating from all sources except the inferior olivary nuclear complex (see Figure 17.13). They include afferents from the spinal cord, posterior column nuclei, trigeminal system, pontine nuclei, and vestibular nuclei.

Climbing fibers originate from the **inferior olivary nuclear complex**. Each neuron within the inferior olivary nuclear complex gives rise to about 10 climbing fibers.

Parallel fibers arise from granule cells and are perpendicular to the plane of the Purkinje cell dendritic tree in the molecular layer.

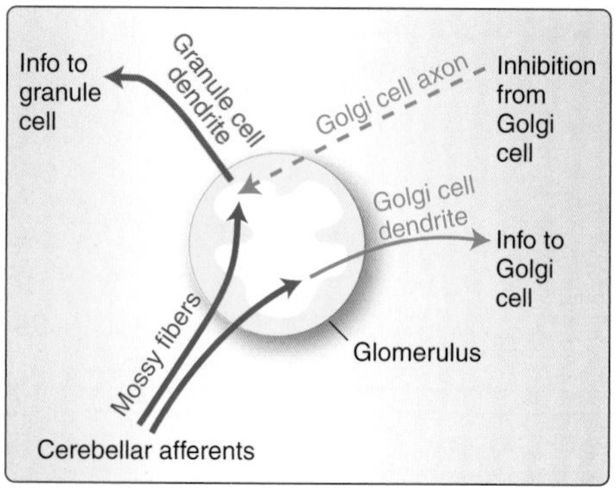

B. Wiring of the cerebellar cortex

All cells of the cerebellar cortex are interconnected in a complex wiring scheme, and a simplified diagram of the key connections is shown in Figure 17.15.

Mossy fiber afferents synapse with Golgi cell and granule cell dendrites in the cerebellar glomerulus. The glomerulus is located in the granular layer and is the first processing station for mossy fiber afferents to the cerebellum (Figure 17.14). The synapse between mossy fibers and granule cells is under the inhibitory control of Golgi cell axons. After this first processing stage, granule cells convey this "prescreened" afferent information to the Purkinje cells (see Figure 17.15).

Figure 17.14
Cerebellar glomerulus.

Individual climbing fibers from the inferior olivary nuclear complex form a direct excitatory synapse with one Purkinje cell. This one-to-one relationship makes this a powerful synapse within the cerebellar circuitry. Climbing fibers provide feedback about the sensory consequence of movement and motor errors. These synapses are key in motor learning.

After receiving processed afferent information from granule cells and direct afferent information from climbing fibers, Purkinje cells send inhibitory output to the deep cerebellar nuclei. Further synapses with the Purkinje cells are from stellate cells and basket cells (both inhibitory). Purkinje cells are the source of all efferent information from the cerebellum.

VI. BLOOD SUPPLY

The blood supply to the cerebellum comes from three vessels from the vertebral basilar system: the **superior cerebellar artery (SCA)**, the **anterior inferior cerebellar artery (AICA)**, and the **posterior inferior cerebellar artery (PICA)** (Figure 17.16). Supply can be variable and specific structures can have overlapping blood supply. Table 17.2 summarizes the most common pattern of blood supply to the cerebellum.

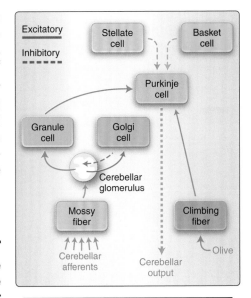

Figure 17.15
Wiring diagram for the cerebellar cortex.

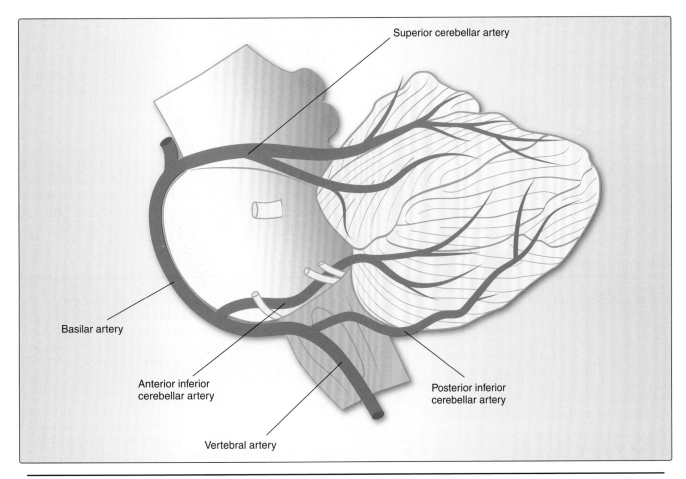

Figure 17.16
Blood supply of the cerebellum in a lateral view.

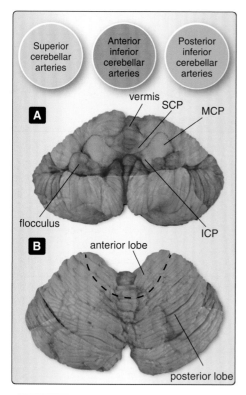

Figure 17.17
Perfusion areas of the cerebellar vessels in anterior (**A**) and superior (**B**) views, respectively.

Table 17.2: Overview of Blood Supply to the Cerebellum

Artery	Region Supplied
Anterior inferior cerebellar artery (AICA)	MCP and ICP Flocculonodular lobe Assists with deep nuclei Some areas of posterior lobe
Posterior inferior cerebellar artery (PICA)	Inferior two-thirds of the posterior lobe Tonsils Nodulus, variable Inferior vermis
Superior cerebellar artery (SCA)	Anterior lobe Superior one-third of the posterior lobe Superior part of the vermis Superior cerebellar peduncle All deep nuclei

A. Superior cerebellar artery

The SCA supplies the anterior lobe, the superior part of the vermis, the superior third of the posterior lobe, as well as the deep cerebellar nuclei and the SCP. It also supplies lateral parts of the rostral pons (Figure 17.17).

B. Anterior inferior cerebellar artery

The AICA supplies the middle portion of the anterior surface of the cerebellum and the flocculonodular lobe. It provides the major supply to the MCP and the ICP. AICA also provides variable contributions to the deep cerebellar nuclei. It also provides blood supply to the lateral parts of the caudal pons (see Figure 17.16). The blood supply to the SCP may come from AICA as well, but this is variable.

C. Posterior inferior cerebellar artery

PICA supplies the inferior two-thirds of the posterior lobe and the inferior part of the vermis. It supplies the cerebellar tonsils and has variable supply to the nodule and ICP. It also provides blood supply to the lateral parts of the medulla (see Figure 17.16 and Chapter 6, "Overview and Organization of the Brainstem") (Table 17.2).

Clinical Case:

Evan's Instability and Gait Problems

Evan is a 6-year-old boy who presents with progressive gait instability. He has a history of normal development and no other preexisting medical conditions. He does not take any medications. There is no history of toxin exposure. There is no family history of coordination or gait problems. He has intermittent occipital headaches with nausea and vomiting, particularly when he coughs or lies down.

On examination, he is alert with normal speech and language function. Fundoscopic examination is normal. Pupillary responses are normal. His visual acuity and visual field testing are normal. He has full range of extraocular movement with no nystagmus. The rest of his cranial nerve examination

is normal. Muscle bulk, tone, strength, and reflexes are normal. Sensory examination is normal. He has a wide-based gait and sways when he walks. He is unable to stand in a heel-to-toe posture (tandem stance) or walk heel to toe (tandem gait). Finger-to-nose and heel-to-shin testing however are normal. There is no dysmetria. Rapid alternating movements of his hands are normal. When sitting on the edge of the bed, he has a mild sway of his trunk.

What type of cerebellar syndrome causes the wide-based gait, truncal instability, and impairment in tandem walking with normal limb coordination?

Flocculonodular lobe syndrome causes instability or imbalance of axial muscles. Truncal ataxia with wide-based gait, swaying, and impairment of tandem gait are features of this syndrome.

Anterior lobe syndrome causes gait ataxia due to impairment of proprioceptive input from the limbs. Impaired heel-to-shin testing, wide-based gait and inability to tandem walk are features of anterior lobe syndrome, but in Evan's case, heel-to-shin testing was normal.

Posterior lobe syndrome causes dysmetria (inability to calculate the trajectory of a target) and dysdiadochokinesia (inability to coordinate agonist—antagonist movements in the extremity).

What is the likely etiology of Evan's symptoms?

Occipital headaches with nausea and vomiting that worsen when lying down are worrisome features of his history and prompt urgent neuroimaging.

An MRI shows a midline mass along the roof of the fourth ventricle (Figure 17.18). Pathology from the surgical resection shows that the tumor is a medulloblastoma, a type of primitive neural ectodermal tumor (PNET). PNETs are the most common group of primary brain tumors in children.

If the medulloblastoma blocks flow of cerebrospinal fluid in the fourth ventricle, obstructive hydrocephalus can occur. Symptoms related to obstructive hydrocephalus include headache and nausea/vomiting. In Evan's case, the increased intracranial pressure had not yet resulted in bulging of the optic disc, which would have been seen in the fundoscopic examination.

Why is it important to ask about medication, toxin exposure, and family history?

Medications such as certain anticonvulsants (phenytoin and carbamazepine), lithium, and some chemotherapeutic agents, as well as toxins such as alcohol and toluene (glue), can cause ataxia.

There are autosomal dominant, autosomal recessive, and X-linked causes of ataxia. Asking about a family history of imbalance, incoordination, and gait issues is therefore important. Spinocerebellar ataxia and Friedrich ataxia are examples of genetic causes of ataxia.

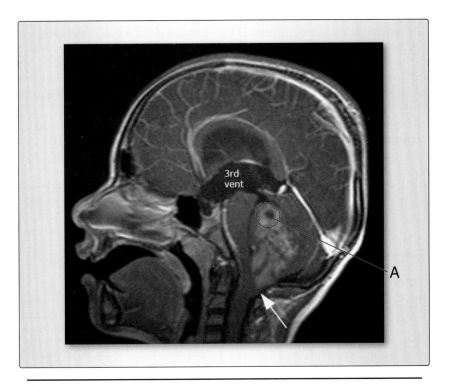

Figure 17.18

Posterior fossa medulloblastoma in a child. Sagittal T1-weighted MRI post-gadolinium image shows a tumor (**A**) within the posterior cranial fossa compressing the fourth ventricle. This heterogeneously-enhancing tumor is typical of medulloblastoma. Note inferior cerebellar tonsillar herniation (white arrow) and dilation of the third ventricle above the tumor level. (Image from Iyer RS, Chapman T. *Pediatric Imaging: The Essentials*. Philadelphia, PA: Wolters Kluwer Health, 2015.)

Chapter Summary

- The cerebellum can be considered the coordinator and predictor of movement. It compares and integrates sensory information from the periphery with the plans for movement received from the cortex. The cerebellum operates through both feedback and feed-forward mechanisms. Feedback is slow, does not require learning, and allows for the correction of errors in movement after they have occurred. Feed-forward is quick, relies on learning from previous experience, and can modulate ongoing movement patterns. The cerebellum also plays a significant role in cognition, particularly language, and helps to coordinate and predict mental concepts.

- The cerebellum is located in the posterior cranial fossa, overlying the fourth ventricle. It is connected to the brainstem by three cerebellar peduncles: inferior, middle, and superior. It can be divided into three lobes: anterior, posterior, and flocculonodular. The dentate, emboliform, globose, and fastigial nuclei lie deep within the white matter.

- Input from the inferior olivary nuclear complex, which helps coordinate motor output, comprises the climbing fibers of the cerebellar cortex. The mossy fibers carry vestibular, proprioceptive, and cortical information. Output from the cerebellum is relayed through the deep cerebellar nuclei.

- The afferents and efferents to and from the cerebellum can be organized into functional loops of information.

- The vestibulocerebellum consists of the vestibular nuclei, flocculonodular lobe, inferior parts of the paravermal area, and fastigial nucleus. Afferents to the vestibulocerebellum provide information about the position of the head in space and help in orienting eye movements. Bilateral efferent projections from the fastigial nucleus provide for adjustment of axial stability and balance.

- The spinocerebellum consists of the anterior lobe, vermis (without the nodule), and superior paravermal area. Afferents to the spinocerebellum include proprioceptive and motor information. The spinocerebellum coordinates truncal movements through outputs from the fastigial nucleus and coordinates limb movement through output from the emboliform and globose nuclei. The coordination of limbs is lateralized, with the cerebellum controlling the ipsilateral side of the body.

- The cerebrocerebellum comprises the lateral aspects of the posterior lobes. Input to the cerebrocerebellum from broad areas of cortex is sent via the pontine nuclei. Output from the cerebrocerebellar cortex is mainly through the dentate nucleus. The cerebrocerebellum coordinates and streamlines motor output from the cortex, is necessary for eye–hand coordination, and predicts the sensory consequence of a movement through comparison with past experience. It is also involved in the planning and automatization of voluntary movements as well as sensory and cognitive skills, including the linguistic coordination and fluidity of language.

- The cells and fibers of the cerebellar cortex (Golgi, granule, and Purkinje cells; stellate and basket cells; mossy and climbing fibers) are interconnected in a complex wiring scheme.

- The blood supply to the cerebellum comes from the superior cerebellar artery, the anterior inferior cerebellar artery, and the posterior inferior cerebellar artery.

Study Questions

Choose the ONE best answer.

17.1 Which of the following statements best describes the fibers traveling through the cerebellar peduncles?

- A. Efferent fibers from the deep cerebellar nuclei exit the cerebellum mainly through the superior cerebellar peduncle.
- B. The middle cerebellar peduncle carries a mix of afferent and efferent fibers.
- C. Proprioceptive information from the spinal cord enters the cerebellum through the middle cerebellar peduncle.
- D. Motor information from spinal border cells enters the cerebellum through the inferior cerebellar peduncle.
- E. Efferents from the cerebellum to the inferior olivary nuclear complex travel through the superior cerebellar peduncle.

Correct answer is A. The middle cerebellar peduncle carries only afferent fibers to the cerebellum. Proprioceptive information from the spinal cord enters the cerebellum through the inferior cerebellar peduncle. Motor information from spinal border cells enters the cerebellum through the superior cerebellar peduncle. Efferents from the cerebellum to the inferior olivary nuclear complex travel through the inferior cerebellar peduncle.

17.2 Which of the following statements about the spinocerebellar loop is correct?

   A. Proprioceptive information from the spinal cord enters the cerebellum and terminates in the posterior lobe.

   B. The coordination of truncal movements happens primarily in the anterior lobe.

   C. The cerebellum coordinates limb movements on the ipsilateral side of the body.

   D. The emboliform and globose nuclei project primarily to the red nucleus.

   E. The fastigial nucleus sends projections only to the ipsilateral vestibular nuclei.

> Correct answer is C. Proprioceptive information from the spinal cord terminates in the anterior lobe of the cerebellum. Truncal movements are processed primarily in the vermis. The emboliform and globose nuclei project primarily to the ventral lateral nucleus of the thalamus. The fastigial nucleus sends bilateral projections to the vestibular nuclei.

17.3 Which of the following statements about the wiring of the cerebrocerebellar loop is correct?

   A. The pontine nuclei receive projections from the contralateral cerebral cortex and, in turn, project to the contralateral cerebellar cortex.

   B. Input to the neocerebellum comes primarily from the vestibular nuclei.

   C. Climbing fibers from the ipsilateral inferior olivary nuclear complex project to the neocerebellar cortex.

   D. Output from the neocerebellum is through the dentate nucleus to the red nucleus and the thalamus.

   E. The neocerebellum has a role in the generation of language and cognition.

> Correct answer is D. The pontine nuclei receive projections from the ipsilateral cerebral cortex and, in turn, project to the contralateral cerebellar cortex. Input to the neocerebellum comes primarily from the pontine nuclei. Climbing fibers from the contralateral inferior olivary nuclear complex project to the neocerebellar cortex. The neocerebellum modulates but does not generate cognition and language.

17.4 In relation to the cytoarchitecture of the cerebellar cortex, which of the following is true?

   A. The Purkinje cell axon is an afferent pathway of the cerebellum.

   B. Mossy fibers synapse on Golgi cell and granule cell dendrites in the cerebellar glomerulus.

   C. Basket cells synapse in a one-to-one ratio with Purkinje cells.

   D. Golgi cells provide excitatory input to the cerebellar glomerulus.

   E. Climbing fibers provide feedback about motor output of the dentate nuclei.

> Correct answer is B. The Purkinje cell axon is the efferent pathway of the cerebellum. Basket cells synapse in a one-to-seventy ratio with Purkinje cells. Golgi cells provide inhibitory input to the cerebellar glomerulus. Climbing fibers constitute a sensory feedback loop indicating motor errors.

18 The Integration of Motor Control

I. OVERVIEW

Movement is the basis of our behavioral output, whether through gross muscle movement, fine movement of our hands, or movement of muscles of facial expression or muscles involved in speech. It is through motor activity that we interact and communicate with the world around us (Figure 18.1).

In the previous chapters, we discussed spinal cord circuitry, descending motor tracts and their influence on lower motor neurons (LMNs) in the spinal cord or brainstem, the modulatory influences of cerebellar and basal ganglia circuits, and the role of sensory input on voluntary motor activity. In this chapter, we bring these concepts together and examine how all of these systems are coordinated and create behavior, or motor output, as summarized in Figure 18.1.

The motor system can be divided into an **upper motor neuron (UMN) system** located in the cortex and the brainstem and an **LMN system** that comprises both spinal cord circuits and brainstem motor nuclei. Together, these systems are the **efferent branch of the CNS**. These two systems receive sensory input, which drives the output. Sensory input to the motor system can be found at all levels.

The cerebellum and basal ganglia play important modulatory roles in motor activity. The **cerebellum** receives sensory input and sends **independent and segregated pathways to the LMN system and the UMN system** of the brainstem and the cortex. The cerebellum coordinates both truncal and limb movement and can be considered the coordinator and predictor of movement. The **basal ganglia**, on the other hand, receive input from widespread cortical areas, integrate that input, and project back to the cortex. The basal ganglia encode for the decision to move, the direction and amplitude of movement, and the motor expression of emotions.

This chapter summarizes and integrates preceding chapters on the motor systems.

II. THE UPPER MOTOR NEURON SYSTEM

The UMN system comprises the cortical output through the corticospinal and corticobulbar tracts and the brainstem output through descending tracts from the vestibular nuclear complex, nuclei in the reticular formation,

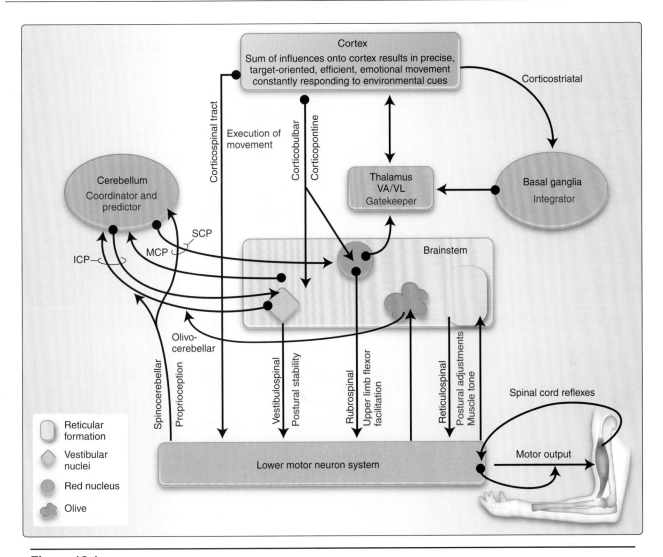

Figure 18.1
Conceptual summary of motor control. VA = ventral anterior nucleus; VL = ventral lateral nucleus; MCP = middle cerebellar peduncle; ICP = inferior cerebellar peduncle; SCP = superior cerebellar peduncle.

and the red nucleus. In other words, the UMN system refers to the sum of all descending influences on the voluntary control of motor activity. In this chapter, we discuss only the corticospinal tract as the exemplar of UMN output from the cortex, but it is important to keep in mind that the same principles apply to the corticobulbar tract (see Chapter 8, "Descending Motor Tracts").

A. Cortical motor system

The cortical motor system is a **network of cortical areas** spanning the primary motor cortex, motor association areas, and the motor area of the anterior cingulate gyrus. The neurons in these cortical areas comprise the UMNs, which send their axons through the corticospinal tract to the LMN system. These UMNs are under the influence of other cortical areas, primarily widespread association areas, and, importantly, the cerebellum and the basal ganglia. As well, output from sensory areas contributes to the descending

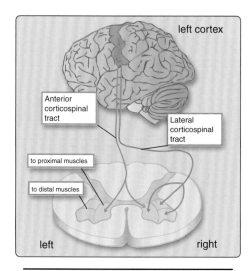

Figure 18.2
Descending motor tracts from the cortex and the spinal cord targets they innervate.

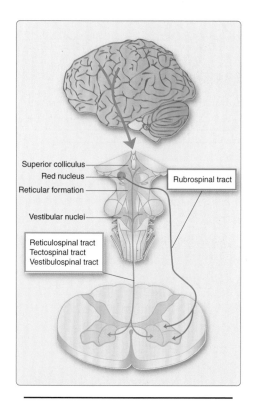

Figure 18.3
Descending motor tracts from the brainstem and the spinal cord targets they innervate. (See Chapter 8, "Descending Motor Tracts" for information on Tectospinal tract.)

corticospinal tract, allowing movements to be guided by touch and other senses. Through the sum of all the inputs to the UMNs, a movement can be **precise**, **target oriented**, and **efficient**; contain **emotional output**; and **respond to changes** in sensory input and the environment.

The cortical motor neurons project to the spinal cord circuits to control limb movement and to the brainstem nuclei, which control truncal movements (Figure 18.2).

B. Vestibular nuclei

The vestibular nuclei project directly to the LMNs of the spinal cord via the lateral and medial vestibulospinal tracts. The lateral vestibulospinal tract strongly facilitates extensor muscles in the ipsilateral limbs, whereas the medial vestibulospinal tract (or descending medial longitudinal fasciculus [MLF]) bilaterally influences muscles of the head, neck, trunk, and proximal parts of the limbs (Figure 18.3). This results in **postural and movement adjustments** and keeps the body oriented with the head when the head position changes.

C. Reticular formation

Nuclei of the reticular formation also have direct influences on the LMNs. These nuclei are responsible for the temporal and spatial coordination of movement and the maintenance of postural stability. When the cerebral cortex initiates a movement via impulses through the corticospinal tract, collaterals of the descending projection go to the reticular formation nuclei. Here, feed-forward mechanisms predict the consequence of this movement for *postural stability*. The reticular formation then activates stabilizing muscle groups in the trunk (see Figure 18.3). Nuclei of the reticular formation are influenced directly by the cerebral cortex (corticoreticular fibers), cerebellum, vestibular nuclei (descending inputs not shown in the figure), and ascending fibers from the spinal cord.

The reticulospinal tract acts on α motor neurons, supplying trunk and proximal limb muscles to influence **locomotion and postural control**, and, through influences on γ motor neurons, modulates **muscle tone**.

D. Red nucleus

The red nucleus functions primarily as a **relay and processing** station for information going from the cerebellum to the thalamus (dentatorubrothalamic tract). It plays a significant role as an integrator of information through its connections with the cerebral cortex, cerebellum, and thalamus. It also receives descending input from motor areas of cortex and sends a descending projection to LMNs in the spinal cord. The small rubrospinal tract originates in the red nucleus, crosses the midline, and descends in the spinal cord with the lateral corticospinal tract (see Figure 18.3). The red nucleus facilitates flexor movements in the contralateral upper limb directly through the rubrospinal tract and indirectly through influences on the medullary reticular formation. However, the direct influence of the red nucleus on the LMNs plays only a minor role in modulation of movement in humans.

Clinical Application 18.1: Upper Motor Neuron Lesions

Lesions of the upper motor neuron (UMN) system result in interruption of the descending influences on the lower motor neuron (LMN) system. They can occur as a result of stroke (sudden vascular lesion) at any level of the system or transection of the spinal cord, resulting in a characteristic set of symptoms. When the lesion is rostral to the pyramidal decussation, the paralysis will be on the contralateral side, while a lesion caudal to the decussation will result in paralysis on the ipsilateral side. The damage typically involves entire limbs or muscle groups.

The symptoms can include the following:

Paresis (partial weakness) or **paralysis** (complete weakness) of voluntary movement is due to loss of cortical input to the LMNs (weaker extensors than flexors in upper, weaker flexors than extensors in lower limbs.)

Spasticity is due to the increased activity of γ motor neurons in the muscle spindle and therefore increased afferent signaling from the muscle spindle, which results in more α motor neuron firing. An additional mechanism appears to be a loss of descending inhibitory input on spinal cord circuit neurons.

Increased muscle tone is due to a disruption of UMN inputs to the reticular formation and therefore a dysfunction in reticulospinal projections, which regulate muscle tone.

Hyperreflexia, or increased tendon reflexes, appears to be due to the loss of descending inhibitory influences on the spinal cord circuit neurons.

Clonus is a sequence of rapid involuntary muscle contractions in response to a sustained stretch stimulus (i.e., testing of reflexes). The underlying cause for clonus is a combination of spasticity, heightened muscle tone, and hyperreflexia.

Extensor plantar response, or Babinski sign, is usually, but not always, present. Under normal conditions, when the lateral part of the sole of the foot is stroked, the big toe plantar flexes (goes down). This reflex depends on the input from the corticospinal tract, and when the corticospinal tract is lesioned, the big toe extends or dorsiflexes (turns up).

There is no acute muscle atrophy, because the LMNs are still active. Over the long term, however, disuse atrophy can be seen.

III. THE LOWER MOTOR NEURON SYSTEM

The LMN is the **final common pathway** for signal transduction to skeletal muscles. The LMN system comprises the **circuitry within the anterior horn** of the spinal cord: the α **motor neurons** that innervate the skeletal muscles, the γ **motor neurons** that innervate the muscle spindles, and the **local circuit neurons** that play a role in spinal reflex pathways. The input to the LMNs will determine their firing rate, which will in turn determine muscle contraction. The α motor neurons that control limb movement receive their input from the cortex (corticospinal) to initiate voluntary movements. The α motor neurons that control balance and posture project mostly to the axial musculature and receive their input from brainstem nuclei. The γ motor neurons that determine muscle tone through their influence on the muscle spindles receive their principal input from the reticular formation. Another set of influences on the LMNs are the spinal reflexes (see Chapter 5, "The Spinal Cord").

Clinical Application 18.2: Lower Motor Neuron Lesions

A lower motor neuron (LMN) lesion is due to damage to the α motor neurons in the spinal cord or cranial nerve nuclei. A lesion of this type can be systemic, as in poliomyelitis, where the virus selectively affects anterior horn motor neurons, or due to the transection of a peripheral nerve. The damage is always limited to the muscles innervated by the affected LMNs.

Typical signs include flaccid weakness or paralysis. This is due to partial or complete loss of motor input to the muscle. It is different from the spastic paralysis seen in upper motor neuron lesions, where the input to the muscle is still intact, but the descending influences on the α motor neuron are lost. The loss of innervation also includes a loss of muscle tone (hypotonia), which can be diagnosed as a diminished resistance to passive stretch. Tendon reflexes are depressed or lost (depending on the extent of the damage), because there is no input to the muscle fibers. Loss of neuronal input also means a loss of trophic factors supplied by the anterior horn neurons to the muscles, which leads to muscle atrophy. Fasciculations (visible muscle twitch) also occur in LMN lesions due to spontaneous discharges of a single motor unit (LMN).

IV. MODULATORY INFLUENCES ON THE MOTOR SYSTEM

Control centers that fine-tune the signal to both the LMNs and the UMNs come from two distinct systems: the basal ganglia and the cerebellum. Whereas the basal ganglia project only to the cortex, the cerebellum projects to all levels of the motor system.

A. Basal ganglia

The basal ganglia are the main integrators of cortical and limbic input to the motor system. They receive input from widespread cortical areas and then project back to the cortex through three functional loops (motor, cognitive, limbic). These loops are not independent of each other. Rather, the information they carry is integrated into our behavior, which comprises motor output, cognition, and emotionality.

The basal ganglia ensure that movements are planned and executed precisely. They encode for the decision to move, the direction of movement, the amplitude of movement, and the motor expression of emotions (Figure 18.4) (see Chapter 16, "The Basal Ganglia," for more information).

The thalamus is under tonic inhibition. This tonic inhibition is modulated through the influence of the direct and indirect pathways from the basal ganglia to the thalamus. Through the basal ganglia circuitry and its influence on the motor system, each movement becomes efficient and target oriented. Superfluous movements are suppressed. The precise firing amplitude and firing intensity of each UMN are determined through these basal ganglia pathways.

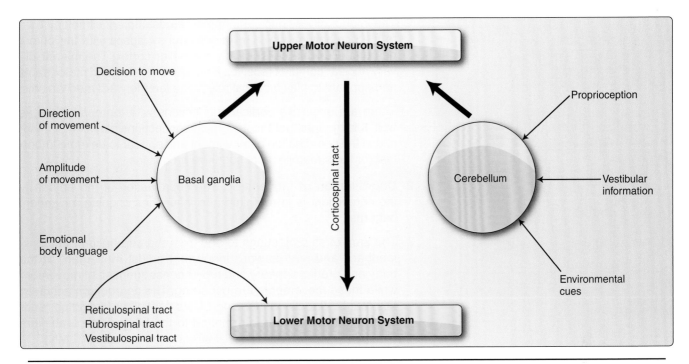

Figure 18.4
Conceptual overview of motor control.

B. Cerebellum

The cerebellum ensures smooth coordinated movements of voluntary muscles. It receives information concerning the position of the body (proprioceptive information from spinocerebellar and cuneocerebellar tracts), muscle tone and activity of spinal reflex arcs (spinocerebellar and cuneocerebellar tracts), position of the head (vestibular information), cortical activity (sensory, motor, and association areas), and the environment (visual and acoustic) (see Figure 18.4).

1. **Connections:** The cerebellum has independent and segregated connections to all levels of the motor system. It influences UMN activity through projections to the cortex and the brainstem tracts. It receives sensory input regarding proprioception from the lower and upper limbs through the posterior spinocerebellar and the cuneocerebellar tracts, respectively. It receives integrated information about motor activity from the spinal border cells through the anterior spinocerebellar tracts. Through these inputs, the cerebellum can compare the motor command with the sensory consequence of the command. Through experience and learning, the cerebellum is able to predict the sensory consequences of movement and feed that information forward to the cortex. Further input to the cerebellum comes from the vestibular nuclei, which convey information about the position of the head. When this information is compared to the position of the body (through proprioception), an accurate representation of the entire body in space can be extrapolated. This information is important for maintaining stable posture as well as for the vestibuloocular reflex.

The inferior olivary nuclear complex sends afferent climbing fibers to the cerebellum, which form powerful synapses with the output neurons (Purkinje cells) of the cerebellar cortex. The inferior olivary nuclear complex receives input from sensory and motor tracts and appears to play a critical role in the feed-forward mechanisms.

There are extensive cortical connections with the cerebellum as well. It is thought that these cortical connections play only a minor role in the coordination of movement and appear to be more important for the streamlining of cognitive processes.

2. **Coordination of movement:** The coordination of movement by the cerebellum is in real time and influences ongoing movement patterns.

The sum of all projections to and from the cerebellum results in feedback and feed-forward loops that adjust movement of the body and predict sensory and motor consequences of movement while these movements are occurring. The cerebellum achieves this feat by directly influencing LMNs as well as providing feedback and feed-forward information to the UMN. The cerebellum can influence both the plan (UMN) and the execution (LMN).

V. LESIONS TO THE MOTOR SYSTEMS

Clinically, the motor system is often split into a **pyramidal system** and an **extrapyramidal system**. The pyramidal system comprises the corticospinal and corticobulbar tracts. Lesions of these descending tracts result in weakness or paralysis resulting in UMN signs. The extrapyramidal system, in general, refers to everything else, that is, all descending influences on motor activity outside of the pyramidal influences. Clinically, the term extrapyramidal refers more specifically to the basal ganglia and related thalamic and brainstem nuclei, and movement disorders resulting from lesions of the basal ganglia are referred to as extrapyramidal disorders.

This division into pyramidal and extrapyramidal systems is not entirely correct from a functional perspective, because the functions of these motor structures are interrelated. In addition, these terms themselves are not precise. For example, it has been shown that a pure lesion of the corticospinal tract does not cause total paralysis, contrary to what one would expect. Rather, when total paralysis occurs, it is always due to interruption not only of the corticospinal tract but also of additional descending influences from brainstem nuclei. The term "extrapyramidal" is equally imprecise. In referring to all descending influences other than those of the pyramidal tract, this term is so broad that it is virtually meaningless. This term becomes more useful if it is subdivided into influences from the basal ganglia and influences from the cerebellum. Lesions of either of these structures result in significant disruption of movement and posture without major paralysis.

Despite these issues of terminology, the division into pyramidal and extrapyramidal systems has been a relatively useful tool clinically, as the deficits seen due to lesions of these systems are distinct. In common usage, the term "pyramidal" is generally synonymous with UMN, and the term "extrapyramidal" is generally synonymous with basal ganglia.

Table 18.1 compares and contrasts LMN, UMN, and basal ganglia lesions.

Table 18.1: Comparison of Lesions to the Motor System

LMN Lesions	UMN Lesions	Basal Ganglia Lesions
Weakness	Weakness	No weakness
Flaccid paralysis	Spastic paralysis	No paralysis Slowed (hypokinetic) or increased (hyperkinetic) involuntary movement
	Spasticity Velocity-dependent resistance to movement "Clasp-knife" phenomenon: resistance greatest on initiation of passive movement and diminishes as movement continues More obvious with quick movements	Rigidity Not velocity dependent Constant resistance throughout range of movement (lead pipe rigidity)
Hypotonia (decreased muscle tone)	Hypertonia (increased muscle tone)	Normal muscle tone
Hyporeflexia (decreased tendon reflexes)	Hyperreflexia (increased tendon reflexes)	Normal reflexes

LMN = lower motor neuron; UMN = upper motor neuron.

Clinical Case:

Patients at a Neurology Clinic

Three individuals present to the medical clinic with impairment in their mobility. All three individuals have normal sensation, vision, and hearing, but each person has a 5-year history of gait impairment related to a different neurological condition. A neurological exam will show signs that are specific to each one of the following disorders: Parkinson disease, middle cerebral artery (MCA) stroke, and bilateral S1 radiculopathy due to severe degenerative disc disease.

Match the following symptoms (1–3) to these neurological conditions: Parkinson disease, MCA stroke, or bilateral S1 radiculopathy.

1. Spasticity with brisk deep tendon reflexes (DTRs)
2. Rigidity with hypokinesia
3. Reduced ankle DTR with plantar flexion weakness

Answers:

1. Spasticity with brisk DTRs matches to MCA stroke. An MCA stroke can cause an upper motor neuron (UMN) lesion. The site of the lesion can either be cortical, in the primary motor cortex, or subcortical in the internal capsule. As this patient is presenting with gait abnormality, the most likely location of the lesion is the internal capsule, where all descending fibers travel. The cortical representation of the lower limb is on the medial surface of the cortex, which is supplied by the anterior cerebral artery. The deep MCA stroke has resulted in contralateral weakness, spasticity, and brisk DTRs.

2. Rigidity with hypokinesia matches to Parkinson disease. Parkinson disease affects the basal ganglia and causes hypokinesia/bradykinesia, rigidity, tremor, and postural

instability due to the degeneration of the dopaminergic neurons in the substantia nigra. The net effect of dopamine on both the direct and indirect pathway is an increase in movement, and loss of these dopaminergic projections will result in a decrease in movement or hypokinesia.

3. Reduced ankle DTR with plantar flexion weakness matches to bilateral S1 radiculopathy. The bilateral S1 radiculopathy is caused by a protruding intervertebral disc, which compresses the motor roots as they emerge from the spinal cord to form a spinal nerve. These axons are lower motor neurons and compression of these LMNs will result in weakness of the muscles innervated by the S1 nerve root (i.e., gastrocnemius muscle, which causes plantar flexion) and absent or reduced DTRs at the ankle.

Both rigidity and spasticity are associated with increased muscle tone. How can one distinguish between rigidity and spasticity?

Spasticity is velocity dependent, which means that there is increased resistance with rapid movement. In addition, spasticity is associated with a clasp-knife phenomenon, where resistance is greatest on initiation of passive movement and diminishes as the movement continues. *Rigidity*, on the other hand, is velocity independent: there is constant resistance throughout the range of movement.

In a neurological examination, five deep tendon reflexes (DTRs) are typically checked:

- Brachioradialis (C5/C6)
- Biceps (C5/C6)
- Triceps (C7/C8)
- Quadriceps (L3/L4)
- Gastrocnemius (S1/S2)

Quadriceps reflex is also called femoral, patellar, or knee reflex. Gastrocnemius reflex is also called Achilles or ankle reflex.

If there is shoulder abduction and elbow flexion weakness with normal strength in elbow extension, wrist flexion/ extension, and finger flexion/extension, which of the above noted DTRs would be reduced?

Brachioradialis and biceps reflexes test the integrity of the C5/C6 nerve root. Elbow flexion is performed by the biceps and brachioradialis muscle and shoulder abduction is performed by the deltoid muscle. All three muscles are innervated by the C5/C6 nerve root. A radiculopathy that affects the C5 and/or C6 nerve root would cause weakness of those three muscles, in addition to other muscles innervated by C5 and/or C6.

A radiculopathy can cause LMN weakness that includes a reduction in the DTRs. A radiculopathy that affects the C5/C6 nerve root will cause reduction or absence of the DTRs of the C5/C6 innervated muscles.

Study Questions

Choose the ONE best answer.

18.1 A 62-year-old previously healthy woman enters the clinician's office with a complaint of problems with her arm and leg. She states that for the past few months, she has noticed that her right arm feels like it is shaking whenever she is sitting quietly. At first, it was barely noticeable, and she ignored it, but lately, it seems more pronounced. In addition, for the past few weeks, she feels that her right leg drags a bit whenever she walks. She also feels that she is moving more slowly than she used to. Examination confirms a tremor in her right hand that is present primarily at rest. When the clinician holds her right arm and bends it at the elbow, constant resistance to passive stretch of her muscles is seen. When she walks, she takes small shuffling steps, particularly with her right leg. Her facial expression and manner of speaking are subdued or depressed. The most likely site of lesion resulting in these signs and symptoms is:

A. The right cerebellum, posterior lobe.
B. The left cerebellum, anterior lobe.
C. The substantia nigra on the right.
D. The substantia nigra on the left.
E. The tegmentum of the rostral midbrain on the right.

Correct answer is D. This is a description of Parkinson disease, which is a motor disorder characterized by hypokinesia or bradykinesia (a decrease in movement) or akinesia (a loss of movement), masked facies, "lead pipe" rigidity (increased muscle tone), and tremor at rest. The effect on movement is due to a degeneration of dopaminergic neurons in the substantia nigra (SN), which influences the direct and indirect pathways in the basal ganglia circuits. Loss of dopaminergic neurons in the SN reduces the excitatory input from the SN to the direct pathway, allowing greater influence from the indirect pathway, thereby resulting in greater inhibitory input to the thalamus and, in turn, less cortical stimulation and cortical output. A lesion of the posterior lobe of the cerebellum on the right will result in loss of fine control of movement on the right side of the body, including intention tremor, dysmetria, and dysdiadochokinesia. A lesion of the left anterior lobe of the cerebellum will result in anterior lobe syndrome, specifically limb ataxia on the left. Lesion of the substantia nigra on the right will result in Parkinson-like symptoms on the left side. A lesion to the tegmentum of the rostral midbrain will affect the red nucleus, interrupting efferent fibers from the cerebellum and resulting in signs of cerebellar disease.

18.2 A 52-year-old, previously healthy man presents in the emergency room with the following symptoms of sudden onset: intention tremor in the left arm, inability to make rapid alternating movements with the left hand, and missing (over- or undershooting) the target when asked to point with the left hand first to his nose and then to the examiner's finger. The most likely site of a lesion resulting in these symptoms is:

A. Anterior lobe of the cerebellum on the left.
B. Anterior lobe of the cerebellum on the right.
C. Posterior lobe of the cerebellum on the left.
D. Posterior lobe of the cerebellum on the right.
E. Flocculonodular lobe of the cerebellum.

Correct answer is C. The posterior lobe of the cerebellum plays a role in regulating fine motor control on the same side of the body. This is a posterior lobe lesion of the cerebellum on the left, resulting in loss of fine motor control on the left side of the body (i.e., intention tremor, dysdiadochokinesia, and dysmetria [past pointing]). The anterior lobe of the cerebellum receives input from the spinocerebellar tracts related primarily to proprioception. A lesion in the anterior lobe would result mainly in ataxia on the same side of the body. The lesion is in the posterior lobe on the left, not the right, as the cerebellum controls motor activity on the ipsilateral side of the body. The flocculonodular lobe of the cerebellum has reciprocal connections with the vestibular nuclei. A lesion of the flocculonodular lobe would result in truncal ataxia and nystagmus.

18.3 A 5-year-old girl is admitted to the hospital. She has been having trouble walking and keeping her balance, and these problems have gotten worse over the past few months. She seems to sway from side to side when standing and walks with a wide-based gait. Further examination reveals nystagmus in both directions. The most likely site of a lesion that would result in these symptoms is:

A. The right cerebellar hemisphere, posterior lobe.

B. The left caudate nucleus.

C. A midline lesion of the cerebellum involving the vermis and paravermal areas.

D. The flocculonodular lobe of the cerebellum.

E. The right substantia nigra.

Correct answer is D. This is a case of a tumor (typically, a medulloblastoma is most common in children) in the roof of the fourth ventricle, causing pressure on the flocculonodular lobe of the cerebellum. This results in truncal ataxia and nystagmus, both due primarily to disruption of the connections between the vestibular nuclei and the flocculonodular lobe. A lesion of the right cerebellar hemisphere, posterior lobe, would result in loss of fine control of movement (i.e., intention tremor, dysmetria, and dysdiadochokinesia) on the right. A lesion of the left caudate nucleus would result in a movement disorder resulting from an imbalance of the direct and indirect pathways. A lesion of the caudate nucleus would disrupt processing of input from the cortex and the substantia nigra. A lesion of the striatum, which includes the caudate, is one of the underlying pathologies in Huntington disease. A midline lesion of the cerebellum involving the vermis and paravermal areas would result in ataxic gait, as midline areas of the cerebellum receive input from the spinocerebellar pathways. A lesion of the right substantia nigra would result in symptoms of Parkinson disease on the left side of the body.

18.4 A patient presents with weakness of the right arm and hand, reduced muscle tone and depressed reflexes in the right arm and hand, and mild atrophy of the muscles of the right arm and hand. The most likely site of a lesion that would produce these symptoms is:

A. The posterior limb of the internal capsule on the left.

B. The left pyramid in the rostral medulla.

C. The right anterior gray horn of the spinal cord in the cervical region.

D. The right corticospinal tract in the cervical region of the spinal cord.

E. Descending reticulospinal fibers to lower motor neurons in the cervical spinal cord on the right.

Correct answer is C. A lesion of the lower motor neurons (LMNs) in the anterior horn of the spinal cord would result in muscle weakness, including reduced muscle tone, depressed reflexes, and muscle atrophy. A lesion in the posterior limb of the internal capsule on the left would result in upper motor neuron (UMN) signs on the right side of the body, including weakness, with increased muscle tone and increased reflexes. A lesion of the left pyramid in the rostral medulla would result in UMN signs on the right side of the body. A lesion of the right corticospinal tract in the cervical region of the spinal cord would result in UMN signs on the right side of the body. Note that the corticospinal tracts have already crossed at the level of the caudal medulla. A lesion of the descending reticulospinal fibers to LMNs in the cervical spinal cord on the right could result in alterations in muscle tone on the right. However, a lesion of the reticulospinal fibers is unlikely to occur in isolation but rather is part of an overall UMN lesion that includes the corticospinal tract and, perhaps, other descending tracts from the brainstem such as the vestibulospinal tract.

The Hypothalamus

19

I. OVERVIEW

The hypothalamus is a small structure (only about 4 g in weight), located inferior to the **thalamus** and forming the walls and floor of the inferior part of the **third ventricle**. Anatomically, the hypothalamus is part of the **diencephalon**, but functionally, it is part of the **limbic system**, as discussed further in Chapter 20, "The Limbic System." Despite its small size, the hypothalamus is critical for life, as it allows us to respond to both the internal and the external environments to maintain **homeostasis** or internal steady state. Together with the limbic system, the hypothalamus plays a role in regulating basic drives (motivated, goal-directed behaviors) as well as emotional or affective aspects of behavior and in many ways functions as the main "effector" of the limbic system. The hypothalamus is also the central regulator of autonomic and endocrine function and of such key functions as food intake and body weight, fluid and electrolyte balance, body temperature, sexual and reproductive behavior, and the sleep–wake cycle and circadian rhythms. Inputs to the hypothalamus provide information about the internal and external environment. These inputs include somatic and visceral sensory information as well as information from limbic structures, the cortex, the thalamus, and the retina. Outputs from the hypothalamus include connections to the pituitary gland for control of endocrine function and descending projections to autonomic nuclei in the brainstem and spinal cord for control of the visceral nervous system. Other outputs include direct and indirect projections to limbic system structures and areas of the cerebral cortex. Importantly, most pathways involving the hypothalamus are reciprocal. By interacting with and linking pathways involved in autonomic, endocrine, emotional, and somatic functions, diverse inputs and outputs are coordinated to allow appropriate behavioral responses to internal and external stimuli (Figure 19.1).

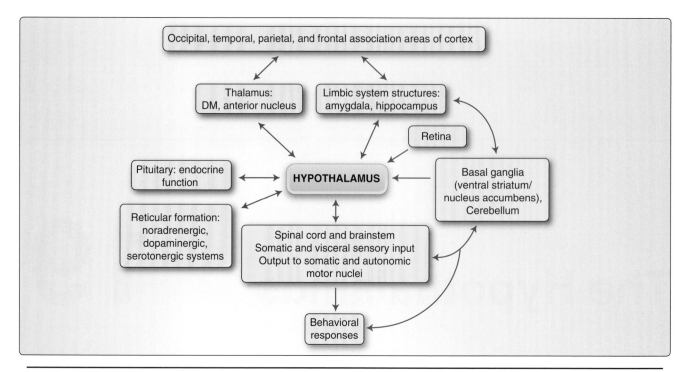

Figure 19.1
Conceptual overview of hypothalamic function. DM = dorsomedial nucleus.

II. ANATOMY

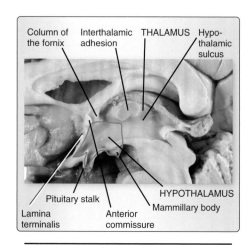

Figure 19.2
Midsagittal view showing the hypothalamus.

The hypothalamus is located inferior to the thalamus and forms the walls and floor of the inferior portion of the third ventricle. In a midsagittal section, the boundaries of the hypothalamus extend from the **lamina terminalis** anteriorly to the posterior edge of the **mammillary bodies**. The lamina terminalis is a thin membrane extending from the anterior commissure inferiorly to the rostral edge of the optic chiasm (Figure 19.2). Superiorly, the hypothalamus is separated from the thalamus by the hypothalamic sulcus (see Figure 19.2), and its lateral boundary is formed by the **posterior limb of the internal capsule** (Figure 19.3). Viewed from the inferior surface, the boundaries of the hypothalamus include the **optic chiasm** anteriorly, the **optic tracts** laterally, and, again, the **mammillary bodies** posteriorly (see Figure 19.3). The inferior surface of the hypothalamus, exclusive of the mammillary bodies, bulges slightly and is known as the **tuber cinereum** (gray swelling). The **infundibulum** (funnel) arises from the tuber cinereum and continues inferiorly as the **pituitary stalk**, which joins the hypothalamus to the pituitary gland (see Figure 19.2). A small swelling of the tuber cinereum, immediately posterior to the infundibulum, is the **median eminence**, which can be considered the anatomical interface between the brain and the anterior pituitary. Releasing and inhibiting hormones produced by hypothalamic neurons are released into capillaries in the median eminence for transport by the hypophyseal portal system to the anterior pituitary.

The hypothalamus can be divided into functional areas of nuclei along a lateral-to-medial axis and along an anterior-to-posterior axis.

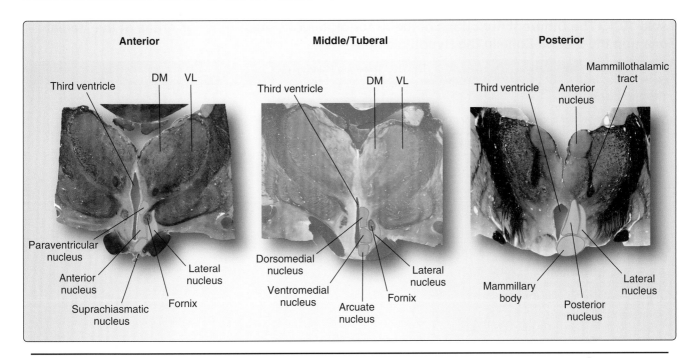

Figure 19.3
Coronal sections through the forebrain at the level of the anterior, middle, and posterior hypothalamus. DM = dorsomedial nucleus; VL = ventral lateral nucleus.

A. Lateral and medial zones

The **columns of the fornix**, which cut through the hypothalamus on their way to the mammillary bodies, divide the hypothalamus into a lateral zone and a medial zone. The **lateral zone** contains diffuse or scattered neurons interspersed among major bundles of fibers, which carry two-way traffic through the hypothalamus, rostrally toward the forebrain and caudally toward the brainstem. In contrast, the **medial zone** contains the majority of the functionally important nuclei of the hypothalamus (Table 19.1).

B. Anterior, middle, and posterior regions

Along the anterior-to-posterior axis, we can identify three functional areas (Figure 19.4; also see Table 19.1). The **anterior area** is the area above the optic chiasm and contains the preoptic, suprachiasmatic, supraoptic (medial and lateral), paraventricular, and anterior nuclei. The **middle** or **tuberal area** is the area above and including the tuber cinereum and includes the dorsomedial, ventromedial, and arcuate nuclei. The **posterior area** is the area above and including the mammillary bodies and contains the posterior nucleus and the mammillary bodies. See Figures 19.3 and 19.4.

III. AFFERENTS AND EFFERENTS

There are numerous afferent and efferent tracts associated with the hypothalamus, which serve to interconnect the hypothalamus with the pituitary, limbic system structures, cortical areas, the spinal cord, and the brainstem. Most tracts are reciprocal, providing loops of information that

Table 19.1: The Different Functions of the Anterior, Middle, and Posterior Areas of the Medial Zone and the Lateral Zone in the Hypothalamus

Anterior Area	Middle/Tuberal Area	Posterior Area
Medial Zone		
Preoptic and anterior nuclei: • Heat loss; response to heat • Parasympathetic activity • Sleep	Dorsomedial nucleus: • Emotional behavior • Role in circadian rhythms	Posterior nucleus: • Heat gain/conservation; response to cold • Sympathetic activity • Arousal/wakefulness
	Ventromedial nucleus: • Inhibits eating and drinking • Satiety	
Suprachiasmatic nucleus: • Circadian rhythms • Secretes releasing/inhibiting hormones	Arcuate nucleus: • Secretes releasing/inhibiting hormones	Mammillary nucleus: • Memory
Supraoptic nucleus: • Secretes hormones (oxytocin and vasopressin, or antidiuretic hormone)		
Paraventricular nucleus: • Magnocellular cells—secrete hormones (oxytocin, vasopressin) • Parvocellular cells—secrete releasing hormones		
Lateral Zone		
Lateral nuclei (diffuse): • Parasympathetic activity • Contains orexinergic neurons	Initiate eating and drinking	

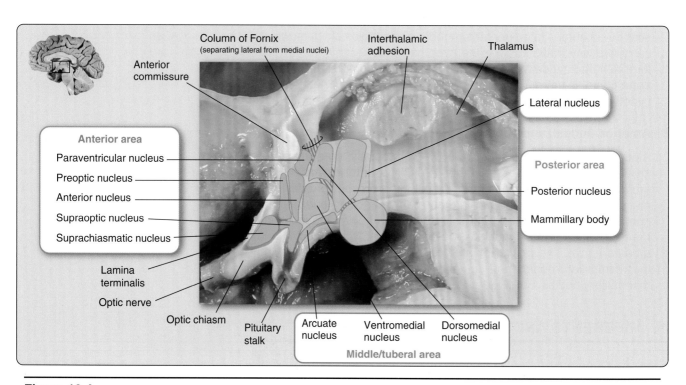

Figure 19.4

Overview of the hypothalamic nuclei.

mediate the complex functions of the hypothalamus and limbic system. It is important to note, however, that lesions to specific hypothalamic pathways are not common, and most hypothalamic dysfunction results from lesions in the hypothalamus itself or in structures or brain areas with which the hypothalamus is connected. A discussion of the major pathways is helpful in understanding the types of information that reach the hypothalamus and the targets to which the hypothalamus projects, which will facilitate an understanding of hypothalamic function.

A. The medial forebrain bundle and the dorsal longitudinal fasciculus

Two major pathways carry a majority of the two-way traffic to and from the hypothalamus, the **medial forebrain bundle** and the **dorsal longitudinal fasciculus**. The medial forebrain bundle is a large fiber bundle that passes through the lateral hypothalamus, traveling rostrally and caudally. It brings information to the hypothalamus from forebrain structures (including the ventral striatum) and from the **reticular formation** of the midbrain and rostral pons and, in turn, sends hypothalamic efferents out to these same areas. The dorsal longitudinal fasciculus carries information primarily between the hypothalamus and the **reticular formation** of the midbrain and rostral pons.

B. Other afferent connections

Afferent inputs to the hypothalamus include sensory (somatic, visceral, and gustatory) information as well as information from limbic structures, the cortex, the thalamus, and the retina.

1. **General somatic, visceral, and gustatory information:** This information comes from the spinal cord and brainstem and reaches the hypothalamus through collateral branches from ascending sensory pathways (mostly related to pain, through the anterolateral system) and the tractus solitarius (gustatory and visceral information), often with relays through the reticular formation (Figure 19.5). The hypothalamus also receives input from monoaminergic cell groups in the reticular formation. Ultimately, all of this ascending information from the spinal cord and brainstem will converge in the dorsal longitudinal fasciculus or medial forebrain bundle.

2. **Limbic afferents:** These come primarily from the **hippocampus** and **amygdala**. The **fornix** is the main fiber tract connecting the hippocampus to the hypothalamus. Afferent information carried in the fornix terminates in the mammillary bodies. The amygdala sends afferents to the hypothalamus through the **stria terminalis** and the **ventral amygdalofugal fibers**, the latter passing beneath the lenticular nucleus (part of the basal ganglia) to enter the hypothalamus.

   An indirect link between the hypothalamus and the limbic circuit of the basal ganglia (see Chapter 16, "The Basal Ganglia") provides one way for the hypothalamus to receive information about and, in turn, to influence emotional, motivational, and affective aspects of behavior.

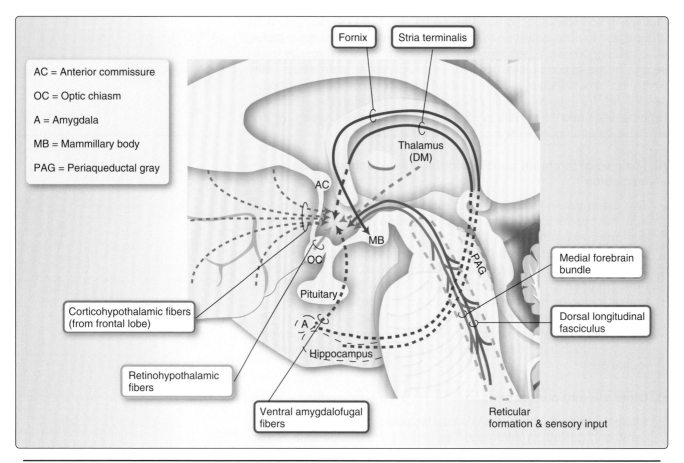

AC = Anterior commissure

OC = Optic chiasm

A = Amygdala

MB = Mammillary body

PAG = Periaqueductal gray

Figure 19.5
Afferent connections of the hypothalamus. DM = dorsomedial nucleus.

3. Afferents from the cortex, thalamus, and retina: The hypothalamus receives numerous cortical (corticohypothalamic) inputs. These come from the **orbitofrontal** and **cingulate** areas, as well as **widespread association** areas. Afferents from the **frontal lobe** travel directly to the hypothalamus and terminate primarily in the lateral area.

Afferents from the **thalamus** come primarily from the DM and anterior nuclei. These terminate, like fibers from the cortex, mainly in the lateral area.

Afferents from the **retina** can arise directly or as collaterals of fibers in the optic tract and terminate in the **suprachiasmatic nucleus (SCN)**.

C. Efferent connections

Most structures projecting to the hypothalamus receive a reciprocal projection back from the hypothalamus. For clarity, we group the hypothalamic efferents according to those that descend to the brainstem and spinal cord and those that ascend to the forebrain (Figure 19.6).

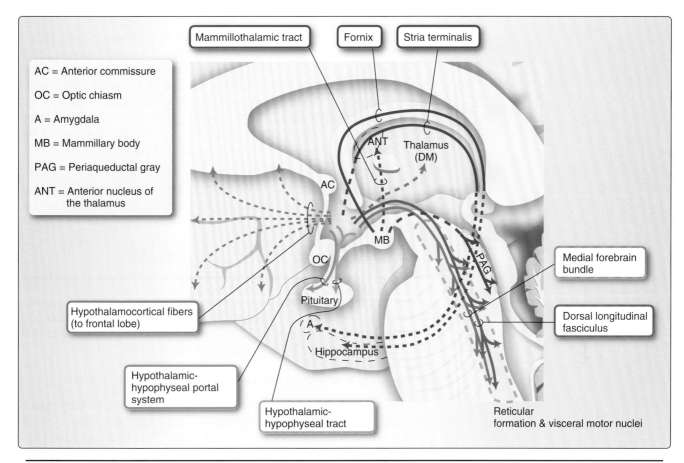

Figure 19.6
Efferent connections of the hypothalamus. DM = dorsomedial nucleus.

In addition, reciprocal connections with the **hippocampus** and **amygdala** enable the hypothalamus to influence limbic system function. Reciprocal connections with **cortical structures** enable the hypothalamus to influence higher cortical functions.

1. **Descending fibers to the brainstem and spinal cord:** The major targets of these descending fibers are the visceral motor nuclei in the brainstem and spinal cord (see below). Descending fibers arise from many different hypothalamic nuclei, descend in the **medial forebrain bundle** and the **dorsal longitudinal fasciculus** to the **periaqueductal gray (PAG)** and **reticular formation** of the midbrain and rostral pons, and then travel in the anterolateral area of the medulla and lateral area of the spinal cord. These fibers terminate on **parasympathetic nuclei** in the brainstem and sacral cord and on **sympathetic nuclei** in the thoracolumbar area of the spinal cord (see below). Other efferents from nuclei in the **medial zone** and the **mammillary bodies** terminate primarily in the PAG of the midbrain. Fibers also relay through the reticular formation and descend to terminate in somatic motor nuclei. The sum of this complex output enables the hypothalamus to influence behavior, in particular, emotional aspects of behavior.

2. **Ascending fibers to the forebrain:** Ascending fibers to the forebrain come from several areas of the hypothalamus (see Figure 19.6). There is a diffuse projection from the hypothalamus to the frontal lobe of the cortex (hypothalamocortical). Fibers from the hypothalamus to the thalamus include the **mammillothalamic tract** that arises in the **mammillary bodies** and terminates in the **anterior nucleus of the thalamus** and fibers from the **lateral zone** of the hypothalamus that project to the **DM nucleus of the thalamus**. These two nuclei then send projections to the cortex, particularly the frontal lobes.

Information from the hypothalamus also reaches the basal ganglia (particularly, the ventral striatum and nucleus accumbens) and cerebellum.

IV. FUNCTIONS

The hypothalamus is responsible for many essential human functions. Through regulation and integration of visceral, somatic, endocrine, and emotional functions, diverse outputs are coordinated to allow appropriate behavioral responses to internal and external stimuli.

A. Regulation of endocrine function

The hypothalamus controls the function of the **pituitary gland** or **hypophysis**. This is accomplished through a neural connection to the posterior lobe of the pituitary (the **neurohypophysis**) and a vascular connection to the anterior lobe of the pituitary (the **adenohypophysis**).

1. **Control of the posterior lobe of the pituitary:** The **supraoptic** and **paraventricular nuclei** contain large neurosecretory cells, known as **magnocellular cells**, which synthesize two peptide hormones (although any given cell produces only one of the two): (1) **oxytocin** causes contraction of the smooth muscles of the uterus and mammary glands and is critical for parturition and lactation, and (2) **vasopressin**, or **antidiuretic hormone (ADH)**, controls water balance by increasing reabsorption of water in the kidneys, thereby regulating the production of urine. These two hormones are then transported down the **hypothalamic–hypophyseal tract**, which consists of axons that arise from the supraoptic and paraventricular nuclei and end in the **posterior lobe of the pituitary** (Figure 19.7). Oxytocin and vasopressin are stored in the axon terminals in the posterior lobe until the appropriate stimuli trigger action potentials that cause release of these hormones into adjacent capillary networks, which carry them into the systemic circulation.

   a. **Vasopressin regulation:** Regulation of vasopressin production in the hypothalamus is through changes in plasma osmolarity. An increase in osmolarity causes increased synthesis in the hypothalamus and increased release from the posterior pituitary. Conversely, a decrease in osmolarity causes decreased synthesis and release.

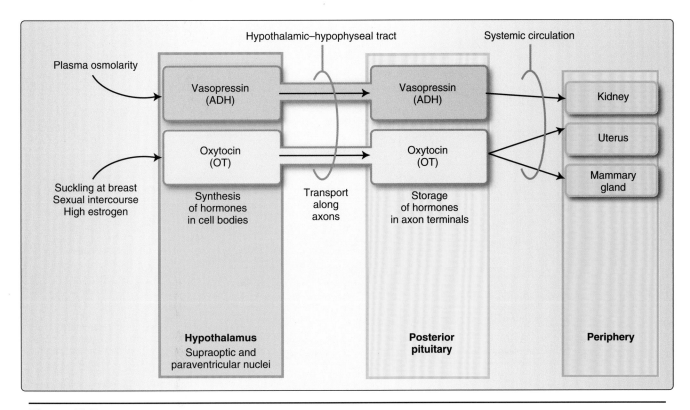

Figure 19.7
Regulation of endocrine function by the hypothalamus and the posterior pituitary.

    b. Oxytocin regulation: Oxytocin is released tonically at very low levels. Increased production of oxytocin in the hypothalamus results from stimuli including suckling at the breast, sexual intercourse, and increased levels of estrogen (see Figure 19.7).

2. **Control of the anterior lobe of the pituitary:** The anterior lobe of the pituitary is controlled by specific **releasing** or **release-inhibiting** (known simply as "**inhibiting**") **hormones** from the hypothalamus. These hormones reach the pituitary through the hypothalamic–hypophyseal portal system. Stimulation of the anterior pituitary by hypothalamic releasing hormones results in the synthesis and secretion of specific trophic hormones. These trophic hormones can then stimulate peripheral endocrine glands to synthesize and secrete their hormones. This entire system is regulated through numerous feedback loops. Two trophic hormones act on peripheral tissues rather than on endocrine glands. In these cases, the regulation of secretion from the anterior pituitary is directly from the hypothalamus through inhibiting hormones.

    The relationship between hypothalamic and pituitary hormones is summarized in Figure 19.8.

    a. The hypothalamic–hypophyseal portal system: The hypophyseal artery forms a primary capillary plexus within the median eminence into which the hypothalamic hormones are released. From here, the hormones are carried by portal veins to a secondary capillary plexus in the anterior lobe of the pituitary, where

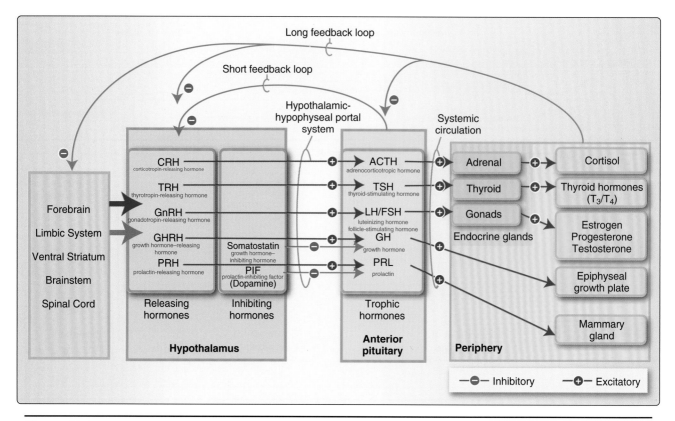

Figure 19.8
Regulation of endocrine function by the hypothalamus and the anterior pituitary.

they are then released to act on the cells of the anterior pituitary. The trophic hormones produced by the anterior pituitary are secreted into a secondary capillary plexus. This arrangement of blood vessels (artery–capillary–vein–capillary) is known as a **portal system**. Pituitary hormones then enter the systemic circulation through the hypophyseal veins and stimulate endocrine glands or target tissues (Figure 19.9).

While the hypothalamus is clearly the driver of pituitary hormone output, the pituitary gland and the portal system cannot be considered as static structures that simply respond to the hypothalamic hormones. Rather, the interrelationship of the hypothalamic neurons and pituitary cells with the capillary networks of the portal system is a dynamic one and can be influenced by an organism's physiological state.

b. **Regulation of the hypothalamic and pituitary hormones:** The hypothalamus is influenced by internal and external stimuli through inputs from the cortex, limbic system, ventral striatum, brainstem, and spinal cord. This is how stress, fear, or physical stimuli, for example, can activate the endocrine system.

Feedback systems regulate endocrine function, resulting in hormonal balance or homeostasis. The hormones produced by endocrine glands feed back at multiple levels to regulate their own production. This occurs through **long feedback loops** to

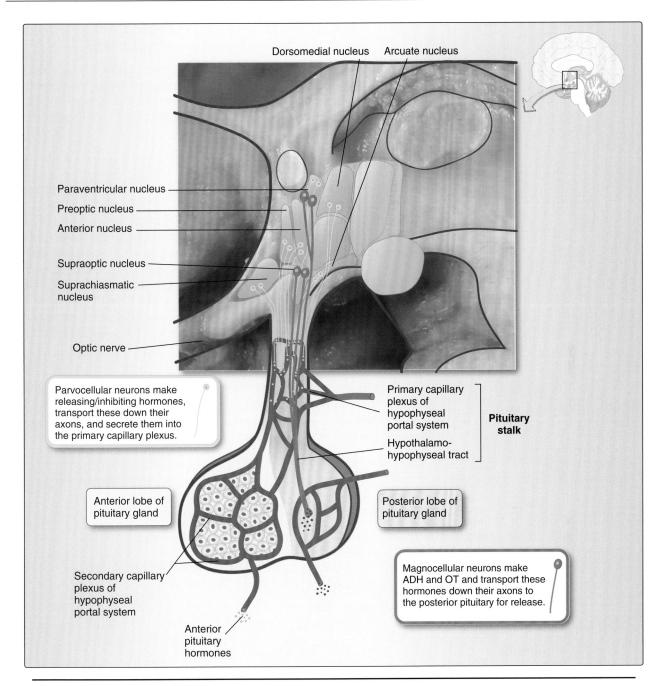

Figure 19.9
Overview of the hypothalamic–pituitary system. ADH = antidiuretic hormone or vasopressin; OT = oxytocin.

the anterior pituitary, the hypothalamus, and forebrain structures (e.g., hippocampus, prefrontal cortex). As plasma hormone levels rise, production and secretion at all levels are inhibited, and as plasma hormone levels fall, production and secretion are increased. In addition, pituitary hormones feedback to the hypothalamus directly through **short feedback loops**.

The production of the pituitary hormones that act directly on target tissues (growth hormone [GH] and prolactin [PRL]) is regulated through a balance of releasing and inhibiting hormones from the hypothalamus (see Figure 19.8).

The specific hypothalamic nuclei responsible for synthesizing and secreting each of the releasing/inhibiting hormones are not fully known. Figure 19.8 summarizes the current knowledge about hypothalamic releasing and inhibiting hormones and their actions.

B. Regulation of visceral function

The visceral motor or autonomic nervous system is regulated by the hypothalamus through a balance in activity between the anterior and posterior areas. The anterior area of the hypothalamus activates **parasympathetic** activity. It sends efferents to brainstem parasympathetic neurons that will travel with cranial nerves III, VII, IX, and X as well as spinal cord parasympathetic (S2–S4) neurons. By contrast, the posterior area of the hypothalamus activates **sympathetic** activity. It sends efferents to sympathetic neurons in the lateral horn of the spinal cord (T1–L2) (Figure 19.10). The visceral nervous system is discussed in detail in Chapter 4, "Overview of the Visceral Nervous System."

C. Regulation of homeostatic functions

Despite the small size of the hypothalamus, many of its nuclei have multiple functions or act in concert with other nuclei. Furthermore, the hypothalamus has an important role in interconnecting and regulating visceral, endocrine, and limbic system functions. Therefore, it is

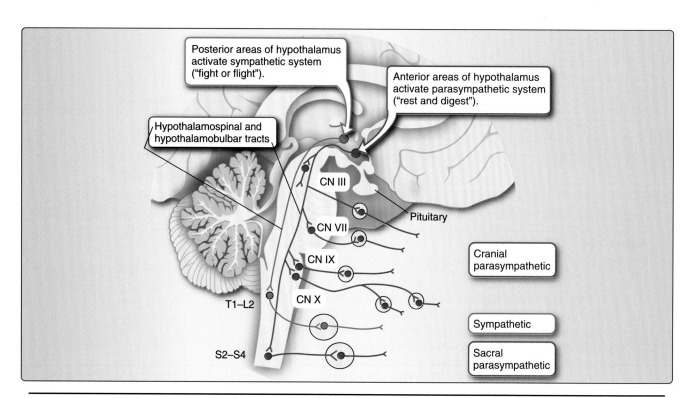

Figure 19.10
The regulation of autonomic function by the hypothalamus. CN = cranial nerve.

difficult to assign a specific homeostatic function to a specific nucleus. Rather, what we see following lesion of the hypothalamus are changes in patterns of response. That is, for most complex homeostatic functions, regulation occurs through activation of anterior versus posterior areas or medial versus lateral areas, with one area counterbalancing or opposing the effects of the other. In the discussion on homeostatic functions below, we generally refer to the anterior, posterior, medial, or lateral areas of the hypothalamus (as defined above) rather than to specific nuclei. In a few instances where we have more specific information, we refer to specific nuclei by name.

Older concepts of the hypothalamus, based primarily on lesion studies, typically referred to "centers" that controlled specific functions. We now know that the "center" concept is too simplistic. Our modern understanding is that there must be "complex integration of complex signals to mediate complex functions." This is discussed in some detail only in relation to the regulation of food intake and satiety, but it is important to realize that this applies to all hypothalamic homeostatic functions that are discussed below.

Of note, for major deficits to be observed in hypothalamic function, bilateral lesions are typically needed.

1. **Temperature regulation:** Maintenance of body core temperature within narrow limits is a major homeostatic function critical for survival. Temperature regulation by the hypothalamus occurs through a balance in activity between the anterior and posterior areas. The anterior area of the hypothalamus contains temperature-sensitive neurons that respond to information from the periphery (environment, skin) and from the body core (temperature of the blood). If body temperature increases, heat loss mechanisms in the anterior area of the hypothalamus are activated, including sweating, cutaneous vasodilation, and a decrease in metabolic rate. In addition, behavioral thermoregulation also occurs; we become consciously aware of the change in temperature and initiate appropriate behavior, such as taking off a sweater or getting a cold drink. Conversely, if body temperature decreases, temperature sensors in the anterior hypothalamus stimulate heat gain/conservation mechanisms in the posterior area of the hypothalamus, which include shivering, cutaneous vasoconstriction, and an increase in metabolic rate.

   Molecular approaches have identified the specific populations of cells that convert temperature information into homeostatic responses. This provides exciting opportunities to study the neural mechanisms of how we keep cool when it is hot and warm when it is cold, which will have clinical implications for pharmacological strategies for treatment.

   a. **Lesions:** A lesion in the anterior area will result in loss of these heat loss mechanisms, resulting in **hyperthermia** in a hot environment or if metabolic rate is high. A lesion in the posterior area of the hypothalamus will result in **hypothermia** in a cold environment or if metabolic rate is low. Interestingly, a large bilateral lesion in the posterior area could result in **poikilothermia**, in which body temperature cannot be regulated and varies with the external environment (just like in lizards). This is

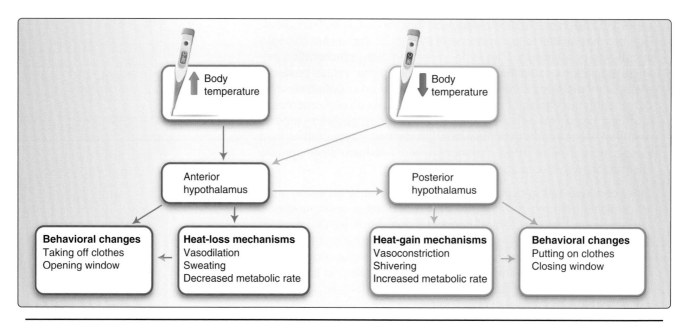

Figure 19.11
The regulation of temperature by the hypothalamus.

due to loss of both the posterior heat gain mechanisms and the descending fibers from the anterior "heat loss" area that pass through the posterior area (Figure 19.11).

b. **Sweat glands:** It may appear somewhat paradoxical that sweat glands are innervated by sympathetic nerve fibers, but sweating is initiated by activation of the anterior area of the hypothalamus, which mediates parasympathetic responses. It has been suggested that, despite the sympathetic innervation of the sweat glands, sweating can be viewed as a function consistent with parasympathetic activity, in that it cools us down. Consistent with this idea, both sympathetic nerve fibers that innervate most sweat glands and parasympathetic nerve fibers innervating the viscera release acetylcholine.

2. **Regulation of food intake:** Regulation of feeding involves a network of nuclei, which receive and integrate multiple peripheral anorexigenic or orexigenic signals. These include circulating nutrients such as glucose and fatty acids, hormones such as leptin and ghrelin, signals from the gut through vagal afferents, and psychological stimuli. In general, food intake is regulated by a balance in activity between the medial and lateral hypothalamic areas.

Hypothalamic nuclei involved in feeding send projections to areas of the limbic system, including the mesolimbic reward circuitry (see Chapter 20, "The Limbic System") that mediates the hedonic control of feeding. In addition, input from the SCN regulates the circadian pattern of food intake.

a. **Lateral hypothalamus:** The **lateral hypothalamus** has been referred to as a "**hunger**" or "**feeding center**" (Figure 19.12). Stimulation of the lateral hypothalamus increases eating and

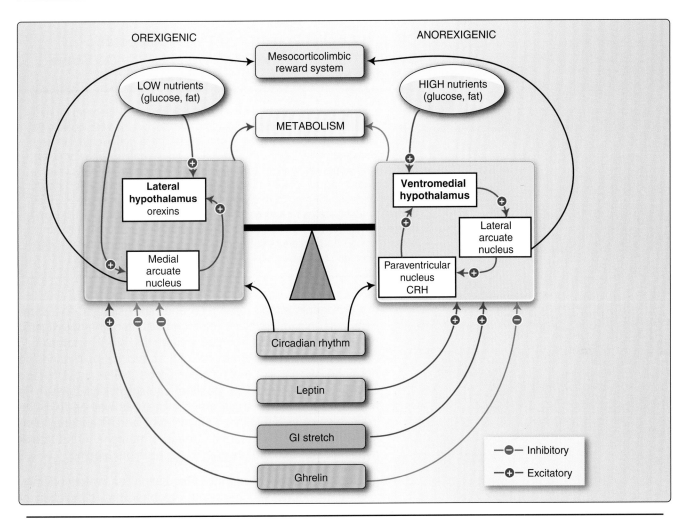

Figure 19.12
The regulation of hunger and satiety by the hypothalamus. CRH = corticotropin-releasing hormone; GI = gastrointestinal.

drinking, whereas a lesion in the lateral area results in aphagia and adipsia or suppression of eating and drinking. The lateral hypothalamus receives diverse input, including gustatory and olfactory information and metabolic signals (glucose, leptin, ghrelin), as well as homeostatic, hedonic (nucleus accumbens), and sensory information.

b. **Ventromedial hypothalamus:** By contrast, the **ventromedial area** has been referred to as a "**satiety center**." Stimulation of the ventromedial area inhibits eating, and lesions of the ventromedial area not only stimulate feeding but also decrease physical activity and alter metabolism, both of which contribute to weight gain. In addition, the arcuate nucleus has been shown to integrate inputs related to energy balance. Neurons in the lateral portion of the arcuate nucleus mediate the anorexigenic signals from leptin and glucose, for example, which reduce food intake and increase energy output. By contrast, nuclei in the medial portion of the arcuate nucleus are inhibited by anorexigenic signals such as leptin and glucose and mediate

Clinical Application 19.1: If This Homeostatic System Can Regulate Body Weight, Why Do Some People Become Obese?

The regulation of body weight depends not only on the "bottom-up" control of food intake but also on "top-down" factors that can override the individual's metabolic state. Recent studies have shown that the lateral hypothalamus is a multifunctional region. While its stimulatory role in eating behavior is well known, it now appears that, surprisingly, inhibition of certain lateral hypothalamic neurons can stimulate eating. Specifically, stimulation of projections to the lateral hypothalamus from cells in the basal forebrain that are involved in reward functions were shown to inhibit lateral hypothalamic activity and cause animals to eat voraciously. It appears that dysfunction in this circuit could override homeostatic control of eating behavior and underlie pathological eating in humans.

the orexigenic effects of ghrelin and other signals. The paraventricular nucleus plays a role in mediating both anorexigenic and orexigenic signals from the arcuate and other hypothalamic nuclei and secretes hormones such as corticotropin-releasing hormone, which, like leptin, reduce food intake and increase energy metabolism. The peptide cholecystokinin, which is released from the GI tract during a meal as well as from the brain, also plays a role in the inhibition of feeding and may contribute to satiety. By contrast, groups of neurons in the lateral hypothalamus secrete orexins, which promote feeding.

3. **Regulation of water balance:** Regulation of water balance involves both neural and hormonal mechanisms. Stimulation of areas in the **lateral hypothalamus** and of **osmoreceptors** in the **anterior region** induces **drinking**, and nuclei in these areas can be considered a "**thirst center**," whereas lesions in the lateral hypothalamus decrease water intake. Hormonal regulation occurs through specialized osmolarity-sensitive neurons in the anterior area of the hypothalamus that monitor osmolarity of the blood. Output from these neurons influences the release of **ADH** from the supraoptic and paraventricular nuclei, which, in turn, influence water resorption in the kidney and the production of urine.

D. **Regulation of circadian rhythms and the sleep–wake cycles**

Similar to the regulation of visceral activity and body temperature, regulation of sleep is mediated by opposing actions of the anterior and posterior/lateral regions of the hypothalamus.

1. **Circadian rhythms:** The **SCN** of the anterior area of the hypothalamus serves as a "**master clock**," controlling both physiological and behavioral circadian rhythms, including the sleep–wake cycle, hormonal secretion, and thermoregulation. The SCN neurons have an approximately 24-hour rhythm of electrical activity, even in the absence of environmental cues. Overriding this intrinsic rhythm, activity of the SCN is regulated by environmental signals and is entrained to the 24-hour light–dark cycle. This entrainment depends on light input from the retina (**retinohypothalamic tract**) as well

as the secretion of **melatonin**, which is secreted by the pineal gland in a circadian pattern. High levels of melatonin secreted at night and low levels secreted during the day play a major role in the regulation of sleep and other cyclical body activities. Secretion of melatonin is controlled by the SCN through projections to the visceral nervous system and, in turn, a sympathetic projection to the pineal gland from the superior cervical ganglion. Interestingly, melatonin signals both the time of day (a "**clock**" function) and the time of the year (a "**calendar**" function) to all tissues of the body. Neurons of the SCN send projections through the hypothalamic nuclei that influence endocrine secretion, visceral function, eating, temperature regulation, and behavior to regulate circadian activity of functions (Figure 19.13).

2. **Sleep–wake cycle:** In addition to regulation of circadian rhythms, the hypothalamus plays a key role in the regulation of sleep and wakefulness. The sleep control system in the hypothalamus interacts with the circadian pacemaker in the SCN and is closely integrated with homeostatic systems. For example, sleep deprivation alters hormone release, increases body temperature, and stimulates appetite. The anterior area of the hypothalamus, particularly the preoptic area, is important for the generation of slow wave sleep (deep sleep, non-rapid eye movement), and lesions of the anterior area are known to produce insomnia in animals and humans. By contrast, the posterior area of the hypothalamus is important in wakefulness, and lesions here produce states ranging from drowsiness to coma. Histamine appears to be a major "wake promoting" neurotransmitter, and lesions in the posterior hypothalamus may produce their effect by inactivating certain subsets of histaminergic neurons. Interestingly, orexins, in addition to their effects in food intake, may have a role in sleep. Studies suggest that deficits in orexin neurotransmission in the lateral hypothalamus may play a role in **narcolepsy**. Narcolepsy

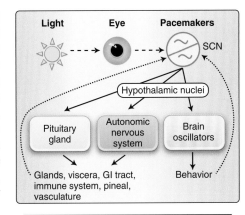

Figure 19.13
The regulation of circadian rhythms by the hypothalamus. SCN = suprachiasmatic nucleus; GI = gastrointestinal.

Clinical Application 19.2: Damage to the Hypothalamus

Typically, for noticeable deficits to occur following hypothalamic lesions, the lesions must be bilateral. Unilateral lesions may not be noticeable or may involve only minimal alterations.

Damage to the hypothalamus can result in neuroendocrine disturbances, autonomic dysfunction, and/or disturbances in homeostatic functions including temperature regulation, water balance, body weight regulation, sleep–wake cycles, and emotional behavior. Importantly, damage rarely occurs through vascular accidents because of a redundant blood supply. Tumors are the most important cause of hypothalamic dysfunction. Hypothalamic tumors are rare; however, pituitary tumors can exert pressure on various hypothalamic nuclei and therefore impair function. With expansion, pituitary tumors can also press on the optic chiasm, resulting in bitemporal hemianopia. Other sources of pressure that can impact hypothalamic function include aneurysms in the circle of Willis and increased intracranial pressure. A fracture at the base of the skull, impacting the hypothalamic–hypophyseal tract, can result in diabetes insipidus (the inability to regulate water balance). Finally, infectious or toxic agents can affect the hypothalamus and alter function in various ways.

Clinical Application 19.3: Melatonin

Melatonin has a major effect on sleep induction. This occurs partly through the inhibition of a circadian wakefulness-generating mechanism in the suprachiasmatic nucleus (SCN) and partly through the lowering of core body temperature as a consequence of peripheral vasodilation. Thus, lesions in the anterior area of the hypothalamus that include the SCN result in sleep disturbances, including fragmentation of sleep and wakefulness, or insomnia. Importantly, the posterior area of the hypothalamus has opposing influences on sleep, such that lesions in the posterior area impair wakefulness, resulting in drowsiness or, at the extreme, coma.

In addition to its effects on sleep induction, melatonin is also known to play a role in phase shifting of the circadian clock. When administered during the evening or the early phase of the night, melatonin phase-advances the circadian clock. In contrast, melatonin delays the circadian rhythm when administered in the second half of the night or in the early part of the morning.

is a chronic sleep disorder, or dyssomnia, characterized by excessive daytime sleepiness in which a person experiences extreme fatigue and may fall asleep suddenly at inappropriate times, such as while at work or at school. Another problem that some narcoleptics experience is cataplexy, a sudden loss of muscle tone, often brought on by strong emotions. This change in tone can range from a slight slackening of the facial muscles to the dropping of the jaw or head, weakness at the knees, or a total collapse.

V. BLOOD SUPPLY

Blood supply to the hypothalamus comes from small arteries that arise from the circle of Willis (Figure 19.14). Branches from the anterior cerebral and anterior communicating arteries supply the anterior areas of the hypothalamus, and branches from the posterior cerebral and posterior communicating arteries supply the middle and posterior areas of the hypothalamus.

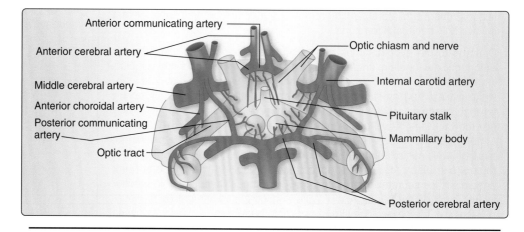

Figure 19.14
Blood supply to the hypothalamus. (Modified from Haines DE. *Neuroanatomy: An Atlas of Structures, Sections, and Systems*, 7th ed. Baltimore, MD: Lippincott Williams & Wilkins, 2007.)

Clinical Case:

Pituitary Adenoma

A 42-year-old female develops galactorrhea (milk discharge from breast). She has a 3-month history of chronic daily headaches and has noticed a reduction in her peripheral vision in both eyes. Her neurological examination is normal except for bitemporal hemianopia.

An MRI of the brain shows an increased FLAIR (fluid-attenuated inversion recovery) signal in the pituitary gland with a large enhancing mass (Figure 19.15A,B). Note that with FLAIR imaging, abnormalities remain bright, but normal CSF is attenuated and appears dark. FLAIR is very sensitive to pathology and makes the differentiation between CSF and an abnormality much easier. Pituitary hormone studies are ordered to determine the cause of this mass in the pituitary, which is greater than 10 mm in size.

Based on the history above, which anterior pituitary hormone would you expect to be elevated?

You would expect to see an elevation in **prolactin (PRL)**; it is essential for milk production by the mammary glands. In nonpregnant women, increased PRL would result in galactorrhea, as seen in this patient.

Other hormones of the anterior pituitary are unlikely to be elevated since the patient is not exhibiting any symptoms related to other hormones. **Thyroid-stimulating hormone (TSH)** stimulates the thyroid gland to secrete T4 (and small amounts of T3); elevated TSH would cause hyperthyroidism. **Growth hormone (GH)** stimulates the epiphyseal growth plates. This would result in acromegaly in adults and gigantism in children. Features of acromegaly include enlarged hands and feet as well as bony changes to the face with brow and lower jaw protrusion and an enlarged nasal bone. **Follicle-stimulating hormone (FSH)** stimulates the maturation of germ cells. In males, FSH induces Sertoli cells to secrete androgen-binding proteins (ABPs) and stimulates primary spermatocytes to undergo the first division of meiosis to form secondary spermatocytes. In females, FSH initiates growth of the ovarian follicles and some secretion of estrogen. **Adrenocorticotropic hormone (ACTH)** stimulates the adrenal glands to synthesize and release cortisol.

Which two hormones are released by the posterior pituitary gland?

Vasopressin or **antidiuretic hormone (ADH)** and **oxytocin** are the two hormones released by the posterior pituitary gland. ADH controls water balance by increasing reabsorption of water in the kidneys, thereby regulating the production of urine. Oxytocin causes contraction of the smooth muscles of the uterus and mammary glands. Uterine contractions are critical for parturition. During breastfeeding, oxytocin stimulates milk letdown.

Why is the pituitary mass or macroadenoma affecting her vision?

The pituitary gland is located in the sella turcica, directly below the optic chiasm. As the adenoma (pituitary tumor) expands from a microadenoma to a macroadenoma (>10 mm), it can compress the optic chiasm from below. The midline crossing fibers within the optic chiasm that carry information from the nasal retinae are typically affected first. These fibers receive input from the temporal visual fields. A lesion (due to compression by a tumor) to these fibers at the optic chiasm will affect information coming from both nasal retinae, resulting in bitemporal hemianopia (see Figure 19.15C).

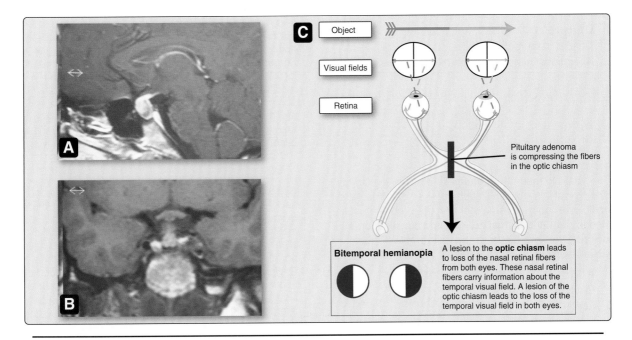

Figure 19.15

A. A sagittal MRI with gadolinium showing an enlarged enhancing (bright white mass) pituitary tumor causing upward compression of the optic chiasm. **B.** Coronal MRI of the same patient. **C.** Bitemporal hemianopia due to lesion of the fibers in the optic chiasm. (From Scher LA, Weinberg G. *General Surgery Board Review*. Wolters Kluwer Health, 2011.)

Chapter Summary

- The hypothalamus is the main regulator of homeostasis in the body. It is a small structure located inferior to the thalamus and lateral and inferior to the third ventricle.

- The hypothalamus receives afferents from somatic and visceral sensory systems, the limbic system, the olfactory system, the cortex, and the retina. It sends efferents to the spinal cord and brainstem and to forebrain structures.

- The hypothalamus regulates endocrine function through the hypothalamic–pituitary system. Releasing and inhibiting hormones regulate the production of trophic hormones by the anterior pituitary, whereas hormones released by the posterior pituitary are actually synthesized in specific nuclei within the hypothalamus. Hormone balance is maintained through feedback to the hypothalamus and higher brain centers.

- The visceral nervous system is regulated by the hypothalamus.

- The hypothalamus regulates body temperature coordinated by the anterior versus the posterior areas, and can initiate behaviors that will result in cooling or warming of the body. Food and water intake are also regulated by the hypothalamus, with multiple signals, including the concentration of nutrients in the blood, coordinated by the lateral versus the medial areas of the hypothalamus to signal hunger/thirst and satiety. The hypothalamus also regulates circadian rhythms through intrinsic activity that is modulated by environmental influences, such as the light–dark cycle and the secretion of melatonin.

- The blood supply to the hypothalamus comes from small arteries directly off the circle of Willis, including branches from the anterior cerebral and anterior communicating arteries as well as the posterior cerebral and posterior communicating arteries.

Study Questions

Choose the ONE best answer.

19.1 Following surgery for a pituitary tumor, a young man experiences pronounced bleeding into the posterior hypothalamus on both sides. What is a likely consequence of this lesion?

    A. Inability to regulate body temperature in a cold environment.

    B. Alterations in food intake resulting in weight gain.

    C. Disruption of the sleep–wake rhythm.

    D. Deficits in memory.

    E. Alterations in the regulation of water balance.

Correct answer is A. Temperature is regulated by opposing actions of the anterior and posterior areas of the hypothalamus. A heat loss mechanism is in the anterior area of the hypothalamus, and a heat gain/conservation mechanism is in the posterior area. A lesion in the posterior area will result in inability to regulate body temperature in a cold environment. A large posterior lesion could result in complete loss of ability to regulate body temperature, because it would include not only the heat gain mechanism but also the descending fibers from the heat loss area that pass through the posterior area. Food intake regulation is complex, but generally, the lateral area mediates feeding, and the ventromedial nucleus mediates satiety. The sleep–wake rhythm is regulated by the suprachiasmatic nucleus in the anterior area of the hypothalamus. Deficits in memory would result from a lesion to the mammillary bodies. Water balance is regulated by the lateral hypothalamus and osmoreceptors in the anterior area.

19.2 A 35 year old man had a traumatic brain injury. On the third day of his hospitalization, he has extreme thirst (polydipsia) and very frequent urination (polyuria). His urine is clear and he has dry skin and constipation. His sodium levels are elevated (hypernatremia). This man has deficiency of which hormone?

    A. ADH

    B. Oxytocin

    C. Leptin

    D. Cortisol

    E. Ghrelin

Correct answer is A. Antidiuretic hormone (ADH) is not being secreted possibly due to a bleed in the area of the hypothalamus where the supraoptic and paraventricular nuclei are located or a lesion of the pituitary stalk. Impairment or absence of secretion of ADH results in diabetes insipidus with polydipsia, polyuria and hypernatremia; this is often transient. Oxytocin is involved in breastfeeding and uterine contractions. Leptin and ghrelin are hormones involved in regulation of food intake and are regulated by the balance between medial and lateral hypothalamic areas. Cortisol is a hormone released from the adrenal gland and is controlled by ACTH. It is not directly involved in fluid balance.

19.3 Which statement below is true regarding the fornix?

A. It carries descending output from the hypothalamus to the brainstem.

B. It connects the hypothalamus with forebrain structures.

C. It connects the hypothalamus and the amygdala.

D. It connects the hypothalamus with the hippocampus.

E. It carries input to the hypothalamus from the thalamus.

Correct answer is D. The fornix carries two-way traffic between the hippocampus and the hypothalamus (mammillary bodies). Descending output from the hypothalamus to the brainstem is carried by two tracts, the medial forebrain bundle and the dorsal longitudinal fasciculus, which bring information to brainstem and spinal cord areas, largely by relays through the reticular formation. Thalamohypothalamic fibers connect the dorsomedial nucleus of the thalamus with the hypothalamus. The amygdala is connected with the hypothalamus through two tracts, the stria terminalis and the ventral amygdalofugal fibers.

19.4 Which statement below is true of hypothalamic regulation of feeding behavior?

A. Corticotropin-releasing hormone provides a signal that stimulates feeding.

B. Nuclei in the lateral area of the hypothalamus play a role in satiety.

C. The ventromedial nucleus plays a role in increasing food intake.

D. Leptin and glucose inhibit neurons in the medial portion of the arcuate nucleus.

E. Ghrelin sends orexigenic signals that stimulate the lateral portion of the arcuate nucleus.

Correct answer is D. Nuclei in the medial portion of the arcuate nucleus are inhibited by anorexigenic signals, such as leptin and glucose, and mediate orexigenic signals from ghrelin. Corticotropin-releasing hormone acts like leptin to reduce food intake and increase energy metabolism. Nuclei in the lateral area of the hypothalamus stimulate food intake, whereas nuclei in the ventromedial nucleus inhibit food intake.

20 The Limbic System

I. OVERVIEW

The term "limbic" was used originally to describe gyri that form a ring or border (*limbus* is Latin for "border") on the medial surface of the cerebral hemispheres, around the corpus callosum and rostral brainstem. Today, we use the term limbic system to describe both cortical areas (the "**limbic lobe**") and **subcortical structures**, located mainly in the medial and inferior regions of the cerebral hemispheres (Figure 20.1). Limbic system structures are interconnected with each other and with the hypothalamus. The limbic system is extremely old from an evolutionary perspective, and for some species in which the neocortex is not highly developed, limbic system structures form the major part of the forebrain. In its connections, the limbic system is interposed between the hypothalamus and the neocortex, thus, providing a bridge connecting endocrine, visceral, emotional, and voluntary responses to the environment. Together with the hypothalamus, the limbic system provides an anatomical substrate for emotional, drive-related, and motivated aspects of behavior.

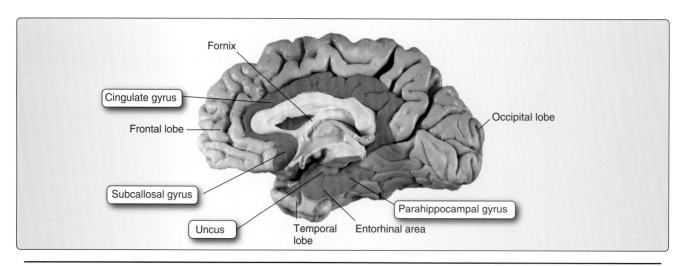

Figure 20.1
Cortical areas of the limbic system.

II. ANATOMY

In this overview of limbic system anatomy, we discuss the limbic lobe, subcortical structures of the limbic system, and the major functionally important interconnections of limbic structures with each other and with other CNS structures.

A. Limbic lobe

As can be seen in Figure 20.1, the limbic lobe is not a true lobe in the way that the frontal, parietal, temporal, and occipital lobes are distinct lobes of the cerebral cortex. Rather, the limbic lobe comprises a ring of cortex on the medial surface of the brain, which spans across aspects of the frontal, parietal, and temporal lobes. It consists of the **parahippocampal gyrus**, the **cingulate gyrus**, and a continuation of the cingulate gyrus anteriorly and inferiorly, called the **subcallosal gyrus** (see Figure 20.1). These cortical areas are interconnected by a subcortical fiber bundle called the **cingulum**. The major **subcortical structures** of the limbic system include the **hippocampus** (major role in learning and memory), the **amygdala** or amygdaloid nuclear complex (major role in emotions and drives), and the **septal nuclei** (associated with reward mechanisms). A swelling in the anteromedial pole of the parahippocampal gyrus is called the *uncus* (Latin for "hook"); it overlies the amygdala and the anterior hippocampus.

The anterior part of the parahippocampal gyrus is known as the **entorhinal cortex**. The entorhinal cortex receives input from widespread cortical association areas, including somatosensory, auditory, visual, taste, and prefrontal areas, and has reciprocal communication with the hippocampus. The entorhinal cortex, in turn, projects information from the hippocampus back to the cortical association areas.

B. Hypothalamus

The hypothalamus (see Chapter 19, "The Hypothalamus") is a functional part of the limbic system insofar as it is intimately interconnected with all limbic system structures and gives rise to efferents that carry limbic system information out to forebrain, brainstem, and spinal cord targets. For example, connections of limbic system structures, such as the amygdala and hippocampus, to the hypothalamus may provide a mechanism by which emotional responses can influence visceral activity (e.g., why anxiety can cause your stomach to "churn" and your palms to get sweaty). Other structures that have major connections with the limbic system include the ventral striatum, the anterior and dorsomedial nuclei of the thalamus, the ventral tegmental area (VTA), the periaqueductal gray, and the prefrontal cortex. In addition, the olfactory system has close interconnections with limbic system structures (Figure 20.2). Olfaction is discussed separately in Chapter 21, "Smell and Taste."

C. Hippocampus

The hippocampus or, more broadly, the **hippocampal formation** is a curved sheet of cortex folded into the medial surface of the temporal lobe, occupying the floor of the inferior or temporal horn of the lateral ventricle. It is a relatively large structure, about 5 cm in length.

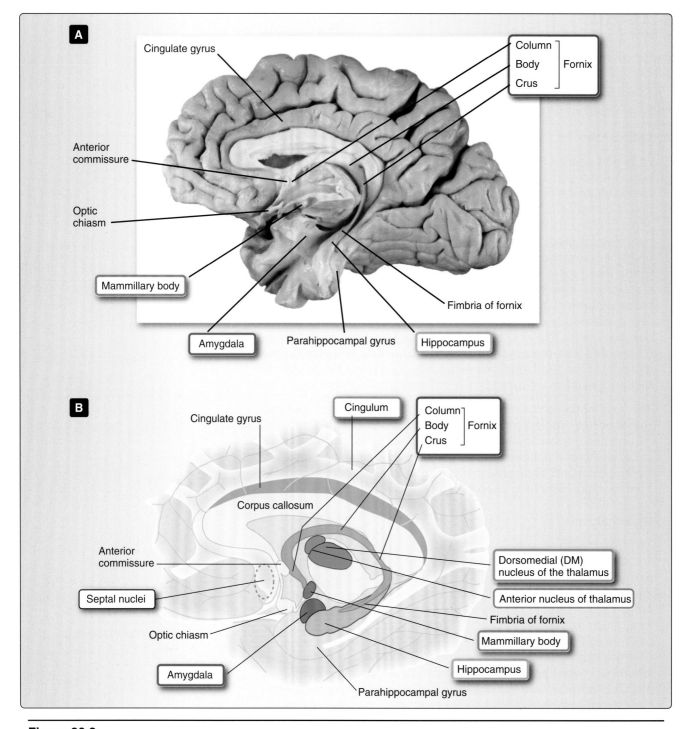

Figure 20.2
A. Midsagittal view of the brain with the inferior horn of the lateral ventricle opened to show the fornix, hippocampus, and amygdala. **B.** A diagrammatic representation of the subcortical structures of the limbic system.

The hippocampal formation consists of three major parts: (1) the **subiculum**; (2) the **hippocampus** proper, also called **Ammon horn**; and (3) the **dentate gyrus** (Figure 20.3). The parahippocampal gyrus comprises the cortex overlying the hippocampal formation (it is "para" or beside the hippocampus).

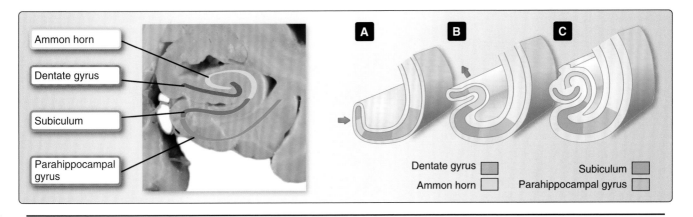

Figure 20.3
Structure of the hippocampus. The *left panel* shows the hippocampus in a coronal section. The *right panel* shows how the structures develop in relation to each other (from **A** to **C**.)

The **subiculum** is a transition zone of cortex, between the hippocampus and the parahippocampal gyrus (see Figure 20.3). The **hippocampus** proper consists of gray matter with an expanded anterior end. Its ventricular surface is covered with ependyma. Fibers arising from cell bodies in the subiculum and hippocampus proper gather into a bundle known as the **fimbria**, which, at the posterior end of the hippocampus, becomes the **fornix**, the most prominent output pathway of the hippocampus. (Note: The fornix also carries reciprocal afferent fibers.) In addition, both the hippocampus proper and the subiculum send outputs directly to the entorhinal cortex, which has reciprocal connections with widespread association areas of the cerebral cortex. The entorhinal cortex thus serves as a "gateway" through which sensory, cognitive, and emotional information can reach the hippocampus, and the hippocampus, in turn, can influence cortical function. The **dentate gyrus** is a notched or "toothlike" band of gray matter. Its toothlike appearance is caused by numerous small blood vessels arising from vessels in the adjacent subarachnoid space that enter the hippocampus along its course and penetrate the dentate gyrus. In coronal section, the dentate gyrus and hippocampus proper take the form of two interlocking letters "*C*" (see Figure 20.3). Interestingly, studies have shown that the dentate gyrus is one of the few regions of the adult brain, other than the olfactory bulb, where **neurogenesis** (i.e., the generation of new neurons) takes place. Neurogenesis takes place in the granule cells of the dentate gyrus. The new cells generated are thought to be fully functional and to play a role in the formation of new memories and possibly in modulating symptoms of stress and depression.

D. Amygdala

The amygdala (amygdaloid nuclear complex) is an "almond-shaped" structure that lies deep to the uncus, just rostral to the hippocampus (Figures 20.2 and 20.4). It is buried in the roof of the inferior horn of the lateral ventricle and abuts the tail of the caudate nucleus. It consists of a collection of morphologically and functionally diverse nuclei that can be divided into three major groups: basolateral, central, and corticomedial. Afferents to the amygdala include sensory

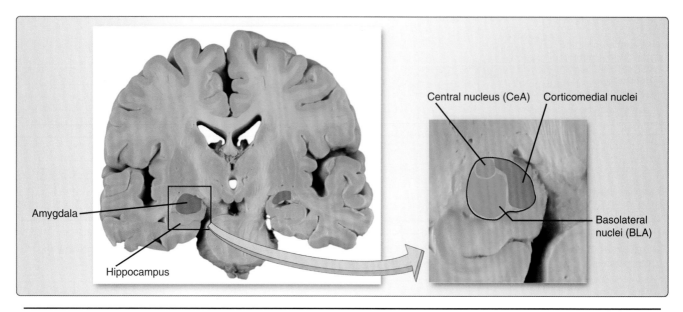

Figure 20.4
The hippocampus and amygdala in coronal section. Inset shows the subnuclei of the amygdala.

information (visual, auditory, somatosensory, gustatory, olfactory), information from the brainstem (raphe nuclei, periaqueductal gray, dorsal motor nucleus of X, nucleus solitarius, and locus ceruleus), input from the dorsomedial nucleus of the thalamus, and information from widespread cortical areas (Figure 20.5). The output from the amygdala travels through two major pathways, the stria terminalis and the ventral amygdalofugal fibers, and projects back to many of the same areas from which the afferents arise. The amygdala can send its output directly to cortical and brainstem areas. Alternatively, output to the brainstem may relay through the hypothalamus, which in turn sends projections to the brainstem (only these latter outputs are shown in Figure 20.5).

1. **Basolateral nuclei:** The **basolateral nuclei** (basolateral amygdala, BLA) are the largest group and well developed in humans. The BLA is thought to play a role in **attaching emotional significance to a stimulus**. It receives information about the modality and particular characteristics of a stimulus through its reciprocal connections with many areas of the cortex, including widespread association areas (prefrontal [mainly orbitofrontal], parietal, temporal), and the cortex of the cingulate and parahippocampal gyri. It also receives inputs from the thalamus (primarily anterior and dorsomedial nuclei) and hippocampus (see Figure 20.5). Efferents from the BLA are sent back to the cerebral cortex as well as the thalamus and the central nucleus of the amygdala.

2. **Central nucleus:** The **central nucleus** (central nucleus of amygdala, CeA) is important in mediating general emotional responses (see Figure 20.5). The CeA has reciprocal connections with visceral nuclei of the brainstem and spinal cord and also receives input from the BLA. The CeA also receives nociceptive inputs and input from the PAG and is a critical part of the pain matrix

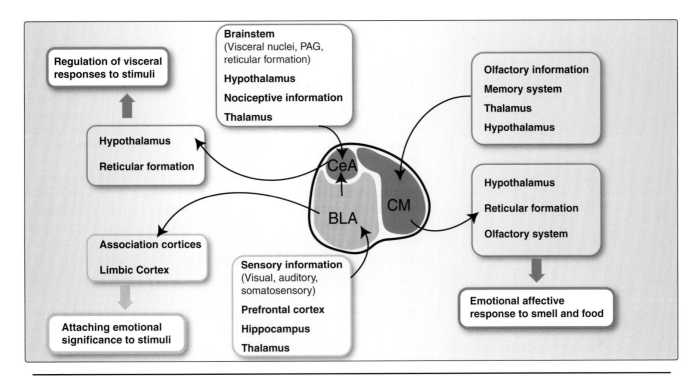

Figure 20.5
The amygdala and its projections. BLA = basolateral nuclei of amygdala; CeA = central nucleus of amygdala; CM = corticomedial nuclei of amygdala; PAG = periaqueductal gray.

(See Chapter 22, Pain). The CeA also has reciprocal connections with cholinergic and aminergic neurotransmitter systems of the brainstem. Through these connections, the CeA plays a role in regulating **visceral responses to emotional stimuli,** including pain.

3. **Corticomedial nuclei:** The **corticomedial nuclei** are not as well developed in humans. They receive olfactory information from the olfactory bulb, gustatory information, and information from the thalamus (dorsomedial nucleus) and have reciprocal connections with the hypothalamus, specifically the ventromedial and lateral areas, which are involved in the regulation of food intake. The corticomedial nuclei may provide information to the hypothalamus about the smell of food and thereby play a role in our **emotional affective responses to food**; pleasant smells stimulate appetite and unpleasant smells suppress appetite.

4. **Stria terminalis:** The **stria terminalis** is a small tract that arises primarily from the **medial nuclei**. It leaves the amygdala, arches over the thalamus, running in the groove between the caudate nucleus and the thalamus, and terminates in the hypothalamus, ventral striatum, and septal nuclei.

5. **Ventral amygdalofugal fibers:** The **ventral amygdalofugal fibers** arise from the basolateral and central nuclei and form a second major efferent pathway. These fibers also terminate in the hypothalamus and septal nuclei. In addition fibers project to the ventral striatum and cortex, including the frontal, prefrontal, cingulate, and inferior temporal cortical areas.

E. The septal nuclei

The **septal nuclei** (see Figure 20.2) are a small group of nuclei in the medial wall of the **frontal lobe**, rostral to the **anterior commissure** and bordering on the anterior horn of the lateral ventricle. The septal nuclei have reciprocal connections with the olfactory bulb, hippocampus (via the fornix), and amygdala (via the stria terminalis and ventral amygdalofugal fibers). In addition, the **medial forebrain bundle** (see Figures 19.5 and 19.6) carries afferents to and efferents from the septal nuclei. The medial forebrain bundle provides a dopaminergic projection to the septal nuclei and connects them with the hypothalamus and the brainstem reticular formation, which in turn projects to the brainstem and spinal cord visceral and motor nuclei. The septal nuclei are one of the few sites in the forebrain that contain cholinergic neurons and send cholinergic projections to the lateral hypothalamus, amygdala, hippocampus, and areas of the frontal cortex.

The clinical importance of the septal nuclei in humans is not well understood. A role in reward and pleasurable feelings has been suggested given the close association of the septal nuclei with the nucleus accumbens and the dopaminergic projections that reach the septal nuclei. Indeed, patients who have received electrical stimulation of the septal region report sexual feelings and feelings of orgasm.

F. The extended Papez circuit or network

In 1937, Dr. James Papez, a neuroanatomist at Cornell University, proposed the idea that the experience of emotion involved reciprocal interactions between the diencephalon and the cerebral cortex. Because emotions reach consciousness, and conscious thoughts can affect emotion, Papez hypothesized that a neural circuit involving the limbic system and specific cortical areas formed the neuroanatomical substrate for emotion: "The hypothalamus, the anterior thalamic nucleus, the cingulate gyrus, the hippocampus and their interconnections constitute a harmonious mechanism which may elaborate the functions of central emotion as well as participate in the emotional expression" (James Papez, 1937). The circuitry involved included (Figure 20.6):

- The **hippocampus**
- Output from the hippocampus through the **fornix**
- Termination of the fornix in the **mammillary bodies**
- Outflow from the mammillary bodies through the **mammillothalamic tract** to the **anterior nucleus of the thalamus**
- Projections from the anterior nucleus to the **cingulate gyrus**
- Output from the cingulate gyrus back to the hippocampus

The concept of a neural substrate for emotion was later expanded to include other areas that are structurally and functionally connected with those described by Papez, and this concept continues to be expanded today as we learn more about this complex neural circuitry. Modern understanding of limbic circuitry includes the following:

- The fornix projects to areas of the hypothalamus beyond the mammillary bodies and to other structures along its route (septal nuclei, hypothalamus, ventral striatum) and carries bidirectional information.

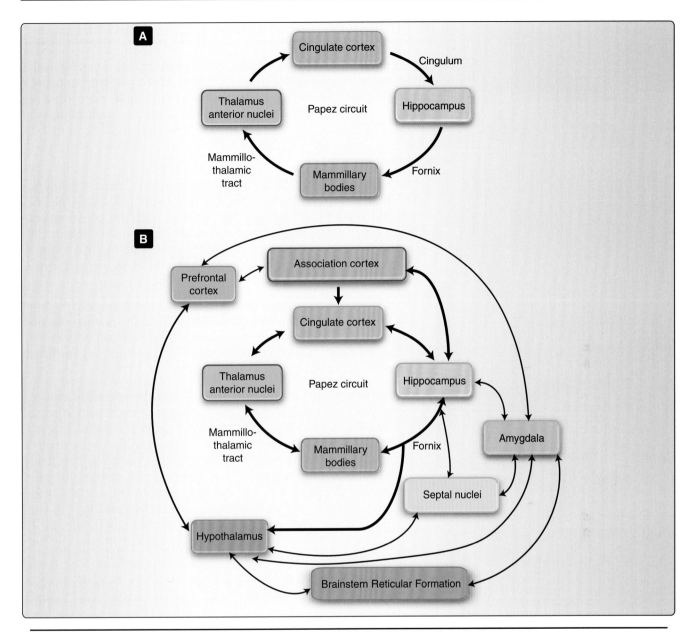

Figure 20.6
The Papez circuit **(A)** and the extended Papez circuit **(B)**.

- The amygdala is a key structure in the expression of emotions, emotional memory, and basic drives.
- Interconnections among and between limbic system structures and the hypothalamus are extensive and complex.
- Association areas of the cortex, and particularly the prefrontal cortex, play key roles.

Figure 20.6 is a diagram showing the expanded structures and connections of the neural circuitry involved in emotion. Notably, this diagram is still a very simplified depiction of this complex neural network. For example, many of these structures also have interconnections with the thalamus (not shown).

III. FUNCTIONS OF LIMBIC SYSTEM STRUCTURES

Limbic system structures include the hippocampus, amygdala, and septal nuclei. These structures are highly interconnected and have both individual and overlapping functions. They play major roles in memory, emotion, emotional learning and behavior, motivation, and reward.

A. Hippocampus

The most important role of the hippocampus in humans is its mediation of learning and the formation of new memories. Intact memory function is critical to everyday life, and disruption of the normal ability to learn, store, and retrieve memories can have an enormously adverse impact on the ability to function. There are multiple forms of memory, each depending on different but overlapping sets of central nervous system structures.

In addition, it should be noted that as part of the limbic system, and in view of its extensive connections with the hypothalamus and other limbic structures, the hippocampus plays a role in visceral and endocrine function and in the expression of emotions and emotional behavior.

1. **Short-term memory: Short-term** or **working memory** involves the short-term maintenance of information in memory and sometimes the manipulation of that information to achieve an immediate goal. The classic example of this is looking up a phone number and keeping the information in mind while picking up the phone and entering the number. Working memory is needed for more complex situations as well, including performing multiple simultaneous tasks, doing calculations, and understanding long written or spoken sentences or paragraphs in a book (Figure 20.7). Working memory involves the hippocampus but depends primarily on the prefrontal cortex.

2. **Long-term memory:** There are two types of long-term memory: **explicit memory**, which involves facts or events, and **implicit memory**, which is not directly accessible to consciousness (see Figure 20.7).

   a. **Explicit or declarative memory:** Explicit or declarative memory involves memories of events or facts that are accessible to consciousness and can be expressed explicitly (i.e., "declared" as remembered events or facts). There are two forms of declarative memory, both of which involve retrieving earlier stored information: episodic memory and semantic memory.

      • *Episodic memory* involves the memory for events and experiences, including information about the time and place of an event and details about the event itself. The context surrounding the event and associated emotions are typically part of the memory. Episodic memory includes the ability to learn, store, and retrieve information about experiences that occur in our everyday lives.

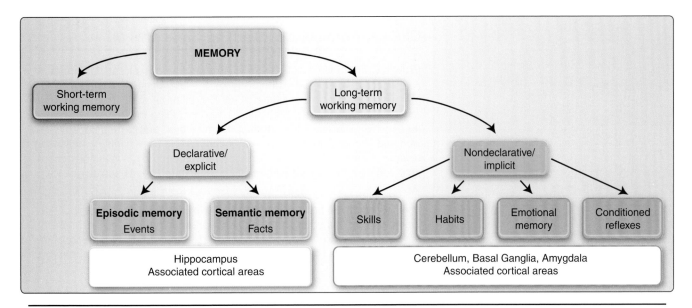

Figure 20.7
The different types of memory.

- *Semantic memory* involves the memory of facts, concepts, and knowledge that have been learned, but are typically independent of personal experience and of the context in which it was acquired. Semantic memory includes common factual knowledge about the external world around us of which the source of the original information is typically not known. Knowledge about categories of objects (e.g., "apples and bananas are fruits"), historical events, capital cities, and mathematical tables are examples of semantic memory.

Explicit memory requires the hippocampus and associated cortical areas (entorhinal cortex, parahippocampal gyrus) and widespread neocortical association areas (which have bidirectional communication with the entorhinal cortex).

b. **Implicit or nondeclarative memory:** Implicit or nondeclarative memory involves memories that manifest as subconscious behavioral or physiological responses to events or stimuli. Implicit memory includes several forms of learning that occur during performance of a task. **Skills** and **habits** such as driving, swimming, and riding a bicycle are examples of implicit memory. Memory of skills and habits depends on the striatum (caudate nucleus and putamen), motor areas of the cortex, and the cerebellum. **Emotional memory** or emotional associations are a second example of implicit memory. Emotional memory involves a change in behavior toward a previously neutral stimulus as a result of experience and depends on the amygdala. An example of this is seeing a red scarf and smiling because you remember your grandfather who always wore one when skating. **Conditioned reflexes** can be considered a third type of implicit memory, and these depend primarily on the cerebellum. The most famous example of this is Pavlov's dog who salivated at the sound of a bell, a stimulus that had been linked to food.

Clinical Application 20.1: The Case of H.M.

Probably, the most famous case illustrating the loss of episodic memory is that of H.M., studied extensively by Brenda Milner and others at the Montreal Neurological Institute. A radical bilateral medial temporal lobe resection, involving both hippocampi, and including the uncus, amygdala, and overlying cortex of the parahippocampal gyri, was carried out on H.M., who had a long history of seizures uncontrollable by maximum medication of various forms. This was an experimental procedure but felt to be acceptable because the patient was totally incapacitated by his seizures.

Although the surgery controlled the seizures, there was "...one striking and totally unexpected behavioral result: a grave loss of recent memory... After the operation, this young man could no longer recognize the hospital staff nor find his way to the bathroom, and he seemed to recall nothing of the day-to-day events of his hospital life. His early memories were apparently vivid and intact."

This memory deficit was persistent and did not improve over time. "He does not know where objects in continual use are kept … [He] has even eaten lunch in front of one of us without being able to name, a mere half hour later, a single item of food he had eaten; in fact, he could not remember having eaten lunch at all." Importantly, intelligence, understanding, and abstract reasoning were intact in H.M., indicating that intelligence and memory can be functionally separated. The memory deficit in H.M. reflects a loss of explicit or declarative long-term memory (memory for events), which depends on the medial temporal lobe. Implicit forms of long-term memory were largely retained, including procedural memory for skills and habits and conditioned reflexes (which depend on the cerebellum) and emotional memory (which depends on the amygdala).

Reproduced from Scoville WB, Milner B. Loss of recent memory after bilateral hippocampal lesions. J Neurol Neurosurg Psychiat 1957;20(*1*):11–21 with permission from BMJ Publishing Group LTD.

Clinical Application 20.2: Alzheimer Disease and Memory

Alzheimer disease is the most common dementia in the world and is characterized by neuronal loss and the presence of neurofibrillary tangles and amyloid plaques. It involves a gradual and progressive loss of both memory and cognitive function. The formation of intracellular neurofibrillary tangles (abnormal tau protein that causes collapse of microtubules) and extracellular amyloid plaques (β amyloid, a protein fragment snipped from an amyloid precursor protein that accumulates to form hard, insoluble plaques) is thought to contribute to the degradation of neurons and the subsequent symptoms of Alzheimer disease. The subiculum and entorhinal cortex are among the first sites where such abnormalities appear, resulting in altered hippocampal function. This may, at least in part, account for some of the memory deficits associated with Alzheimer disease. Loss of cognitive function occurs through neuronal abnormalities in more widespread areas of cortex.

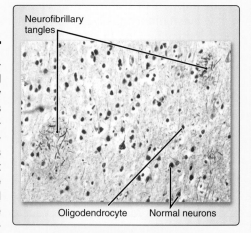

Neurofibrillary tangles characteristic of Alzheimer disease are visible in this microscopic study.

B. Amygdala

With connections to both the hypothalamus and, indirectly, the prefrontal cortex, the amygdala is positioned to play a role in both drive-related behaviors and the processing of emotions that are related to these behaviors (Figure 20.8).

1. **Emotional learning and memory:** The amygdala plays a key role in emotional learning and memory (an **implicit memory** function). The emotional significance of inputs from various cortical areas is assessed primarily by the **basolateral nucleus** (see Figure 20.5). Efferents from there to the **hypothalamus** then activate the appropriate visceral and motor responses. At the same time, the amygdala sends outputs via the **dorso-medial nucleus of the thalamus** to the **orbitofrontal cortex**, which mediates the conscious perception of emotions. Thus, the **amygdala is involved in linking perception with visceral and behavioral responses and with memory**. In relation to this, it is well known that events or facts associated with strong emotions are more likely to be remembered than emotionally neutral events or facts. People remember emotional events such as a wedding day in great detail. Traumatic events can also be remembered easily: People who grew up in the 1960s remember the assassination of John F. Kennedy and the events surrounding that day in minute detail. Similarly, for those who witnessed the events of September 11, 2001, these events will always be linked to memories of "where we were when we heard the news." It is the link of the emotion with the event that solidifies the memory. The amygdala works together with, and enhances the function of, the memory system of the hippocampus and associated cortical areas (medial temporal lobe memory system) during memory formation (Figure 20.9). The link between the amygdala and the medial temporal lobe memory system in the consolidation of emotional memories has been confirmed

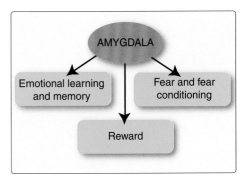

Figure 20.8
Functions of the amygdala.

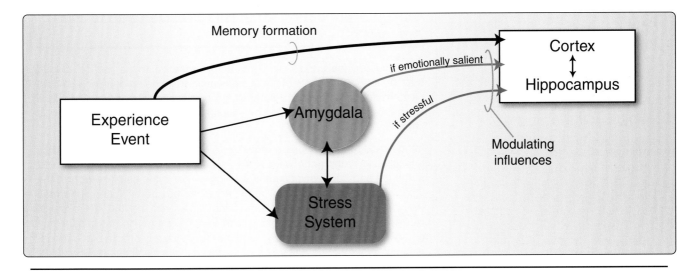

Figure 20.9
The modulation of memory formation by emotion and stress.

in functional magnetic resonance imaging (fMRI) studies in humans. If the amygdala is damaged, the normal facilitation of attention to and memory of emotional stimuli is greatly reduced or absent entirely.

2. **Fear and fear conditioning:** The amygdala is part of a neural system that detects and responds to threats. Brain imaging studies have shown that when people are exposed to threatening stimuli, neural activity in the amygdala increases and both behavioral (freeze or jump back) and physiological (increased heart rate, sweating) responses occur. At the same time, people may experience feelings of fear. However, activity in the amygdala does not mean that fear will be experienced. Conversely, when the amygdala is damaged, stimuli that were previously threatening are no longer viewed as threatening. Thus, defensive responses to threat and fearful feelings can be separated in the brain. The amygdala is involved in immediate behavioral and physiological responses to threats, while experience or feeling of fear results from recognition and interpretation of an event through extensive connections between the amygdala and brain areas involved in cognitive function.

Fear conditioning is a form of emotional learning in which a neutral stimulus (the conditioned stimulus) becomes associated with an aversive event (the unconditioned stimulus) such that, similar to conditioned reflexes (described above), presentation of the conditioned stimulus alone can elicit defensive behavior and the appropriate visceral and endocrine responses (see Figure 20.8). Importantly, in humans with damage to the amygdala, processing of social signals of fear and anger is severely impaired, regardless of the input modality. In particular, visceral responses to fear appear to be mediated by the CeA through its connections with the midbrain periaqueductal gray, the reticular formation, and the hypothalamus.

Clinical Application 20.3: Bilateral Damage to the Amygdala

While the amygdala and its circuitry are not essential to the "feeling" of fear, they are key components of the system involved in the recognition and cognitive evaluation of emotional behaviors or signals from others. The close link between the amygdala and the medial temporal lobe memory system is important in this function.

In one case report, a woman with severe epilepsy that was uncontrolled by antiepileptic drugs underwent a series of stereotaxic operations targeted at the left and right amygdalae. Following surgery, she could recognize the faces of people who were familiar to her before the operation and performed well on a face-matching task involving faces with neutral expressions. However, her interpretation of facial expressions of emotion was impaired. In particular, recognition of fear was differentially severely affected, and recognition of anger and, to a lesser extent, disgust was also impaired. Similarly, she showed impaired perception of vocal affect, with intonation patterns related to fear and anger particularly affected.

IV. REWARD CIRCUITRY

Reward is a key factor for driving incentive-based learning, appropriate responses to stimuli, and the development of goal-directed behavior. Dopamine plays a key role in reward, and we now know that the medial forebrain bundle carries **dopaminergic fibers** projecting from the **VTA** of the midbrain to the **nucleus accumbens**. VTA dopaminergic projections also influence the hippocampus, amygdala, septal nuclei, and prefrontal cortex. In turn, the prefrontal cortex can provide feedback to the VTA either directly or through the nucleus accumbens. Additional modulatory connections fine-tune the system. All of these structures ultimately communicate with the hypothalamus to initiate neuroendocrine and visceral responses to reward (see Chapter 19, "The Hypothalamus"). Together, these structures and their interactions form a neural substrate for reward (Figure 20.10). See Chapter 12, "Brainstem Systems and Review," for more information. In addition, cortical and subcortical structures interact to form a complex network that mediates adaptive behaviors, allowing motivation and reward to be combined with a strategy and action plan for obtaining goals.

A. Addiction

The complex reward circuitry forms a neural substrate for **addiction**. Although substances of abuse are structurally and functionally diverse and produce a variety of behavioral effects, they can all modulate the brain reward system. Substances of abuse, similar to natural rewards, activate the **dopaminergic neurons** that travel in the medial forebrain bundle. Acutely, all substances of abuse increase dopaminergic transmission from the **VTA** to the **nucleus accumbens** but, with chronic intake, the dopaminergic system is impaired. Dopamine D_2 receptors are down-regulated, and dopamine function is reduced, particularly in the nucleus accumbens and

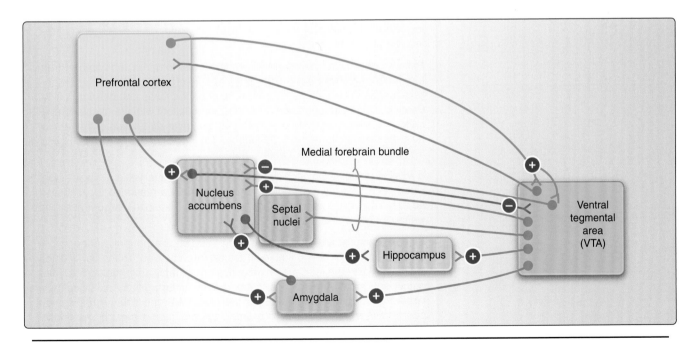

Figure 20.10
Reward circuitry. Shown are VTA projections to the nucleus accumbens, hippocampus, amygdala, septal nuclei, and prefrontal cortex. Feedback from VTA is direct or through the nucleus accumbens.

ventral striatum, of patients addicted to a variety of different substances. The result of dopamine receptor down-regulation is that normally rewarding or naturally reinforcing stimuli are less effective in eliciting dopamine transmission. This may, in part, underlie the negative emotionality seen during **withdrawal**. On the other hand, the dopaminergic system becomes sensitized to substances and their cues, resulting in a more intense response to these stimuli (see Figure 20.10). Substances of abuse typically block the dopamine transporter in the brain reward circuitry, allowing dopamine to remain in the synapse for a long time, resulting in a large and lasting reward, despite the reduced number of receptors.

1. **Dopamine and saliency:** The reason dopamine is so important in addiction is that it signals saliency (i.e., something important or worth paying attention to). In addiction, nonsubstance stimuli take on reduced saliency, whereas stimuli associated with substances of abuse have markedly increased saliency. Substance-associated stimuli activate the prefrontal cortex and increase glutamatergic drive to the nucleus accumbens. It is this increased prefrontal drive that results in the greatly increased salience to substances and substance-associated stimuli, with a corresponding increase in craving and substance-seeking behavior (see Figure 20.11).

Altered emotional learning and memory circuitry may also play a role in the increased saliency of substances and their cues and the reduced saliency of otherwise pleasurable stimuli (e.g., food, sex).

Clinical Application 20.4: Temporal Lobe Epilepsy

A seizure involves abnormally synchronized and high-frequency firing of neurons in the brain. Epilepsy is a disorder in which there is a tendency for unprovoked recurrent seizures. Seizures can be partial (focal or local), involving a localized or specific area of the brain, or can be generalized, involving the entire brain.

Patients who have seizures localized to medial temporal lobe limbic structures (the hippocampus, amygdala, and overlying cortex) often report "experiential" phenomena. The term "experiential" was first used by Wilder Penfield. During an operation on a conscious patient with intractable temporal lobe seizures, Penfield observed that electrical stimulation of the temporal cortex evoked a memory "flashback." Writing about this in his 1975 book, *The Mystery of the Mind*, Penfield noted:

"I was incredulous. On each subsequent occasion I marveled ... I was astonished each time my electrode brought forth such a response. How could it be? This had to do with the mind! I called such responses 'experiential' and waited for more evidence."

Since that time, many studies have shown that such experiential phenomena do exist but occur only if limbic structures are involved. These can include feelings of extreme fear and anxiety, illusions of familiarity (the *déjà vu* phenomenon), memory recalls or flashbacks, visual or auditory hallucinations, visceral sensations ("butterflies" in the stomach), and strange unpleasant odors. In view of the discussion of limbic system functions, such experiential phenomena should no longer seem astonishing.

From Penfield W. *Mystery of the Mind: A Critical Study of Consciousness and the Human Brain.* Princeton, NJ: Princeton University Press, 1975.

Association of a substance-induced pleasurable experience with increases in dopamine will result in strong conditioning not only to the substance but also to the stimuli that predict the substance (e.g., the house of the dealer, the neighborhood, syringes). This could contribute to the enhanced responses to the substance and related stimuli that then overshadow the response to natural rewards (Figure 20.11).

2. **Dopamine and stress:** The dopamine reward circuitry is also responsive to stress. Stress and substances of abuse act similarly to activate structures involved in the reward pathway, including dopamine neurons in the VTA. Stress can also facilitate the reward associated with initial substance exposure and can increase craving and relapse into substance-seeking behavior. Corticotropin-releasing hormone in the hypothalamus and in extrahypothalamic structures including the amygdala plays a role in mediating the effects of stress on the reward circuitry.

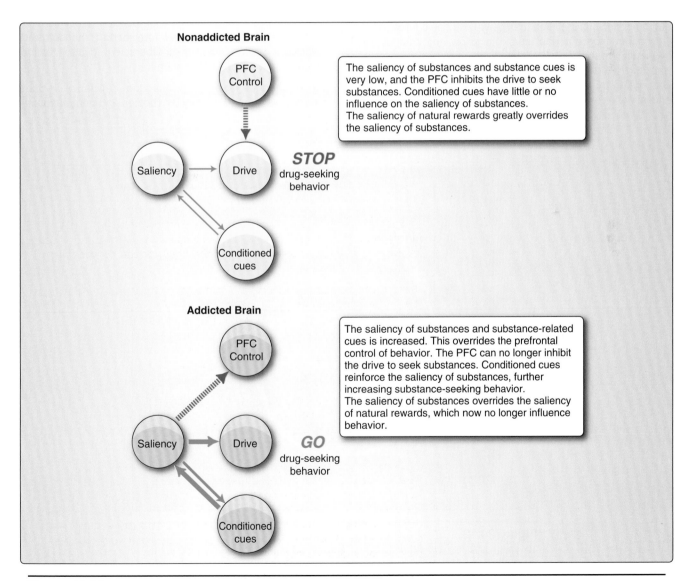

Figure 20.11
Saliency of drugs and drug-related cues drive addictive behavior. PFC = prefrontal cortex.

Clinical Case:

Wernicke Encephalopathy

A 56-year-old man presents with a sudden onset of vertigo, nausea, and unsteady gait. He has no tinnitus (ringing in his ears) or hearing loss. He has no headache, neck pain, diplopia, swallowing/speech difficulties, or any focal motor/sensory symptoms. He has no infectious symptoms and shows no signs of trauma or constitutional symptoms (weight loss, fever, or night sweats). He has been forgetful and disoriented for the past 2–3 days but had no prior difficulties in his memory or cognition. He drinks 5–7 bottles of beer per day and skips meals. His diet consists mainly of starchy and sweet food, with little protein or fruit.

On examination, he was oriented to his name but not to the date or location (place). His speech was generally normal, and he did not confabulate (speech with fabricated, distorted, or misinterpreted memories about oneself or the world, without the conscious intention to deceive). He could follow one-step commands but was unable to follow any complex or sequential commands. His immediate recall of three words was normal, but his delayed recall was zero out of three, and cueing by the physician did not improve his recall. His cranial nerve examination was normal except for changes in

eye movements: There was some disruption of normal eye movements during smooth pursuit, saccades were hypometric (saccades undershoot or fall short of their target), and nystagmus was observed. His muscle bulk, tone, strength, and deep tendon reflexes were normal. There was no tremor. Sensory examination was normal. He had mild dysmetria on finger-to-nose testing and slightly slow random alternating movements (dysdiadochokinesia). He had ataxia on heel-to-shin testing with a wide-based stance and gait.

His MRI (Figure 20.12) showed signal changes in the mammillary bodies and the midbrain. His erythrocyte transketolase level was low and serum pyruvate level was high (both are markers for abnormal serum thiamine levels). He was diagnosed with Wernicke encephalopathy due to thiamine deficiency and was given thiamine replacement and education regarding diet and alcohol consumption.

Conditions associated with thiamine deficiency include alcohol consumption, hyperemesis gravidarum (nausea and vomiting in pregnancy), chemotherapies, gastrointestinal disorders, anorexia, and lymphoma.

Signal changes related to thiamine deficiency include hemorrhage into the mammillary bodies and periaqueductal

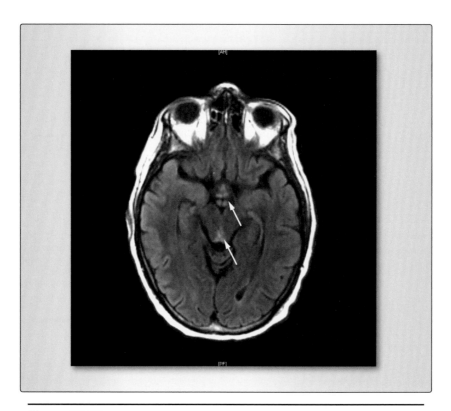

Figure 20.12

This is an MRI, T2 weighted image - axial view. Increased T2 signal seen in the mammillary bodies (top arrow) and midbrain tectum (bottom arrow) in Wernicke encephalopathy due to thiamine deficiency. (From Mayer SA. Rowland LP, Louis ED. *Merritt's Neurology*, 13th ed. Wolters Kluwer Health, 2015.)

gray of the midbrain, which can be seen on MRI, as seen in Figure 20.12.

What is the triad for Wernicke encephalopathy?

The triad consists of confusion, ophthalmoplegia, and ataxia.

1. Confusion: over days to weeks with inattention, apathy, disorientation, and memory impairment

2. Ophthalmoplegia: weakness of extraocular movement with horizontal nystagmus

3. Ataxia: especially truncal and gait ataxia due to cerebellar damage

Other symptoms that can occur include hypothermia and postural hypotension due to involvement of the hypothalamus and brainstem autonomic pathways.

If there were confabulation, what disorder could this man have had in combination with Wernicke encephalopathy?

Korsakoff syndrome. Bilateral temporal lobe dysfunction, particularly with damage to the hippocampus, causes Korsakoff amnestic syndrome with confabulation and both anterograde and retrograde amnesia that are the predominant cognitive deficits. Attention and other cognitive domains are relatively preserved. Confabulation is a prominent feature in the early stages of Korsakoff.

Chapter Summary

- The limbic system, together with the hypothalamus, provides an anatomical substrate for emotional, drive-related, and motivated aspects of behavior. The limbic system comprises both cortical and subcortical structures. The cortical structures or "limbic lobe" consist of the parahippocampal, cingulate, and subcallosal gyri, which are interconnected by the cingulum. The major subcortical structures include the hippocampus, amygdala, and septal nuclei. The hippocampus (hippocampal formation) consists of the subiculum, hippocampus proper (Ammon horn), and the dentate gyrus.

- The hippocampus and associated cortical areas are critical for explicit or declarative memory, that is, memory for facts and events. Implicit memory on the other hand depends on the amygdala, cerebellum, and prefrontal cortex. The hippocampus also plays a role in visceral and endocrine function and in the expression of emotions and emotional behavior. The amygdala (amygdaloid nuclear complex) consists of three major groups of nuclei: the basolateral, central, and corticomedial. The amygdala plays a key role in emotional learning and memory and behavioral and physiological responses to threats and, through its extensive connections to brain areas involved in cognitive function, influences emotional learning and memory and plays a role in attention, perception, and processing of the emotional content of social interactions.

- The extended Papez circuitry links limbic system structures with each other and with specific cortical areas and provides a neural substrate for emotional behavior. Dopaminergic projections from the ventral tegmental area to the nucleus accumbens, hippocampus, amygdala, septal nuclei and prefrontal cortex, and modulatory interactions among these structures form a neural substrate for reward. The reward circuitry forms the neural substrate for addiction. Dopamine is critical in addiction insofar as it signals saliency. In addition, nondrug stimuli take on reduced saliency, whereas stimuli associated with drugs have markedly increased saliency.

Study Questions

Choose the ONE best answer.

20.1 In limbic system circuitry:

    A. The cingulum connects the septal nuclei with the hippocampus.

    B. The entorhinal cortex projects information from the hippocampus to cortical association areas.

    C. Ventral amygdalofugal fibers project to the hippocampus.

    D. The medial forebrain bundle carries cholinergic projections to the septal nuclei.

    E. The fornix arises from the dentate gyrus and carries information from the hippocampus to the hypothalamus.

Correct answer is B. The entorhinal cortex receives input from widespread cortical association areas, has reciprocal connection with the hippocampus, and sends output from the hippocampus back to the cortex. The cingulum is a subcortical bundle of fibers that interconnects areas of the limbic lobe with each other. The ventral amygdalofugal fibers project from the amygdala to the hypothalamus. The medial forebrain bundle carries dopaminergic projections from the ventral tegmental area to the nucleus accumbens and the septal nuclei, among other structures. The fornix arises from the subiculum and the hippocampus proper and terminates in the mammillary bodies of the hypothalamus.

20.2 A 43-year-old woman suffers from relapsing–remitting multiple sclerosis. In one of her relapsing episodes, she complains about problems with her memory. You take her history to pinpoint what type of memory and which structures are affected. Which one of the following statements on memory is correct?

   A. Short-term or working memory depends on the amygdala.
   B. Episodic memory involves the memory of facts such as historical events.
   C. Memories involving learned skills are an example of implicit memory.
   D. Conditioned reflexes are an example of implicit memory that depends primarily on the striatum.
   E. Emotional memory involves a change in behavior toward a previously neutral stimulus and depends on the prefrontal cortex.

Correct answer is C. Implicit memory involves memories that manifest as subconscious behavioral or physiological responses to events or stimuli. Learned skills such as driving are an example of implicit memory. Short-term or working memory depends on the hippocampus and prefrontal cortex. Episodic memory involves memory of events or experience in everyday life. Knowledge of facts is an example of semantic memory. Conditioned reflexes are an example of implicit memory that depends primarily on the cerebellum. Emotional memory depends on the amygdala.

20.3 A physician treating patients with addictions to drugs of abuse counsels patients about the effects of the addiction on the function of their brains. Which one of the following statements best describes the effects of addiction and drug use?

   A. Drugs generally up-regulate dopamine D_2 receptors initially but down-regulate these receptors with chronic intake.
   B. Stress can suppress activity of the dopamine neurons in the ventral tegmental area.
   C. Drugs can activate the dopamine transporter in key limbic structures.
   D. Drug-associated stimuli can activate the prefrontal cortex and increase prefrontal inhibition of the nucleus accumbens.
   E. Drugs can potentiate the rewarding effects of natural rewards such as food and sex.

The correct answer is A. Acutely, drugs increase dopamine transmission from the ventral tegmental area (VTA) to the nucleus accumbens, but with chronic drug intake, the dopaminergic system is down-regulated. Stress activates neurons in the VTA and can facilitate the rewarding effects of drugs. Drugs suppress the dopamine transporter, keeping dopamine in the synaptic cleft for longer periods of time. Drug-associated stimuli activate the prefrontal cortex, which in turn activates the nucleus accumbens. Drugs become highly salient in addiction, whereas natural rewards have greatly diminished salience.

20.4 An elderly patient with Alzheimer disease is diagnosed with degenerative changes to the amygdala on both sides. Which one of the following statements about the amygdala is true?

   A. The emotional significance of inputs from the cortex is assessed primarily by the central nucleus.
   B. The stria terminalis connects the central and basolateral nuclei with the hypothalamus.
   C. In emotional learning, outputs from the amygdala go via the ventral posterior nucleus of the thalamus to the prefrontal cortex for conscious perception of emotion.
   D. The basolateral amygdala plays a role in the regulation of visceral responses to stimuli.
   E. Lesions can impair the ability to process social signals of fear and anger.

Correct answer is E. The amygdala plays a key role in the recognition and interpretation of emotional behaviors in oneself and in others. The emotional significance of inputs from the cortex is assessed primarily by the basolateral nuclei. The stria terminalis arises primarily from the medial nuclei. In emotional learning, information from the amygdala reaches the cortex via the dorsomedial nucleus of the thalamus. The central nucleus of the amygdala plays a role in the regulation of visceral responses to stimuli.

Smell and Taste

21

I. OVERVIEW

The senses of smell (**olfaction**) and taste (**gustation**) are part of the chemical senses in the body. They detect molecules and transduce chemical into electrical stimuli that are then interpreted by the brain as a smell or a taste. Together, these senses can bring us feelings of euphoria or disgust, and they can even warn us of danger (e.g., gas leaks or contaminated food). The inability to smell is termed anosmia, and the inability to taste is called *ageusia*. Some chemicals are not detected by taste and smell but rather through sensory nerve endings, which are sensitive to irritating substances such as camphor. In this chapter, we will discuss all of these types of chemoreception.

II. THE OLFACTORY SYSTEM

The olfactory system is made up of the **olfactory epithelium** containing **receptor neurons**, the **olfactory bulb**, **olfactory tract**, and **cerebral cortex** (Figure 21.1). It permits a conscious appreciation of chemicals (**odorants**) inhaled while breathing or those drifting up to the epithelium from the oropharynx. Although olfaction is not as critical for humans as it is for some animals, we are nevertheless able to discern many odorants that can be pleasant or unpleasant (i.e., noxious) and, ultimately, lifesaving. Olfaction plays a very large role in how we appreciate food. Consider how swollen nasal passages from a head cold affect the ability to taste. The epithelium also receives sensory innervation from V_1 of the trigeminal nerve for the appreciation of noxious stimulants (see below).

The olfactory system is unique in that the primary sensory receptors are **bipolar cells** in a specialized area of epithelium lining the upper part of the nasal cavity. Additionally, these cells undergo continuous replacement throughout life. These primary receptors synapse with secondary olfactory neurons in the olfactory bulb. From the olfactory bulb, signals are sent directly to the cortex, bypassing the thalamus. All other sensory inputs synapse in the thalamus before reaching the cortex.

A. Olfactory epithelium

The olfactory epithelium is a specialized mucosa in the roof of the nasal cavity. The epithelium extends from the upper lateral wall of the nasal cavity, across the **cribriform plate** of the ethmoid bone and

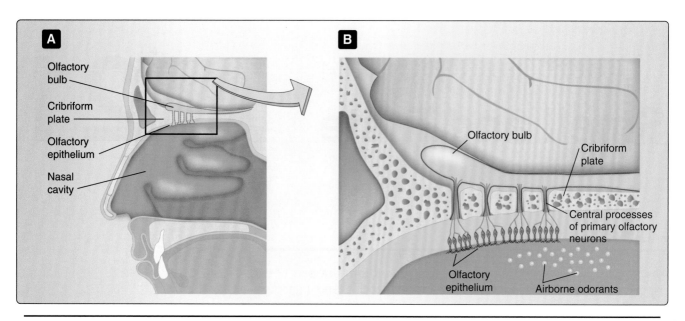

Figure 21.1
Organization of the olfactory system in the nose and inhaled odorants. **A.** Sagittal section of nasal cavity. **B.** Closer sagittal section at the level of the cribriform plate.

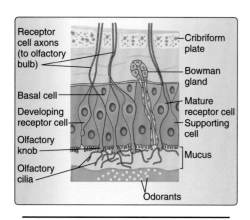

Figure 21.2
Structure of the olfactory epithelium.

partway down the medial wall or **nasal septum** (see Figure 21.1). It is moistened by mucous cells located within the epithelium. Inhaled odorants are dissolved in this moist mucus.

1. **Cells of the olfactory epithelium:** Olfactory receptor neurons are **bipolar neurons** each with a single dendrite that extends to the epithelial surface where it expands into an **olfactory knob** with cilia. The cilia contain the molecular receptor sites for odorant detection. The central processes of these neurons gather together into 20 or more bundles of filaments to traverse the cribriform plate of the ethmoid bone and synapse on the cells of the secondary olfactory neurons in the olfactory bulb (Figure 21.2).

   a. **Basal cells: Basal cells** are situated on the basement membrane and generate new receptor cells. This is one of the few types of cells in the central nervous system (CNS) that can continuously regenerate throughout life.

   b. **Supporting cells: Supporting cells** (sustentacular cells) commingle with the sensory cells and play a supporting role similar to that of glial cells.

   c. **Secretory cells: Secretory cells** in the olfactory glands (**Bowman glands**) release a fluid that contains odorant-binding proteins. The dendritic endings of the receptor cells and their cilia are bathed in this fluid that acts as a solvent for the odorants, enabling a widespread diffusion to the sensory receptors, thereby increasing the efficiency of odor detection.

2. **Olfactory processing and coding:** In the olfactory epithelium, odorants bind to specific receptors on the **primary olfactory neurons**. This leads to the activation of a **G protein**, which acti-

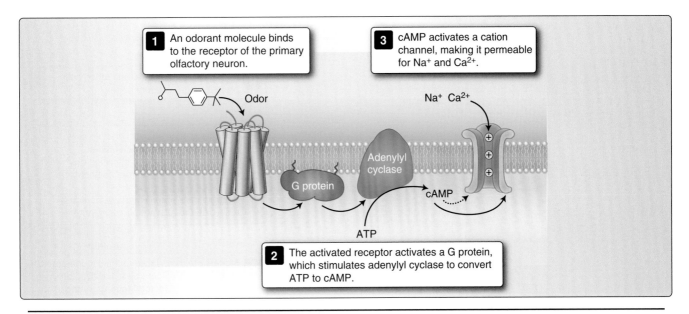

Figure 21.3
Signal transduction in primary olfactory neurons. cAMP = cyclic adenosine monophosphate; ATP = adenosine triphosphate.

vates adenylyl cyclase, which in turn produces cyclic adenosine monophosphate (cAMP) from adenosine triphosphate (ATP) and activates an ion channel that is permeable to cations (Na⁺ and Ca²⁺). This leads to a shift in membrane potential leading to depolarization of the neuron and the production of action potentials (Figure 21.3). See Chapter 1, "Introduction to the Nervous System and Basic Neurophysiology," for more detail.

B. Olfactory bulb

Primary olfactory neurons project to the olfactory bulb. The olfactory bulb and its tracts are the most rostral component of the olfactory system, lying directly over the cribriform plate of the skull.

There are four cell types in the olfactory bulb: **mitral, tufted, periglomerular,** and **granule** cells. The mitral and tufted cells are functionally similar and together carry olfaction from the olfactory bulb to the CNS. The periglomerular and granule cells are interneurons that integrate and modulate afferent olfactory information.

In the bulb, the olfactory receptor neurons synapse with **mitral** and **tufted** cells. The synapses occur in specialized areas called **glomeruli.** Considerable convergence takes place between input of olfactory receptor neurons and the dendrites of the tufted and mitral cells. Olfactory receptor neurons detecting the same odorant are scattered throughout the olfactory epithelium and converge onto the same glomerulus (Figure 21.4). This increases the sensitivity of the olfactory system. In some glomeruli, receptor neurons detecting different odorants converge, indicating that a first level of sensory processing occurs at this early synapse in the olfactory pathway. **Periglomerular cells** mediate contacts between glomeruli, and **granule cells** mediate contacts between two mitral or two tufted cells originating from different glomeruli. In addition, granule cells receive input from cortical

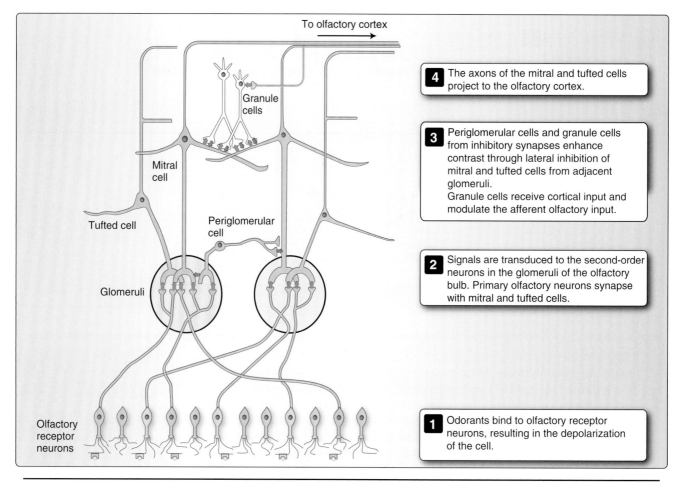

Figure 21.4
Olfactory processing and coding in the olfactory bulb. Inhibitory (*red*) and Excitatory (*green*). Granule cells receive excitatory input from and provide inhibitory input to mitral cells.

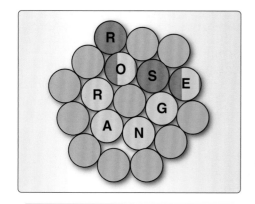

Figure 21.5
Pattern encoding by glomeruli in the olfactory bulb. (From Hatt H. Molecular and cellular basis of human olfaction. *Chem Biodivers* 2004; 1(12): 1857–1869, with permission of John Wiley and Sons.)

areas and thereby modulate the afferent input to the olfactory system based on cortical input. This integration and modulation is thought to help fine-tune the signal and enhance the contrast through lateral inhibition. This helps in the recognition of patterns of glomerular activation (see Figure 21.4).

A scent is often the combination of various odorant molecules, which activate various glomeruli. The pattern of these combinations (Figure 21.5) results in the identification of a specific scent ("rose" versus "orange").

C. Central projections of the olfactory pathway

The axons of the mitral and tufted cells travel in the **olfactory tract**. Fibers then pass laterally to form the **lateral (primary) olfactory tract** and terminate in several areas of the medial temporal lobe and basal frontal lobe for the conscious appreciation of smell. The primary olfactory cortex consists of the cortices of the **uncus** and **entorhinal area** (anterior part of the parahippocampal gyrus), the **limen insulae** (the point of junction between the cortex of the insula and the cortex of the frontal lobe), and part of the **amygdala**. The uncus, entorhinal area, and the limen insulae collectively are called the **piriform** ("pear-shaped") **area** (Figure 21.6).

Some collateral branches of the axons from the olfactory bulb neurons terminate in a small group of cells called the **anterior olfactory nucleus**, a collection of nerve cell bodies located along the olfactory tract. From this nucleus, fibers project to the contralateral olfactory bulb via the **anterior commissure**. Their role is believed to be inhibitory, thereby enhancing the more active bulb and providing directional clues to the source of the olfactory stimulation. The significance of this connection in humans is minor.

1. **Downstream relays of the olfactory cortex:** From the primary olfactory areas of the cortex, profuse projections go directly to the olfactory association area in the entorhinal cortex, sometimes referred to as the **secondary olfactory cortex**. In addition, significant projections go to the amygdala, the hippocampus, and the striatum; these connections mediate the behavioral responses to odors and the direct formation of emotional memories related to smell. Indeed, the links among smell, memory, and behavior are strong: it takes just one odor to be transported to a childhood memory. The interactions between the various cortical and subcortical areas in response to smell enable us to subjectively appreciate olfactory stimuli, adapt our behavior, and form or retrieve related memories.

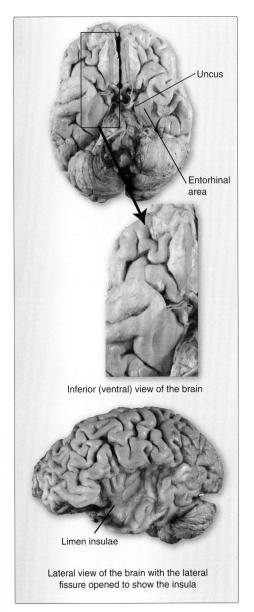

Inferior (ventral) view of the brain

Lateral view of the brain with the lateral fissure opened to show the insula

Figure 21.6
Central projections of the olfactory system. Olfactory bulb, olfactory tract, and primary olfactory cortex shown in green.

Clinical Application 21.1: Anosmia Following Traumatic Contrecoup Brain Injury

Following a fall from his mountain bike, a cyclist hit the back of his head on a large boulder. He lost consciousness briefly and, when examined later in hospital, was found to have a fracture of his skull in the occipital bone and the cribriform plate. He was hospitalized overnight and in the morning complained that he could not smell anything and that his breakfast had no taste to it. He also had a clear discharge from his nose. In this case, the force of the blow to the back of his head caused the brain to be propelled forward, acting to tear the processes of the olfactory receptor neurons passing through the cribriform plate. This is called a contrecoup injury and explains his loss of smell. Because what we smell greatly influences our sense of taste, his breakfast had no flavor. The clear discharge from his nose was cerebrospinal fluid (CSF) leaking into the nasal cavity from a dural tear caused by the fracture to the cribriform plate. The leakage of CSF can recover as the meninges heal; if it does not recover, surgery may be required. After such an injury, many patients never recover their sense of smell. This has widespread implications as our memories and our sense of pleasure derived from food are mediated through connections to the limbic system. Eating new food might make these patients nauseous, and they may refuse to eat anything for which they have no memory. People without a sense of smell rely on memories to accept a food as "safe." As there are no memories connected to new foods, ingestion can lead to an aversive or neophobic response. This neophobic response is likely a protective mechanism to prevent the ingestion of unknown, potentially toxic foods.

III. THE GUSTATORY SYSTEM

Taste detects nutritionally relevant stimuli and elicits behaviors and reactions, which will lead to either consumption or avoidance of food. The sense of taste is therefore directly linked with brainstem systems that will initiate salivation (when accepting food) or gagging (when rejecting food) as well as higher cortical areas through the thalamus that will initiate more complex behaviors.

Taste is a chemical sense that depends on **taste receptors** on cells clustered in **taste buds**. Taste buds are located in the tongue and also in the soft palate as well as in parts of the pharynx and larynx. In the tongue, taste buds are housed in folds called **papillae**. **Fungiform** papillae are found at the front of the tongue, **foliate** papillae at the side, and **vallate** papillae at the rear of the tongue. We are able to detect five different qualities of taste: sour, salty, sweet, bitter, and *umami* ("savory").

It is important to realize that what we appreciate as taste and flavor is a combination of olfaction, taste, and somatosensory input.

A. Taste buds

There 2,000–5,000 taste buds throughout the oral cavity. Taste buds are located on fungiform, foliate, and vallate papillae (Figure 21.7A). Taste buds are modified epithelial cells in an ovoid shape, somewhat like garlic cloves within a garlic bulb. The taste receptor cells are spindle shaped with microvilli on their apical ends that extend through a small opening, the **taste pore**, where they are exposed to ingested chemical stimuli (see Figure 21.7C). Each taste bud contains glial-like support cells, sensory **taste receptor cells**, and basal cells that provide for the continual replacement of taste receptor cells that have a turnover time of 10–14 days. At the base of the taste bud, afferent fibers form the postsynaptic element of a chemical synapse.

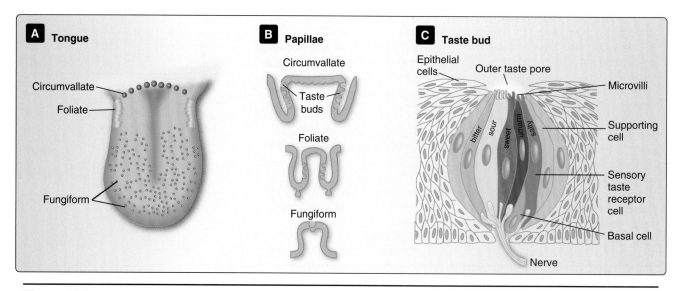

Figure 21.7
General organization of the gustatory system in the tongue. **A.** Tongue, **B.** Papillae, and **C.** Taste bud

1. **Fungiform papillae: Fungiform papillae** are so called because they look like button mushrooms scattered over the anterior tongue as little red bumps. They appear red because of their very rich blood supply. They usually have dozens of taste buds per papilla and are innervated by the facial nerve (CN VII) via the **chorda tympani** branch. It has been estimated that there are approximately 1,100 fungiform papillae per tongue. Fungiform papillae can also be found on the palate and the epiglottis; these are innervated by the vagus nerve (CN X) via the greater petrosal nerve.

2. **Foliate papillae:** Approximately 5–6 **foliate papillae** line each edge of the tongue just anterior to the circumvallate line. They contain up to hundreds of taste buds each and are primarily innervated by the glossopharyngeal nerve (CN IX), with some innervation via chorda tympani (CN VII).

3. **Circumvallate papillae: Vallate** or **circumvallate papillae** are sunken papillae surrounded by a trough. The taste buds lie deep within the trough on the walls of the papillae (see Figure 21.7B). There are usually 8–12 papillae per tongue, but each one contains hundreds of taste buds. These are also innervated by the glossopharyngeal nerve (CN IX).

B. Signal transduction for taste

Taste receptor cells respond to a very large variety of molecules and encode for the five basic qualities of taste (sour, salty, sweet, bitter, and umami). Each cell responds predominantly to one taste quality, but a single cell can express different receptors and respond to more than one taste. Cells that respond to these five taste qualities are found throughout the oral cavity; there is no map for these tastes on the tongue. In fact, a single taste bud will often contain cells for all five qualities. There is however some evidence that the sensitivities to the five tastes are regionally variable.

Taste receptor cells exist in three basic subtypes: (1) **Type I cells** are glial-like support cells. They limit the spread of transmitters released by type II and type III cells. (2) **Receptor (type II)** cells respond to sweet, bitter, and umami input via a G-protein–coupled mechanism. Each receptor cell is specific for a taste quality. Receptor cells do not form specific synapses, but rather secrete ATP onto afferent fibers and adjacent presynaptic cells. (3) **Presynaptic (type III)** cells respond to sour stimuli and can be indirectly stimulated through ATP from the receptor cells. These cells form synapses and release serotonin, norepinephrine, and GABA as neurotransmitters. Tastants can act on either type II cells or type III cells. We now look into how these taste receptor cells can detect and then encode for the different taste qualities.

1. **Sour:** Sour is perceived as an aversive taste unless it is combined with other flavors and tastes. The taste quality "sour" probably evolved to prevent us from eating unripe fruit or rotting foods ("which have gone sour"). Sour compounds contain acids and these acids have the ability to permeate type III taste receptor cells (presynaptic cells) and lower the intracellular pH, which results in

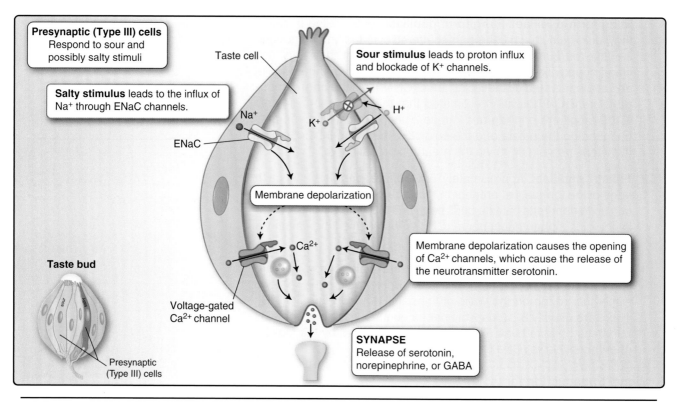

Presynaptic (Type III) cells
Respond to sour and possibly salty stimuli

Taste cell

Sour stimulus leads to proton influx and blockade of K⁺ channels.

Salty stimulus leads to the influx of Na⁺ through ENaC channels.

Na^+

K^+ H^+

ENaC

Membrane depolarization

Taste bud

Ca^{2+}

Membrane depolarization causes the opening of Ca^{2+} channels, which cause the release of the neurotransmitter serotonin.

Voltage-gated Ca^{2+} channel

SYNAPSE
Release of serotonin, norepinephrine, or GABA

Presynaptic (Type III) cells

Figure 21.8
Signal transduction in type III cells for salty and sour. ENaC = epithelial sodium channel.

signaling (Figure 21.8). The exact intracellular target that initiates the signaling, which results in synaptic transmission, is unknown.

2. **Salty:** The prototypical salty stimulus comes from table salt: NaCl. The Na⁺ ion is thought to act directly on ion channels and causes a rise in intracellular Na⁺, which results in depolarization and signal transduction (see Figure 21.8). Several different ion channels have been discussed as candidates for this sodium detection, such as the epithelial sodium channel[1] (ENaC) and the vanilloid receptor V1R. It is still unclear which taste cells respond to Na⁺ taste; there is some evidence that this may be via type III or type I cells. In addition, activation of these Na⁺ channels is thought to amplify the taste signals elicited by other stimuli in a taste cell.

3. **Sweet:** There are a multitude of molecules that evoke the taste "sweet," and their molecular structures are very different, ranging from sugars to amino acids and peptides to alcohols and certain salts. These compounds act on G-protein–coupled receptors on type II cells (Figure 21.9). These receptors are protein dimers that have a large extracellular domain, which can bind the large variety of sweet compounds and initiate a signal transduction cascade (see below).

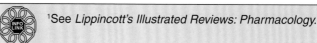 [1]See *Lippincott's Illustrated Reviews: Pharmacology.*

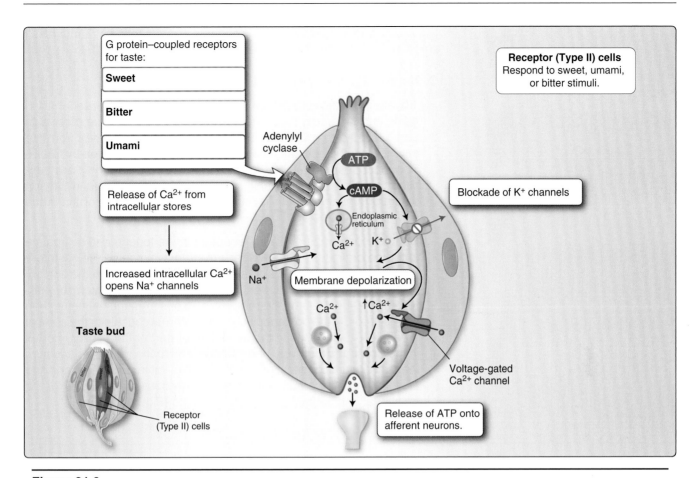

Figure 21.9
Signal transduction in type II cells for sweet, bitter, and umami. cAMP = cyclic adenosine monophosphate; ATP = adenosine triphosphate.

4. **Bitter:** Compounds that elicit a bitter taste sensation far outnumber those that elicit any of the other tastes. Although most bitter tastes are unpleasant (the bitter taste likely evolved to protect us from toxins), some bitter compounds are perceived as very pleasant, especially in combination with other tastes and flavors (caffeine in coffee, bitter compounds in fruits and vegetables). These bitter compounds act on the G-protein–coupled receptors in type II cells (see Figure 21.9). Its extracellular domain can bind the various bitter compounds and initiate a signaling cascade (see below).

5. **Umami:** The prototypical compound that elicits the savory "umami" taste is monosodium glutamate (commonly known as "MSG"). The umami receptor is also a protein dimer that contains an extracellular domain specialized for the detection of amino acids; it is found on type II cells (see Figure 21.9).

6. **Signal transduction in type II receptor cells:** Activation of G-protein–coupled receptors initiates a signaling cascade for bitter-, sweet-, and umami-initiated signals: The G protein can activate cAMP, which ultimately results in the blockade of K⁺ channels, or it can result in a second messenger–mediated increase in intracellular calcium (released from the endoplasmic reticulum), which

causes the opening of Na⁺ channels. Both of these streams result in the depolarization of the cell and the secretion of ATP.

C. Neuronal pathways for taste

Taste is carried by cranial nerves (CNs) VII (facial), IX (glossopharyngeal), and X (vagus). The anterior two-thirds of the tongue receive its nerve supply from CN VII. The posterior third is separated from the anterior two-thirds by the circumvallate line and is supplied by CN IX. The most posterior part of the tongue and the oropharynx are supplied by CN X. There are also sensory receptors for taste in the soft palate and pharynx, but they are thought to be less important in the appreciation of taste.

Central processes from taste receptors in the tongue and soft palate enter the brainstem in the **solitary tract** to synapse on the gustatory nucleus in the rostral part of the **solitary nucleus**. From the solitary nucleus, the ascending fibers project ipsilaterally to the **ventral posteromedial nucleus of the thalamus**. Axons from the thalamus then project through the posterior limb of the internal capsule to the cortical area for taste, situated in the most inferior part of the sensory cortex in the **postcentral gyrus**, and extending on to the **insula.** In addition, there are direct connections between the solitary nucleus and the amygdala and hypothalamus. These connections are the basis for the emotional and behavioral reactions to taste (Figure 21.10).

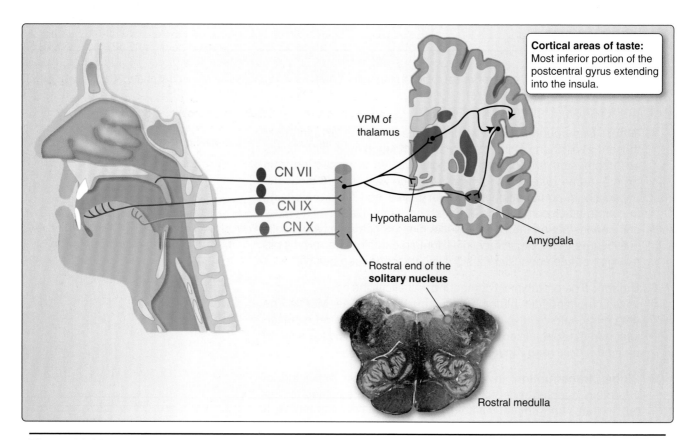

Cortical areas of taste:
Most inferior portion of the postcentral gyrus extending into the insula.

VPM of thalamus

CN VII

CN IX

CN X

Hypothalamus

Amygdala

Rostral end of the **solitary nucleus**

Rostral medulla

Figure 21.10
Neuronal pathway for taste. CN = cranial nerve; VPM = ventral posteromedial nucleus of the thalamus.

IV. TRIGEMINAL CHEMORECEPTION

The trigeminal nerve (CN V) is responsible for the sensory innervation of the face. It carries discriminative touch, vibration, and proprioception as well as pain and temperature. Chemicals perceived as painful are detected by the trigeminal system through its pain pathway. **Polymodal C fibers** respond to chemical irritants and elicit a pain response. Some chemicals are pleasant in low doses and "taste" good but are painful at high doses. Examples of these are ethanol that in high concentrations gives a burning feeling and chili peppers that can "burn." Low concentrations of acid (e.g., in a salad dressing) can be pleasantly sour but in high concentrations can be painful. The trigeminal chemoreception system is part of the pain system, and it protects us from noxious stimuli. We will discuss the pain pathway in detail in Chapter 22, "Pain."

Clinical Case:

Frontal Meningioma

A 62-year-old woman has a 6-month history of loss of taste and smell. She states that her food does not "taste" the same; however, she is still able to detect sweet, sour, and salt in her foods. She also has a history of progressive frontal headache (pressure that is mild to moderate in intensity) with nausea and occasional vomiting. There is no facial numbness or weakness. There is no diplopia (double vision), dysphagia (trouble swallowing), or dysarthria (trouble speaking). There is no limb weakness and no numbness in her body. She is otherwise healthy and takes no medications.

On examination, she is unable to detect the smell of coffee grounds, lavender, or orange.

Fundoscopic examination shows slight blurring of her optic discs on both sides (papilledema). There is no optic disc pallor, relative afferent pupillary defect (RAPD) or impairment in her visual acuity. She has intact visual fields. Pupils are 3 mm and equally reactive to light and accommodation. She has full range of extraocular movement with no diplopia or nystagmus. Tearing is normal. Facial sensation and strength are normal. She is able to taste sour, bitter, sweet, and salty on the taste test. Her hearing is normal. Speech and language are normal. The rest of her cranial nerve and neurological examination is normal.

Why does she report impaired taste when her taste test is normal?

Recognition of food flavors requires input from the olfactory system. Individuals assign the complex flavors detected by olfaction to taste. The five basic taste qualities detected by the taste cells do not result in what is commonly referred to as a taste without olfactory input, particularly in the context of food consumption.

An MRI of the brain shows a large enhancing tumor along the floor of the anterior fossa. She undergoes a complete resection of the tumor. The pathology is consistent with a meningioma (tumor of the meninges) that is typically benign and is more common in women than men.

If a meningioma is typically benign, then why was the tumor resected?

In this case, there was clinical evidence of increased intracranial pressure with headaches and papilledema. Meningiomas are typically slow growing and benign. However, if they expand and cause mass effect on the brain, then a resection should be considered.

How did the meningioma affect her sense of smell?

The olfactory bulb is located in the anterior cranial fossa and the olfactory nerves pass through the meninges and the cribriform plate into the nasal cavity. Her olfactory bulb was likely compressed by the meningioma.

Chapter Summary

- Smell and taste are two sensory systems that detect chemicals in our environment.

- The sense of smell detects airborne molecules through receptor cells in the olfactory epithelium in the nose. From the olfactory receptor cells, the smell is encoded into electrical signals, which are relayed to the olfactory bulb, where the first level of processing occurs. Pooling of inputs in glomeruli and lateral inhibition of adjacent glomeruli enhance contrast and facilitate pattern recognition. Every smell has a distinct pattern of glomerular activation, which is relayed to the olfactory cortex on the inferior surface of the brain.

- The sense of taste detects molecules on our tongue and palate through receptor cells in taste buds. Taste receptors are distributed in different areas of the tongue and can discern five basic qualities: sour, salty, sweet, bitter, and umami. Signal transduction occurs either through direct depolarization through changes in pH or Na^+ concentration or through a G-protein–coupled mechanism. Cranial nerves VII, IX, and X have afferents that carry taste and synapse in the nucleus solitarius. From the nucleus solitarius, fibers travel to the ventral posteromedial (VPM) nucleus of the thalamus, from which the axons of postsynaptic neurons travel to the primary taste area in the cortex in the most inferior part of the postcentral gyrus and the insula.

- The trigeminal nerve also detects chemical stimuli in the nose and mouth and works in conjunction with the senses of smell and taste. Free nerve endings of the trigeminal nerve are only stimulated when the chemical concentrations reach noxious levels. This trigeminal signal is part of the pain pathway.

Study Questions

Choose the ONE best answer.

21.1 A 43-year-old woman suffers a head injury during a motor vehicle accident, resulting in shear injuries of the brain in the anterior cranial fossa. Three months later, she comes to the office complaining about a diminished sense of taste and smell, probably resulting from a lesion to which one of the following cranial nerves?

    A. Olfactory nerve.
    B. Trigeminal nerve.
    C. Facial nerve.
    D. Glossopharyngeal nerve.
    E. Vagus nerve.

Correct answer is A. The sense of smell is carried by cranial nerve I, the olfactory nerve. The distinct flavors we appreciate in our food are due to our sense of smell, the loss of which would greatly depreciate one's sense of taste. The facial, glossopharyngeal, and vagus nerves all carry taste from the tongue and the larynx, but they are not at all involved in smell. A select lesion to any one of these nerves would be compensated for by the other nerves. The trigeminal nerve is not involved in taste or smell. It carries general sensory afferents from the oral cavity and helps us appreciate the texture of food. During a traumatic brain injury, the nerves passing through the cribriform plate from the nasal cavity to the olfactory bulb can be sheared off.

21.2 Olfactory epithelium is unique in that it is continuously regenerated throughout life. The cell responsible for this is the:

    A. Primary sensory receptor.
    B. Support cell.
    C. Secretory cell.
    D. Basal cell.
    E. Periglomerular cell.

Correct answer is D. These cells are situated on the basement membrane of the epithelium and continuously regenerate new cells that will mature into mature bipolar olfactory neurons. The support cell is a supporting cell, and the secretory cell secretes a fluid containing odorant-binding proteins. Periglomerular cells are located in the olfactory bulb, where they are involved in the processing of olfactory information.

21.3 In the olfactory bulb, the primary olfactory neurons synapse with mitral and tufted cells in specialized areas called:

    A. Olfactory trigone.
    B. Lateral olfactory stria.
    C. Glomeruli.
    D. Anterior olfactory nucleus.
    E. Olfactory knob.

Correct answer is C. Axons of primary sensory neurons synapse with mitral and tufted cells in specialized areas called *glomeruli* where considerable convergence takes place. The trigone is an expansion of the olfactory tract. The lateral olfactory stria is made up of the fibers traveling to the primary olfactory area in the cortices of the uncus and entorhinal area, limen insulae, and amygdala. The anterior olfactory nucleus is a collection of nerve cell bodies along the olfactory tract. The olfactory knob is the part of the mature receptor cell as it abuts into the mucous layer.

21.4 A patient comes to the walk-in clinic complaining about a metallic and bitter taste to all foods he eats. His medical history has no previous instances of these symptoms. The patient reports having eaten pine nuts 3 days earlier, and shortly thereafter, the symptoms started. Which of the following statements about the taste quality "bitter" is true?

A. Bitter compounds act on T1R2 receptors.
B. Activation of bitter receptors directly opens Na^+ channels.
C. Bitter receptors are located at the tip of the tongue.
D. Bitter compounds act on T2R receptors.
E. Bitter compounds lead to the opening of K^+ channels.

Correct answer is D. Bitter receptors are located at the back of the tongue. Bitter compounds bind to T2R receptors. Binding to these receptors leads to a G-protein–coupled signaling cascade, which ultimately leads to the blockade of K^+ channels and a rise in intracellular Ca^{2+}, which, in turn, opens Na^+ channels. This causes the depolarization of the cell and release of serotonin and adenosine triphosphate. Pine nuts have been implicated in transient dysgeusia, likely related to pine oils found in the nuts that have a very bitter taste. This usually resolves within a few days.

22 Pain

I. OVERVIEW

The International Association for the Study of Pain defines pain as, "an unpleasant sensory and emotional experience associated with actual or potential tissue damage, or described in terms of such damage." Pain is a condition that we all experience throughout our lives, and it is a leading cause of disability and suffering.

Pain is a multifactorial and multisystem phenomenon. In previous chapters, we discussed peripheral receptors and ascending sensory systems and the cortical representations of sensory sensations. We then discussed how the limbic system and hypothalamus influence our emotional experience and our behavioral responses to environmental stimuli.

In this chapter, we bring all of these systems together to see how they interact to mediate the perception and modulation of pain. We then look at chronic pain conditions and the maladaptive mechanisms that underlie chronic neuropathic pain.

II. NOCICEPTORS

The perception of pain, or **nociception**, depends on specialized receptors in the periphery that can respond to extreme temperatures (heat and cold), intense mechanical stimulation, and chemical stimuli. Nociceptors are free nerve endings, whose cell bodies are located in the spinal ganglia for the body and the trigeminal ganglia for the face.

Nociceptors are only activated when the stimulus reaches a noxious threshold. They respond progressively, according to the intensity of the stimulus. Whereas other receptors show adaptation (see Chapter 1, "Introduction to the Nervous System and Basic Neurophysiology"), nociceptors do not. Continued stimulation can decrease the threshold at which nociceptors respond, a phenomenon termed **sensitization**.

A. Nociceptive fibers

There are two major classes of fibers associated with nociceptors: **Aδ fibers** and **C fibers**. The myelinated Aδ fibers are responsible for the localized, sharp "first" pain and respond to intense thermal (extreme hot or cold) and mechanical (pinching) stimuli. Unmyelinated C fibers, conversely, mediate the poorly localized, diffuse "second" pain. They are polymodal in that they respond to thermal, mechanical, and chemical stimuli. The neurotransmitters used by nociceptive fibers are **glutamate**, **substance P**, and **calcitonin gene–related peptide (CGRP)**. These neurotransmitters are produced in the cell body within the spinal and trigeminal ganglia. When the nociceptor is activated, these neurotransmitters are released centrally at the synapse with the second-order neuron in the spinal cord or brainstem and, interestingly, peripherally at the site of injury. In the periphery, these neurotransmitters lead to redness, swelling, and tenderness, the classic indicators of pain. They also increase the activation of nociceptors and lower their activation threshold (sensitization). In addition, peripheral neurotransmitters activate so-called **"silent" nociceptors**, which expand the receptive field for the painful stimulus.

B. Nociceptor activation

Activation of nociceptors by extreme temperature, intense mechanical stimulation, or an array of chemicals occurs through specific receptors (Figure 22.1). This results in the opening of cation channels (mainly Na^+), which will result in **membrane depolarization** (see Chapter 1, "Introduction to the Nervous System and Basic Neurophysiology") and the generation of action potentials (APs).

1. **Temperature:** Heat can be perceived as a painful stimulus, with the threshold in humans at about 43°C. Subtypes of both Aδ fibers and C fibers can be activated by heat, and some are activated at temperatures below or above the 43°C threshold. One of the receptors responsive to hot thermal stimulation is the **TRPV1 receptor**. This

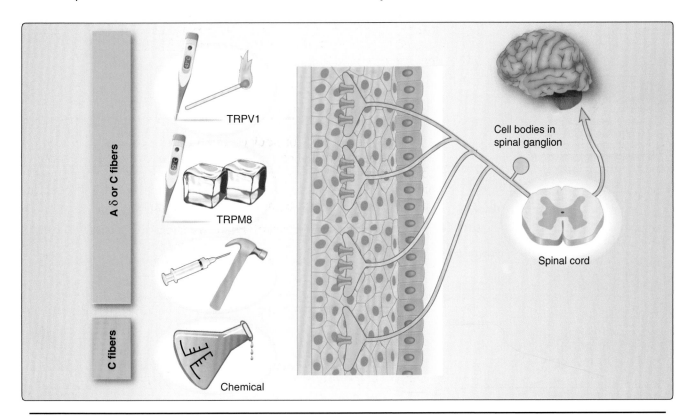

Figure 22.1
An overview of nociceptor activation (from top to bottom, Heat, Cold, Mechanical, Chemical). TRPV1 = Transient receptor potential type V1; TRPM8 = Transient receptor potential type M8.

is the receptor for **capsaicin**, the main "hot" compound found in chili peppers. The TRPV1 receptor belongs to the family of **transient receptor potential (TRP) channels**, and its physiological properties are similar to those of voltage-gated channels. At rest, the pore is closed and, when activated, there is an influx of Na+ and Ca2+ ions.

Cold can also be perceived as painful, with the threshold at about 25°C. Cold sensitivity is mediated through a receptor that also responds to **menthol**, the **TRPM8 receptor**, another type of TRP channel.

Activation of these thermally sensitive receptors results in behaviors that will either cool the body (in response to heat stimulation) or warm the body (in response to cold stimulation). These behavioral responses are mediated through projections to the hypothalamus (see Chapter 19, "The Hypothalamus").

2. **Mechanical activation:** Mechanically activated receptors can be found in both Aδ fibers and C fibers. These receptors cannot be activated by light touch. They have a high threshold and are activated only when the stimulus is noxious and may result in tissue damage.

3. **Chemical activation:** Free nerve endings of C fibers can be activated by a range of chemicals, which can be external irritants or substances released during tissue damage, either by the nociceptors, as discussed above, or through **inflammation**.

C. Sensitization of peripheral receptors

Tissue damage and inflammation result in the release of inflammatory molecules, such as **bradykinin** and **prostaglandins**, which sensitize peripheral nociceptors (Figure 22.2). In addition, when a

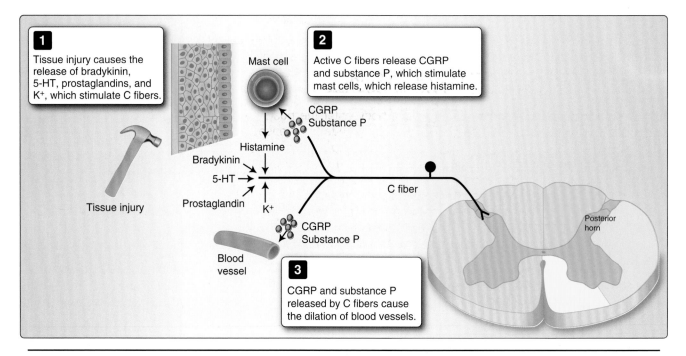

Figure 22.2
Principles of peripheral sensitization. 5-HT = serotonin; CGRP = calcitonin gene–related peptide. Bradykinin stimulates C fibers through stimulation of mast cells.

noxious stimulus is detected by a nociceptor, free nerve endings will release substance P and CGRP. These two neuropeptides contribute to the inflammatory response at the site of tissue injury. Bradykinin stimulates mast cells to release histamine. CGRP induces vasodilation, which results in further release of inflammatory molecules from peripheral immune cells.

1. **Activation threshold:** All of these processes lower the **activation threshold** of nociceptors such that they will then be activated by stimuli that would normally not be perceived as painful. Anyone who has had a sunburn can relate to the sensitivity of that area of skin. Light touch is perceived as very painful and a warm shower is perceived as painfully hot. This increased sensitivity is due to the heightened activity of the skin nociceptors and a loss of the specificity of the stimulus in that sunburned area of skin. In addition, the enhanced sensitivity of peripheral nociceptors due to exposure to the inflammatory mediators can cause spontaneous discharges, which increase the perception of pain.

   When the response to a normally painful stimulus is heightened (more pain felt than usual), it is referred to as **hyperalgesia**. When a normally nonpainful stimulus is perceived as painful (light touch on sunburn), it is referred to as **allodynia**.

2. **Silent nociceptors:** The release of inflammatory molecules and neurotransmitters in the periphery not only results in redness, swelling, and tenderness but also recruits the silent nociceptors. These receptors further amplify the signal to the posterior horn by increasing the temporal and spatial summation of the incoming signal. The interesting thing about the silent nociceptors is that they only signal in response to the molecules secreted by other activated nociceptors and are not activated by any noxious stimulus per se.

III. PAIN PROCESSING IN THE SPINAL CORD

All fibers carrying sensory information from the body enter the posterior horn of the spinal cord. Aδ and C fibers, which encode for intense thermal, mechanical, and chemical stimuli, synapse in the posterior horn before the signal is relayed to higher centers. We explore the synapses in the spinal cord because they are the first step in the nociceptive pathway. Some pain medications target the spinal cord, and modulation of pain occurs at this first synapse through descending pathways (see below). An understanding of the processes in the spinal cord can provide insight into the maladaptive mechanisms that underlie chronic and neuropathic pain.

Pain processing in the spinal trigeminal nucleus is analogous to the mechanisms in the spinal cord; this nucleus receives nociceptor input from the head.

In Chapter 5, "The Spinal Cord," we discussed the laminar organization of the gray matter in the spinal cord. Aβ fibers (carrying discriminative touch, proprioception, and vibration) project mostly to the deep layers of the posterior horn, whereas Aδ and C fibers project mostly to laminae I and II and to lamina V (to influence wide dynamic range neurons).

A. Synaptic targets

The synaptic targets of Aδ and C fibers are either **nociceptive-specific (NS)** cells, which synapse only with Aδ and C fibers, or **wide dynamic range (WDR)** neurons, which receive synaptic input from all types of sensory fibers (Figure 22.3). Whereas NS neurons will encode only for painful stimuli and project to higher centers, the WDR cells can encode for a range of stimuli, both painful and nonpainful.

1. **Neurotransmitters:** The synapse in the posterior horn is excitatory. The neurotransmitters released by the afferent nociceptive fibers are glutamate, which acts mainly on α-amino-3-hydroxy-5-methyl-4-isoxazole propionic acid (AMPA) and N-methyl-D-aspartic acid (NMDA) receptors (see below); substance P, which acts on the neurokinin 1 (NK1) receptor; and CGRP, which also has an excitatory effect via the CGRP receptor.

2. **Wide dynamic range neurons:** The firing pattern of WDR cells depends on the stimulus intensity, which is encoded by the frequency of C-fiber signaling. The more painful the stimulus, the higher the frequency of C-fiber discharge and the greater the WDR neuron response. The WDR neuron can then amplify this signal through a mechanism called "**wind-up.**"

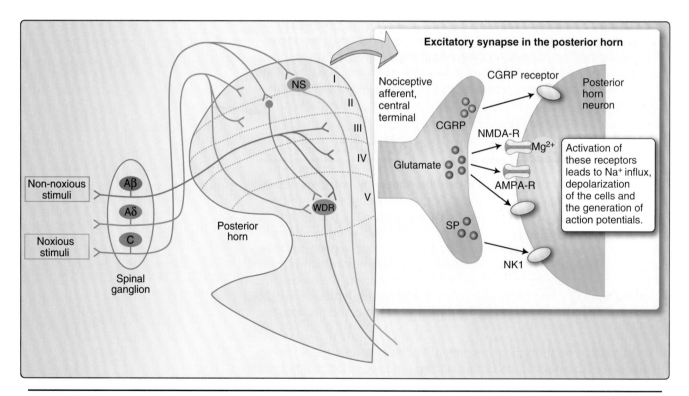

Figure 22.3

The projection of fibers into the spinal cord. NS = nociceptive-specific; WDR = wide dynamic range neurons; CGRP = calcitonin gene–related peptide; AMPA-R = α-amino-3-hydroxy-5-methyl-4-isoxazole propionic acid receptor; NMDA-R = N-methyl-D-aspartic acid receptor; CN = Cranial nerve; SP = substance P; NK1 = neurokinin 1.

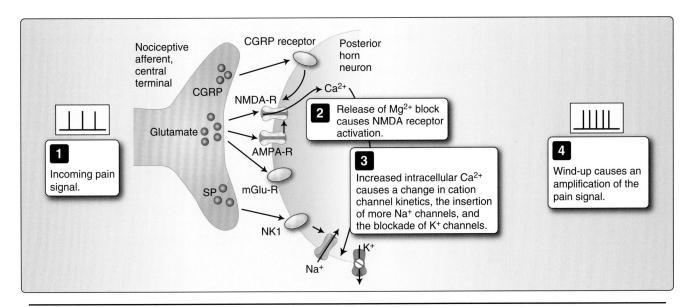

Figure 22.4
Principles of central sensitization or "wind-up." CGRP = calcitonin gene–related peptide; AMPA-R = α-amino-3-hydroxy-5-methyl-4-isoxazole propionic acid receptor; NMDA-R = N-methyl-D-aspartic acid receptor; SP = substance P; NK1 = neurokinin 1. Certain metabotropic glutamate receptors (mGluRs) increase posterior horn activity and are involved in pain processing.

B. Wind-up

Central sensitization, or "wind-up," in the spinal cord is mediated through neurotransmitter release at the postsynaptic neuron in the posterior horn. Wind-up is essentially an amplification system within the spinal cord to respond to the cumulative nociceptive input from C fibers. Wind-up results from the repetitive excitatory stimulation of the WDR neuron through glutamate acting mainly on AMPA receptors. The increased AP frequency and sustained membrane depolarization that then occur result in activation of NMDA receptors. The NMDA receptor is usually inactive due to blockade of channels by Mg^{2+} ions. Sustained depolarization releases this Mg^{2+} block, and the NMDA receptor can then be activated by glutamate. Significantly, because the NMDA receptor is Ca^{2+} permeable, Ca^{2+} influx into the cell changes the electrophysiological signaling properties of the WDR neuron (Figure 22.4).

C. Substance P and CGRP

The release of **substance P** by C fibers also appears to play an important role in the activity of WDR neurons through their NK1 receptors. Activation of the NK1 receptor causes long-lasting depolarization that contributes to the temporal summation of the nociceptive input from C fibers. **CGRP** contributes to sensitization through its downstream enhancement of NMDA receptor activity.

D. Increased excitation

Together, these mechanisms contribute to the augmentation of the excitatory signal in the spinal cord. Increases in intracellular calcium in spinal cord neurons through the activation of NMDA receptors lead to postsynaptic changes that, in turn, lead to a reduction of threshold, a change in cation channel kinetics, and the insertion of more receptors in the postsynaptic membrane. These changes increase the sensitivity of the postsynaptic neurons, which now fire APs more readily (see Figure 22.4).

E. Sensitization

Repeated C-fiber activity can lead to wind-up within seconds, and the sensitization that occurs through wind-up can last for a long time. In the short-term, this can be beneficial in a number of ways. Amplification of the nociceptive signal gives it "priority salience" on its way to the cortex. Continued hyperalgesia of the affected area can protect the injured body part because the person typically shields the injured area from further trauma. On the other hand, sensitization can become maladaptive in the long-term, persisting long after an injury has healed (see below).

IV. ASCENDING NOCICEPTIVE PATHWAYS

The posterior horn neurons that receive input from the peripheral nociceptors send projections to the **thalamus**, which then sends projections to the cortex. This ascending system gives off collaterals that activate brainstem centers, which, in turn, will send descending modulating fibers to the spinal cord (Figure 22.5). We discuss the **spinothalamic tract** first and then examine the collaterals of the ascending system in the brainstem. The sum of all ascending nociceptive pathways and collaterals is collectively referred to as the **anterolateral system** due to its location in the spinal cord, in the anterior part of the lateral column (see Chapter 5, "The Spinal Cord").

Generally speaking, there are two main qualitative components of the nociceptive pathways: (1) a **lateral sensory-discriminative component**, which projects to the primary somatosensory cortex, and (2) a **medial affective-motivational component**, which projects to various cortical association areas and the limbic system.

A. Lateral sensory-discriminative pathway

This phylogenetically newer pathway encodes for the location, intensity, and quality of pain and comprises the **neospinothalamic tract**. Fibers originate in the posterior horn from both NS and WDR neurons. The main input to these neurons appears to come from **Aδ fibers**. The second-order neurons cross the midline in the **anterior white commissure**, ascend in the anterior part of the lateral column of the spinal cord and brainstem, and project to the ventral posterolateral (VPL) nucleus of the thalamus (fibers from the trigeminal system (from spinal trigeminal nucleus) project to the ventral posteromedial [VPM] nucleus of the thalamus), where they synapse with third-order neurons. From the thalamus, axons project to the **somatosensory cortex** (see Figure 22.5). The tract is organized somatotopically and allows for the localization of pain in the primary somatosensory cortex. The lateral sensory-discriminative pathway mediates the **"first" pain**, the sharp, well-localized sensation relayed rapidly to the cortex: "I have a sharp pain in my left arm!"

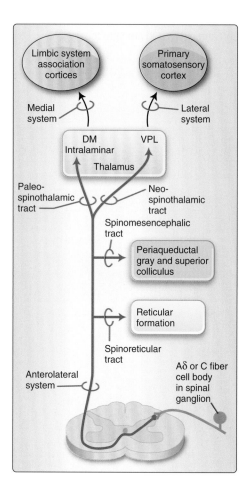

Figure 22.5
Conceptual overview of ascending pain pathways. DM = dorsomedial nucleus; VPL = ventral posterolateral nucleus.

B. Medial affective-motivational pathways

This phylogenetically older collection of pathways comprises the projections to the **reticular formation**, **midbrain**, **thalamus**, **hypothalamus**, and **limbic system**. Together, these pathways influence the emotional and visceral responses to pain as well as the descending modulation of pain. Each of these pathways originates in the posterior horn of the spinal cord and carries afferents from NS and WDR neurons. They all cross in the anterior white commissure and travel rostrally as part of the anterolateral system.

The medial affective-motivational component of the system is the **"second" pain**, the dull, throbbing poorly localized type of pain. This component includes the emotional response to pain: "Ouch! That hurts. I don't like it!"

1. **Spinomesencephalic tract:** The **spinomesencephalic tract** projects to midbrain structures: the **periaqueductal gray (PAG)** and the **superior colliculus**. The PAG sends projections directly back to the posterior horn, where it modulates pain at the level of both the NS and the WDR neurons. The PAG has reciprocal connections with the limbic system and receives input from the cortex and hypothalamus. It is an important *integrator and modulator of the pain experience*. Projections to the **superior colliculus** are thought to influence our eye movements and direct our gaze to the site of injury (see Chapter 9, "Control of Eye Movements").

2. **Spinoreticular tract:** One set of ascending fibers projects to the reticular formation (**spinoreticular tract**), where it influences the motor response to pain (see Chapter 12, "Brainstem Systems and Review"). Fibers also project to the **noradrenergic locus ceruleus** and the **serotonergic raphe nuclei** as well as to the **rostral anteromedial medulla**, an area that is particularly rich in **opioid receptors**. All of these target areas send descending projections that modulate nociception (see Chapter 12, "Brainstem Systems and Review"). Activation of the reticular formation also leads to widespread activation of cortical areas through the ascending monoaminergic pathways.

3. **Paleospinothalamic tract:** The **paleospinothalamic tract** projects to the **dorsomedial** (connections to limbic areas of cortex) and **intralaminar** (role in cortical arousal) **nuclei of the thalamus**, which, in turn, project to cortical association areas (see Figure 22.5). The activity in this pathway is thought to mediate a more diffuse response to pain and is key in the affective processing of the pain experience.

4. **Other tracts:** There are additional direct and indirect projections that bring sensory information to the hypothalamus and the limbic system. These projections influence the **neuroendocrine** and **visceral** responses to pain, which result in the release of stress hormones (cortisol and epinephrine) and the activation of the visceral nervous system. The emotional response to pain comes through the projections to limbic structures such as the anterior cingulate gyrus, the amygdala, and cortical association areas. These pathways can also influence our level of alertness by activation of noradrenergic neurons through projections from the amygdala.

Clinical Application 22.1: Referred Pain

Referred pain is pain perceived in an area of the body that is not directly innervated by the neurons exposed to the noxious stimulus. Typically, visceral pain is experienced as referred pain and is experienced in the dermatomes associated with the spinal cord levels of the sympathetic fibers that supply those viscera. For example, pain resulting from inflammation of the gastrointestinal tract is felt in the dermatomes related to the abdomen. Visceral pain travels along sympathetic fibers and enters the spinal cord where those sympathetic fibers originate. Pain can also be referred from one somatic area to another: Irritation of the diaphragm, which is innervated by the phrenic nerve (C3, C4, C5), is referred to the dermatomes of C3, C4, and C5 around the shoulder. It appears that when there is no sensory reference point for a part of our body, pain is interpreted by the CNS as originating from an area for which it does have a somatic reference point. This mechanism is important to keep in mind during the clinical assessment of pain.

The figure shows the cutaneous areas to which visceral pain is referred.

If all special tests are negative yet the individual continues to feel pain at a specific site, it may be referred pain. (From Anderson MK. *Foundations of Athletic Training*, 6th ed. Wolters Kluwer, 2017.)

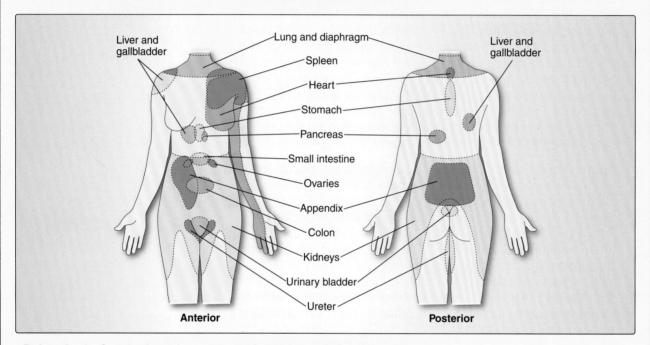

Referred pain. Certain visceral organs can refer pain to specific cutaneous areas.

V. CORTICAL PAIN MATRIX

Given the multitude of nociceptive pathways and their various targets in the cortex, it is apparent that there is no single "pain center" in the brain. Pain is a multifactorial, multipathway system that projects to many areas of the cortex that are involved in the perception of pain. These cortical areas are collectively referred to as the **cortical pain matrix**. These cortical regions include the primary somatosensory cortex, which is critical for the localization of pain as well as cortical areas related to the limbic system, and various association areas. There are reciprocal connections between these cortical areas and subcortical structures such as the amygdala, the hippocampus, and the hypothalamus. The global experience of pain involves complex interactions among cortical, subcortical, and brainstem systems.

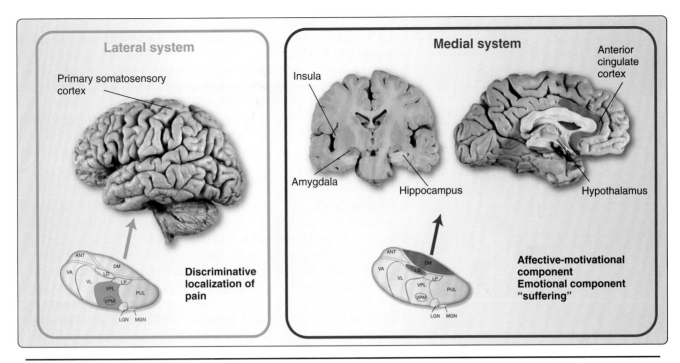

Figure 22.6
The pain matrix. DM = dorsomedial nucleus of the thalamus; VPL = ventral posterolateral nucleus of the thalamus; VPM = ventral posteromedial nucleus of the thalamus. LD = lateral dorsal nucleus of the thalamus

The neuronal populations that comprise the pain matrix are not necessarily specific to pain but may be activated by many other stimuli including somatosensory stimuli, visual and auditory cues that may be perceived as threatening, memory of pain, and, possibly, anticipation of pain. Some neurons in the pain matrix are thought to be part of the **mirror neuron system** that lets us empathize with the pain of others (see Chapter 13, "The Cerebral Cortex").

In parallel with the nociceptive pathways, the cortical representation of pain has classically been divided into two systems: the **lateral pain system** and the **medial pain system** (Figure 22.6). We now know that these two systems are highly integrated and both are necessary for the full experience of pain, which includes both the sensory-discriminatory and the emotional components.

A. Lateral pain system

The **lateral pain system** includes the projections from the VPL and VPM of the thalamus to the **primary** and **secondary somatosensory cortices**. Localization of pain and interpretation of pain information carried by the neospinothalamic pathway ("sharp pain in my left arm") has been attributed to these cortical areas. The primary somatosensory area is somatotopically organized and represents information coming from the contralateral side of the body. This is where the *localization* of pain occurs.

B. Medial pain system

The **medial pain system** includes the more medial cortical areas, the **anterior cingulate cortex (ACC)**, **insula**, **amygdala**, and

hypothalamus. These areas are responsible for the affective-motivational, emotional component of pain, that is, the suffering associated with pain ("Ouch! That hurts. I don't like it!"). The ACC is active both during the actual perception of pain and during imagined pain as well as during the observation of someone else being subjected to pain. From the ACC, direct behavioral reactions to pain are initiated. The **insula** is thought to play a role as a relay station to the limbic system, where pain-related learning and memory are processed, and to the **hypothalamus**, from which the visceral response to pain is initiated.

C. Saliency

Recent evidence has shown that the pain matrix is not necessarily specific to nociceptive input. It is now thought that these brain areas are active during **salient sensory input**. Saliency of an input depends on how much that stimulus contrasts with the background activity or contrasts with past experience. Nociceptive inputs, by their very nature, will always have high saliency and activate cortical areas that detect saliency (see further discussion below). The detection of a "different," "interesting," and, possibly, "threatening" stimulus allows us to react quickly to a situation through the activation of behaviors that will result in avoidance or possibly exploration. During nociception, avoidance behaviors are routinely activated: (1) the **withdrawal reflex** at the spinal cord level (see Chapter 5, "The Spinal Cord") and (2) a voluntary reaction at a cortical level through the activation of motor systems involving the basal ganglia and cerebellum. Through the medial pain system, limbic structures such as the amygdala are also brought into play and contribute to the saliency of a pain stimulus.

VI. PAIN MODULATION

In the previous sections, we explored the nociceptive pathways and how nociceptive signals can be amplified through sensitization. Equally important is the ability to modulate pain so that we can react to a potentially hazardous situation and continue to function appropriately. The control of pain is not primarily through a pain suppression system but is part of a broader central nervous system function in which competing stimuli and needs are prioritized. Surprisingly, these descending influences on posterior horn neurons can be both inhibitory and facilitatory. The balance between inhibition and facilitation can be altered to meet different behavioral, emotional, and pathophysiological needs.

Famous and powerful examples of **pain inhibition** are the battlefield stories of injured soldiers who manage to get themselves, and even others, out of a life-threatening situation before even noticing the pain from a severe injury. Although not everyone has such an experience, we all, nevertheless, modulate pain when we experience it. When we rub a sore spot or when we try to get our mind off pain, we are, effectively, modulating the pain experience. These are examples of **descending inhibition** of the incoming signal. On the other hand, emotional states, inflammation, and chronic drug use can exacerbate the experience of pain. These are examples of **descending facilitation** of the incoming signal.

In this section, we look at the neuroanatomical pathways and the mechanisms that underlie the modulation of pain at the level of the spinal cord, the brainstem, and the cortex. As noted, the first synapse in the pain pathway is in the posterior horn of the spinal cord, and it is the first site where modulation of the incoming nociceptive signal occurs.

A. Gate control theory of pain

Sensory input to the posterior horn of the spinal cord activates both projection neurons and a system of interneurons, most of which are inhibitory. One of the most basic modulatory systems involves influencing the balance between the nociceptive input (through Aδ and C fibers) and the input through Aβ fibers, which encode for touch. One of the things we do when we hurt ourselves is touch or rub that area in order to alleviate the pain. The underlying mechanism is that the balance of inputs is shifted toward touch (through Aβ-fiber activation) and away from pain (through Aδ-and C-fiber activation). This mechanism has been termed the **gate control theory of pain** (Figure 22.7).

Nociceptive fibers have excitatory synapses with the posterior horn neurons, as do mechanosensitive Aβ fibers. The gate control theory involves **local circuit interneurons**, which are **inhibitory** and synapse with the posterior horn WDR neurons. These interneurons are inhibited by nociceptive input. This *inhibition of the inhibition* results in *more* signaling of the posterior horn neurons. The interneurons also receive excitatory input from the Aβ fibers. In this case, excitation of the inhibition results in *more inhibition* of the posterior horn neurons and *less signaling*. Hence, there is a balance: When the Aβ fiber predominates, there is *less* signaling; and when the nociceptive fiber predominates, there is *more* signaling.

B. Descending influences from the brainstem

As Figure 22.8 shows, other modulatory systems come from a number of areas in the brainstem, which receive input from the ascending nociceptive pathways (see above). These structures include the **PAG** and the **reticular formation**, which include the **locus ceruleus** and the **raphe nuclei** (Figure 22.9). From these structures, descending fibers project back down to the posterior horn, where they influence NS and WDR neurons through interneurons. Noradrenergic projections from the locus ceruleus appear to play a role in the inhibition of mostly inflammatory pain. The descending serotonergic pathways from the raphe nuclei can either inhibit or enhance nociceptive input in the posterior horn. Manipulating these neurotransmitter systems is the basis for some pharmacological approaches to the treatment of pain (see below).

1. **Descending cortical input:** These brainstem nuclei are tightly linked to the thalamus, limbic system, hypothalamus, basal ganglia, and cortical areas that are part of the pain matrix (see Figure 22.8). Thus, brainstem activity depends not only on the ascending nociceptive input from the spinal cord but also on the descending cortical input. The various components of the pain matrix exert influence on the descending brainstem systems as well as on the thalamus, which gates the information to the cortex.

2. **Nociceptive modulation:** These descending influences should not be considered an analgesic system that indiscriminately blocks

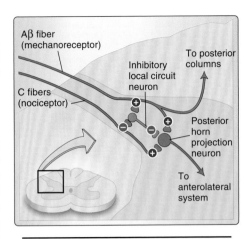

Figure 22.7
The gate control theory of pain. Input from Aβ fibers dampens the nociceptive input of Aδ and C fibers.

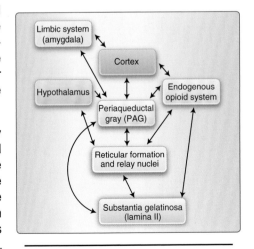

Figure 22.8
Conceptual overview of pain modulation networks.

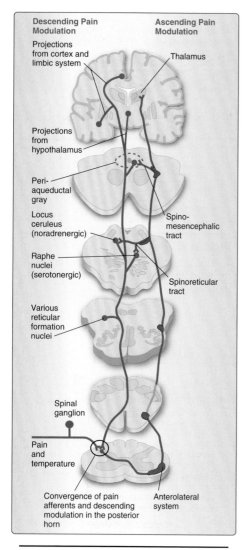

Figure 22.9
Anatomical overview of pain modulation pathways.

incoming nociceptive signals but rather a modulator that can both inhibit and facilitate nociception. Together with the cortex, these structures and pathways act as a **salience filter**, as described above. If the signal is deemed important and behavioral responses need to be initiated, nociception is facilitated. If it is deemed as distracting from other more important behavioral needs, nociception is inhibited.

C. Endogenous opioid system

The **endogenous opioid system** provides another modulatory influence on cortical pain processing. Opioid receptors can be found at all levels of the pain system: in various parts of the **pain matrix**, particularly in the **medial pain system** (cingulate cortex and insula); in the **descending brainstem systems**; and in the **posterior horn of the spinal cord**. All levels of the pain system that are involved in the modulation of pain are responsive to opioids. The powerful analgesic effect of opiate drugs is likely due to the fact that they influence the pain system at every level. Opioid receptors are physiologically activated by a group of endogenous molecules that comprise **enkephalins**, **endorphins**, and **dynorphins**. These opioid peptides act as neurotransmitters or neuromodulators and can produce potent analgesic effects.

1. **Opioid receptors:** Three classes of opioid receptors have been identified: μ, δ, and κ. Activation of these receptors results in the inhibition of voltage-gated Ca^{2+} channels and/or the opening of K^+ channels, which results in **hyperpolarization** and less neuronal excitability (Figure 22.10).

   Opioid receptors have been found on **peripheral nerves**. They are up-regulated in response to inflammation, which occurs in response to injury. Receptor agonists may come from immune cells, which release cytokines that are precursors of endorphins and enkephalins, possibly providing peripheral analgesic effects. This is clinically relevant because peripheral application of morphine can relieve pain, as has been shown in local application of opioids during surgery. Morphine can, however, delay the wound-healing process significantly due to the suppression of proinflammatory molecules that initiate tissue healing.

2. **Descending inhibition:** In the spinal cord, **enkephalin receptors** are found in **circuit neurons** of the posterior horn. These circuit neurons receive input through descending modulatory pathways and act as powerful inhibitors of the second-order neuron, which receives the nociceptor input.

   The brainstem areas with high opioid receptor density are the **PAG** and the **rostral anteromedial medulla**. These areas send the most potent descending inhibition to the posterior horn. Interestingly, it is necessary for these brainstem areas to interact with each other and the spinal cord in order to achieve a maximal analgesic effect.[1]

 [1]See *Lippincott's Illustrated Reviews: Pharmacology.*

VII. CHRONIC PAIN

Chronic pain can be divided into two broad categories: chronic nociceptive pain and chronic neuropathic pain.

A. Chronic nociceptive pain

Chronic nociceptive pain occurs when nociceptors are chronically activated, in chronic inflammatory diseases such as osteoarthritis, for example. The therapeutic approach is to prevent further nociceptor activation by treating the underlying inflammation or tissue damage.

B. Chronic neuropathic pain

Chronic neuropathic pain is defined as pain that arises as a consequence of a lesion of the somatosensory system and persists after the injury has healed or is described in terms of such damage. It is characterized by spontaneous pain without any apparent stimulus as well as **hyperalgesia** (heightened reaction to low-intensity noxious stimuli) and **allodynia** (painful reaction to nonnoxious stimuli). **Summation**, another sign of neuropathic pain, occurs when repeated applications of a low-intensity noxious stimulus lead to a worsening pain experience. The affected area can display **paresthesias**, which are spontaneous tingling sensations, or **dysesthesias**, which are painful experiences characterized by burning, electrical sensations, or shooting pain, in the absence of an external stimulus.

Neuropathic pain can be classified into **peripheral** and **central syndromes**, **complex pain disorders** (such as complex regional pain syndrome), and **mixed pain syndromes** (in which a combination of stimulus-induced [nociceptive] and stimulus-independent pain is present).

The mechanisms underlying neuropathic pain comprise sensitization or wind-up, as discussed above. Sensitization is a mechanism that serves to amplify the noxious stimulus in order to evoke a quick and effective behavioral response. When these mechanisms become maladaptive, they can lead to permanent changes in the wiring of the pain pathways and, through that, to a persistent experience of pain. Glia also have a role in sensitization, as discussed below.

C. Maladaptive peripheral sensitization

This type of sensitization can lead to changes in the peripheral free nerve endings of both Aδ and C fibers, which include overexpression of Na+ channels. Overexpression of Na+ channels in clusters on the peripheral branches of the Aδ and C fibers can lead to spontaneous discharges and lower the threshold for signaling. Overexpression of capsaicin receptors (TRPV1) can lead to a burning feeling in the affected area.

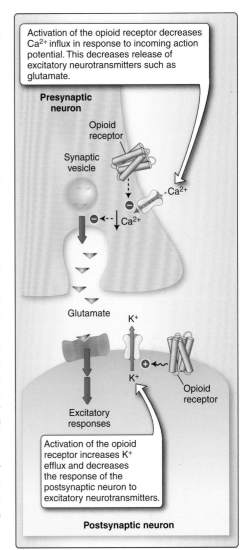

Activation of the opioid receptor decreases Ca^{2+} influx in response to incoming action potential. This decreases release of excitatory neurotransmitters such as glutamate.

Presynaptic neuron

Opioid receptor

Synaptic vesicle

Ca^{2+}

$\downarrow Ca^{2+}$

Glutamate

K^+

K^+

Opioid receptor

Excitatory responses

Activation of the opioid receptor increases K^+ efflux and decreases the response of the postsynaptic neuron to excitatory neurotransmitters.

Postsynaptic neuron

Figure 22.10
The action of opiates at the synapse. Receptors are on both the pre- and postsynaptic neuron.

D. Maladaptive central sensitization

This type of sensitization, conversely, can lead to changes in the synapses of the posterior horn of the spinal cord. Changes in glutamate receptor physiology and the insertion of more glutamate receptors in postsynaptic posterior horn neurons lead to increased excitability of these neurons. The overexpression of ion channels that result in inward currents (Na^+ channels) and the reduction of ion channels that result in outward currents (K^+ channels) will produce a lowered threshold and facilitated synaptic transmission. A reduction in the release of inhibitory neurotransmitters can lead to the loss of inhibition by inhibitory interneurons at the synapse with projection neurons (**disinhibition**). Collectively, this leads to increases in spontaneous activity (paresthesias, dysesthesias) and a reduced threshold for pain perception (**hyperalgesia**). In addition, some NS neurons can be converted to WDR neurons that now receive input from sprouting Aβ fibers, which typically occurs following nerve injury. This will result in nonnoxious stimuli being interpreted as pain (**allodynia**).

E. Glial signaling

In neuropathic pain, there is a loss of descending inhibition from brainstem centers, especially a decrease in serotonergic and noradrenergic fibers from the reticular formation. The role of glia in neuropathic pain should not be underestimated. Both astroglia and microglia can be activated by inflammation and, interestingly, by direct neuron-to-glia signaling. Following nociception, both astroglia and microglia can secrete proinflammatory signaling molecules, which in turn have lasting effects on synapse formation and connectivity.

1. **Astroglia:** Astroglia are significantly involved in synapse formation, function, and plasticity. They can influence the level of neurotransmitter receptors expressed by the second-order neurons and can guide sprouting Aβ fibers, possibly to the wrong targets. Astroglia down-regulate their glutamate reuptake and secrete additional glutamate, which increases excitatory neurotransmitter levels in the synaptic cleft and may increase the excitability of adjacent neurons. In addition, astroglia secrete a variety of other molecules that enhance the excitability of the postsynaptic neurons, which leads to more pain signaling.

2. **Microglia:** Microglia are the immune cells in the CNS, and their activation in response to chronic nociceptor signaling includes the release of transmitters that will inhibit the GABAergic (inhibitory) signaling of interneurons in the posterior horn. Reactive microglia express various neurotransmitter receptors; their stimulation leads to further secretion of proinflammatory molecules, which perpetuates the cycle of inflammation leading to neuronal excitation.

Clinical Application 22.2: Principles of Pain Therapy

The treatment of pain can occur at the various stages of the pain pathway, from the peripheral nociceptor to the processing of pain in the pain matrix.

Ideally, pain therapy is best managed in **multidisciplinary pain clinics** that address not only the pharmacotherapy but also the psychological counseling, physiotherapy, and community support of these patients.

Peripheral nociceptors: The peripheral nociceptor can be influenced to reduce firing by several of the following mechanisms.

- **Nonsteroidal anti-inflammatory drugs (NSAIDs)** decrease tissue inflammation and, thereby, decrease the noxious stimuli at the free nerve endings of Aδ and C fibers.

- **A blockade of Na$^+$ channels** in the peripheral free nerve endings blocks the generation and transmission of action potentials (APs). This is achieved through the local application of a Na$^+$ channel blocker such as the local anesthetic **lidocaine**.

- Stimulated free nerve endings also secrete neurotransmitters, which then amplify the noxious signal, termed **peripheral sensitization** (discussed above). The repeated application of **capsaicin**, a potent TRPV1 agonist (see above), causes the secretion of substance P from the free nerve endings, resulting in a depletion of substance P stores, which effectively desensitizes the affected nerve endings causing analgesia in that area.

- Calcitonin gene–related peptide (CGRP) contributes to both peripheral and central sensitization. Novel CGRP antagonists and monoclonal antibodies against CGRP or CGRP receptors are currently being developed for acute or preventive migraine treatment.

Posterior horn: In the posterior horn of the spinal cord, the first synapse of the pain pathway offers several possibilities for therapeutic intervention.

- APs in the afferent nociceptive fibers cause the opening of **Ca^{2+} channels** near the synapse, which induces the release of neurotransmitters into the synaptic cleft. A blockade of these Ca^{2+} channels effectively blocks the release of neurotransmitters. This can be achieved through a the use of gabapentin that binds to the alpha 2 delta subunit of the Ca^{2+} channels. Gabapentin has also been shown to inhibit synaptogenesis, which could be an effective mechanism to prevent the misdirected sprouting of Aβ fibers, which results in allodynia.

- One of the mechanisms of action of **opioids** is the opioid receptor–mediated blockade of presynaptic Ca^{2+} channels, which also prevents neurotransmitter release. Opioids also act on the postsynaptic side of the synapse by opening K$^+$ channels, which results in the efflux of K$^+$, hyperpolarization, and a decrease in the response of the postsynaptic neuron to excitatory neurotransmitters.

- Central sensitization at the posterior horn synapse is due to wind-up, which is caused by N-methyl-D-aspartic acid (NMDA) receptor activity (see above). Antagonists to NMDA receptors may be a potent way to block this central sensitization and treat neuropathic pain. The NMDA receptor antagonist **ketamine** is used in low doses for pain treatment. High doses of ketamine are used as an anesthetic and cause the patient to be unconscious and pain free. The opioid **methadone** also acts on the NMDA receptor and is used in the treatment of neuropathic pain.

Brainstem: A number of modulating influences originate in brainstem structures.

- Opioid receptors are abundant in the PAG of the brainstem, where opioids appear to block the facilitation of pain transmission.

- Noradrenergic and serotonergic fibers from the brainstem provide descending influences on the posterior horn circuitry. Here, they excite inhibitory interneurons, resulting in a net inhibition of the transmission of the nociceptive signal. Antidepressants, such as **serotonin** and **norepinephrine reuptake inhibitors** and **tricyclic antidepressants**, all increase the neurotransmitter at the synapse, which has been shown to be an effective way to enhance the descending inhibition.

Cortex: The experience of pain is processed in the pain matrix in the forebrain. Pain **management** strategies that target the psychological processing of the pain experience include **cognitive–behavioral therapy**, **mind–body therapy**, **relaxation techniques**, and **coping strategies**.

Clinical Application 22.2: Principles of Pain Therapy continued

Musculoskeletal system: Pain often results in, or is caused by, changes in the function of the **musculoskeletal** system. Pain will lead to postural changes and underuse of muscles. This can lead to muscle pain, spasms, and a shortening of muscles.

- **Botulinum toxin** blocks the release of acetylcholine at the neuromuscular junction, which blocks neuromuscular transmission. It has been shown to be effective in decreasing spasticity. Onabotulinum toxin A (Botox) is currently used in the treatment of chronic migraine.

- **Therapeutic exercise** is an important part of pain therapy. Exercise will help to return muscles to their normal length and correct postural changes that can lead to secondary problems. Exercise causes the release of endorphins, which can act to relieve pain as well.

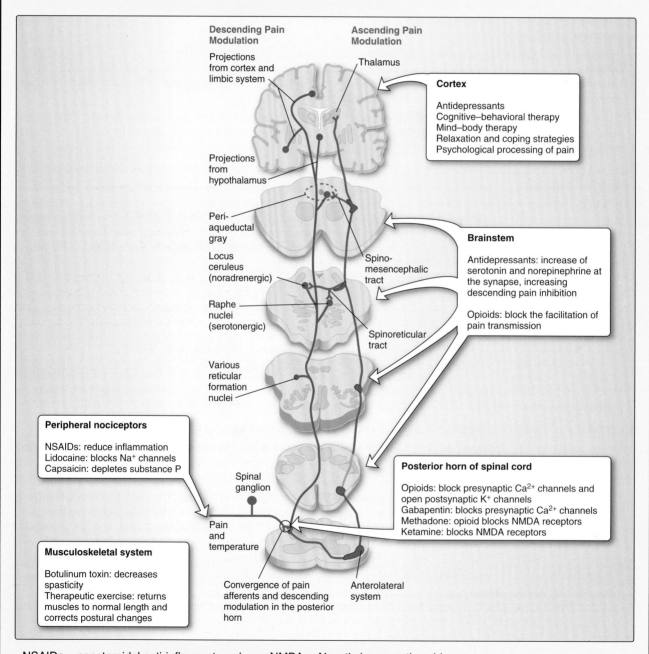

NSAIDs = nonsteroidal anti-inflammatory drugs; NMDA = *N*-methyl-D-aspartic acid.

Clinical Case:

Migraine Headache

A 32-year-old woman presents with a 10-year history of intermittent moderate-degree headaches that last 12–48 hours without medication. They occur every month prior to her menstrual period. She has light and noise sensitivity with her headaches and notices that physical activity during her headaches increases her pain. A few times per year, her headaches will be very severe with nausea and vomiting. Her headaches are migraines.

What are the typical clinical features of migraines?

Migraines can be divided into four phases: (1) premonitory phase, which includes changes in mood and activity; (2) aura (occuring only in some patients) is a fully reversible, short neurological deficit that can be sensory or motor; (3) headache phase; 4) postdromal phase, with changes in mood and activity. Migraines are moderate to severe headaches that last 4–72 hours without treatment. They are typically worse with activity and improve with rest. Pain that starts on one side of the head and is throbbing in quality is a common feature of migraine. Light and noise sensitivity or nausea and/or vomiting are associated with migraine.

Migraine can be associated with sensitivity to gentle touch. What is the name of this nonpainful stimulation that causes pain?

Allodynia is a term used to describe a painful reaction to nonnoxious stimulation (light, noise, or gentle touch). Normally, nonnoxious stimulation is not painful; however in migraine, patients can develop allodynia and are sensitive to these stimuli.

Dysesthesia is a term used to describe painful experiences such as burning, electrical sensation, and shooting pain in the absence of an external stimulation. *Hyperalgesia* is a term used to describe a heightened reaction to low-intensity noxious stimulation. *Paresthesia* is a term used to describe spontaneous tingling sensation that is not painful.

Chapter Summary

- Pain is detected by the free nerve endings of myelinated Aδ and unmyelinated C fibers. These fibers respond to noxious thermal, mechanical, and chemical stimuli. Aδ fibers mediate the "first pain," a sharp, well-localized pain. C fibers mediate the "second pain," a dull, poorly localized throbbing sensation. Both Aδ and C fibers project to the posterior horn of the spinal cord, cross the midline, and ascend in the anterolateral system.

- The ascending nociceptive pathways can be divided into two components. The lateral sensory-discriminative component comprises the neospinothalamic tract, which projects to the ventral posterolateral nucleus and ventral posteromedial nucleus of the thalamus and, from there, to the primary and secondary somatosensory cortices. It allows for the localization of pain. The medial affective-motivational component comprises all other tracts of the anterolateral system. It projects to brainstem structures, where it influences the descending modulation of pain, and to the limbic system, hypothalamus, and association areas of cortex, where it mediates the affective-motivational and emotional response to pain.

- A pain matrix, which includes the lateral somatosensory cortices and medial limbic structures, is activated in response to the perception of pain.

- Sensitization of pain occurs both peripherally and centrally. Peripheral sensitization is due to neurotransmitters released by stimulated nociceptors and to substances released during tissue injury and inflammation. These substances exacerbate the stimulation of nociceptors and can recruit so-called silent nociceptors, which expand the receptive field. Central sensitization occurs in the spinal cord due to the continued activation of the posterior horn neurons. One of the main components of this process, called "wind-up," is the activation of *N*-methyl-D-aspartic acid receptors. This has a number of downstream effects that lower the activation threshold of the postsynaptic neuron.

- Modulation of pain occurs both at the level of the spinal cord and through descending fibers from the brainstem reticular formation and the periaqueducdal gray.

- The modulation can be explained through the gate control theory, which describes a shift in the balance of input through nociceptive fibers and touch fibers resulting in either more or less pain transmission. The descending fibers from the brainstem can either inhibit the nociceptive signal or facilitate the signaling to the cortex, depending on the behavioral needs of the situation. Another pain-modulating system is the endogenous opioid system.

- The nociceptive system can be amplified through peripheral and central sensitization, which increases the saliency of the signal and gives it priority for a behavioral response. When sensitization is maladaptive, either centrally or peripherally, neuropathic pain (continued pain after nerve injury) can result. Sprouting of Aβ fibers onto nociceptive-specific neurons in the posterior horn results in nonnoxious stimuli being interpreted as painful. Many of these changes can result in permanent rewiring of the nociceptive pathways.

Study Questions

Choose the ONE best answer.

22.1 A patient was admitted to the emergency room with severe burns to his hand. The pathway by which the nociceptive stimulus reached a level of awareness in his cortex is through the:

    A. Spinohypothalamic pathway.
    B. Spinomesencephalic tract.
    C. Spinoreticular tract.
    D. Spinothalamic tract.
    E. Spinocerebellar tract.

The next three questions are based on this scenario:

An elderly lady noticed that the right side of her face had become extremely painful with a burning sensation like a **bad** sunburn. The hypersensitive area then developed a rash with blister-like nodules. Her doctor told her that she had developed herpes zoster, more commonly known as shingles, which is caused by the same virus that causes chickenpox. Unfortunately, after treatment for the disease and the disappearance of the rash, her painful burning sensation continued. When pain persists after the shingles outbreak has disappeared, it is called postherpetic pain for which there is no cure; however, treatments are available. The herpes virus attacks the nerve cell bodies in the ganglia of sensory neurons and is carried to the areas of skin supplied by the peripheral processes of those neurons.

22.2 The nerve cell bodies of which ganglion are involved in this patient's case?

    A. Spinal ganglion T12.
    B. Glossopharyngeal.
    C. Inferior vagal.
    D. Geniculate.
    E. Trigeminal.

22.3 After treatment for the disease and the disappearance of the rash, her painful burning sensation continued. This phenomenon is referred to as:

    A. Hyperalgesia.
    B. Allodynia.
    C. Paresthesia.
    D. Dysesthesia.
    E. Summation.

The correct answer is D. The spinothalamic tract is part of the anterolateral system and is responsible for sensory discrimination and affective appreciation of pain. It reaches consciousness through the thalamus to the cortex. The spinohypothalamic pathway is a projection to the hypothalamus and can be involved in the neuroendocrine–visceral response to pain. The spinomesencephalic tract projects to the PAG and the superior colliculus to modulate pain and direct our gaze. The spinoreticular tract is a projection to the reticular formation and mediates activation of the reticular formation. The spinocerebellar tract is a tract carrying nonconscious proprioception to the cerebellum.

The correct answer is E. The trigeminal ganglion contains the nerve cell bodies of the sensory nerves of the face. The spinal ganglion is associated with spinal nerves. The glossopharyngeal ganglion is the sensory ganglion of cranial nerve (CN) IX that is sensory to the pharynx. The inferior vagal ganglion is concerned with sensory input from the pharynx and larynx. The geniculate ganglion is the sensory ganglion of CN VII associated with taste from the anterior two-thirds of the tongue.

The correct answer is D. Dysesthesia is a burning, electrical sensation or shooting pain in the absence of an external stimulus. Hyperalgesia is a heightened reaction to low-intensity noxious stimuli. Allodynia is a painful reaction to nonnoxious stimuli. Paresthesia is spontaneous tingling sensations. Summation is when repeated application of a low-intensity noxious stimulus leads to a worsening pain experience.

22.4 A likely mechanism underlying the persistence of pain in this patient is

   A. A down-regulation of peripheral *N*-methyl-ᴅ-aspartic acid receptors.

   B. A loss of Na⁺ channels in the postsynaptic membrane of the second-order neuron.

   C. Altered synaptic circuitry in the posterior horn of the spinal cord resulting in central sensitization.

   D. An up-regulation of opioid receptors on peripheral nerves in the face.

   E. A change in the balance of inputs to the face favoring Aβ fibers over Aδ and C fibers.

The correct answer is C. Postherpetic pain is an example of chronic neuropathic pain. It is most likely due to central sensitization or wind-up, which includes a lowered pain threshold, disinhibition, and spontaneous activity of nociceptive neurons. *N*-Methyl-ᴅ-aspartic acid (NMDA) receptor activation plays a role in wind-up and a down-regulation of NMDA receptors would result in less excitability. During sensitization, more Na⁺ channels are inserted into the membrane of the postsynaptic neuron, which leads to heightened excitability. Down-regulation rather than up-regulation of opioid receptors could play a role in increasing pain. However, the pain being experienced by this patient is likely due to changes in central mechanisms. Increased pain is likely to involve a shift in the balance of inputs favoring Aδ and C fibers.

22.5 In the processing of pain, which statement is true?

   A. Synaptic targets of Aδ and C fibers include both nociceptive-specific and wide dynamic range neurons.

   B. Descending inputs to posterior horn cells from brainstem areas are exclusively inhibitory.

   C. Afferent nociceptive fibers release primarily norepinephrine.

   D. Hot thermal stimulation activates primarily TRPM8 receptors.

   E. The lateral sensory-discriminative pathway mediates the emotional response to pain.

The correct answer is A. Aδ and C fibers target both nociceptive-specific and wide dynamic range neurons. Descending input to the posterior horn cells can be either inhibitory or facilitatory. Afferent nociceptive fibers release glutamate, calcitonin gene–related peptide, and substance P but not norepinephrine. Cold sensitivity is mediated through TRPM8 receptors, whereas hot thermal stimulation activates TRPV1 receptors. The medial affective-motivational pathway mediates the emotional response to pain. The lateral pathway is important in perception and localization of pain.

Index

Note: Page numbers in *italics* denote figures; those followed by b denote boxes; those followed by t denote tables.